PROCEEDINGS OF THE HERBRAND SYMPOSIUM
LOGIC COLLOQUIUM '81

STUDIES IN LOGIC

AND

THE FOUNDATIONS OF MATHEMATICS

VOLUME 107

NORTH-HOLLAND PUBLISHING COMPANY
AMSTERDAM · NEW YORK · OXFORD

PROCEEDINGS OF THE HERBRAND SYMPOSIUM

LOGIC COLLOQUIUM '81

Proceedings of the Herbrand Symposium
held in Marseilles, France, July 1981

Edited by

J. STERN
Université de Caen
Caen, France

1982

NORTH-HOLLAND PUBLISHING COMPANY
AMSTERDAM · NEW YORK · OXFORD

ISBN: 0 444 86417 2

Published by:

North-Holland Publishing Company – Amsterdam · New York · Oxford

Sole distributors for the U.S.A. and Canada:

Elsevier Science Publishing Company, Inc.
52 Vanderbilt Avenue
New York, N.Y. 10017

Library of Congress Cataloging in Publication Data

Logic Colloquium (1981 : Marseille, France)
Logic Colloquium '81.

(Studies in logic and the foundations of mathematics ; v. 107)
1. Logic, Symbolic and mathematical--Congresses.
I. Stern, J. (Jacques), 1949- . II. Herbrand, Jacques, 1908-1931. III. Series.
QA9.A1L63 1981 511.3 82-6433
ISBN 0-444-86417-2 AACR2

PRINTED IN THE NETHERLANDS

A LA MEMOIRE

DE JACQUES HERBRAND

PREFACE

Fifty years after the death of Herbrand, about two hundred people attended the colloquium which was held in Marseilles from July 16 through July 24, 1981.

The conference has tried to encompass fifty years of logic; this could be considered as a very ambitious goal, but because of the enthusiasm of the invited speakers and of the audience, our aim was achieved to a large extent.

During the opening ceremony, the participants could hear a short address by Professor A. Guinier who knew Herbrand personally; messages from Professor C. Chevalley and Professor J. Dieudonné, who also knew Herbrand, were read.

A large part of the program was devoted to invited lectures on the work of Herbrand and on the role of Herbrand's ideas in the subsequent development of logic. These lectures appear in the first part of the proceedings. The other invited lectures dealt with other topics of current research in mathematical logic (set theory, recursion theory, model theory, proof theory, computer science).

In his opening address, Professor Guinier wondered about the theorems that Herbrand would have proved if he had not met with his tragic fate. This question cannot be answered. On the other hand, it is clear that the death of Herbrand somehow delayed the firm establishment of logic in France. For this reason, I want to thank warmly those who have chosen to write their contributions in the French language as a tribute to the memory of Herbrand.

The symposium was sponsored by the CNRS (Centre National de la Recherche Scientifique) and was the summer meeting of the European branch of the Association for symbolic logic. Financial support was also given by the Société française de logique and both Universities in Marseilles.

I would like to close this preface by thanking all those who helped with the organization of this meeting and the preparation of these proceedings.

J. STERN

MEMBERS OF THE PROGRAM COMMITTEE

S. Feferman	Stanford
R. Fraïsse (Chairman)	Marseille I
H. Gaifman	Jerusalem
J.Y. Girard	Paris VII
M. Guillaume	Clermont II
T. Jech	Los Angeles
D. Lascar	Paris VII
G.H. Müller	Heidelberg
L. Pacholski	Wroclaw
J.B. Paris	Manchester
J. Stern	Caen

MEMBERS OF THE ORGANIZING COMMITTEE

G. Blanc	Marseille - Luminy
M.R. Donnadieu	Marseille - Luminy
R. Fraïsse	Marseille I
M. Guillaume	Clermont II
A. Preller (Chairman)	Marseille - Luminy
C. Rambaud	Marseille - Luminy
R. Smadja	Marseille - Luminy

TABLE OF CONTENTS

PROCEEDINGS OF THE HERBRAND SYMPOSIUM
LOGIC COLLOQUIUM '81, J. Stern (editor)

JACQUES HERBRAND

par C. Chevalley

Jacques Herbrand était d'origine belge. Son père, négociant en tableaux anciens, vint s'établir à Paris où Jacques Herbrand continua des études secondaires, qui n'avaient pas trop bien commencé du fait des difficultés qu'il rencontrait... en mathématiques ! Faut-il voir là un cas d'éclosion très soudaine d'un talent mathématique jusqu' alors caché, ou plutôt un exemple du manque total de discernement d'un enseignant à qui Herbrand - avec la franchise parfois terrible avec laquelle il s'exprimait - aurait fait part des insuffisances de ses leçons.

Quoiqu'il en soit, il fut reçu - premier, si mes souvenirs sont exacts - à l'Ecole Normale Supérieure, en 1925 ; c'est là qu'entré un an après lui je le rencontrai. Sans aller suivre les cours donnés à la Sorbonne ou à l'école même par le directeur - ce qui lui semblait une perte de temps - il se consacra tout de suite aux questions qui l'intéressaient, ce qui explique en partie qu'il ait pu produire en si peu de temps une oeuvre considérable.

Il ne s'intéressait pas seulement aux mathématiques, mais aussi à la philosophie et à la poésie. Il me fit découvrir et aimer Valéry et surtout Mallarmé pour lequel il sut me faire partager son enthousiasme. En philosophie, il était attiré par l'idéalisme absolu de Hamelin, - la France n'avait alors rien de mieux à offrir à un jeune esprit ardent ; il en critiqua cependant la pensée, et notamment la déduction a priori des concepts de la spatialité, dont il perçut l'insuffisance en la comparant à la méthode axiomatique. Peut-être d'ailleurs était-il moins attiré par l'idéalisme de la doctrine que par son caractère absolu. C'était en effet un des traits de son esprit que de pousser toutes choses jusqu'à leur limite extrême et de mépriser toutes les demi-mesures. Sa pratique du sport même était follement risquée : expédition d'alpinisme seul et sans guide par temps de brouillard dans les Pyrénées, traversée de l'entrée du port de La Rochelle alors qu'il savait à peine nager ... Par une amère ironie du destin il devait périr dans un accident qui ne résultait d'aucune imprudence de sa part : un rocher sur lequel il était assis s'est subitement détaché et l'a entraîné dans sa chute.

C'est le goût de l'aventure intellectuelle qui le porta vers la logique. Il était séduit par le caractère grandiose de l'oeuvre de Hilbert et il crut un temps - c'était avant Gödel - que le problème de la décision était sur le point d'être résolu.

Outre la logique mathématique, qui fut toujours la première de ses préoccupations, Jacques Herbrand s'intéressa à la théorie des nombres algébriques. Peut-être l'attira-t-elle par cette double circonstance que cette branche des mathématiques était totalement inconnue en France et que la théorie du corps de classes passait à juste titre pour l'une des plus difficiles en mathématiques. La découverte de la loi générale de réciprocité de Artin n'en avait pas encore éclairé les abords et c'est à travers le grand mémoire d'exposition de Hasse qu'il s'y initia. Le fascicule du "Mémorial des sciences mathématiques" fut rédigé en vue de fournir au public mathématique de langue française un accès à la théorie ; il se signale par une grande clarté d'exposition plutôt que par des résultats nouveaux. Herbrand s'intéressait d'ailleurs moins aux fondements de la théorie qu'aux applications qu'on pouvait en faire. Dans son mémoire "Sur les classes des corps circulaires", il utilisa la théorie du corps de classes pour redémontrer certains résultats de Kummer, pour en prouver d'autres dont aucune démonstration n'avait été publiée, et surtout pour en obtenir de nouveaux qui précisent considérablement ceux de Kummer; de plus, il raffine également les critères de validité pour certains exposants du théorème de Fermat. Son mémoire "Théorie des groupes de décomposition, d'inertie et de ramification" a pour but de déterminer ces groupes pour une extension galoisienne K/k à partir de leur connaissance pour un sur-corps L de K, également galoisien en k. Ces résultats peuvent s'étendre au cas non galoisien, comme l'ont montré les méthodes de Krasner.

Vous allez entendre durant ce colloque des exposés sur les travaux de Herbrand en logique. Les quelques phrases qui précèdent n'ont pour ambition que de rappeler que Jacques Herbrand était ouvert à tous les vents de l'esprit et serait probablement devenu non seulement un grand mathématicien mais un de ces hommes profondément cultivés dans tous les domaines dont s'honore leur siècle.

Claude Chevalley

PROCEEDINGS OF THE HERBRAND SYMPOSIUM
LOGIC COLLOQUIUM '81, J. Stern (editor)

JACQUES HERBRAND ET LA THEORIE DES NOMBRES

par J. Dieudonné

Académie des Sciences

Entré premier à l'Ecole Normale supérieure à 17 ans, Herbrand était élève externe et fréquentait peu l'Ecole. Il dédaignait les cours de la Sorbonne et s'instruisait lui-même ; on le voyait cependant à certains cours du Collège de France et notamment au Séminaire d'Hadamard. Selon une coutume assez répandue, il avait voulu se débarrasser des 4 certificats de Licence réglementaires dès sa première année d'Ecole ; il y parvint sans peine sauf pour le certificat de Mécanique rationnelle, qu'il avait entièrement négligé de préparer, persuadé qu'il ne s'agissait que d'applications triviales de théorèmes d'Analyse ; mais à la session de Juillet le problème d'examen comportait un petit piège, bien connu de tous ceux qui avaient suivi quelques travaux pratiques, mais ignoré d'Herbrand, qui dut repasser en Octobre.

J'ai donc très peu connu Herbrand à l'Ecole, où j'étais entré un an avant lui, mais sa réputation n'avait pas tardé à se répandre. Il était déjà docteur ès sciences un an après sa sortie de l'Ecole, alors que je commençais à peine à m'initier à la recherche ; j'ignorais entièrement les domaines des mathématiques dans lesquels il travaillait, et qui me paraissaient alors inaccessibles ; aussi m'inspirait-il une admiration un peu craintive. Après sa thèse, il passa l'année 1930-1931 en Allemagne, où je le rencontrai peu avant sa mort ; sans abandonner pour autant la logique mathématique, il consacra ses efforts pendant cette année à la Théorie des Nombres ; l'Allemagne était alors la Mecque de cette discipline, et la profondeur et la nouveauté de ses idées firent une grande impression sur E. Artin, Hasse et E. Noether, trois des principaux représentants de cette école. En très peu de temps, il obtint des résultats très originaux dans trois directions différentes : la théorie du corps de classes, la théorie des corps de nombres de degré infini, et celle des corps cyclotomiques.

I. LA THEORIE DU CORPS DE CLASSES

C'est le développement de conjectures émises par Hilbert et H. Weber entre 1896 et 1902, et qui a été au centre des travaux des arithméticiens dans la première moitié du XXe siècle. L'idée centrale de Hilbert peut se décrire en disant qu'il interprète le groupe des classes d'idéaux d'un corps de nombres k, à isomorphie près, comme groupe de Galois d'une extension abélienne de k ; Weber, à la suite de résultats particuliers de Kronecker, avait conjecturé indépendamment un résultat plus général, associant de même une extension abélienne de k à des groupes de classes d'idéaux restreintes (deux idéaux d'une même classe au sens usuel n'étant considérés comme équivalents que s'ils satisfont en outre à certaines congruences). On donna à ces extensions abéliennes le nom de corps de classes.

Après qu'en 1907 Furtwängler, un élève de Hilbert, eut prouvé l'existence du corps de classes dans le cas particulier envisagé par ce dernier, le mathématicien japonais Takagi, en 1920, est parvenu, non seulement à démontrer les conjectures de Weber dans le cas général, mais aussi à établir que toute extension abélienne de k est un corps de classes pour un groupe de classes restreintes bien déterminé. Ses résultats furent complétés sur un point important par la loi de réciprocité de E. Artin (1927), qui définit un isomorphisme canonique entre un groupe de classes d'idéaux restreintes et le groupe de Galois du corps de classes correspondant.

La théorie du corps de classes pouvait donc sembler achevée ; mais les méthodes de Takagi étaient extrêmement compliquées et paraissaient très artificielles. Après le mathématicien allemand F.K. Schmidt, Herbrand et Chevalley s'attaquèrent à la recherche de méthodes plus simples et qui feraient mieux comprendre la structure de la théorie. Après la mort d'Herbrand, Chevalley continua seul cette tâche, jalonnée par sa thèse de 1933 [1] et le mémoire de 1940 [3], où il donna un essor nouveau à la théorie des nombres algébriques par l'introduction de la notion d'idèle et l'utilisation de la topologie, devenue prépondérante dans les exposés actuels.

Il n'est évidemment pas étonnant que les contributions de Herbrand à ce renouveau des conceptions relatives au corps de classes soient demeurées fragmentaires, mais deux d'entre elles ont une portée plus générale et sont encore d'un emploi courant :

1) Ce qu'on appelle le "lemme de Herbrand" ; c'est un résultat technique de la théorie des groupes finis, que Herbrand ne semble pas avoir publié, mais qui est présenté sous ce nom par Chevalley dans sa thèse ([1], p. 375), comme une propriété de deux endomorphismes d'un groupe fini satisfaisant à certaines conditions. De nos jours la forme de ce lemme s'est transformée en un énoncé de la théorie cohomologique des groupes finis, qui a de nombreuses applications ([13], p. 143, prop. 9).

2) Le mémoire [5] est consacré à la théorie des groupes de ramification, définis

par Hilbert pour les extensions galoisiennes. Si K est une telle extension d'un corps de nombres k, et P un idéal premier de K, on associe à P une suite décroissante $(G_i)_{i \geq 0}$ de sous-groupes du groupe de Galois G de K sur k : G_o, appelé groupe de décomposition, est formé des $\sigma \in G$ tels que $\sigma(P) = P$; pour $i \geq i$, G_i est formé des $\sigma \in G_o$ tels que $\sigma(x) - x \in P^i$ pour tout entier x de K non dans P ; G_1 est le groupe d'inertie de P, et les G_i pour $i \geq 2$ les groupes de ramification. Le problème que résoud Herbrand concerne un corps L intermédiaire entre k et K et galoisien sur k ; à l'aide des G_i et du groupe de Galois de L sur k, il détermine les groupes d'inertie et de ramification de l'idéal premier $P \cap L$ de L. Le résultat est devenu classique et d'un usage constant dans la théorie des corps locaux ([13], p. 101, cor. 3).

Un autre résultat de Herbrand concerne les unités d'une extension galoisienne K d'un corps de nombres k [4]. Le groupe de Galois G de K sur k opère naturellement sur le groupe abélien E des unités de K, d'où (en prenant les logarithmes des unités et l'espace vectoriel V sur $\mathbb{Q}$ qu'ils engendrent) une représentation linéaire de G dans l'espace vectoriel $V \otimes_{\mathbb{Q}} \mathbb{C}$. Herbrand détermine cette représentation et en déduit des propriétés du groupe E. Moins important actuellement que les deux résultats précédents, il a néanmoins été utilisé par Chevalley dans sa thèse, ainsi que plusieurs méthodes imaginées par Herbrand dans les articles [6] et [10] en vue de simplifier les démonstrations dans la théorie du corps de classes.

II. LES CORPS DE NOMBRES DE DEGRE INFINI

Ce sont les extensions algébriques du corps $\mathbb{Q}$ des rationnels, dont le degré sur $\mathbb{Q}$ est infini. L'étude de l'anneau des entiers d'un tel corps, et notamment de ses idéaux premiers, avait été commencée dans la période 1920-1930 par Stiemke et Krull. Herbrand, dans les mémoires [8] et [9], se proposa d'examiner les généralisations possibles des groupes de décomposition, d'inertie et de ramification dans la situation où les corps k et $K \supset k$ sont de degré infini sur $\mathbb{Q}$. Le principal intérêt de ces mémoires est que Herbrand y inaugure une méthode nouvelle basée sur les notions de limite inductive et de limite projective, qu'il introduit systématiquement pour la première fois en mathématiques, et qui sont devenues fondamentales de nos jours, aussi bien en Algèbre qu'en Topologie. Il considère un corps de nombres de degré infini comme limite inductive d'une suite de sous-corps de degré fini, et fait une étude systématique du "passage à la limite" pour diverses notions liées à ces corps. Quant à la notion de limite projective, elle se présente naturellement quand on considère une extension galoisienne K (de degré infini) d'un corps de nombres k (de degré fini ou non sur $\mathbb{Q}$), comme limite inductive d'extensions galoisiennes K_n de k de degré fini, car le groupe de

Galois de K sur k est alors limite projective des groupes de Galois des K_n sur k.

Il n'est peut-être pas sans intérêt de rappeler que c'est en prolongeant ces travaux de Herbrand que Chevalley, pour étendre la théorie du corps de classes aux corps de nombres de degré infini, introduisit la notion d'idèle [2].

III. LES CORPS CYCLOTOMIQUES

Le mémoire de Herbrand sur les classes d'idéaux des corps cyclotomiques [7] est celui qui maintenant est considéré comme la plus originale de ses contributions à la Théorie des nombres, et qui a eu le plus de répercussions sur les travaux actuels. Soit $K = \mathbb{Q}(\zeta)$ le corps des racines p-èmes de l'unité, où p est premier impair et $\zeta = \exp(2\pi i / p)$. Kummer a prouvé que le nombre de classes d'idéaux de K est divisible par p si et seulement si l'un des nombres de Bernoulli b_k, où k est <u>pair</u> et $2 \leq k \leq p-3$, est divisible par p. On peut chercher à améliorer ce résultat de la façon suivante. Dans le groupe des classes d'idéaux de K (écrit additivement), on considère le sous-groupe C des classes x telles que $p.x = 0$. Le groupe de Galois G de K sur $\mathbb{Q}$ est cyclique d'ordre $p-1$, et on l'identifie au groupe multiplicatif $\mathbb{F}_p^*$ du corps à p éléments $\mathbb{F}_p = \mathbb{Z}/p\mathbb{Z}$; il opère sur C de façon naturelle, de sorte que C (qu'on peut considérer comme espace vectoriel sur $\mathbb{F}_p$) se décompose en somme directe de sous-espaces C_h, où $h \in \mathbb{Z}/(p-1)\mathbb{Z}$, et C_h est le sous-espace formé des $x \in C$ tels que $s_t . x = t^h x$, $s_t \in G$ correspondant à $t \in \mathbb{F}_p^*$ par l'identification précédente. Le problème est alors de trouver des critères pour que C_h ne soit pas réduit à 0 ; il avait été considéré par Pollaczek en 1924 [11], qui avait donné des conditions nécessaires faisant encore intervenir les nombres de Bernoulli (loc.cit., p. 26, Satz VIII). Herbrand ne paraît pas avoir connu le travail de Pollaczek, qu'il ne cite pas dans son mémoire ; mais alors que Pollaczek n'avait utilisé que les résultats de Furtwängler sur le corps de classes de Hilbert, Herbrand arrive à des résultats plus complets grâce aux théorèmes de Takagi ; il obtient comme condition nécessaire pour que $C_{1-k} \neq 0$ (où k est <u>pair</u> et $2 \leq k \leq p-3$) que b_k soit divisible par p, et prouve que cette condition est suffisante quand on suppose en outre que $C_h = 0$ pour h <u>pair</u> (on ne connaît aucun nombre premier p pour lequel cette propriété n'a pas lieu). En 1976, Ribet a montré que la condition de Herbrand est suffisante dans tous les cas, par une ingénieuse méthode nouvelle utilisant les fonctions modulaires [12].

Je remercie J.P. Serre pour les précieuses indications qu'il m'a fournies concernant les répercussions des travaux de Herbrand sur les recherches actuelles de Théorie des nombres.

REFERENCES

[1] C. Chevalley, Sur la théorie du corps de classes dans les corps finis et les corps locaux, Journ. Fac. Sci. Univ. Tokyo, 2 (1929-34), p. 365-476 (Thèse).

[2] C. Chevalley, Généralisation de la théorie du corps de classes pour les extensions infinies, Journ. de Math., (9), 15 (1936), p. 359-371.

[3] C. Chevalley, La théorie du corps de classes, Ann. of Math., 41 (1940), 395-418.

[4] J. Herbrand, Nouvelle démonstration et généralisation d'un théorème de Minkowski, C.R. de l'Acad. des Sci., 191 (1930), p. 1282-1285.

[5] J. Herbrand, Sur la théorie des groupes de décomposition, d'inertie et de ramification, Journ. de Math., (9), 10 (1931), p. 481-498.

[6] J. Herbrand, Sur les théorèmes du genre principal et des idéaux principaux, Hamb. Abh., 9 (1931), p. 84-92.

[7] J. Herbrand, Sur les classes des corps circulaires, Journ. de Math., (9), 15 (1932), p. 417-441.

[8] J. Herbrand, Théorie arithmétique des corps de nombres de degré infini, I, Math. Ann., 106 (1932), p. 473-501.

[9] J. Herbrand, Théorie arithmétique des corps de nombres de degré infini, II, Math. Ann., 108 (1933), p. 699-717 (rédigé par C. Chevalley d'après des papiers de Herbrand ; le dernier paragraphe est une addition de Chevalley).

[10] J. Herbrand et C. Chevalley, Nouvelle démonstration du théorème d'existence en théorie du corps de classes, C.R. de l'Acad. des Sci., 193 (1931), p. 814-815.

[11] F. Pollaczek, Über die irregulären Kreiskörper der ℓ- ten und ℓ^2- ten Einheitswurzeln, Math. Zeitschr., 21 (1924), p. 1-38.

[12] K. Ribet, A modular construction of unramified p-extensions of $\mathbb{Q}(\mu_p)$, Inv. Math., 34 (1976), p. 151-162.

[13] J.P. Serre, Corps locaux, Paris, Hermann, 1962.

PROCEEDINGS OF THE HERBRAND SYMPOSIUM
LOGIC COLLOQUIUM '81, J. Stern (editor)

UN JOUR DE JUILLET 1931, DANS LE MASSIF D'OISANS ...

par A. Guinier

Académie des Sciences

Je ne connaissais pas Jacques Herbrand quand il est venu en juillet 1931, à la Bérarde rejoindre un petit groupe de camarades de l'Ecole Normale, Jean Brille, Pierre Bellair et moi-même, pour faire des courses en montagne. Il en avait fait les années précédentes avec Jean Brille, qui était de sa promotion ; ce dernier n'ayant pas pu venir aujourd'hui, le Comité du Colloque m'a demandé de le remplacer pour évoquer devant vous l'accident qui mit fin si tôt à la carrière pleine de promesses du grand savant qu'était déjà Jacques Herbrand.

La première ascension que nous avions projetée était celle des Bans. Dès l'arrivée de J. Herbrand, nous avons quitté la vallée pour le refuge et, le lendemain, suivant l'itinéraire normal, nous avons traversé le glacier avant d'escalader l'arête rocheuse conduisant au sommet. Nous avons encore la photographie de J. Herbrand au sommet du glacier et une photo de la cordée à la pointe des Bans : ce devait être la dernière image de J. Herbrand.

Au cours de la descente, pour franchir un passage assez raide, nous avons posé un rappel. P. Bellair le passa le premier, puis ce fut mon tour ; J. Herbrand, en attendant, s'était assis sur une dalle. Pendant que je descendais, je sentis que le piton autour duquel était enroulée la corde de rappel cédait : je fus retenu par la corde d'assurance, mais le bloc sur lequel était assis J. Herbrand bascula et je le vis projeté dans le vide. Ensuite ce fut le silence : nous étions seuls dans la montagne qui s'était embrumée. Nous ne voyions pas où la chute de J. Herbrand avait pu s'arrêter : son corps devait avoir disparu dans une crevasse du glacier d'où émergeait le rocher des Bans. Nous descendimes ; le refuge était vide et c'est dans la nuit que nous atteignimes La Bérarde pour donner l'alerte.

Le lendemain, des guides partirent pour localiser le corps et ensuite le redescendre à La Bérarde. La mise en bière eut lieu en présence des parents de J. Herbrand, de sa fiancée et de Claude Chevalley.

Pendant l'hiver suivant, M. et Mme Herbrand me demandèrent plusieurs fois de venir les voir. Ils étaient très aimables et très éloignés du milieu dans lequel avait vécu et travaillé leur fils. En juillet 1932, M. Herbrand tint, bien qu'il ne fût pas du tout un habitué de la montagne, à se faire conduire jusqu'à la crevasse où l'on avait retrouvé le corps de son fils ; je l'ai accompagné dans cette course très dure pour lui.

Cinquante ans après la disparition de J. Herbrand voici des spécialistes réunis pour discuter des travaux qu'il avait effectués en moins de quatre ans. Si le destin n'avait pas interrompu sa vie prématurément, il est non seulement probable mais quasi-certain qu'il aurait accompli une oeuvre considérable de grande originalité. Il vient alors une question à l'esprit : quelle serait la place dans l'ensemble de nos connaissances actuelles de cette oeuvre inaccomplie ?

A une telle question, transposée dans divers domaines de la création humaine, les réponses seraient très variables. Pour prendre deux exemples extrêmes, d'abord nous avons tous la conviction que personne ne pouvait et ne pourra composer la Dixième Symphonie que Beethoven aurait écrite s'il en avait eu la possibilité matérielle. Par contre, les inventions techniques surgissent bien souvent en plusieurs endroits presque simultanément, et sous des formes très voisines : si une réalisation ne se fait pas ici, elle se fera là bientôt. On a l'impression de l'avancée continue du front des connaissances sous les efforts d'individus si nombreux qu'au bout de quelques années seulement, les percées individuelles ne se perçoivent plus guère.

La situation pour les sciences expérimentales est assez semblable. Nous sommes naturellement persuadés que, si Pasteur n'avait pas existé, nous aurions quand même, depuis longtemps déjà, vaccins et asespsie. En physique, on pourrait citer maints exemples allant dans le même sens : le progrès considèrable qu'a constitué l'introduction de la mécanique quantique a débuté par les travaux indépendants et presque simultanés, de L. de Broglie et W. Heisenberg.

Mais comment les mathématiciens placent-ils leur science dans l'éventail des possibilités que nous avons évoquées ? Est-ce qu'un individu hors du commun a un rôle irremplaçable, ou bien l'essentiel est-il fait de petites avancées réalisées par des équipes nombreuses ?

Evoquer la mémoire de Jacques Herbrand, c'est se demander si les découvertes qu'il n'a pas pu faire à cause de son tragique destin manquent encore, et manqueront peut-être toujours, à la communauté des mathématiciens.

A. Guinier.

PROCEEDINGS OF THE HERBRAND SYMPOSIUM
LOGIC COLLOQUIUM '81, J. Stern (editor)

COMPUTATIONALLY IMPROVED VERSIONS OF HERBRAND'S THEOREM

Wolfgang Bibel

Technische Universität München
Germany

Herbrand's Fundamental Theorem provides a straightforward mechanical proof procedure which, however, is blown up with such a lot of redundancy that it is of no use in practice. We therefore develop in this paper in a sequence of refinements a version of this theorem which from this computational point of view is substantially more efficient. The resulting proof method is superior to any other known proof method.

INTRODUCTION

Herbrand's Fundamental Theorem [7] is "the central theorem of predicate logic" which has a variety of applications of great importance (for details see the Introduction in [6]). One of these applications is its use as a constructive tool for providing effective proof procedures, which beginning in the mid-fifties has extensively been pursued in the field of Automated Theorem Proving (ATP).

Soon it turned out, however, that Herbrand's original version, applied for this purpose in a straightforward way, yields a proof procedure blown up with such a lot of redundancy that in reality it fails even for rather simple theorems by quickly exhausting the available computational resources. Of course, this is of no surprise, since Herbrand in the late twenties could hardly think of such an application. In detail, there are the following 4 major sources of reduncancy.

a. The required *expansion* of the given formula P explicitly involves several variants of P which all encode essentially the same information.

b. The number of these variants of P used in the expansion is not minimal and, for instance by appropriate *splitting*, often may be decreased.

c. Expansion and the tautology test of the resulting quantifier-free disjunction is treated separately, which results in a stupid behavior.

d. Classical tautology test methods are inefficient even in relatively simple cases.

Some of these redundancies have been eliminated with the *resolution* proof method, developed by J.A. Robinson in the early sixties on the basis of various related results (cf. section IV. 13 in [4] for some historical details). In particular, resolution offers an efficient interplay between the operation of expansion and the - considerably improved - tautology test via *unification*, thus coping with (c) and (d) above.

But resolution has realized only one among many possible steps. For instance, it does not care about (a) and (b), and for this reason may wildly generate vast amounts of redundant new clauses. But even w.r.t. (c) and (d) it is far away from offering an optimal solution. In one respect, it is even less advanced than Herbrand's theorem suggests since it requires the formula presented in normal form, a new source of sometimes enormous redundancy (cf. [1]).

A lot of energy has been invested to overcome some of these drawbacks of resolution resulting in refinements, structure-shared representations, and numerous proposals of detail. The problem with these proposals is that they are to be understood within a complex procedural environment and that they are supported by a few experiments rather than by intuitive mathematical statements like Herbrand's theorem, for which reason their value may often be doubted.

Motivated by this unsatisfactory situation, the author, in an independent approach, has tried to gain a deeper understanding of the problem as a whole by carefully studying efficient ways of determining derivations of given theorems in Gentzen-like formal systems as developed by Schütte [9]. These efforts, lasting more than a decade, have produced theorems which afterwards turned out to be refined versions of Herbrand's theorem. They support a proof method, called the *connection* method, which copes with all 4 kinds of redundancies listed above, and in that respect is more efficient than any other known proof method [4].

A comprehensive treatment of this approach is contained in [4], while a short informal and intuitive introduction into the connection method may be found in [3]. The goal of the present paper is the development of our computationally most advanced version of Herbrand's theorem which first was proved in [2] (although its main ideas appear in earlier publications of the author). Here, we present it as the last in a sequence of 5 Herbrand-type theorems, $H_0, H_1, \ldots, H_4$, each being a computationally refined version of the preceding one. Even H_4 neglects a number of issues, listed in section (IV.11) from [4], the incorporation of which would extend this list, thus opening a wide field for further logical research with immediately useful applications.

For space limitations the detailed proofs (contained in [4]) cannot be given here. But for readers well familiar with first-order logic our step-wise refinement approach will to some extent provide intuitive evidence, which is further supported by our representational form of formulas as *matrices*, which are defined exclusively in terms of set-theoretic notions and which can be nicely illustrated in the plane w.r.t. the notions relevant for theorems like *paths, connections*, and others. Also for any further algorithmic or other details, discussions, illustrations, references, etc., the reader once more is referred to [4].

In each of the subsequent sections one further of these versions H_i will be presented, with a summary of all 5 versions listed in a table concluding the paper.

1. A CHARACTERIZATION OF VALIDITY IN PROPOSITIONAL LOGIC

Logicians with interest in ATP tended towards ignoring the propositional features of proof procedures, which are important in practice, however. We therefore begin with an appropriate characterization of validity on this *ground level* which might be regarded as a degenerated version H_0 of Herbrand's theorem. At the same time this introduces our set-theoretic representation of formulas.

Propositional variables are denoted by P,Q,R; *occurrences* are denoted by r .

A (*ground*) *literal* is a pair (n,P) with $n \in \{0,1\}$. Literals are denoted by K,L,M . For L=(n,P) and $m \in \{0,1\}$, we define ${}^{m}L = (n+m \bmod 2,\ P)$. By convention we write P for (0,P) and $\neg P$ for (1,P) .

(Propositional) matrices, denoted by D,E,F, are defined inductively by (m1),(m2) .

m1. Any pair (L,r) , shortly written L^r , or even L if r is understood, is a matrix.

m2. If $F_1,\ldots,F_n$ with $n\geq 0$ are matrices then $\{F_1,\ldots,F_n\}$ is a matrix.

$\{\{L\},\{^1L\}\}$ and $\{\{\},K,\{\{K\},M\}\}$ are two examples of such matrices. They represent propositional formulas obtained by alternatively interpreting the set constructor $\{\}$ by $\vee$ and $\wedge$. If in this translational process we begin with interpreting the outermost set constructor as $\vee$ $(\wedge)$, then we speak of the *positive* (*negative*) representation. In this paper we always have in mind the positive representation. Thus the 2 previous matrices represent the formulas $L\vee\neg L$ and *true* $\vee$ $K \vee (K\wedge M)$, respectively. In fact, each represents more than one formula, for instance, the first one also represents $\neg L\vee L$, $L\rightarrow L$, $\neg L\rightarrow\neg L$, etcetera, which truth-functionally all are equivalent.

A *path through* a matrix F is a set of (occurrences of) literals which is defined inductively by (p1) through (p3).

p1. If $F = \emptyset$ then the only path through F is $\emptyset$.

p2. If $F = L^r$ then the only path through F is $\{L^r\}$.

p3. If $F = \{L_1,\ldots,L_m,F_1,\ldots,F_n\}$ where $0\leq m,n$, $1\leq m+n$, and F_i is not a literal, $i=1,\ldots,n$, then for any matrix $E_i\in F_i$ and any path p_i through E_i, $i=1,\ldots,n$, the set $\bigcup_{j=1}^{m}\{L_j\} \cup \bigcup_{i=1}^{n} p_i$ is a path through F .

The paths through a matrix are best illustrated in a two-dimensional display where the matrices of the given matrix are displayed left-to-right, their matrices top-down at a smaller distance, and so on, always alternating left-to-right and top-down, and decreasing the distance. If this loop terminates after the first round, we speak of *normal* form matrices, which is the case for the matrix $E1 = \{L,\{K,\{^1L\}\},\{^1K,M\},\{^1M\}\}$, thus displayed as

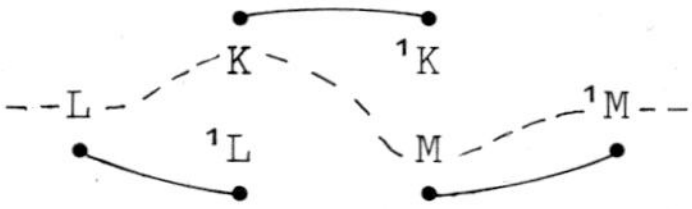

(for the moment ignore the lines). Note that in our representation left-to-right corresponds to disjunction and top-town to conjunction. Now, a path through any such matrix is obtained by crossing the matrix left-to-right and selecting among the top-down items, as illustrated with the dashed line in the previous picture. It should now be obvious that these paths represent the conjuncts of the formula obtained by transformation into conjunctive normal form of the formula represented by the given matrix. With this in mind, H_0 stated after the following notions is in fact an obvious result.

A *connection* in F is a two-element subset of a path through F of the form $\{L,\neg L\}$.

A set W of connections in F is called *spanning for* F if for each path p through F there is an $w \in W$ with $w \subseteq p$.

F is called *complementary* if there exists a spanning set of connections in F .

THEOREM H_0 . (Any formula represented by) a matrix F is valid iff F is complementary.

The previous matrix E with its 3 connections thus is in fact valid.

The value of this result lies in the computational use which we can make of it. In accordance with the purpose of the present paper we will not go into any details, except for an illustration mainly provided by figure 1. It shows a deduction of E in the *connection calculus*. The important point to note is the fact that the matrix itself never changes during the deduction. Rather the changes are encoded by pointers and a dashed line, called a *structure* for the matrix, while the connections have been added only for a more intuitive illustration.

Thus formally we are considering a (complete and consistent) calculus with rules of the form (F,S) $\vdash$ (F,S'), where the structure S varies but the formula remains fixed, further with special initial and

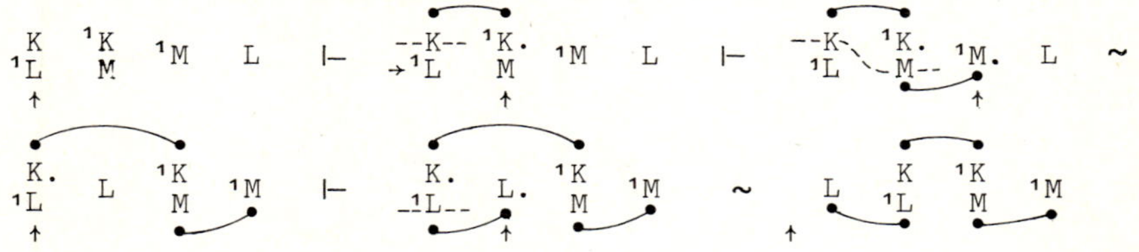

Figure 1. A connection deduction for the matrix E1

terminating structures. Each structure in the simplest case of normal form matrices essentially consists of 3 boolean functions which in practice may be encoded with a few additional bits. One might even think of a realization in hardware.

Besides this obvious representational advantage, this calculus has been developed upto a point where it is provably more efficient than any other known calculus [4]. In particular, it applies to arbitrary matrices (not only normal form ones). Moreover, these features may be *lifted* to the level of first-order logic. We, however, will continue to focus our interest to the aspects connected with lifting and refining the theoretical basis H_0 of this calculus.

2. A BASIC VERSION OF HERBRAND'S THEOREM

The set-theoretic representation of formulas introduced in the previous section for propositional logic may be easily generalized to the first-order level. For instance, the matrix
$^{\{c\}\{x,y\}}\{\{Pc,Pfx\},{}^{1}Py\}$ represents the formula $\forall c\exists x\exists y((Pc\wedge Pfx)\vee\neg Py)$.
Though we always will think in set-theoretic terms and thus take for granted the resulting comfort as explained in the previous section, we will not bother the reader with this unfamiliar notation and rather explicitly use the traditional one, with the following stipulations reflecting the set-theoretic background.

We use the logical symbols $\neg,\wedge,\vee,\forall,\exists$, where $\neg$ occurs in literals (such as $\neg Py$) only. In the case of $\forall c$ we speak of the *bounded constant* c , and in the case of $\exists x$ of the *bounded variable* x . The formulas are to be taken only modulo associativity and commutativity both in their propositional structure and within sequences of quantifiers, as illustrated by the previous example. We take scarce use of parentheses if possible, such as in Pfx which actually means P(f(x)) . With all denotations we follow fixed conventions which will be clear by the context.

Substitutions are defined as usual. Our notation is, e.g., $\sigma = \{x\backslash t1, y\backslash t2\}$ to denote the substitution of the variable x by the term t1 and y by t2 ; thus $P(x,y,c)\sigma = P(t1,t2,c)$ where we write σ behind the literal as usual in ATP.

There are two ways to generalize the notion of a path to the first-order level. In the first one, subformulas determined by an outermost quantifier are treated just like literals, in which case we speak of *propositional paths*. As in section 1 this notion implies those of *propositional connections*, *spanning* sets of propositional connections, and *propositionally complementary* formulas (such as $Px \vee \neg Px \vee \forall c Qc$). A subformula E in F is called a (*propositionally*) *minimal* subformula if $E \in p$ for some propositional path through F.

For the second way yielding the general notion of a *path*, the definition of section 1 is applied thereby ignoring all quantifiers. This notion will not be used until the next section.

We now define a *standard procedure* SP applicable to any formula F and based on some fixed enumeration $t_1, t_2, \ldots$ of all terms.
STEP0. Let $i \leftarrow 1$ and attach 1 to any existential quantifier as an upper index ($\exists^1$).
STEP1. If F is propositionally complementary then return "valid".
STEP2. If there is no minimal subformula of the form $\exists^k x E_0$ for $k \leq i$ or $\forall c E_0$ in F then let $i \leftarrow i+1$, and if there are still no minimal subformulas of the form $\exists^i x E_0$ in F then return "invalid".
STEP3. Select a minimal subformula E in F of the form $\forall c E_0$ or $\exists^k x E_0$ for $k \leq i$; in the case $E = \forall c E_0$ substitute E by $E_0\{c \backslash c_j\}$ in F and go to STEP1, where j is the smallest index such that c_j does not occur free in F; in the case $E = \exists^k x E_0$ substitute E by $E_0\{x \backslash t_j\} \vee \exists^{k+1} x E_0$ in F (*expansion*) and go to STEP1, where j is the smallest index such that t_j has not been used before in a substitution of $\exists^{\tilde{k}} x E_0$ with $\tilde{k} < k$, and no variable or constant in t_j is bound in E_0 by some quantifier.

This tool SP is quite a familiar one except that here SP leaves the propositional structure of the original formula completely untouched. This is possible because of our version of theorem H_0 from the previous section. Otherwise the proof of the following result goes as usual (with several straightforward inductions and König's lemma).

THEOREM A. A formula is valid iff SP, with input F, terminates after a finite number of steps returning "valid".

This theorem lays the ground for all our results. For instance, H_1 below is simply a corollary. For its formulation, we call, for any formula $F = \forall c_1 \ldots \forall c_m \exists x_1 \ldots \exists x_n F_0$ in *Skolem normal form*, $F_1 \vee \ldots \vee F_k$ a *compound instance* (or an *expansion*) *of* F if

$F_i = F_o\{x_1\backslash t_{i1},\ldots,x_n\backslash t_{in}\}$ for any *Herbrand terms* $t_{i1},\ldots,t_{in}$ of F , $1\le i\le k$, $k\ge 1$.

COROLLARY H_1. A formula in Skolem normal form is valid iff it has a complementary compound instance.

In this formulation, it is actually a generalized version of what is known as the *Skolem-Gödel-Herbrand* (or *expansion*) *theorem*, the generalization being captured in our notion "complementary". But its features which are of interest for ATP are really those due to Herbrand. In this sense we may regard it as a version of Herbrand's Fundamental Theorem.

The connection calculus developed on its grounds and generalized from that on the propositional level is illustrated in figure 2 with a deduction of the formula $\forall c\exists xy(\neg Px \vee \neg Qy \vee (Pfy \wedge Qgc))$ in the same style as in figure 1. Here the structures in addition involve substitutions which are determined by *unification*. The example, because of its simplicity, does not demonstrate the features of *expansion* and of *backtracking*, which, in contrast to the propositional level, are intrinsic on the first-order level. Both are sources of redundancy, for which reason we are looking for refined versions of H_1 .

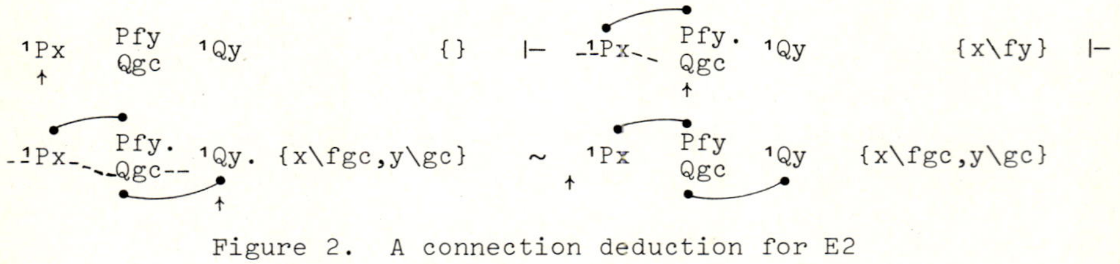

Figure 2. A connection deduction for E2

3. ENCODING EXPANSION WITH INDEXING

The redundancy caused by generating copies of subformulas by expansion can be easily eliminated by a simple indexing idea. For illustration, consider the "father-grandfather" formula $\forall ab\exists xyzuv(\neg Fau \vee (Fzy\wedge Fyx\wedge\neg GFzx) \vee Gfvb)$, which cannot be proved without expansion. With 2 copies we easily obtain a spanning set of connections and an appropriate substitution as shown in figure 3. The same information apparently is contained in

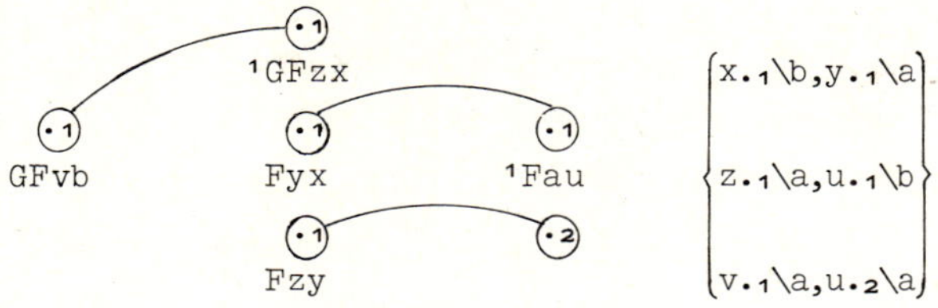

where the 2 copies are distinguished by 2 indices $._1$ und $._2$. The index encoding affects our basic notions in the following way.

For any formula F a *multiplicity* μ is a function assigning to each occurrence r of a subformula of the form $\exists x_1 \ldots x_n F'$ for some F' in F a natural number $\mu(r) \geq 1$ (determining the number of copies to be considered). F together with μ will be written F^μ.

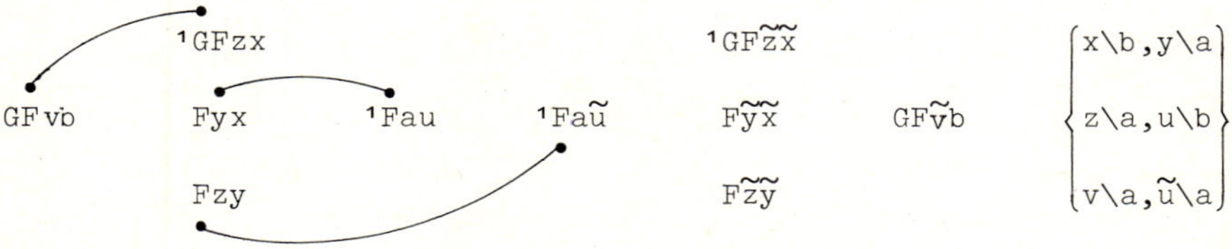

Figure 3. A proof for the father-grandfather formula

In the previous section we have already introduced the notion of a *path* (ignoring quantifiers). An *indexed path* $p_{.i}$, $i \geq 1$, *through* F in Skolem normal form is a path through F where each variable x occurring in any of its literals L is indexed by i, i.e. $x_{.t}$. We also write $L_{.i}$.

For any formula $F = \forall c_1 \ldots c_m \exists x_1 \ldots x_n F_0$ in Skolem normal form with multiplicity μ, a *path through* F^μ is a set $\bigcup_{i=1}^{\mu(r)} (p_i)_{.i}$ where r denotes the occurrence of the subformula $\exists x_1 \ldots x_n F_0$ and p_i is any path through F which is then indexed by i, $i=1,\ldots,\mu(r)$.

In the previous example we have had $\mu=2$. Note that these definitions are expressed with a tendency towards arbitrary formulas to be considered in the next section.

A (*first-order*) *connection* in a Skolem normal form formula F^μ is a 2-element subset of a path through F^μ of the form

$\{(Ps_1 \ldots s_n)_{.i}\ ,\ (\neg Pt_1 \ldots t_n)_{.j}\}$. A set of connections in F^μ is called *spanning* if each path through F^μ contains at least one of these connections. With these notions it is obvious that H_1 can now be rewritten in the following form.

COROLLARY H_2. A formula F in Skolem normal form is valid iff for some (constant) multiplicity μ there is a spanning set U of connections in F^μ and a substitution σ such that $u\sigma$ is a propositionally complementary pair of literals, for any $u \in U$.

With this result the connection calculus will now operate on a single copy of the formula to be proved, as explained for the propositional level in section one.

4. GENERALIZATION TO ARBITRARY FORMULAS

So far our results are restricted to formulas in Skolem normal form, which can be easily achieved via *skolemization*. For various reasons however, it is desirable to get rid of this restriction which will be carried out in the present section.

The idea for this generalization can be obtained by noticing the equivalent of skolemization in the standard procedure SP from section 2. In the formula $\exists x \forall c \exists y (Px \vee Py \vee \neg Pc)$ by skolemization c is substituted by a Skolem function $f_c(x)$ which prevents considering the dashed connection since the 2 terms x and $f_c(x)$ are never unifiable. This central feature of skolemization in SP is achieved via the sequence of selections of subformulas in conjunction with the conditions on constants in STEP3. Namely, the subformula $\exists x \ldots$ necessarily has to be selected *before* the one $\forall c \ldots$, and thus x must not be substituted by c . This suggests to consider an ordering $<$, which captures this notion of *before*, and appropriately embed the restrictions on substitutions, as follows.

For $<$ we consider the tree ordering of the given formula (among its occurrences) with the root being the least element, by convention often expressed in the form $x<c$ and $c<y$ in case of the previous formula.

For any substitution σ substituting terms for variables, an equivalence relation $\sim$ and a relation $<\!\cdot$ is defined by (r1) through (r3).

r1. If $x\backslash y \in \sigma$ then $x \sim y$ holds.
r2. If $x\backslash t \in \sigma$ and t is no variable, then $y \lessdot x$ $(c \lessdot x)$ holds for any variable y (constant c) occurring in t, respectively.
r3. If $x \sim y$ and $c \lessdot x$ hold then $c \lessdot y$ holds.

By $\lhd$ we denote the transitive closure of the union of $<$ and $\lessdot$.

The previous and further examples illustrate that the restrictions above equivalently can be expressed by forbidding *cycles* in $\lhd$. This, however, has now to be carried out for formulas F^{μ} with arbitrary multiplicity μ.

For any formula F and any of its occurrences r, by $F_{:r}$ we denote the subformula occurring at r. Now, for F^{μ} let $\Omega(F)$ denote the set of all the nodes in the formula tree of F. Then the set of its *index nodes* $\Omega(F^{\mu})$ is a set of pairs (r,κ), or shortly r^{κ} such that (i1) and (i2) holds.
i1. $r \in \Omega(F)$.
i2. If the set $\{r_1,\ldots,r_n\} \subseteq \Omega(F)$, $n \geq 0$, contains all positions r_i in F such that $r_1 < r_2 < \ldots < r_n \leq r$ and, for all i, some F_i, and some x_{ij}, $j=1,\ldots,n_i$, $F_{:r_i} = \exists x_{i1} \ldots x_{in_i} F_i$ holds, then we have $\kappa = j_1.j_2 \ldots .j_n$ and $1 \leq j_i \leq \mu(r_i)$ for $i=1,\ldots,n$.

In F^{μ}, $r1_{.\kappa 1} < r2_{.\kappa 2}$ is defined by requiring $r1 < r2$ in F and $\kappa 1.\kappa = \kappa 2$ for some κ.

For any literal L^r in F^{μ} and for any index κ with $r_{.\kappa} \in \Omega(F^{\mu})$, the *indexed* literal $L_{.\kappa}$ is the literal $L\sigma$ where σ contains all pairs $z\backslash z_{.\kappa 1}$ such that z is a variable or constant occurring in L and quantified at some position r1 in F with $r1_{.\kappa 1} \in \Omega(F^{\mu})$ and $r1_{.\kappa 1} < r_{.\kappa}$.

A *path* p *through* F^{μ} is a path through F^{μ} w.r.t. the empty index, where p *w.r.t.* (any) *index* κ is a set of (occurrences of) indexed literals in F, defined inductively by (p1) through (p5).
p1. If $F=L^r$ then $\{L^r_{.\kappa}\}$ is the only path through F w.r.t. κ.
p2. If $F = \vee(F_1,\ldots,F_n)$, $n \geq 0$, then for any path p_i through $F_i^{\mu_i}$ w.r.t. κ, where μ_i is μ restricted to F_i, $i=1,\ldots,n$, $\bigcup_{i=1}^{n} p_i$ is a path through F^{μ} w.r.t. κ.
p.3. If $F = \wedge(F_1,\ldots,F_n)$, $n \geq 0$, then any path p_i through $F_i^{\mu_i}$ w.r.t. κ, $i \in \{1,\ldots,n\}$, is a path through F^{μ} w.r.t. κ.
p.4. If $F = \exists^r x_1 \ldots x_n F_0$ with $\mu(r)=m$, then for any path p_i

through $F_o^{\mu o}$ w.r.t. $\kappa.i$, for $i=1,\ldots,m$, $\bigcup_{i=1}^{m} p_i$ is a path through F^μ w.r.t. κ .

p5. If $F = \forall c_1 \ldots c_n F_o$ then any path p through F_o^μ w.r.t. κ is a path through F^μ w.r.t. κ .

A *connection* in F^μ is a 2-element subset of a path through F^μ of the form $\{(Ps_1 \ldots s_n)^{ro}_{.\kappa o}, (\neg Pt_1 \ldots t_n)^{r1}_{.\kappa 1}\}$. *Spanning* is defined as before.

The definition of $\sim, <\cdot$, and $\lhd$ given above is literally the same in the present case with indexed variables and constants. With all these notions we are now able to formulate a version H_3 of Herbrand's theorem for arbitrary formulas.

THEOREM H_3. A formula is valid iff for some multiplicity μ there is a spanning set U of connections in F^μ and a substitution σ such that $u\sigma$ is a (propositionally) complementary pair of literals for any $u \in U$, and the relation $\lhd$ determined by σ and F^μ has no cycles.

In comparison with the version H_2 we see that skolemization has been replaced by a test for cycles. Now we know from unification theory that a test for cycles is involved in all fast unification algorithms anyway. Therefore, H_3 suggests a more direct solution for ATP since the detour of introducing Skolem functions is avoided without causing any kind of additional complication.

5. A VERSION FOR SPLITTING BY NEED

In the previous sections we have provided the tools by which the redundancy caused by explicit expansion may be eliminated. Nevertheless, an increase of the values of the multiplicity, still may complicate the search for a proof considerably since the number of indexed paths may increase exponentially. It is therefore desirable to find a proof with as small a μ as possible.

There is a rule, in ATP known as *splitting*, by which the values of μ may be decreased. For instance, $\exists x \neg Px \vee Pa \wedge Pb$ represents a proof with $\mu=2$ for the occurring $\exists$. However, this formula apparently may be splitted into two independent formulas $\exists x \neg Px \vee Pa$ and

$\exists x \neg Px \vee Pb$. Moreover, this splitting has not to be carried out explicitly rather with some proviso it appears to be reasonable to regard

$\exists x \neg Px \vee Pa \wedge Pb$ with $\mu=1$ as an acceptable proof. Of course, the example is too simple to demonstrate the gains. But the reader may easily think of a large formula instead of $\neg Px$ in order to believe that the smaller index in fact may have a substantial effect. Therefore we are looking for a general method which in such cases allows for the simpler proofs with smaller μ-values whenever possible. Our last version H_4 provides a flexible solution to this problem.

In order to describe the idea behind this solution, we consider our standard procedure in a modified form, say SP', such that E to be selected there in STEP3 may also be of the form $C \wedge D$ resulting in 2 independent subcases and for the previous 2 quantifier-cases for E it is now required that E is not subformula in a subformula of the form $C \wedge D$ in F . In other words, SP' is the standard procedure usually considered in logic which also includes a transformation into conjunctive normal form.

For the previous example SP' may select among the left and the right disjunct, but splitting occurs only if the right one is preferred, that is, carried out *before* handling the left one. As in the previous section we again encounter here such *before*-relation which will be formalized by extending the relation $<\!\cdot$ to include also pairs of occurrences of $\wedge$-symbols and existentially quantified variables x in the form $\wedge <\!\cdot\, x$. For instance, $\wedge <\!\cdot\, x$ is all we need in the previous proof with $\mu=1$ as the proviso mentioned there. With the following admittedly messy details we are doing nothing more than incorporating this simple idea in general terms into our previous notions.

For any formula F with the set $\Omega(F)$ of its positions and for any such position $r \in \Omega(F)$, we let λr denote the lable at this position in the formula tree.

A formula F is called a *contracted* formula if for any two positions r and q from $\Omega(F)$, such that q is a successor of r , the following 2 properties hold:

c1. λr and λq are not of the same logical sort, that is, not both $\wedge$, nor both $\vee$, nor both (sets of) constants, nor both (sets of) variables.

c2. If $\lambda r = \wedge$ then λq is not a (set of) variable(s).

For any contracted formula F with multiplicity μ and for any binary relation $<\cdot$ on the set $\Omega(F^\mu)$ of indexed positions in F^μ, the transitive closure of $< \cup <\cdot$ is denoted by $\lhd$. Then $<\cdot$ is called a *skeleton ordering* if the conditions (s1), (s2) and (s3) hold.

s1. For any 2 indexed positions $r1_{.\kappa 1}$ and $r2_{.\kappa 2}$ from $\Omega(F^\mu)$, $r1_{.\kappa 1} <\cdot_{.\kappa 2}$ implies the properties (i), (ii), and (iii).

i. One of the following two possibilities holds:

i.1. There is some indexed node $r_{.\kappa}$ in F^μ, labeled with an $\vee$, such that $r1_{.\kappa 1}$ and $r2_{.\kappa 2}$ occur in different subtrees of $r_{.\kappa}$ w.r.t. $<$ (which implies $\kappa i = \kappa.\kappa i'$ for some $\kappa i'$, $i=1,2$).

i.2. There is some indexed node $r_{.\kappa}$ in F^μ, labeled with some variable x, that is, $x \in \lambda r$, such that $r_{.\kappa} \leq ri_{.\kappa i}$ and $\kappa i = \kappa.\kappa i'$ for some $\kappa i'$ with $\kappa 1' = \kappa 2'$ holds, for $i=1,2$.

ii. The label of $r1$ is an $\wedge$ or a constant, and that of $r2$ is a variable, that is, $\lambda r1 = \wedge$ or $a \in \lambda r1$, and $x \in \lambda r2$, for some a and x.

iii. For no indexed position $r_{.\kappa}$ in F^μ, $r1_{.\kappa 1} \leq\cdot\, r_{.\kappa} < r2_{.\kappa 2}$ or $r1_{.\kappa 1} < r_{.\kappa} \leq\cdot\, r2_{.\kappa 2}$ holds, that is $<\cdot$ in this sense is a minimal relation.

s2. $\lhd$ has no cycles, that is, for no indexed position $r_{.\kappa}$ in F^μ, $r_{.\kappa} \lhd r_{.\kappa}$ holds.

s3. If for 2 indexed positions $ri_{.\kappa i} \in \Omega(F^\mu)$, $i=1,2$, both $r1$ and $r2$ are labeled with variables and $\kappa i = \kappa.mi$ for some index κ and some number $mi \geq 1$ holds for $i=1,2$, then $r1_{.\kappa 1} \lhd r2_{.\kappa 2}$ implies $m1 < m2$.

A *skeleton* is a triple $(F^\mu, U, <\cdot)$ where F^μ is a formula F with multiplicity μ, U is a spanning and non-empty set of connections in F^μ, and $<\cdot$ is a skeleton ordering for the contracted F.

For any skeleton $(F^\mu, U, <\cdot)$ the following concepts are defined.

U is called *locally consistent* if for each connection $u \in U$ there is a substitution σ such that $u\sigma$ is a (propositionally) complementary pair of literals. For notational simplicity, we may think in the sequel of U as of such a locally consistent set.

For any unordered pair of indexed occurrences of terms a function DIFF is defined inductively as follows.

d1. $\mathrm{DIFF}\{(t, r1_{.\kappa 1}), (t, r2_{.\kappa 2})\} = \emptyset$.

d2. If $s = fs_1 s_2 \ldots s_n$ and $t = ft_1 t_2 \ldots t_n$ then

$DIFF\{(s,r1_{.\kappa 1}),(t,r2_{.\kappa 2})\} = \bigcup_{i=1}^{n} DIFF\{(s_i,r1_{.\kappa 1})(t_i,r2_{.\kappa 2})\}$.

d3. $DIFF\ (s,r1_{.\kappa 1}),(t,r1_{.\kappa 2})\} = \{\{(s,r1_{.\kappa 1}),(t,r2_{.\kappa 2})\}$ in any other case not covered by (d1) and (d2).

For any connection $u=\{(Ps_1\ldots s_n)^{r1}_{.\kappa 1},(\neg Pt_1\ldots t_n)^{r2}_{.\kappa 2}\}\in t$ $DIFF(u) = \bigcup_{i=1}^{n} DIFF\{(s_i,r1_{.\kappa 1}),(t_i,r2_{.\kappa 2})\}$ where the terms are taken from the *indexed* literals. For the whole set U, $DIFF(U)=\bigcup_{u\in U} DIFF(u)$.

For any $u\in U$, an "ignorance" function IGN_u determines a subset in $DIFF(U)$ such that $\{(s,r1_{.\kappa 1}),(t,r2_{.\kappa 2})\}\in IGN_u(U)$ holds if for some indexed position $r_{.\kappa}\in\Omega(F^{\mu})$, labeled with an $\wedge$, and for the indexed position $ro_{.\kappa o}$ of one of the literals in u the following conditions (i1), (i2) and (i3) are satisfied.

i1. $r_{.\kappa} < ro_{.\kappa o}$

i2. $ro_{.\kappa o}$ and $r1_{.\kappa 1}$ occur in different subtrees of $r_{.\kappa}$.

i3. t is a variable x indexed with ι such that $x\in\lambda q_{.\iota}$ for some node $q_{.\iota}\in\Omega(F^{\mu})$ with $r_{.\kappa}\lhd q_{.\iota}$.

Let $PART_u(U) = DIFF(U)\backslash IGN(DIFF(U))$, for any $u\in U$; for such a set of pairs of indexed occurrences of terms a substitution is called a mgu if it is a mgu of the set of pairs of the terms themselves (without the indexed positions).

Then U is called *unifiable w.r.t.* $\lessdot$ if for each $u\in U$ there is a mgu σ_u of $PART_u(U)$ such that the relation determined by σ_u is a subrelation of $\lhd$. (If a set of connections is not locally consistent then unification is never possible, with or without splitting; hence the restriction to a locally consistent U is not serious, but simplifies the situation.)

The only difficult part of this sequence of definitions is the one concerned with the ignorance function which for any connection $u\in U$ determines a certain set of term-pairs. The reader should instantiate this definition for the case of the previous example. Its proof involved two connections, $u1=\{{}^1Px^r,Pa^p\}$ and $u2=\{{}^1Px^r,Pb^q\}$, with $\mu=1$ and $\wedge\lessdot x$, where r,p,q denote the nodes for these literals and the indices are left out for simplicity. Thus one obtains $DIFF(\{u1,u2\}) = \{\{x^r,a^p\},\{x^r,b^q\}\}$, $IGN_{u1} = \{\{x^r,b^q\}\}$, and $IGN_{u2} = \{\{x^r,a^p\}\}$.

The central notion of unifiability w.r.t. $\lessdot$ given above then says that unification has not to be done for all connections in U at the same time, rather the pairs of terms may be partitioned into the subsets $PART_u(U)$, $u \in U$, for independent unification, which captures exactly our idea of implicit splitting as described at the beginning of this section. With these notions one can now prove our most refined version H_4 of Herbrand's theorem stated as follows.

THEOREM H_4. A formula F is valid iff there exists a skeleton $(F^\mu, U, <\cdot)$ such that U is unifiable w.r.t. $<\cdot$.

As a further example the reader may check the following formula

$$\forall abcd(\exists x Payx \wedge_1 \exists z Pbzx) \vee (\exists u \neg Puuc \wedge_2 \exists v \neg Pvbd))$$

is valid on grounds of its 4 connections and of $<\cdot$ determined by $\wedge_2 <\cdot x$, $\wedge_1 <\cdot u$, and $\wedge_1 <\cdot v$.

In more complicated examples it is certainly difficult for humans to decipher such a connection proof. Therefore it is important to note that skeletons actually encode the essential information for deductions in a Gentzen-like system GS (similar to one used by Schütte [9]) which may be regarded as the *generative* type version of SP' introduced further above. In other words it is straightforward to print out a connection proof in a natural-deduction-like setting which is expressed in the following result.

COROLLARY. For any skeleton $(F^\mu, U, <\cdot)$ such that U is unifiable w.r.t. $<\cdot$, a derivation δ of F in GS can be computed from $(F^\mu, U, <\cdot)$ in time linear in the size of δ .

6. SUMMARY

The main result of this paper is the theorem H_4 from the previous section. It has been obtained by successively refining versions $H_0, \ldots, H_3$ of Herbrand's Fundamental Theorem. In order to provide an illustrative comparison of their respective refinements, these altogether 5 versions are listed in the table below. In the case of H_4 the formulation of the characterizing properties has been adjusted in an obvious way for reasons of uniformity. For instance, in terms of the notions from the previous section we have
$V = \{PART_u(U) \mid u \in U\}$.

Version	Restriction on formula F	Properties characterizing validity of F
H_0	propositional	$\exists U$ s.t. - U is spanning set of connections in F
H_1	Skolem normal form $\forall a_1 \ldots a_m \exists x_1 \ldots x_n F_0$	$\exists n\ \exists U\ \exists\sigma$ s.t. - $n \geq 1$ - U is spanning set of connections for $F_0\tau_1 \vee \ldots \vee F_0\tau_n$ where $\tau_i = \{x_1 \backslash x_{1i}, \ldots, x_n \backslash x_{ni}\}$, $1 \leq i \leq n$, and - σ is substitution s.t. $u\sigma$ is complementary for any $u \in U$
H_2	Skolem normal form	$\exists\mu\ \exists U\ \exists\sigma$ s.t. - μ is constant multiplicity for F - U is spanning set of connections for F^μ - σ is substitution s.t. $u\sigma$ is complementary for any $u \in U$
H_3	none	$\exists\mu\ \exists U\ \exists\sigma$ s.t. - μ is multiplicity for F - U is spanning set of connections for F^μ - σ is substitution s.t. $u\sigma$ is complementary for any $u \in U$ and $\lhd$ has no cycles
H_4	none	$\exists\mu\ \exists U\ \exists\mathcal{V}\ \forall V\ \exists\sigma$ - μ is multiplicity for F - U is spanning set of connections for F^μ - $\mathcal{V}$ is a certain set of subsets of U - $V \in \mathcal{V}$, i.e. V is subset of U - σ is substitution such that $u\sigma$ is complementary for any $u \in V \subseteq U$ and, among several more properties $\lhd$ has no cycles

Table. A comparative display of $H_0, H_1, \ldots, H_4$

This sequence of refinements, which may be continued in various ways, has been developed in view of improving the efficiency of proof procedures for first-order logic. In fact, the connection method based on this approach is computationally more efficient than any other known proof method.

However, it seems to be obvious that cur results may also be applied for problems concerning decidability and complexity, since, for certain formula classes, H_4 together with its corollary may be used to determine a shortest possible derivation in GS for a given formula.

REFERENCES

[1] Andrews, P.B.: Theorem proving via general matings. JACM 28, 193-214 (1981)

[2] Bibel, W.: The complete theoretical basis for the systematic proof method. Bericht ATP-6-XII-80, Projekt Beweisverfahren, Institut für Informatik, Technische Universität München (1980)

[3] Bibel, W.: Matings in matrices. Proc. of German Workshop on Artificial Intelligence (J. Siekmann, ed.), Informatik Fachberichte 47, Springer, (1981) 171-187

[4] Bibel, W.: Automated theorem proving. Vieweg, Verlag, Wiesbaden (book in preparation)

[5] Gentzen, G.: Untersuchungen über das logische Schließen. Mathem. Zeitschr. 39 (1935) 176-210, 405-431

[6] Goldfarb, W.D. (ed.): Jaques Herbrand, logical writings. Reidel, Dordrecht (1971)

[7] Herbrand, J.J.: Recherches sur la théorie de la démonstration. Travaux Soc. Sciences et Lettres Varsovie, Cl. 3 (Math., Phys.) (1930) 128pp. English translation in [6].

[8] Robinson, J.A.: A machine-oriented logic based on the resolution principle, JACM 12, 23-41 (1965)

[9] Schütte, K.: Proof theory. Springer Verlag (Berlin, 1977)

Typescript: A. Bußmann

PROCEEDINGS OF THE HERBRAND SYMPOSIUM
LOGIC COLLOQUIUM '81, J. Stern (editor)

HERBRAND'S THEOREM AND PROOF-THEORY

Jean-Yves Girard
Université Paris VII

Herbrand's "Théorème Fondamental", one of the milestones of mathematical logic, is also one of the few basic results of proof-theory, only challenged (and presumably overtaken) by Gentzen's Hauptsatz.

We shall investigate the importance of Herbrand's theorem for the proof-theorist of 1981, 50 years after Herbrand's death. We shall of course consider direct uses or extensions of Herbrand's result, but also general ideas in proof-theory that are still alive and which can be- at least partly - ascribed to Herbrand : we shall analyse the main ideas in Herbrand's theorem -as they appear fifty years later - and try to see their posterity in recent work, usually not connected at all with Herbrand's theorem. The parallel with Gentzen's Hauptsatz, when possible, will provide a natural counterpoint.

It will be convenient to recall Herbrand's theorem in a particular case : assume that A is a prenex formula $\exists x \forall y \exists z \forall t\ B(x,y,z,t)$, B quantifier-free ; let f and g be new function letters of one and two arguments, and let $\underline{t} = t_1,\dots,t_n$, $\underline{u} = u_1,\dots,u_n$ be finite sequences of terms (involving f and g) ; one definies $A_{\overline{H}}^{\underline{t},\underline{u}}$:

$$A_{\overline{H}}^{\underline{t},\underline{u}} = B(t_1,f(t_1),u_1,g(t_1,u_1)) \vee \dots \vee B(t_n,f(t_n),u_n,g(t_n,u_n)).$$

The implication $\mathrm{cl}(A_{\overline{H}}^{\underline{t},\underline{u}}) \rightarrow A$ is obviously provable in predicate calculus. Herbrand's theorem states that, if A is provable, so is $A_{\overline{H}}^{\underline{t},\underline{u}}$ for some sequences $\underline{t}$ and $\underline{u}$; in fact, we have (a little) more : $A_{\overline{H}}^{\underline{t},\underline{u}}$ is a propositional tautology. Throughout the paper, we shall always refer to this particular instance of Herbrand's theorem.

1. BOUNDING THE LOGICAL COMPLEXITY

One of the applications of Herbrand's theorem is that if A is provable in predicate calculus, then it is provable by means of formulas of bounded complexity (here complexity means "number of quantifiers") : if A is provable in predicate calculus, $A_{\overline{H}}^{\underline{t},\underline{u}}$ is provable in propositional calculus (hence without quantifiers) for some sequences t and u ; it suffices to look carefully to the standard proof of $cl(A\frac{t}{H},\frac{u}{}) \longrightarrow A$ to get a reasonable bound on the complexity of formulas involved in it : this complexity depends on the actual choice of axioms and rules for predicate calculus : with our choice of A, all formulas will have, say, less than 10 quantifiers. So we have obtained a proof of A by restricted means, i.e. using a subsystem of predicate calculus with a limited number of quantifiers.

This result has been obviously superseded by Gentzen's subformula property, a standard corollary to the Hauptsatz : if A is provable in predicate calculus, then it is provable by means of sequents consisting only of subformulas of A (in Gentzen's sense $A(t)$ is a subformula of $\exists xA(x)$ and $\forall xA(x)$ for all terms t). Indeed, a close inspection would show that the subformula property is exactly the same thing as the application of Herbrand's theorem just mentioned but we must acknowledge that Gentzen's formulation is much more elegant and flexible. (*)

Typical application of bounded logical complexity is the so called reflexion schema for predicate calculus, which was originated by Kreisel (1). As often with Kreisel, the point is to find the positive content of a general property : in that case one starts from the fact that all theorems of predicate calculus are true, which is essential for the philosopher, but trifling for the mathematician, the positive content being a particular corollary of this fact, of mathematical interest. The idea of the reflection scheme is just a formalized version in arithmetic :

$$PA \vdash Thm_{PC}(A(\overline{x})) \longrightarrow A(x)$$

where PC stands for the predicate calculus. If one tries to formalize the trivial proof of this fact (i.e. by induction on proofs), then one has a problem because of Tarski's theorem saying that there is no truth predicate in arithmetic. But there are truth predicates for formulas of bounded complexity : hence the use of Herbrand's theorem or of Gentzen's subformula property to bound the logical complexity in a proof of $A(\overline{x})$ permits to prove easily the reflection scheme. It is well-known that the result subsists when predicate calculus is replaced by finite subsystems of Peano arithmetic. A standard corollary is that PA is not finitely axiomatizable, but there are less trifling applications : one can say without exageration that the reflexion scheme is one the essential tools of everyday proof theory.

It is very natural to try to extend the bounded logical complexity property to more general situations than predicate calculus. The obvious thing to generalize is not Herbrand's theorem, but Gentzen's Hauptsatz.

Schütte (2) proved the generalization of Hauptsatz for ω-logic, and it follows

(*) See Van Heijenoort (this volume) for a detailed comparizon.

that ω-logic enjoys the subformula property. The point is that arithmetic can be imbedded into ω-logic, so in some sense, we have a subformula property for arithmetic. Reflection properties can be extended to ω-logic : this leads to the important point of <u>proofs by transfinite induction</u> that is considered in part 2 below.

The solution of Takeuti's conjecture [3] semantically by Tait [4] and syntactically by myself [5] deserves little attention, simply because the concept of a cut-free proof à la Takeuti gives no usable subformula property for second order logic. In fact it can be shown [6] that many formulas of second order logic enjoy a trivial normalization theorem. [In fact this is true as soon as the formula is true : see [7]]. Takeuti's conjecture has shared the common fate of many similar problems : it was too ambitions compared with its solutions.

The situation of theories of inductive definitions is more satisfactory ; roughly speaking, one can determine three cut-elimination procedures :

(i) The translation of the inductively defined set in second order arithmetic, gives, via Takeuti's conjecture a cut-elimination theorem, see for instance Martin-Löf [8]. This theorem gives no subformula property, except for Σ^0_1 sentences.

(ii) Considerable progress has been made by Pohlers [9], Buchholz [10] and Sieg [11], simplifying and improving earlier work by Takeuti [12] Howard [13], and Tait [14]. Only partial cut-elimination is achieved (negative occurences of the inductively defined predicate are forbidden), but, for cut-free proofs, we have a genuine subformula property ; compared with (i), the range of application of the theorem has been switched from Σ^0_1 to Π^1_1 sentences. I don't know whether (i) + ω-logic would do the same thing.

(iii) In [15] I introduced a new method, based on "Π^1_2-logic". This gives full cut-elimination as well as full subformula property. The success of (iii) is presumably due to the fact that Π^1_2-logic is complete w.r.t. inductive definitions, and this is comparable to the remark that ω-logic is complete w.r.t. the integers.

The cut-elimination theorem for the Π^1_2-comprehension axiom is now pending. The natural approach is obviously the generalization of Π^1_2 - logic to "Π^1_3 - logic". However, pertubations coming from known facts of set theory may destroy the expected symmetry, and in order to restore it, the appeal to extra axioms of set theory is perhaps necessary. (*)

2. FUNCTIONAL INTERPRETATIONS

It is evident that A is not at all equivalent with the generalized disjunction of the formulas $A_H^{\underline{t},\underline{u}}$. But A is equivalent to its <u>no counterexample interpretation</u> introduced by Kreisel [16]. Let A_K be the formula

$$\exists T \exists U \forall f \forall g\, B(T(f,g),\ (T(f,g)),\ U(f,g),\ g(T(f,g),\ U(f,g))$$

then A is equivalent to A_K, the point being that T and U can be chosen among recursive continuous functionals of type 2. The no counterexample is a natural and straightforward extension of Herbrand's theorem ; obviously $cL(A_H^{\underline{t},\underline{u}}) \longrightarrow A_K$ (let $\theta(f,g)$ be the smallest i such that $B(t_i, f(t_i), u_i, g(t_i, u_i))$ and let $T(f,g) = t_{\theta(f,g)}$, $U(f,g) = u_{\theta(f,g)}$). In fact, the no-counterexample interpretation is the natural extension of Herbrand's theorem to ω-logic, just as Schütte's cut-elimination extends Gentzen's Hauptsatz to ω-logic. In many situations, the two approaches are theoretically equivalent, but, contrarily to what happens for predicate calculus, the use of no-counterexample is very often more manageable than cut-elimination. Perhaps it is the right place to indulge in a short discussion in the general interest of such generalisations :

(i) In generalizations from predicate calculus to infinitary logics, something is irremediately lost ; of course generalization is made necessary by the wider range of applications (not all theorems of mathematics can be expected to be proved in predicate calculus) ; but, usually the original theorem permits deeper results : a typical example may be found in Kreisel's analysis of Roth's theorem [17] : using Herbrand's theorem, one can get a bound on the number of rationals $^p/_q$ such that $|\alpha - {}^p/_q| < q^{-2-1/n}$, α being an algebraic irrational. The analysis depends on the very choice of the terms involved in the Herbrand interpretation of Roth's proof : it is only in exceptional cases that proof-theory can give bounds for Σ_2^o - formulas.

(ii) However, the point of a good generalization is that some usable corollaries can still be enunciated : typically, cut-elimination for ω-logic as well as no-counterexample yield bounds for provable Π_2^o formulas of arithmetic. For instance, it is possible to extract the bounds implicit in the Furstenberg $\wedge$ Weiss proof of Van der Waerden's theorem, simply because this is a Π_2^o statement, by use of no counterexample [18].

(*) Remark (Oct. 81) : I have now solved this problem : extra axioms of set theory are not needed ; the method applies to Π_n^1 - CA as well.

One of the major defects of Herbrand's theorem (the same is true for no counter-example) is the bad behaviour with respect to implication. A more regular notion with respect to this problem is obtained by Gödel's functional interpretation [19], which can be thought as an unwinding of the no-counterexample interpretation by means of functionals of finite type. Gödel's interpretation has deserved a general uninterest, due to untolerable complexity, and scarce applicability, of the interpretation. The extensions to analysis deserve a fortiori the same appreciation [20], [5]. However, the case of Spector's interpretation is a bit apart, because, although this work is simply of no use as it stands, it is connected with the very important ideas of Bar-induction and Bar-recursion, which under more refined forms, can be one of the main tools of proof theory in the next years. The works of Howard and Kreisel [21], [22] have permitted to raise part of the veil, at least in the case of Bar-recursion of type 0 and 1. In fact, the idea of transfinite induction on recursive well-orders which is another formulation of Bar-induction of type 0 and 1, appears naturally when one works with ω-logic, or with the no-counterexample interpretation : this indicates surely that this principle is central among generalizations of Herbrand's viewpoint.

Bounded complexity still holds for ω-logic. From this it follows that, whenever A is provable in arithmetic, then A is provable by means of transfinite induction on formulas of bounded complexity (close to that of A) and up to an ordinal $\alpha < \epsilon_0$. This was proved by Kreisel [23]. The case of Π^0_1 formulas deserves a special mention. According to a French mathematician "Gentzen est l'homme qui a prouvé la cohérence de l'induction jusqu'à ω au moyen de l'induction jusqu'à ϵ_0", so the need of a positive content of Gentzen's or Schütte's consistency proofs is obvious. In [24], Kreisel gave such a content, namely, if A is a provable Π^0_1 statement of arithmetic, then A can be proved by means of the following transfinite induction scheme : (f is a primitive recursive function)

$$\frac{B(0,\underline{Z}) \quad B(X,\underline{Z}) \rightarrow B(X+1,\underline{Z}) \quad B(X[f(X,\underline{Z})],\ \underline{Z}) \wedge X \text{ limit} \rightarrow B(X,\underline{Z})}{B(\alpha,\underline{Z})}$$

where α is (the code of) an ordinal $< \epsilon_0$, B quantifier-free.

The question of a positive content for the new cut-elimination procedure of [15], which yields a new cut-elimination procedure for arithmetic, was solved by Schmerl [25] ; if A is provable Π^0_1 sentence of arithmetic, then A can be proved by means of the scheme above, with f not depending of X, but the ordinal α can go up to the Howard ordinal η_0. The result of Schmerl is very important because it restricts the possible descending sequences : we have presumably reached here the last word on the subject, at least for Π^0_1 and Π^0_2 sentences. The theorem is connected (even technically) with my theorem of comparizon of hierarchies :

$\gamma_{\eta_o} = \lambda_{\epsilon_o}$ [26], γ and λ standing for the slows and fast hierarchies of recursive functions. Before these results (which where all obtained in the 6 last years), there was a common agreement (of course one must exclude Kreisel of this - and of any - common agreement) that God had liberally bestowed ordinals to theories, and people endeavoured to find "the" ordinal of Π^1_2-CA, nay set-theory. This simple-minded view is no longer tenable : whilst ϵ_o will continue to be the ordinal of arithmetic for a certain number of problems (recently, the work of Solovay and Ketonen [27], is a new evidence that ϵ_o is not dead), it is evident that η_o plays a presumably equally important role w.r.t. arithmetic, and that the uses of η_o in relation to PA will be numerous. However, one must say that the relation of arithmetic to ϵ_o and η_o is of a different nature : the <u>order</u> ϵ_o can be connected with PA, whereas η_o is connected to PA only when dressed with its fundamental sequences : the question of the relation of the "naked" ordinal η_o to PA is still open.

Existing ordinal analyses yield proofs by transfinite induction of bounded complexity, of their arithmetic, nay Π^1_1, theorems. Schmerl's result will be adapted easily for Π^o_1 formulas, and for the biggest of the theories, (*) the ordinal will be unchanged.

Let us come back to functional interpretation and to higher type Bar-recursion : Π^1_2-logic sheds a new light on this subject, for the following reasons :

(i) Objects of Gödel's T appear as hereditarily functors over the ordinals, commuting to direct limits and to pull-backs ([29]). Compared to known models, such as Kreisel's HRO and HEO [30] which do not say anything non-trivial on type 1 objects, we have now all the fine structure of dilators [26] that can be used, for instance, ordinals can be naturally associated to objects of type 1 of Gödel's T : if t is of type 1, let T be the associated dilator, and define $|t| = T(\omega)$. $|t| < \eta_o$.

(ii) <u>Induction on dilators</u> ([26], ch. 3) is a generalization of transfinite induction, presumably equivalent to Bar-induction of type 2. Formulas of logical complexity greater than Π^1_1, but not than Π^1_2, will certainly be provable by means of induction on dilators. A close question is to find the analogue of Schmerl's fine result for Σ^o_2 or Π^o_3 sentences ; presumably some kind of induction on dilators with very elementary sequences will give the solution.

(iii) Bar recursion of finite type can certainly be replaced by operations on

(*) For instance the Jäger and Pohlers analysis of the 1st rec. inacc. [28]

ptyxes of finite type, which generalize dilators [29], similar to Λ ([26], chap. 5). If one compares this with existent models of Bar-recursion of finite type, due to Scarpellini [31], one sees that commutation to direct limits contain the continuity properties of these functionals, but commutation to pull-backs and functoriality are a new kind of information, and an essential simplification is that ordinals are no longer type 2 objects, but of type 0.

3. FINITARY APPROXIMATIONS

One of the most characteristic ideas in Herbrand's work is that of "champ fini" (finite domain) ; roughly speaking, the finite domain consists of the points $t_1 \ldots t_n, u_1, \ldots, u_n$. So this does not mean "finite model" ; the meaning is elsewhere. Any attempt to explain this in familiar terms would be deceiving. Closely connected is the asymetry in the treatment of quantifiers (observe what happens when one replaces A by $\neg A$). The trouble of giving a satisfactory explanation of "champs finis" is perhaps the indication that there is here a direction that has not been sufficiently well understood. In order to make some progress, we shall move to Gentzen's Hauptsatz, and see which kind of finitary semantics one can derive. Cut-free systems are characterized by the independant behaviour of positive and negative occurences of connectives and quantifiers in the rules for sequent calculus. The natural semantics for such a situation is a 3-valued semantics ([32] : semi-valuations, partial valuations, [7]), or equivalently by a schizophrenic distinction between strong form !A (A is true) and weak form (A is not false) ?A, of all predicates, and of course, even the range of the quantifiers may obey to this dissociation. The cut rule expresses that $?A \rightarrow\ !A$ (i.e. not false $\rightarrow$ true) in other terms that the value " undeterminated" is absent. Of course, a theory having only infinite models can have finite 3-valued models ; if this remark is not an idle triviality, then the finite 3-valued models may serve to give explicit bounds in some questions. The typical use of this is connected with theories of inductive definitions and gives results concerning generalized recursion : Let Φ be a positive operator, and let $I\Phi^\alpha$, $D\Phi$ stand for the αth iterate of Φ and the complete iteration of Φ ; given any formula A involving $D\Phi$, one can replace, α being an ordinal and F a recursive dilator :

- negative occurences of $D\Phi$ by $I\Phi^\alpha$
- positive occurences of $D\Phi$ by $I\Phi^{F(\alpha)}$

The result of this substitution is denoted $?A_{F,\alpha}$. In [12] I proved that $A \leftrightarrow \exists F\ \forall\alpha\ \ ?A_{F,\alpha}$ and this result looks very close to Herbrand's theorem and to the no-counterexample. As a corollary, one can reduce α^+-recursion to usual recursion, i.e. by considering the extensions by direct limits functors from integers to integers ; Van de Wiele [33] has obtained similar results in the same line.

Work by Masseron [34], Ressayre [35], Simpson [36] gave proofs of these results by direct recursion theoretic arguments.

Another kind of use of infinitary constructions is the search of equivalents for comprehension axioms ; the principles of Bar-induction provide such equivalents, but they are usually unmanageable, see Kreisel ∧ Howard [21].

I would make the following conjecture : for each n, define an "exponential" $\Lambda^n \in PT^{n+1}$; then

$$\Pi^1_n \text{-CA} \Longleftrightarrow \Lambda^n \in PT^{n+1} \quad (*)$$

is formally derivable in PRA^2. For $n = 0$, and $n = 1$

$$\Pi^0_1 - CA \Longleftrightarrow 2^{Id} \in PT^1$$

$$\Pi^1_1 - CA \Longleftrightarrow \Lambda^n \in PT^2$$

Have already been proved [15] ; the problem, for $n = 2$, would certainly give "the ordinal" of Π^1_2 -CA. For the case $n = 0$, we have another equivalent, proved by Friedman [37] : namely König's lemma ; the search of natural combinatorial equivalents (in particular for $n = 1$) seems to be an essential question.

So the idea of interpretation of the infinite objects by means of finitary approximations, that would have seemed obsolete (if not paranoiac) a few years ago, seems to be central again for the research : in this spirit, we are perhaps closer to Herbrand than our fathers.

REFERENCES :

[1] Kreisel G., Levy A. : Reflection principles and their uses for establishing the complexity of axiom systems. Zeitschrift math. Logik ∧ Grundlagen der Mathematic 14 (1968).

[2] Schütte K. : Beweistheorie. Berlin 1960.

[3] Takeuti G. : On a generalized logical calculus. Jap. Jour. Math. 23, 1953.

[4] Tait W.W. : A non constructive proof of Gentzen's Hauptsatz for second order predicate calculus. Bull. Ann. Math. Soc. 72, 1966.

[5] Girard J.Y. : Une extension de l'interprétation de Gödel à l'analyse et son application à l'élimination des coupures dans l'analyse et la théorie des types. Proc. 2nd. Log. Scand. Symp. ed. Fenstad, North-Holland, Amsterdam, 1971.

[6] Kreisel G., Takeuti G. : Formally self-referential propositions for cut-free classical analysis and related systems, Dissertations Mathematicae CXVIII, 1974

(*) This question seems to be solved by the cut-elimination procedure for Π^1_n -CA.

[7] Girard J.Y. : Three-valued logic and cut-elimination : the actual meaning of Takeuti's conjecture. Dissertationes Mathematicae CXXXVI, 1976.

[8] Martin-Löf P. : Hauptsatz for the intuitionistic theory of iterated inductive definitions. Proc. 2nd. Scand. Log. Symp., ed. Fenstad, North-Holland, Amsterdam, 1971.

[9] Pohlers W. : Beweistheorie der iterierten induktiven Definitionen. Habilitationschrift München 1977.

[10] Buchholz W. : Three contributions to the conference "Recent advances in proof theory", handwritten, 1980.

[11] Sieg W. : Trees in metamathematics Ph.D. Thesis Stanford, 1977.

[12] Takeuti G. : Consistency proofs of subsystems of classical analysis, Ann. of Math. 86, 1967.

[13] Howard W.A. : A system of abstract constructive ordinals, JSL 37, pp. 355-374

[14] Tait W.W. : Applications of the cut-elimination theorem to some subsystem of classical analysis. Intuitionism and Proof theory, ed. Kino, Myhill, Vesley, North-Holland, Amsterdam, 1970.

[15] Girard J.Y. : A survey of Π^1_2 -logic. Proceedings of the Hannover conference, North-Holland.

[16] Kreisel G. : On the interpretation of non-finitist proofs, I : J.S.L. 16, 1951, II : J.S.L. 17, 1952.

[17] Kreisel G. : In Springer lecture Notes 125 (1968)

[18] Girard J.Y. : L'analyse du théorème de Van der Waerden et de sa démonstration topologique à l'aide de la théorie de la démonstration manuscript, 1980.

[19] Gödel K. : Uber eine bisher noch nicht Erweiterung des finiten Standpunktes. Dialectica 12 (1958).

[20] Spector C. : Provably recursive functionals of analysis : a consistency proof by means of principles formulated in current intuitionistic mathematics in recursive function theory, Proc. Symp. Pure Math., vol. V, AMS Providence RI, 1962.

[21] Howard W.A. $\wedge$ Kreisel G. : Transfinite induction and Bar-induction of type 0 and 1 and the role of continuity in intuitionistic mathematics. J.S.L. 31, 1966.

[22] Howard W.A. : Functional interpretation of Bar-induction by Bar-recursion. Compositio Mathematica, vol. 20 (1968).

[23] Kreisel G. : Mathematical logic, Lectures in modern mathematics, vol. III, ed. Saaty, Wiley, N.Y. 1965.

[24] Kreisel G. : Mathematical significance of consistency proofs. J.S.L. 23 (1958)

[25] Schmerl U. : Number-theory and the Bachmann-Howard ordinal, manuscript 1981.

[26] Girard J.Y. : Π^1_2 - logic, part I, to appear in Ann. Math. Log.

[27] Ketonen J. and Solovay R. : Rapidly growing Ramsey functions. Annals of Mathematics, 1981.

[28] Jäger G. et Pohlers W. : Eine beweistheorische Untersuchung von $(\Delta^1_2 - CA) + (BI)$ und verwanater Systeme der Mengenlehre, to appear in "Sitzungsberichte der Bayerischen Akademie der Wissenschaften".

[29] Girard J.Y. : Introduction to Π^1_2 - logic. Manuscript to appear in Proceedings of the Florence conference.

[30] Kreisel G. : Course notes 1958-1959.

[31] Scarpellini B. : A model for Bar-recursion of higher type. Compositio Mathematica, vol. 23, 1971.

[32] Schütte K. : Syntactical and semantical properties of simple type theory. J.S.L. 25 (1960).

[33] Van de Wiele J. : Dilatateurs récursifs et récursivités généralisées, thèse de $3^{\text{ième}}$ cycle. Paris, 1981.

[34] Masseron M. : Majoration des fonctions ω_1^{ck}- recursives par des ω-échelles, thèse de $3^{\text{ième}}$ cycle, Paris, 1980.

[35] Ressayre J.P. : Bounding generalized recursive functions of ordinals by effective functors : a complement to the Girard theorem. (to appear).

[36] Simpson S.G. : Private communication.

[37] Friedman H. :

PROCEEDINGS OF THE HERBRAND SYMPOSIUM
LOGIC COLLOQUIUM '81, J. Stern (editor)

FINITENESS THEOREMS IN ARITHMETIC: AN APPLICATION OF HERBRAND'S THEOREM FOR Σ_2 - FORMULAS

G. Kreisel

Ohio State University and Stanford University, U.S.A.

Except for (derivations of) $\forall_2$ and hence: $\forall_1 \rightarrow \forall_2$, formulas Herbrand's disjunctions are generally too convoluted for practical use. The application below derives bounds on the number of elements of the set X from simple arithmetic conditions on the disjunction of a Σ_2 formula expressing that X is finite, or, more precisely, that the size of its elements is bounded. In this way some new (proofs of) effective bounds are found, but also previously ad hoc unwindings in the literature can be interpreted as instances of the general result just mentioned. Its analysis leads to criteria for useful extensions or refinements of Herbrand's Theorem. Those criteria are then shown to correspond to conclusions derived from two extremes of logical research: from foundational discussions of aims in vogue fifty years ago, and from sifting the body of results obtained in proof theory since then. Finally, a parallel from model theory is discussed briefly.

INTRODUCTION

Herbrand's Theorem, HT for short, is familiar for Σ_1 -formulas, say $\exists x A_o$ with quantifier - free A_o, both in its model - theoretic and proof - theoretic forms. (Since free variables are admitted, the results extend to $\forall_2$-formulas.) Specifically,

> if $\exists x A_o$ is valid (or formally derived by d) then some finite disjunction $W A_o[x/t_n]$ is valid (and, for current formalizations, its formal structure is bounded by an elementary function of d);

the terms t_n are built up by superposition of those in A_o . Thus, proved existential formulas possess remarkably uniform realizations inasmuch as the t_n are independent of the relation parameters and, of course, of the range of variables involved in A_o; 'proved' because, in general,

$$(\exists x A_o) \rightarrow W A_o[x/t_n]$$

is not valid (even) for the infinite disjunction of all t_n. - Hint: consider the language of first order arithmetic with the defining equations for + and $\times$ as ($\forall_1$ -) axioms, and a diophantine equation which has non-standard, but no standard solutions. - HT for Σ_1 formulas was exploited successfully in the fifties, when Σ_2 formulas were considered too, for example, in [8], but quickly dropped: for one thing there were many more rewarding topics for logicians.

Another reason for the neglect goes back to the original purpose of HT, of eliminating all non-constructive, and certainly all non-recursive operations from proofs. Taken literally, this purpose is hard to satisfy in the case of Σ_2 formulas. These tend to be proved by use of logically compound instances of induction whose Skolem functions are generally non-recursive. So, to apply HT (for elementary logic), either complex Herbrand disjunctions of some theorem with more alternations of quantifiers enter into the picture or such non-recursive auxiliaries as the Skolem functions which are needed to reduce the instances of induction to $\forall_1$ form. The application below indicates why, even for effective estimates, not all is lost provided the original aim behind HT is appropriately modified. Warning. Though the modification is in accordance with common sense and mathematical practice, it involves a radical ideological change; roughly speaking, it applies constructivity requirements to operations but not to proofs.

Reminders. A Herbrand disjunction of $\exists x \forall y A_o(x, y)$ has the form

$$A_o(t_1, b_1) \vee A_o(t_2, b_2) \vee \dots \vee A_o(t_k, b_k)$$

where the b are new variables, and the term t_i does not contain any b_j with $i \leq j$. The disjunction is to be read as:

Either t_1 satisfies $A_o(t_1, b_1)$ for all b_1 or, if

not and $\neg A(t_1, \bar{b}_1)$ then $\bar{t}_2$, that is, $t_2(\bar{b}_1)$

satisfies $A_o(\bar{t}_2, b_2)$ for all b_2 or ...

Universal premisses are absorbed since an implication $\forall z\, B_o(z) \rightarrow \exists x \forall y A_o(x,y)$ is reduced to $\exists x \forall y [A_o(x, y) \vee \neg B_o(x)]$. NB. The latter may have much simpler Herbrand disjunctions than $\exists x \forall y\, A_o(x, y)$; for example, if $B_o(z)$ is $A_o(t, z)$.

FINITENESS THEOREMS

As is well-known even quite simple proofs of the finiteness of some set X - tacitly: of natural numbers - need not provide any effective bound on the number, let alone on the size of the elements of X. One way of finding out whether a proof does provide either kind of bound is to inspect it by the light of nature. Occasionally, it is more convenient to apply HT to the proof, and to check whether the terms satisfy appropriate conditions, such as those listed below, which ensure a bound (on the number of elements). Evidently, the details depend on the particular formula used to express that X is finite (for the usual meaning of the symbols, here assumed to be determined by $\forall$-axioms). We take the

Σ_2 formula

$$\exists n \forall m [n < m \to \neg X(m)].$$

with the Herbrand disjunction: $\ldots \vee [t_i < b_i \to \neg X(b_i)] \vee$, for $1 \le i \le k$. It seems open whether there are useful conditions for the application of HT when finiteness in the sense of: the cardinal of X is bounded (by p), is used. The obvious corresponding formula uses a richer vocabulary of finite sequences s of length $\bar{s}$ with elements s_i $(1 \le i \le \bar{s})$

$$\exists p \forall s \{[p \le \bar{s} \wedge (\forall i \le \bar{s})\, X(s_i)] \to (\exists j < i \le \bar{s})(s_i = s_j)\}$$

Arithmetic growth conditions. Suppose there are effectively computable c_i: $1 \le i \le k$, s.t. for the terms t_i in the Herbrand disjunction above

$$t_1 \le c_1 \text{ and, for } 1 < i \le k \text{ and all } b,$$
$$t_i(b_1, \ldots, b_{i-1}) < c_i + \max(b_1, \ldots, b_{i-1}).$$

Then the number of elements in X is bounded by Σc_i.
NB. It is not assumed that the t_i are (known to be) computable. The result extends to sets X depending on some parameter π if the c_i are computable functions of π.

To verify the observation above, define simultaneously sets X_i and numbers $\bar{b}_i$ as follows:

$X_1 = X \cap \{b: b \le t_1\}$; if $X - X_1$ is empty, so are all X_j for $j > 1$; otherwise let $\bar{b}_1$ be the least element of $X > t_1$.
Suppose $X - X_1 - \ldots - X_i$ is not empty, and $\bar{b}_1, \ldots \bar{b}_i$ are defined. Then $X_{i+1} = X \cap \{b: \bar{b}_i < b \le \bar{t}_{i+1}\}$ where $\bar{t}_{i+1} = t_{i+1}(\bar{b}_1, \ldots, \bar{b}_i)$.
If there is an element of X beyond $\bar{t}_{i+1}$, let $\bar{b}_{i+1}$ be the least one.
Since each X_i contains at most c_i elements and $X = \cup X_i$, the result follows.

Corollary. The conclusion applies unchanged if the finiteness of X is proved from a true $\forall_1$ statement $\forall z\, B(z)$ because all disjuncts $\neg B(t_i)$ appearing in the relevant Herbrand disjunction are false.

Remark. When HT is applied to other branches of mathematics, for example, topology or function theory, useful additional conditions will, of course, involve: not the rate of growth, but continuity or analyticity (of the functions defined by terms in Herbrand's disjunctions).

Diverse comments: (i) An obvious area for applying the observation on arithmetic growth conditions is the literature on bad approximations of algebraic numbers α by rationals (in their lowest terms), X being the set of exceptionally good approximations in the sense of the theorem considered; for example, in Roth's Theorem, for a given α, X depends on n :

$$\{ p/q: |\alpha - p/q| \le q^{-2-n^{-1}} \} .$$

As it stands, Roth's proof derives a contradiction from assuming X to be infinite: bounds on the number of elements are obtained by applying HT, and checking growth conditions. Nothing would be gained in this way in the cases of Thue's, Siegel's, Dyson's or Schneider's earlier theorems because, for any α, they provide an obvious list of, say k, exceptional approximations and derive a contradiction from assuming that there are 2 more: so the bound k + 1 stares us in the face. For earlier remarks on the topic, see p. 110 of [7]. Correction. Baker's result on $\sqrt[3]{2}$ was vaguely described there as an instance of his later work. But the special properties of hypergeometric series, sufficient for his result on $\sqrt[3]{2}$, since improved in [1], are not used for his general results on linear forms of logarithms. (ii) In constrast to the use of HT for Σ_1 or $\forall_2$ theorems in the fifties, at least so far the analysis of finiteness theorems in (i) has not provided the first nor, even temporarily, best bounds, but answered different questions: Can bounds be extracted from given proofs at all? and if so: Which improvements need novel tricks? In a general way this kind of theoretical explanation may be compared to physical theory which explains why there is no perpetuum mobile, without replacing the ingenuity needed to spot cheap sources of energy; cf. also the unwinding of informal proofs below. (iii) The earliest mathematical applications of HT were made by Herbrand himself, to settle some socalled solvable cases of the decision problem for elementary logic, where the classes of formulas considered are not selected by their meaning (for some reinterpretation in mainstream mathematics), but by their syntactic form. Correspondingly, the techniques used by Herbrand, besides HT, came from certain corners of combinatorial mathematics explained in [2].

Remark: Goldfarb, the second author of [2], brought HT into his talk at this colloquium to manufacture formally undecided $\forall_2$ sentences. This gives an unsatisfactory idea of the possibilities of HT because (a) the principal property of Herbrand's rules used by Goldfarb holds for all recursive rules (namely: domination of the provably recursive functions by some recursive function got from their r.e. set of indices) and (b) no memorable interpretation of the particular $\forall_2$ sentence arising from HT (and satisfied by the dominating function in (a) above) was given in the vernacular. The talk belonged to the tradition which asks: What can be done at all if one knows HT, and little else? not: Where is HT more effective than, say, modern alternatives, including methods more or less related to HT such as Kripke's n-fulfillability? cf Quinsey's thesis.

EFFECTIVE USES OF HT : PRACTICAL EXPERIENCE AND FORMAL THEORY

Broadly speaking, there are two sides to the use of HT. First, as illustrated by the examples above and in the fifties, corollaries are derived from Herbrand's disjunctions of suitable formulas A , and significant features - of the disjunctions, not merely of A itself - are isolated which make the corollaries important for the branch of mathematics to which they belong. All this is largely a matter of experience. The second step is to indicate the structural properties of proofs of A which determine those significant features, by introducing significant measures. The following reminders should be sufficient to make the broad points concrete enough for practical use.

Suitable formulas. HT for $\Sigma_1, \forall_1 \to \exists_1, \forall_2$ formulas is not only rewarding in number theory, but also in axiomatic algebra. For example, in the theory of fields, a theorem T which holds for characteristic 0 or under real closure, may be sharpened by giving bounds p and d (depending on the parameters in T) s.t. T holds for every characteristic $\geq p$, resp. for any Euclidean field in which polynomials of odd degree $\leq 2d+1$ have a zero.

Significant features. In the finiteness theorem in the last section, the significant feature of the Herbrand disjunction involved is Σc_i for the growth parameters c_i.

In applications to sums of squares in the fifties, it was the complexity of the contraction of the disjunction (over i) of: $t = \sum_j p_{ij}^2$.

Reminder: if $t = \sum_j R_j^2 \vee t = \sum_k S_k^2$ and $T = \sum_j R_j^2 + \sum_k S_k^2$ then

$$t = \frac{[\,t^2 + (\sum_j R_j^2)(\sum_k S_k^2)\,] \cdot T}{T^2};$$

thus t is a sum of squares of fractions, T being a sum of squares.

Significant measures: $\forall_1$-lemmas. As observed in the last section, for several classes of formulas A , a valid Herbrand disjunction is obtained from the so-called end piece of any derivation (of A) in which all subderivations of $\forall_1$-formulas are suppressed. So only the end piece, not the whole derivation is significant for determining the formal structure of that disjunction (but, in general, not enough for an unconditional proof of the latter). As a corollary: traditional measures, which apply to the whole derivation, are largely irrelevant for the use of HT. Reminder: The formal structure of the disjunction is generally also only loosely connected with its significant features, as in the case of Σc_i above.

A more quantitative analysis of the defects of traditional measures follows from:

Lower bounds on HT; cf. [9]. The formal complexity of Herbrand disjunctions grows very fast with that of proof figures measured in the traditional way by cut degree, number of inferences, and the like. Far from establishing that HT is practically useless, this piece of complexity theory shows that the traditional measures are superficial: they ignore actual constraints which are satisfied in cases of recognized interest (such as the simplicity of the end piece) and are significant for the use of HT. Put differently, not arbitrary proofs, only those which have a low significant measure, are relevant.

As an obvious corollary: even in the case of elementary logic there are Σ_1-theorems for which constructive analysis is demonstrably unrewarding, namely, when the size of the corresponding disjunction and that of its terms are both significant and unacceptably large, but the derivation of the theorem is acceptable. The corollary ratifies a general conviction; specifically, about the need for abstract methods (in fact, even those that are eliminated by HT).

Extensions and refinements of HT. In view of this corollary, no obvious direction of extending HT to systems going beyond logic, for example, to so-called strong subsystems of analysis, is promising. Furthermore, it runs the risk of simply establishing even larger lower bounds. Of course just working out those bounds does no intrinsic 'harm'. But it is liable to obscure a philosophical point:

> If rate of growth is relevant at all, much smaller lower bounds may be sufficient; and if it is not then the glamor of 'huge' lower bounds is a sham.

In both cases the worry about precise bounds draws attention away from the more central questions of locating the critical lower bounds, resp. the properties which, unlike rate of growth, are relevant. A more promising direction is to refine HT, for example, by establishing more recondite significant measures than the trivial one above in terms of the 'endpiece'.

Discussion. Historically, extensions of HT preceded refinements. There are many parallels in other parts of mathematics, for example, the theory of functions of a real variable. Corresponding to stronger systems, one has larger classes of functions on $\mathbb{R}$. Here the other extreme is algebraization where the class of functions admitted as solutions, is restricted but their domains are extended. In this way the problem is solved for a wider class of structures than $\mathbb{R}$ (though often the greater simplicity of the solution is the more rewarding result of algebraization). Incidentally, the uniformity property of Σ_1 or $\forall_2$

theorems proved from algebraic $\forall_2$ axioms, as established by HT, expresses this advantage of algebraization more formally.

The positive side of the parallel lies in the fact that choices between extensions and refinements, and of particular refinements had to be made in the past, and some were successful. The negative side is that the proper choice of algebraic notions is notoriously delicate. It is therefore particularly satisfying that, in the present case of developing HT, some of the basic options, though not the details of any solution, follow directly from elementary foundational ideas, of the kind that were in the air when HT was discovered.

FIFTY YEARS AGO: EXAGGERATIONS AND SECOND THOUGHTS

The background for HT was Hilbert's presentation of so-called finitist foundations for (all of) mathematics. Taken literally the claims made were not only simple but almost staggeringly simple-minded. This combination had popular appeal but, as will be documented below, little attraction for the logically fastidious. As is well-known, two elements of Hilbert's presentation go back to the great logical discoveries of the 19th century, formalization of informal proofs and the evidently elementary character of formal derivability (which Hilbert compared to solubility of diophantine equations). The remainder of his presentation is actually well understood without closer examination of what is meant by 'finitist'; in fact, it becomes more interesting if presented as a general reductionist scheme; here: for reducing all of (formalized) mathematics M^+ to its finitist part M^-.

Put this way, the failure of finitist foundations for their intended purpose, as an analysis or theory of mathematical reasoning, is not at all surprising. But also, it is a fact that finitist foundations had some, albeit limited successes, for example, with formal systems M for the successor or addition in arithmetic. So experience with the subject should come in useful for building up a flexible technology of reductions by skillful choice of M^+ and M^-, preferably - and contrary to an almost universal temptation - far removed from the original, unsuccessful choices. Generally speaking, the value of the technology for solving mathematical problems is easier to judge than its use for analysing mathematical reasoning. There it is enough if M^+ and M^- represent correct proofs relevant to the problems, while here they must represent correctly the phenomena exhibited by the (informal) proofs considered.

Below is a list of the main lessons from fifty years ago which are relevant to HT.

Normal forms for finitist assertions: $\forall_1$ formulas, or, equivalently as we know now, insolubility of diophantine equations. Since only the most elementary operations are involved, such formulas isolate the principal new element of modern foundations: restrictions on proofs (and not only operations as in the case of radicals for solving quintics). Reminder: for the mathematical uses of HT considered above, restrictions on proofs of $\forall_1$-formulas are demonstrably irrelevant.

Consistency: autonomy of finitist mathematics. Under well-known conditions, a proof in M^- of the consistency of M^+ ensures that all $\forall_1$-theorems T of M^+ are also theorems of M^-; in fact, a derivation of T in M^- is easily got from one in M^+ together with the consistency proof. Reminders: the obvious irrelevance of consistency to more complicated formulas than $\forall_1$ made a bigger impression on critics than the surprising fact that consistency was good for anything at all. For the consistency of elementary logic itself, HT adds nothing to the one-element model (of logic).

Finitist sense of non-finitist proofs. Here to every formula A of M^+ is associated an interpretation in M^-, usually WA_n, and to a derivation in M^+, one in M^- of $A_{\nu(d)}$. To avoid loss of information, A is required to follow from A_n, with an obvious extension in the case of deduction, when A is proved in M^+ from hypotheses B. Reminder: HT can be regarded as a paradigm for such interpretations; in the case of deductions d of A from B, in place of a monadic ν, we have: $A_{\nu(n,d)}$ follows from B_n.

M^+: a manner of speaking (about M^-). Here the so to speak 'theoretical' requirement above, on ν being a finitist function, is sharpened to: a practically simple function, inasmuch as a (mere) manner of speaking must be easy to eliminate. Reminder: by the large lower bounds given in [9], HT does not establish that elementary logic constitutes a mere manner of speaking, *façon de parler*, about free variable logic formalized by M^+, resp. M^-. Here it should be added that if 'manner of speaking' is to be taken literally, not merely as a metaphor for a simple transformation, the conclusion must be restated:

> either current formalizations are wildly unrealistic representations of logical reasoning or the use of their non-finitist parts is not merely a manner of speaking.

Formal representations in science and technology. For a so-called basic or fundamental scientific study the essential elements - or, in popular terms, the true nature - of the phenomena considered must figure in the representation, for example, the atomic structure of matter (involved in the phenomena). As a corollary: the conclusions claimed must be in proportion to the accuracy of the data represented. This warning applies, for example, to questions of unwinding if the latter is taken in its realistic sense, as an operation on proofs as ordinarily understood (and more or less easily identified from current texts or formalizations). Much weaker requirements are enough when specific, limited purposes are at issue, as in any technological problem. For example, suppose bounds are to be extracted by use of HT, and two derivations seem trivially different (representations of the same proof), but have significantly different Herbrand disjunctions (in the sense explained in the last section). Then we have simply learnt to revise first impressions on the nature of HT; not of the laws of thought, to use one of Hilbert's or Boole's favorite expressions: for example it is not necessary to decide whether the additional information provided by HT is 'trivial compared to the total content' of the proof in the first place, while this question would be most natural if we were concerned with the laws of thought.

Points of detail. (i) While Hilbert's consistency program caught the imagination of the public, (the irascible) Zermelo was contemptuous of formalization generally, Gödel complained sharply in [5] that consistency did not even ensure the truth of proved Σ_1-theorems in arithmetic, and Gentzen, who on p. 564 of [4] considered the assignment of a finitist sense to logically compound theorems as the principal achievement of [4], was dissatisfied because that sense was not related simply to the form of the theorem. Evidently, this was in aid of the claim that those theorems constituted a manner of speaking, to be eliminated as easily as points at infinity by (von Staudt's) bundles of parallel straight lines. Incidentally, HT was known to both Gödel, who lectured on it in Hahn's seminar soon after its discovery, and to Gentzen. But neither of them saw its relevance to their complaints. (ii) Herbrand himself did not pin-point the interest of HT in connection with weaknesses of Hilbert's program, for example, in terms of some such notion as: interpretation. He obviously 'felt' the interest, as reflected in the name théorème fondamental, perhaps to be compared to Gentzen who also could not make up his mind on the precise interest of his discovery which he called Hauptsatz (though he toyed with the idea that the subformula property was essential, by giving it a name). Their failure to analyse their principal aims is in sharp contrast to Gödel's practice in almost contemporary publications on completeness and incompleteness, and in harmony with the long delay before HT or, for that matter, Gentzen's work was widely used; 'in harmony with' and not 'the cause of' to avoid a dubious relation of cause and effect here. (iii) As far as I know, historical studies of Herbrand neglect the question whether, like Godel and Gentzen, he saw the logical atrocities committed in Hilbert's presentation of his program, and, at least vaguely, thought of HT as a correction. This kind of broad question tends to be obscured by minutiae, of Herbrand's rules or mistakes, in the fundamentalist tradition of biblical scholarship. Besides, the rules are superseded; and something like the false lemmas ought to be true for non-finitist methods to be only a manner of speaking!

Speculation about Hilbert's favorite quotation (from Gauss) on removing the scaffolding when the building is completed. Point (iii) above, but especially the obviously permanent value of the broad general lessons in this section raises the question: When it comes to the pair, those lessons and formal logical theorems (consolidating each other), which is the building and which the scaffolding?

THE LAST FIFTY YEARS

As already discussed at the end of the last section but one, developments of HT followed a well-trodden path: obvious extensions to stronger systems preceded refinements. This fitted in with the project, begun in the fifties and described as a 'calculated risk', of formalizing various stronger kinds of proof (than elementary logical reasoning) considered in the foundational literature, like finitist and predicative. Contrary to first impressions the principal problem with those foundational classifications of proofs was not vagueness, nor instability, but lack of any broad significance. This situation has since been compared to the categories used in astrology: they are perfectly precise, binding on the faithful (who presumably regard themselves as predestined to respect their astrological charts), but not at all well suited for stating, let alone understanding the facts we encounter.

A more convincing aim for developing, in fact, for refining HT was found in the last 5 or 10 years. Computers had become cheap enough to implement the following idea. Proofs of $\forall_2$- not of arbitrary! - theorems are regarded as programs, so to speak in some 'higher' programming language, for a function which realizes the $\forall_2$ theorem proved (either by constructive or non-constructive means); 'straight code' for an algorithm defining that function is then extracted mechanically by some proof - theoretic transformation like HT. Here the obvious first concern is to establish how the algorithm, the outcome, depends on - the details of - the data, that is, the formal presentation of the proof, and on the particular extraction processes available. Some of the earlier experience with unwinding proofs by hand is incorporated in the new work. Reminder: as already mentioned in the last section, on science and technology, requirements for success are modest here; for example, it is enough if programmers learn to get easily a formal presentation from the general idea of the proof one has in mind; it is not necessary to have a theory for the passage.

Functional interpretations: recursiveness, continuity and higher types. The idea of 'reducing' an arbitrary assertion, first illustrated by Herbrand's disjunctions, was extended to arithmetic in the no-countereexample-interpretation (n.c.i.) which, in contrast to HT, introduces higher types. If $\exists f \forall n \neg A_o$ is the Skolem normal form of $\neg A$ then, for obvious reasons,

$\forall f \exists n A_o$ is called the n.c.i. of A, and the map:

$f \mapsto \mu_n A_o$ is recursive and continuous in f(for the product topology).

N.B. The Skolem functions of A need not be recursive even if A is provable in predicate logic. It is an easy excercise, in hierarchy theory, that arithmetic theorems A cannot be generally interpreted by purely universal arithmetic formulas with recursive constants. In contrast, the n.c.i. of A is $\forall f A_o[n/N(f)]$, where N is a recursively continuous functional of f. For formulas A in the language of second-order arithmetic with variables over ω and $\omega \mapsto \omega$, one uses certain continuous functions of all finite types over ω. Again, a simple hierarchy argument shows that this sort of passage to higher types is needed. - Remark. Though the classification by quantifier complexity is adequate for such simple negative results it is wholly inadequate for analysing the gain achieved by functional interpretations: at the cost of a 'mild' increase in types, one has a dramatic restriction on (the higher type) operations, for example, in place of the non-recursive Skolem functions of A, a narrow subclass of recursive functionals in the n.c.i. of A. Exaggerating a bit: the categorization of formulas by their type, also called: level of 'abstraction', has some of the flavor familiar from astrological categories.

Stability results for interpretations and other transformations of given derivations. As a hangover from the unrealistic project of formalizing proofs or, more generally, thoughts correctly, there was the aim of showing that the intended unwinding was faithfully rendered by the available transformations, under the slogan: stability of E-theorems. But the same work has also an important consequence for the more realistic technological project described above. Since lower bounds for any one of these transformations are very large, they are all equally, so to speak, systematically defective. For practical success one has to separate efficient from inefficient procedures. This obvious, but decisive conclusion was implemented in Goad's pruning (of redundancies). - Reminder: The work in question illustrates two general points: first, the weakness of stability results without safeguard against systematic oversights; secondly, the use of instability or sensitivity as a means for choice (between the good and the bad).

Big effects of small changes: effects on the result of transforming different derivations which formalize 'essentially the same key idea' of a proof. As soon as people began to look for this phenomenon, many isolated facts and promising problems were found, but so far no general quantitative meta-mathematical results.

For most readers of this volume probably the issue is best illustrated by the following question from Ramsey theory: Let Π_∞ be an economical proof of RT_∞, Ramsey's theorem for partitions of the (infinite) set of all natural numbers, and let Π_F and Π_A be the - mathematically trivially different - compactness arguments which derive from RT_∞: Ramsey's own finite version RT_F, resp. its arithmetic variant RT_A due to Paris and Harrington.

The two proofs:

Π_∞ followed by Π_F, resp. by Π_A

differ only 'slightly', but the unwinding processes are 'complicated'. So it seems wide open whether the bounds obtained by unwinding these two proofs differ little or much. - NB. The derivation Π_∞, stated in an appropriate formalism, may not be long (though it will certainly be boring to read); I believe that J. Ketonen uses < 30 lines in his 'proof-checker'.

More mathematically inclined readers would prefer examples from topological dynamics, such as the proof of von der Waerden's theorem given by Furstenberg and Weiss and analysed by Girard, or from function theory such as Landau's proof of $L(1,\chi) \neq 0$ by deriving a contradiction from assuming that

$$L(s,\chi)\,\zeta(s)/\zeta(2s) \text{ has no pole at } s = 1,$$

and its modification which uses the analytic continuation, across $s = 1$, of

$$L(s,\chi)\zeta(s)/\zeta(2s) - L(1,\chi)\,\zeta(s)/\zeta(2)\,.$$

<u>Recycling traditional proof theory</u>: how to get efficiency in areas of obvious interest by going beyond them, for example, in geometry by going beyond dimensions 4 or 5. The idea, often repeated and sometimes valid, is that simplifications of proofs and constructions are needed to succeed at all in the far beyond; afterwards the improvements can be brought down to earth to get efficient solutions where they are really wanted. A promising candidate in the present context seems to be the
normalization for ZF presented in Powell's abstract at this meeting.

His own aim seems to have been the extension from the (already hopelessly 'strong') theory of species to the stronger system ZF. But another side is the use of <u>set-theoretic</u> language, with the following consequence. His scheme can be specialized to quite weak subsystems (where bounds are manageable), and applied, for example, to proofs of $\forall_2$ theorems directly. Without his scheme, the subsystems would have to be embedded in appropriate type theories, in which - the translations of-the set-theoretic axioms have to be proved, and then used as <u>cut formulas</u> of relatively high complexity. When, as here, the growth rate is high, this kind of detour may make the difference between failure and success.

Remark on a neglected contrast. The large lower bounds on HT provide, as already mentioned, a simple formal ratification of a general conviction which specialists have derived from experience: the practical need for abstract methods. So those bounds are of limited interest to specialists, and of great interest to outsiders for whom a simple proof of so to speak a point of principle replaces painful experience. Almost the opposite applies to recent results on the (high) rate of growth of solutions for specific problems in finite combinatorics which follow easily from Ramsey's theorem, Higman's on well partial orderings, Kruskal's on sequences of trees. As a matter of historical fact, experienced specialists have had trouble guessing even roughly just which corollaries do or do not have, say, rapidly growing bounding functions or straightforward direct proofs. NB. The relation of this work, on bounds, to traditional proof theory is delicate. The work sometimes uses the ordinal notations introduced for the sake of proof - theoretic analysis; but it is rewarding just because it separates them from other, irrelevant features of that analysis.

Sundry observations: (i) In the first edition of Hilbert - Bernays, vol. 2, HT is proved for (valid) prenex formulas by reduction to (the validity of) Σ_1-formulas. In this auxiliary step new function symbols f are introduced which have a very clear meaning in terms of the n.c.i.; they have to be eliminated, most simply by substituting new variables a_t for terms beginning with f, in order to get Herbrand's own form of HT, which is much less memorable. Unquestionably, the restriction to prenex formulas A is essential if HT is to be used to meet Gentzen's demand, mentioned earlier, for a 'finitist sense' which is related simply - tacitly, for us - to the form of A. This is perfectly consistent with the increase in the computational complexity of Herbrand's disjunctions that must be expected from a semantically trivial passage to prenex form, yet one more instance of the big difference between information processing by us and by current computers. (Thus we remember shapes and descriptions of substitutions separately as shown most clearly in experiments on preattentive perception.) (ii) Sensitivity of the n.c.i. to the choice of representations, for example, of so-called real number generators: a neglected point in the literature. Let F be a (recursively) continuous map of the circle into itself, and let ξ represent points of the circle; subscripts n denote the n th approximation. Then

$\forall N \exists \xi (|F(\xi) - \xi|_{\bar{N}} < \bar{N}^{-1})$ is the n.c.i. of $\exists \xi \forall n (|F(\xi) - \xi|_n < n^{-1})$,

Brouwer's fixed point theorem for F, where N ranges over continuous maps : $\xi \mapsto \omega$, and $\bar{N} = N(\xi)$. Specializing N to constants reduces the n.c.i. to

$$\forall n \exists \xi (|F(\xi) - \xi|_n < n^{-1}),$$

which merely says that, for some, possibly quite irregular sequence $\xi_n : F(\xi_n)$

and ξ_n get close. But the effect of the reduction is delicate. If the ξ were reals with N respecting the usual equivalence, N would be constant anyway. If the ξ are oscillating decimals, they form a compact though not connected space, and N can have different, but only finitely many values. It seems to be open whether the n.c.i. provides useful additional information if the coordinates ξ_i of ξ are represented by Dedekind cuts in the sense used by the topos people, that is,

pairs of enumerations of the rationals $< \xi_i$, resp. $> \xi_i$.

NB. This sense is to be distinguished from the traditional one where, as in the case of continued fractions, we have a characteristic function for the set of rationals $r: r < \xi_i$. (This sense is not suitable here since few functions F which are continous in the unit circle also have a continuous representation for the appropriate topology on those characteristic functions.) (ii) Readers may wish to compare mechanical unwinding of proofs with the type of <u>ad hoc</u> analysis used informally, for example, for Euclid's and Euler's proofs showing that there are infinitely many primes p. -(a) $1 + \Pi p: p \leq n$, is not divisible by any $p \leq n$, and so there is a prime between n and $1 + \Pi p: p \leq n$. <u>Case</u> 1. Without any further information about $p \leq n$, we have $(\Pi p: p \leq n) \leq n!$, and so by Stirling's formula, $p: 2 < n < p < 1 + (2\pi n)^{1/2}(n/e)^n$. <u>Case</u> 2. Using all that is known about $p \leq n$ (for large n, since $\theta(n) \sim n$), we have: $(\Pi p: p \leq n) \sim e^n$, and so a prime $p: n < p < e^n$. (b) Euler's proof, stripped of detail, uses the identity:

$$\prod_{p\leq n}(1-p^{-1})^{-1} = \Sigma' \frac{1}{m} \quad \text{where } \Sigma' \text{ ranges over } m: \forall p(p \mid m \rightarrow p \leq n).$$

<u>Case</u> 1. Without any further information about $p \leq n$, we have

$$\prod_{p\leq n}(1-p^{-1})^{-1} \leq \prod_{m\leq n}(1-m^{-1})^{-1}, = n.$$

But also

$$\sum_{m\leq x} \frac{1}{m} = \log x + \gamma + O(\tfrac{1}{x}), \text{ where } \gamma, = 0.577\ldots, \text{ is Euler's constant.}$$

So there is a $p: n < p < e^{n-\gamma}$. <u>Case</u> 2. By Mertens formula, for large n,

$$\prod_{p\leq n}(1-p^{-1})^{-1} \sim e^{\gamma} \log n;$$

So there is a $p: n < p < \delta^{-1} n^{\delta}$ where $\delta = e^{\gamma}$ and so $1.75 < \delta < 2$. - In short, Euler's proof without any additional information about the primes $\leq n$, actually gives a better bound than Euclid's proof together with an exact estimate for $\Pi p: p \leq n$ provided n is sufficiently large.

HOW TYPICAL IS THE FATE OF HT FOR OTHER ELEMENTARY LOGIC?

As is evident from the style of this article, the developments arising from HT seem to me to illustrate some very general points which, sub specie aeternitatis, may be as memorable as the specific contributions of HT to (elementary) logic or its other known applications. Without any attempt to assess the range of validity of those general points, they will now be stated quite briefly, and illustrated by just one other simple theorem, also around 50 years old: what used to be called finiteness, and now more often, compactness theorem for elementary logic.

First and foremost, the continued effective use of HT involves combination with facts and methods in highly developed branches of mathematics. This is in harmony with everyday experience where the value of a (simple) piece of technology like HT goes up when it has access to rich resources. In the case of HT the "entry fee" is knowledge of the resources offered by those rich branches of mathematics. The principal skill needed is to spot a good use for what we have already, rather than to manufacture costly new components, a point known elsewhere as dadaism. A convincing use, of this kind, of the finiteness theorem was recently made by van den Dries and Wilkie in [10], answering Gromov's question whether the bounds for his theorem on finitely generated groups of polynomial growth are recursive. They observed that the group-theoretical properties involved in the theorem could be expressed by recursive sets of first-order sentences. - Remark: Gromov himself had applied a (topological) compactness argument, but without paying attention to the language used. Incidentally, the term finiteness theorem seems preferable to 'compactness theorem' because its principal competitor in practice, as in the case of Gromov himself, is the introduction of a suitable topology on the objects studied, using compactness (of the space) but without attention to the logical character of the propositions about those objects.

The tortuous path from HT to refinements via extensions to stronger systems has a convincing analog in various generalizations of the finiteness theorem to languages with infinitely long expressions (in a more or less simple-minded manner: $L_{\kappa\kappa}$ for large cardinals κ, resp. to $L_{\omega_1\omega}$), with additional quantifiers and the like, occasionally turning up a trick or two applicable to elementary logic. The natural parallel to the refinements of HT considered in this article is, in the case of the finiteness theorem, its use for getting models with properties adapted to particular branches of mathematics; for example, not merely some non-standard model of arithmetic, but special ones which we can learn to visualize almost as well as the geometry of the plane $\mathbb{C}$ on which the success of analytic number theory depends (where ω is embedded in $\mathbb{C}$). - Remark. It is perhaps significant that mathematicians were so to speak instinctively attracted to proofs of the finiteness theorem by means of ultrapowers where special algebraic and other additional properties suggest themselves automatically, in contrast to the case of, say, term models.

Finally, we have the use of HT for improving - as Bertrand Russell put it: from within - Hilbert's formulation of his idea of finitist foundations; removing the particular blemishes of the consistency program which offended Godel, Gentzen and us other logically delicate souls before and after them. Speaking for myself, I find this contribution of the essence, however weak the finitist philosophy may be: the latter is so natural that, in an age of intellectual affluence, we want to put it in its place, not merely to put it out of mind. For the finiteness theorem, the background corresponding to Hilbert's foundational scheme, is a preoccupation with elementary logic, as the sum total of our means of precise definitions. This idea was quite wide spread fifty years ago; its most embarrassing form being Skolem's so-called paradox or relativity theorem. The ideal of formal precision had been clearly expressed, in connection with (model theoretic) validity of elementary formulas, by the completeness property; in modern terms, the recursive enumerability of the set of valid formulas. But here one is bound to ask, consciously or not:

What about consequence?

especially, from an infinite set of assertions. The finiteness theorem is the perfect answer since it permits a reduction to finite (conjunctions of) formulas, and hence to the case of validity. This way of looking at the finiteness theorem must have occurred to people at the time, just as it occurred to me when I tried to popularize the subject in the fifties; cf. pp. 169-170 of [6]. - Remark. These kinds of foundational uses retain their value for the common understanding, even after the grand original claims have been cut down to size. Of course, neither finitist methods nor first-order definitions are the only valid ones. But they are among those that are-trivially: valid, and - worth attention; how much attention (or refinement) is determined by research: finitist assertions seem here to stay, while the category of finitist proofs is unconvincing.

REFERENCES :

[1] Choodnovsky, G.V., Formules d'Hermite pour les approximants de Pade de logarithmes et de fonctions binomes et mesures d'irrationalité. C. r. Acad. Sc. Paris, Ser. A. 288 (1979) 965-967.

[2] Dreben, B. and Goldfarb, W.D., The decision problem (Addison Wesley Publ. Co. 1979).

[3] Fürstenberg, H. and Weiss, B., Topological dynamics and combinatorial number theory, Journal d'Analyse, 34(1978) 61-86.

[4] Gentzen, G., Die Widerspruchsfreiheit der reinen Zahlentheorie, Math. Ann. 112 (1936) 493-565.

[5] Gödel, K., Diskussion zur Grundlegung der Mathematik, Erkenntnis 2 (1931/32) 147-151.

[6] Kreisel, G., Some uses of metamathematics, Brit. Jrnl. Phil. Sc. 7 (1956)161-173.

[7] Kreisel, G., What have we learnt from Hilbert's second problem? Proc. Symposia Pure Math., 28 (1976) 93-130.

[8] Lightstone, A.H., and Robinson, A., On the representation of Herbrand functions in algebraically closed fields, J.S.L. 22 (1957) 187-204.

[9] Statman, R., Lower bounds on Herbrand's theorem, Proc. Amer. Math. Soc. 75 (1979) 104-107.

[10] van den Dries, L. and Wilkie, A.J., On Gromov's Theorem concerning groups of polynomial growth, Jrnl. of Algebra (to appear).

PROCEEDINGS OF THE HERBRAND SYMPOSIUM
LOGIC COLLOQUIUM '81, J. Stern (editor)

L'OEUVRE LOGIQUE DE JACQUES HERBRAND ET SON CONTEXTE HISTORIQUE

Jean van Heijenoort
Brandeis University

En 1929, à l'âge de vingt-et-un ans, Jacques Herbrand obtint des résultats de grande importance pour la théorie de la quantification. Celui qu'on peut mentionner en premier lieu peut s'énoncer ainsi : Etant donnée une formule quelconque $\underline{F}$ de la théorie de la quantification, il est possible d'engendrer de manière mécanique (et relativement simple) une suite infinie de formules sans quantificateurs, $\underline{F}_o$, $\underline{F}_1$,..., telle que $\underline{F}$ est démontrable en théorie de la quantification (c'est-à-dire dans l'un quelconque des systèmes équivalents qui cristallisent cette théorie) si et seulement s'il existe un nombre $\underline{k}$ tel que $\underline{F}_{\underline{k}}$ est démontrable en logique propositionnelle ou, ce qui est équivalent, est valide par les tables de vérité. Il existe plusieurs manières d'engendrer une telle suite de formules sans quantificateurs. C'est ainsi que l'on peut considérer soit ce que l'on appelle les expansions de Herbrand, soit ce que l'on appelle les disjonctions de Herbrand. En outre, bien des détails peuvent varier. Dans ce qui suit, ce sont les expansions de Herbrand que nous aurons en vue.

Un second résultat de Herbrand, lié au premier, est que la formule $\underline{F}$ peut se récupérer à partir de la formule $\underline{F}_{\underline{k}}$ à l'aide de trois règles de caractère bien particulier. Ces règles sont la règle généralisée d'existentialisation, la règle généralisée d'universalisation et la règle généralisée de simplification. Les deux premières permettent d'introduire, dans des conditions bien précises, un quantificateur, existentiel ou universel, dans une formule, et la troisième permet de supprimer, dans une formule, le second terme d'une disjonction lorsqu'il ne diffère du premier que par ses variables liées. Chacune de ces trois règles a une seule prémisse. Donc, si nous considérons la formule $\underline{F}_{\underline{k}}$, supposée propositionnellement valide par les tables de vérité, comme un axiome, les applications que nous faisons successivement des trois règles pour retrouver la formule $\underline{F}$ constituent une démonstration de $\underline{F}$ qui, regardée comme un arbre, n'a aucune bifurcation. En outre, lors de l'application de l'une des trois règles, la conclusion ne contient aucun connecteur propositionnel, aucun symbole de prédicat, aucun symbole de fonction qui ne soit contenu dans la prémisse.

Ces propriétés des trois règles permettent de retrouver la démonstration qui va de

F_k à F ; car, si l'on est déjà remonté de F jusqu'à une certaine formule, il y a, pour chacune des trois règles, un nombre fini d'applications qui aient pu donner cette formule comme conclusion et pour chacune de ces applications il est possible de présenter la prémisse. Nous pouvons donc, après un nombre fini d'essais, reconstruire la démonstration. Tel ne serait pas le cas si, par exemple, nous avions à employer dans la démonstration la règle de détachement après avoir déjà employé la règle d'existentialisation ou la règle d'universalisation. Si la formule B est la conclusion d'une application de la règle de détachement aux formules A et $A \supset B$, alors A, étant une formule quelconque prise dans un ensemble infini, est irrémédiablement perdue dans l'application de la règle. Cette possibilité de remonter dans la démonstration lorsque celle-ci s'opère selon des règles jouissant de propriétés analogues à celles des règles de Herbrand est aujourd'hui utilisée à fond dans la méthode des arbres de falsifiabilité. Le système basé sur les trois règles de Herbrand est, historiquement, le premier exemple de ce qu'on appelle aujourd'hui les systèmes sans coupure ; il jouit aussi de ce qu'on appelle la propriété des sous-formules.

Soit Q le système particulier que nous avons choisi pour la théorie de la quantification, un système du genre de ceux de Frege, de Russell ou de Hilbert-Ackermann. Nous appelons maintenant Q_H le système dont les axiomes sont les formules sans quantificateurs de Q qui sont valides (par les tables de vérité) et dont les règles sont les trois règles que nous venons de mentionner. Le théorème de Herbrand peut s'énoncer ainsi : Q et Q_H sont équivalents. La démonstration de ce théorème est constructive et les fonctions introduites au cours de celle-ci sont récursives primitives (et non simplement récursives).

Pour obtenir ces résultats, Herbrand introduisit une notion importante, celle d'une suite de champs finis. Soit F une formule du système Q, et plaçons-nous au point de vue de la satisfaisabilité, plutôt qu'à celui de la validité. Les quantificateurs universaloïdes (existentialoïdes) de F sont ceux qui deviendraient universels (existentiels) si F était mise sous forme prénexe. Dans la matrice de F on remplace chaque variable existentialoïde y par un terme fonctionnel dont les arguments sont les variables universaloïdes supérieures à y (la variable x est supérieure à y si le quantificateur liant y est dans l'étendue du quantificateur liant x, et l'on compte les variables réelles parmi les variables universaloïdes) ; la formule ainsi obtenue est la forme fonctionnelle stricte de F pour satisfaisabilité, F_{ffss} (la forme fonctionnelle aurait été non stricte si l'on avait conservé les quantificateurs universels). Pour construire la suite des champs, on part d'un domaine initial non vide, C_o, et l'on obtient C_{k+1} en adjoignant à C_k les termes obtenus en remplaçant, dans les termes fonctionnels de F_{ffss}, les arguments par des éléments de C_k, l'un au moins de ces éléments étant un élément de $\underline{C}_k - \underline{C}_{k-1}$. La formule $\underline{F}$ est vraie dans le champ C_k si et seulement si chacun des cas particuliers de $\underline{F}_{ffss}$ obtenus en employant des éléments de $\underline{C}_k$ a, par les tables de vérité, la valeur logique v pour certaines valeurs logiques assignées aux formules atomiques. Remarquons que certains de ces cas particuliers contiennent des éléments de $\underline{C}_{k+1}$, et cette vérité dans un champ fini doit être nettement distinguée de la satisfaisabilité dans un ensemble fini, c'est-à-dire celle obtenue par la transformation des quantificateurs universels en conjonctions et des quantificateurs existentiels en disjonctions. Les deux notions se rejoignent à l'infini, pour ainsi dire : si la formule $\underline{F}$ est vraie dans chacun des champs finis d'une suite infinie, elle est satisfaisable dans un ensemble dénombrable. Les champs finis peuvent être considérés comme des approximations successives de ce domaine infini. Mais ceci, bien entendu, requiert une démonstration.

Dans quelle situation se trouvait la théorie de la quantification quand Herbrand présenta ses résultats ? Frege avait, en 1879, introduit pour la théorie de la quantification des règles formelles qui permettaient de reconnaître mécaniquement si un object syntactique est, oui ou non, une démonstration. Ces règles engendrent donc un ensemble récursivement énumérable de théorèmes, et les différents systèmes proposés par Frege, Russell ou Hilbert-Ackermann avaient pour objet de retrouver cet ensemble récursivement énumérable de théorèmes. Ce qu'on demandait à ces systèmes, c'était de respecter la fiabilité et d'atteindre, si possible, la complétude. Mais, au-delà de ces exigences, seules de vagues considérations esthétiques permettaient de préférer un système à un autre. On manquait de critères objectifs pour distinguer deux démonstrations différentes du même théorème.

Chez Frege, les axiomes et les règles de la théorie de la quantification étaient encastrés dans un vaste système, car, avide qu'il était de définir la notion de nombre naturel, il lui fallait quantifier sur les fonctions. La logique, telle que Frege l'envisageait, devait prétendre à englober l'ensemble des mathématiques et peut-être même davantage (voir la préface de _Begriffsschrift_). Ce devait être un système universel, dans le sens précis qu'au niveau le plus bas les quantificateurs s'étendaient à tous les objets. Cette construction grandiose, qui révélait des ambitions de reconstruction ontologique de l'univers, laissait perplexes les mathématiciens professionnels. Les paradoxes de la théorie des ensembles inquiétaient aussi. Aiguillonné par eux, Russell avait bâti un système indéfiniment stratifié. Il avait aussi déjà introduit, par rapport à Frege, un certain relativisme, en ce sens que pour lui, sur le plan logique au moins, les objets du type le plus bas ne se distinguent que par leur place relative dans la hiérarchie. A la différence de la construction fixe de Frege, la hiérarchie des types n'est pas, chez Russell, ontologiquement ancrée à un certain point et elle peut, si l'on peut dire, se mouvoir vers le haut ou vers le bas. Comme l'écrit Russell (_Whitehead et_

Russell 1910, page 169, ou 1925, page 161), 'seuls les types relatifs sont pertinents'.

Un mathématicien comme Hilbert allait encore plus loin dans la voie de cet affranchissement à l'égard des présupposés ontologiques. Il en était venu, pour traiter des problèmes qui se présentent dans l'étude des fondements des mathématiques, à considérer l'arithmétique formulée en théorie de la quantification. Sous cette influence, et aussi à la suite des travaux de Löwenheim et de Skolem, la théorie de la quantification était devenue, dans le milieu des années vingt, un système digne d'une étude indépendante et, d'ailleurs, bientôt codifié dans Hilbert et Ackermann 1928.

On ne trouve dans les écrits de Herbrand aucune mention du nom de Frege et, s'il sut quelque chose de lui, ce ne fut, apparemment, qu'à travers les écrits de Russell ou de Hilbert. Le premier ouvrage de logique qu'il lut, ce fut, semble-t-il, Zaremba 1926, qui fut publié en français à Paris en 1926 et auquel il emprunta certains termes. En mai 1927 il se mit à lire Principia mathematica, au moins le premier volume. Ce qui nous frappe, c'est que, encore plongé dans Principia mathematica, avant même que le livre de Hilbert et Ackermann soit accessible à Paris, Herbrand isole, comme objet de ses travaux, la théorie de la quantification. Par quoi fut-il guidé ? Par une certaine méfiance philosophique à l'égard de la théorie des types, méfiance qu'il pouvait avoir puisée dans l'écrit de Zaremba ? Par certains articles de Hilbert qu'il avait alors peut-être déjà lus ? Par son instinct mathématique ?

Bien que l'oeuvre de Russell et Whitehead soit plongée dans ce courant de la logique pour lequel l'obtention de théorèmes soit plus importante que l'étude des propriétés des systèmes, il y a dans Principia mathematica un paragraphe, le paragraphe *9, d'ailleurs intitulé 'Extension de la théorie de la déduction d'un type inférieur à un type supérieur de propositions', qui établit un métathéorème pour un certain système de la théorie de la quantification. Herbrand fut arrêté par ce paragraphe *9, il l'étudia de près et, comme l'a déjà noté Goldfarb (1971, page 5) y trouva sans doute l'inspiration qui le conduisit à certaines de ses notions fondamentales.

Dans le sommaire qui présente *9 au lecteur et essaie de justifier son existence, Russell écrit que la négation et la disjonction (donc aussi les autres connecteurs propositionnels, puisqu'ils sont définis à partir de ces deux-ci) ont, appliqués à des formules sans quantificateurs, un sens différent de celui qu'ils ont lorsqu'ils sont appliqués à des formules qui contiennent des quantificateurs. Ce qui l'amenait à cette affirmation, c'est la stratification des valeurs de vérité. La formule $\forall \underline{p}(\underline{p} \vee \sim \underline{p})$, par exemple, a ce qu'il appelle une vérité seconde, alors que la formule $\underline{p}$ a, si elle est vraie, une vérité première. Cette distinction ne s'applique pas, à proprement parler, à la théorie de la quantification, où les quantificateurs ne régissent que des individus. Mais, quels qu'aient été ses soucis philosophiques, Russell ne se posait pas moins le but suivant : si l'on admet pour les connecteurs propositionnels des axiomes (obtenus à partir de schémas) où seules apparaissent des formules sans quantificateurs, établir un métathéorème qui dirait que sont démontrables les formules obtenues à partir de ces mêmes schémas au moyen de formules quelconques. Il découvre alors qu'il ne peut démontrer la formule

(1) $(\exists \underline{z}\varphi \underline{z} \vee \exists \underline{z}\varphi z) \supset \exists \underline{z}\varphi z,$

bien que le système contienne le schéma d'axiomes

$$(\underline{A} \vee \underline{A}) \supset \underline{A},$$

à moins d'admettre, parmi les schémas d'axiomes qui régissent les quantificateurs, le schéma suivant :

*9.11 $(\varphi \underline{x} \vee \varphi \underline{y}) \supset \exists \underline{z}\varphi \underline{z},$

et il donne un argument pour établir l'impossibilité de démontrer dans le système qu'il considère alors, (1) sans *9.11. Dans le paragraphe suivant de Principia mathematica, *10, il abandonne le système de *9 et adopte un système dans lequel des formules quelconques peuvent être utilisées dans les axiomes qui régissent la négation et la disjonction.

A lire la thèse de Herbrand, on s'aperçoit combien ce paragraphe *9 de Principia mathematica avait retenu son attention. La Section 4 du Chapitre 2 de la thèse s'occupe du 'lien' que la théorie de la quantification a avec la logique propositionnelle. Herbrand y énonce et démontre le théorème suivant, qui correspond tout à fait aux préoccupations de Russell dans *9 : si, dans une formule démontrable en logique propositionnelle, on remplace les symboles propositionnels par des formules de la théorie de la quantification, on obtient une formule qui est démontrable en théorie de la quantification.

La portée de ce théorème vient évidemment de ce que, dans les axiomes adoptés par Herbrand pour les connecteurs propositionnels, seules sont admises des formules sans quantificateurs. Dans la démonstration du théorème (et aussi occasionnellement dans les Sections 5, 6 et 7 du Chapitre 2) Herbrand utilise ce qu'il appellera plus tard, dans le Chapitre 5 de sa thèse, une identité normale. Etant donnée une formule $\underline{F}$ de la théorie de la quantification, soit F' une forme prénexe de $\underline{F}$. Dans $\underline{F'}$ nous supprimons chaque quantificateur existentiel et remplaçons la variable qu'il liait soit par une variable réelle (se trouvant ou non dans $\underline{F}$), soit par une variable universelle liée par un quantificateur qui, dans le préfixe de $\underline{F}'$, précédait le quantificateur supprimé. S'il existe une forme prénexe de $\underline{F}$ telle que la matrice de la formule ainsi obtenue soit propositionnellement démontrable, $\underline{F}$ est dit être une identité normale.

La considération des identités normales est liée à la règle d'existentialisation. En vertu de cette règle on peut passer de

(2) $\underline{A}(\underline{x},\underline{x})$

à

(3) $\exists \underline{y}\, \underline{A}(\underline{x},\underline{y})$,

certaines occurrences de $\underline{x}$ dans (2) ayant été remplacées par des occurrences de $\underline{y}$. Si l'on essaie de faire marche en arrière, on passe évidemment de (3) à (2) en supprimant le quantificateur $\exists \underline{y}$ et en remplaçant les occurrences de $\underline{y}$ par des occurrences de $\underline{x}$, de sorte que, si $\underline{A}(\underline{x},\underline{x})$ est démontrable, $\exists \underline{y}\, \underline{A}(\underline{x},\underline{y})$ l'est également. Nous avons là le germe d'une idée qui revient souvent dans l'oeuvre de Herbrand : remonter dans les démonstrations.

Toute identité normale est démontrable et la classe des identités normales est décidable. Pour aller au-delà des identités normales, Herbrand introduit la notion de propriété A. Soit $\underline{PM}$ une forme prénexe de la formule $\underline{F}$, $\underline{P}$ étant le préfixe et $\underline{M}$ la matrice. Soit $\underline{F}_{\underline{m}}$ la formule

$$\underline{P}'(\underline{M}_o \vee \underline{M}_1 \vee \ldots \vee \underline{M}_{\underline{m}-1}),$$

où $\underline{m}$ est un nombre naturel, $\underline{M}_o, \underline{M}_1, \ldots, \underline{M}_{\underline{m}-1}$ sont des variantes alphabétiques de $\underline{M}$ et $\underline{P}'$ est un préfixe ainsi obtenu : soit Π l'arbre des quantificateurs qui se trouvent dans $\underline{P}$; soit Π' un arbre que l'on obtient à partir de Π en répétant certaines branches (ou des segments terminaux de certaines branches) et en renommant des variables ; on obtient $\underline{P}'$ en aplatissant (linéarisant) Π'. Si, pour un certain $\underline{m}$, $\underline{F}_{\underline{m}}$ est une identité normale, alors $\underline{F}$ est dite avoir la propriété A et est démontrable en théorie de la quantification.

Le problème auquel répond la propriété A est celui que se pose Russell dans *9 : retrouver les formules démontrables dans la théorie de la quantification bien que les schèmas d'axiomes propositionnels ne soient utilisés qu'avec des formules sans quantificateurs. Pour atteindre ce but, Russell introduit le schéma *9.11, alors que Herbrand introduit la propriété A. Cette idée de considérer une disjonction de variantes de la formule (ou de parties de la formule) se retrouve dans la notion de propriété C, laquelle est nécessaire et suffisante pour qu'une formule soit démontrable et qui est au centre même des résultats de Herbrand. Dans la recherche de la propriété C, l'identification des variables, en vue d'obtenir la formule propositionnellement valide, se fait à l'aide d'une suite de champs finis, une notion dont Herbrand a trouvé une esquisse dans la démonstration par Löwenheim de son théorème bien connu. Le théorème de Herbrand, c'est l'enfant de *9 et de la démonstration de Löwenheim.

Le seul point précis où Herbrand puisa son inspiration dans Principia mathematica, c'est *9. Pour le reste, l'influence de cet ouvrage resta limitée. Ce que Herbrand en retint, c'est que les énoncés mathématiques peuvent se traduire en un langage artificiel, aux règles de formation relativement simples, et que dans ce langage les démonstrations peuvent s'effectuer à l'aide de règles formelles, aussi relativement simples à énoncer. 'Ce qui fait qu'on peut, en un certain sens, considérer ce système de signes comme équivalent à l'ensemble des mathématiques' (Herbrand 1930, page 1, ou 1968, page 35). Et en 1929 il écrivait : 'Il résulte des recherches de Russell et Whitehead que les signes que nous allons indiquer suffisent pour traduire toutes les propositions mathématiques' (1931, page 16, ou 1968, page 170). Mais ce système de Russell et Whitehead, Herbrand le voit sous un jour particulier. Dans son premier écrit logique, 1928 (page 1275, ou 1968, page 22), il considère, et il fut sans doute le premier à le faire, la théorie des types comme un système de la théorie de la quantification à plusieurs sortes. Dans 1931 (page 41, ou 1968, page 203) il introduit le système R + Inf Ax + Mult Ax, qui est la théorie simple des types, avec une version de l'axiome de l'infini et une de l'axiome de choix, formulée en théorie de la quantification, avec plusieurs sortes, et il écrit que dans ce système 'on peut refaire toutes les mathématiques classiques, jusqu'à Cantor'. Si Herbrand pouvait ainsi considérer la théorie des types comme un système de la théorie de la quantification, c'est qu'il n'y a pas chez lui un intérêt soutenu pour les soubassements philosophiques de la construction de Russell. Les difficultés causées par le paradoxe de Russell, la stratification infinie, la ramification, les problèmes qui se rattachent à l'axiome de l'infini ou l'axiome de réducibilité, rien de tout cela ne semble retenir son attention.

La raison de cette attitude, c'est que Herbrand ne partage pas la conception que Russell a des rapports de la logique et des mathématiques, mais a adopté celle de Hilbert. En 1930 Herbrand indique bien où il voit les limites de l'oeuvre de Russell : 'Nous n'avons jusqu'ici fait que remplacer le langage ordinaire par un autre, plus commode, certes; mais cela ne nous avance en rien quant aux problèmes touchant les principes des mathématiques' (1930a, page 248, ou 1968, page 160). Et la phrase suivante indique la voie à suivre : 'Hilbert a cherché à résoudre les questions que l'on pouvait se poser en s'attachant à l'étude des ensembles de signes qui sont la traduction des propositions vraies dans une théorie déterminée.'

La contribution de Russell, c'est finalement d'avoir préparé la voie à Hilbert. De la formalisation des mathématiques Herbrand écrit : 'Tel était le premier pas - d'ailleurs nullement dû à Hilbert, qui a ici tout emprunté à ses devanciers - qu'il fallait accomplir pour poser de manière nette les problèmes fondamentaux de la métamathématique' (1930a, page 247, ou 1968, page 159). Et le mérite de *Principia mathematica*, c'est de nous avoir apporté une 'certitude expérimentale' quant à la possibilité de formaliser les mathématiques (*ibidem*).

Herbrand arriva en Allemagne en octobre 1930, donc bien longtemps après avoir écrit sa thèse. Là, il rencontra von Neumann et Bernays. Mais avant d'aller en Allemagne, encore isolé à Paris, il avait lu les articles que, de 1917 à 1928, Hilbert avait consacré aux problèmes des fondements des mathématiques ; il avait lu aussi Ackermann 1924 et von Neumann 1927. Il lut Hilbert et Ackermann 1928 peu de temps après sa parution. Son condisciple à l'Ecole Normale, André Weil, avait traduit en français Hilbert 1925 immédiatement après sa parution en allemand. C'est dans le contexte de ces écrits que Herbrand se mit à considérer les problèmes de non-contradiction, de complétude, de décision. Des problèmes qui, soit dit en passant, sont absents de l'oeuvre de Russell. Herbrand adopta le point de vue métamathématique que Hilbert avait préconisé et, plusieurs fois dans ses écrits, on en trouve des caractérisations qui se répètent en des termes presque identiques. Le titre même de sa thèse, *Recherches sur la théorie de la démonstration*, montre bien cette filiation.

Mais il y a dans l'adhésion de Herbrand aux conceptions de Hilbert une limite. C'est ainsi qu'il écrit : 'Jusqu'à quel point cette théorie [de Hilbert] atteint le fond des choses, ce n'était pas ici le lieu de le discuter ; on verra par l'analyse que nous en donnerons qu'elle peut prétendre au positivisme le plus strict et à la plus parfaite rigueur, mais qu'elle s'interdit aussi de considérer certaines questions appartenant à la théorie de la connaissance, et c'est là peut-être que gît son insuffisance au point de vue philosophique' (1931a, page 186, ou 1968, page 209). Herbrand était un jeune homme extrêmement doué qui avait certainement des vues sur bien des questions philosophiques. Dans leur notice biographique sur lui, Claude Chevalley et Albert Lautman écrivent : 'Il aimait en effet extrêmement la philosophie, la philosophie des sciences tout d'abord, mais aussi et surtout celle qui traite abstraitement des sentiments et des désirs de l'âme' (Chevalley et Lautman 1931, page 67, ou Herbrand 1968, page 14). Toutefois, cette effervescence intellectuelle n'a guère laissé de traces dans les écrits de Herbrand et on ne trouve pas dans ceux-ci, par delà son adoption, avec les réserves indiquées, du point de vue métamathématique, une opinion tant soit peu précise sur la nature des mathématiques ou sur la connaissance en général.

En 1930 Herbrand écrivait : 'Mais il ne faut pas se cacher que le rôle des mathématiques est peut-être uniquement de nous fournir des raisonnements et des formes, et non pas de chercher quels sont ceux qui s'appliquent à tel objet' (1930a, page 253, ou 1968, pages 164-165). Cette dernière tâche incomberait, selon Herbrand, au physicien ou au philosophe, mais non pas au mathématicien. Chevalley, dans une conférence faite en 1934, voit dans ces remarques de Herbrand une indication que la pensée de celui-ci 'n'était pas d'inclination platonicienne'. Et il ajoute : 'C'est dire, et c'est, je crois, ce que pensait Herbrand, que l'objectivité ne s'atteint que dans la symbolique pure, c'est-à-dire en vidant complètement les symboles de toute signification' (Chevalley 1934, page 100, ou Herbrand 1968, page 19). Herbrand avait cependant écrit (1930a, page 250, ou 1968, page 162) qu'une théorie mathématique, pour ne pas être 'un vain jeu de symboles', 'doit être la traduction de quelque chose de réel ; elle doit s'appliquer à des objets réellement concevables par l'entendement'. Certes, il avait fait précéder cette remarque de 'On considère souvent que ...' ; mais, par la suite, il n'avait nullement rejeté ce qu'il venait de présenter.

Tout ceci ne nous permet pas de serrer une conception bien précise. Peut-être Herbrand, dans les conversations et dans des remarques faites çà et là, se plaçait-il parfois à des points de vue différents pour en éprouver la valeur. Plutôt qu'à une conception philosophique bien définie, il faut sans doute attribuer l'adoption du point de vue finitiste de Hilbert par Herbrand à la précision des problèmes techniques que ce point de vue permettait de poser. Cette précision attirait l'algébriste qu'était Herbrand.

Herbrand lui-même note (1930, page 97, ou 1968, page 125) 'l'étroite relation' qui existe entre les termes fonctionnels qu'il utilise pour former la forme fonctionnelle d'une formule et les termes considérés dans le calcul epsilon introduit par Hilbert en 1925 (après diverses ébauches dans les années antérieures). Etant donnée une formule $\underline{A}(\underline{x}_o, \underline{x}_1,\ldots,\underline{x}_{k-1})$ de la théorie de la quantification, contenant les $\underline{k}$ variables libres $\underline{x}_o, \underline{x}_1,\ldots,\underline{x}_{k-1}$, pour une de ces variables, disons x_o, on associe à la formule un terme, $\varepsilon_{\underline{x}_o} \underline{A}(\underline{x}_o, \underline{x}_1,\ldots,\underline{x}_{k-1})$. Si $\underline{k} = 1$, toute interprétation du calcul dans un domaine $\underline{D}$ associe à ce terme un élément de $\underline{D}$; si $\underline{k}>1$, l'interprétation associe au terme epsilon une fonction $(\underline{k}-1)$-aire définie dans $\underline{D}$.

Pour comparer la méthode de Hilbert à celle de Herbrand, supposons que nous veuillons discuter le problème de la satisfaisabilité de la formule $\forall\underline{x}\exists\underline{y}\,\underline{A}(\underline{x},\underline{y})$, où $\underline{x}$ et $\underline{y}$ sont les seules variables (libres ou liées) de $\underline{A}(\underline{x},\underline{y})$. La forme fonctionnelle non stricte de $\underline{F}$ pour satisfaisabilité est $\forall\underline{x}\,\underline{A}(\underline{x},\underline{f}(\underline{x}))$. Dans le calcul epsilon, l'élimination du quantificateur existentiel nous donne la formule $\forall\underline{x}\,\underline{A}(\underline{x},\varepsilon_{\underline{y}}\underline{A}(\underline{x},\underline{y}))$ et, sémantiquement, le terme $\varepsilon_{\underline{y}}\underline{A}(\underline{x},\underline{y})$, tout comme le terme fonctionnel $f(x)$, dénote, dans une interprétation quelconque, une fonction de x. Certes, le terme epsilon a une plus grande complexité syntactique que le terme fonctionnel de Herbrand, lequel est formé à l'aide d'un symbole nouvellement introduit. Mais apparaît bientôt une différence plus profonde, car, à l'étape suivante, les chemins divergent. Herbrand passe à la forme fonctionnelle stricte pour satisfaisabilité, $A(x, f(x))$, dans laquelle x est désormais une variable libre, tandis que Hilbert considère le quantificateur universel comme la négation d'un quantificateur existentiel et répète la même opération, de sorte qu'il obtient la formule $A(\varepsilon_x \sim A(x, \varepsilon_y A(x,y)), \varepsilon_y A(x,y))$. Sémantiquement, cette formule est déjà moins transparente que $A(x, f(x))$, et pourtant notre exemple était l'un des plus simples que nous eussions pu prendre. Dès que le nombre des quantificateurs superposés dans une formule augmente, la traduction de la formule dans le calcul epsilon devient vite sémantiquement opaque.

En entreprenant d'éliminer, d'une démonstration dans l'arithmétique, toutes les variables, liées ou réelles, au profit de chiffres déterminés, Hilbert traite les deux quantificateurs de la même manière, alors qu'il y a, tout au long des écrits logiques de Herbrand, un fil rouge, qui est la façon différente dont il traite les variables universaloïdes et les variables existentialoïdes d'une formule.

Herbrand étudie la logique pure, passant ensuite à des théories particulières en ajoutant des hypothèses. Le système initial de Hilbert est déjà l'arithmétique ; ainsi, dans la liste de ses axiomes, les axiomes sur le successeur, l'addition et la multiplication précèdent les axiomes sur les quantificateurs. Dans 1929 Hilbert opéra un passage inverse à celui de Herbrand et va de l'arithmétique à la logique ; il abandonne le successeur et, écrit-il (page 8), 'ceci signifie essentiellement que nous faisons abstraction du caractère ordonné du système des nombres et le considérons comme un système arbitraire de choses'. L'aboutissement de cette marche en arrière, c'est que dans la seconde moitié des années trente, déjà bien après la publication des travaux de Herbrand, Bernays prend comme son point de départ la théorie de la quantification, pour laquelle il établit les deux théorèmes epsilon. Mais, ce faisant, il abandonne la façon unique de traiter les deux quantificateurs, il introduit entre eux une dissymétrie, rejoignant ainsi Herbrand (Bernays 1936, pages 93-98 ; Hilbert et Bernays 1939, pages 149-163, ou 1970, pages 149-169). Il y a entre la méthode de Herbrand et celle de Hilbert revue par Bernays une analogie assez étroite, que la présentation de Bernays fait bien ressortir (bien qu'il se limite à des formules prénexes). Mais, néanmoins, Hilbert et Bernays étaient passés de l'arithmétique à la logique, alors que Herbrand était allé de la logique à l'arithmétique. Cette différence de perspective explique, par exemple, pourquoi le premier théorème epsilon a, pour Hilbert et Bernays, une importance particulière qu'il n'a pas pour Herbrand. Si l'on en vient maintenant aux applications qui concernent l'arithmétique avec quantificateurs, les différences s'estompent. La démonstration, basée sur le théorème de Herbrand, de la non-contradiction de cette arithmétique (Scanlon 1973) est du même degré de complexité que celle qui suit les idées de Hilbert (Ackermann 1940) ; même plus, les deux démonstrations ont de profondes ressemblances. Enfin, dans leur démonstration du théorème de Herbrand, Dreben et Denton (1966) adaptent à la logique pure la notion de résolvante, introduite par Ackermann dans sa démonstration (1940) de la non-contradiction de l'arithmétique.

Herbrand indique, parmi les écrits qui lui ont servi lors de la rédaction de sa thèse, Löwenheim 1915. On ne trouve, ni dans cette thèse ni dans ses autres écrits, aucune remarque sur la notation si particulière utilisée dans cet article et empruntée, comme on sait, à Schröder. Mais Herbrand lut certainement l'article, car, lorsque dans son 1931 il s'occupe du problème de la réduction, pour la théorie de la quantification, à des prédicats tout au plus binaires, il utilise des symboles ('R', 'H' et 'V') introduits par Löwenheim dans son traitement de ce problème. C'est donc, on peut le dire avec assurance, dans l'article de Löwenheim que Herbrand trouva l'argument sémantique qui le conduisit à introduire ses suites de champs finis.

Herbrand adresse à Löwenheim un certain nombre de reproches. Certes, l'article de Löwenheim contient des lacunes dans les définitions et les démonstrations (voir, par exemple, van Heijenoort 1967, pages 228-232), mais les griefs que Herbrand formule contre Löwenheim ont un tour particulier qu'il nous faut examiner. Ces griefs peuvent se ramener à deux : premièrement, Löwenheim donne à la notion de validité 'un sens intuitif' et, par suite, son théorème 'n'a aucun sens précis' ; deuxièmement, sa démonstration manque de rigueur et reste insuffisante. Certes, Löwenheim n'énonce pas de définitions pour les notions sémantiques qu'il utilise, mais la façon dont il les manie montre bien, une fois qu'on a maîtrisé son langage, qu'il n'y a aucun malentendu dans son esprit à leur sujet. Ces notions restent pour lui, il est vrai, 'intuitives', car il les emprunte à une théorie naïve des ensembles. Quant à sa démonstration, elle souffre réellement de deux insuffisances : il fait un détour par l'infini, c'est-à-dire considère des formules infiniment longues, au lieu de se servir de l'axiome de choix, et, lorsqu'il s'agit d'obtenir le contre-modèle infini à partir de la suite d'assignations finies, il emploie une conjonction infinie (le texte de Löwenheim est tel qu'il n'est pas facile de dire s'il passe là indûment du fini à l'infini ou s'il a un véritable argument en tête).

Les griefs de Herbrand semblent donc justifiés si l'on s'en tient au sens littéral des mots : il y a dans les notions de Löwenheim de l''intuitif' et dans ses démonstrations des lacunes. Mais, derrière ces mots, Herbrand avait en vue des reproches bien différents de ceux que nous venons de rappeler. Ce qu'il pense, ce n'est nullement que Löwenheim aurait dû donner des définitions explicites des notions sémantiques qu'il utilise, mais qu'il aurait dû abandonner ces notions mêmes ; et non pas combler les lacunes de sa démonstration, mais prouver un théorème différent.

Si, au lieu de dire 'pour tout nombre *r* la *r*-ième expansion conjonctive de Herbrand de la formule *F* est satisfaisable (par les tables de vérité)', nous convenons, avec Herbrand, de dire 'F est vraie dans un champ infini', nous pouvons énoncer ses résultats fondamentaux sous la forme suivante (voir *Herbrand 1929a*, page 1077, ou *1968*, page 28) :

Théorème I. Si une formule *F* est démontrable en théorie de la quantification, ∿*F* ne peut être vraie dans un champ infini.

Théorème II. Si une formule *F* n'est pas démontrable en théorie de la quantification, on peut construire un champ infini dans lequel ∿*F* est vraie.

Herbrand voit dans le Théorème II un 'résultat analogue' (*1930*, page 118, ou *1968*, page 143) au théorème de Löwenheim sur l'équivalence de la validité à la validité dans un ensemble dénombrable. Le résultat de Löwenheim, mis sous une forme comparable à celle du Théorème II, serait :

Théorème II*. Si une formule *F* n'est pas valide, ∿*F* est satisfaisable dans un ensemble dénombrable.

Löwenheim n'a ni axiomes ni règles d'inférence, il laisse de côté la notion de démonstrabilité dans un système donné. Donc les notions sémantiques appartiennent non seulement à ses arguments, mais aussi à ses résultats mêmes. L'intérêt fondamental de ces résultats, c'est qu'ils sont, pour ainsi dire, une réduction du non-dénombrable au dénombrable, alors que ceux de Herbrand établissent un pont entre la théorie de la quantification et la logique propositionnelle. La prémisse du Théorème II contient 'démontrable' tandis que celle du Théorème II* contient 'valide', ce qui montre bien que les deux théorèmes appartiennent à des domaines différents de la logique. Mais, comme Herbrand n'accorde pas aux notions sémantiques de statut propre, il voit dans la validité une ébauche grossière de la démonstrabilité ; les deux théorèmes ont donc pour lui la même prémisse, le premier sous une forme exacte, le second sous une forme confuse. Il peut à la fois parler d'analogie entre les deux théorèmes et critiquer Löwenheim pour son manque de rigueur.

Quant au Théorème I, Herbrand reproche à Löwenheim, 'et ce reproche est le plus grave' (*1930*, page 118, ou *1968*, page 143), de le considérer comme évident. Si nous faisons subir au Théorème I la transposition qui nous a permis de passer du Théorème II au Théorème II*, nous obtenons :

Théorème I*. Si une formule F est valide, ∼F n'est pas satisfaisable dans un ensemble dénombrable.

Ce qui, du point de vue adopté par Löwenheim, découle immédiatement des définitions mêmes. Mais, ce point de vue sémantique, Herbrand se refuse précisément à la considérer. Il a été visiblement séduit par l'argument sémantique qui est au coeur de la démonstration de Löwenheim. Mais, cet argument, Herbrand le modifie de trois manières, liées entre elles par le fait qu'il rejette, dans les études logiques, la notion d'ensemble infini. Premièrement, il ne considère pas l'ensemble infini dont les champs finis sont les approximations successives ; il mentionne plusieurs fois l'existence de cet ensemble, mais il ne le fait pas figurer dans ses arguments. Deuxièmement, après avoir ajouté au vocabulaire de son système comme constantes individuelles les noms des éléments de ces champs (ou ces éléments eux-mêmes s'ils sont censés être déjà des noms), il ne considère que des disjonctions ou conjonctions finies, passant ainsi du plan sémantique au plan syntactique. Troisièmement, l'hypothèse de validité faite par Löwenheim, Herbrand la remplace par celle de démonstrabilité dans un système qu'il spécifie. Herbrand lui-même considérait son théorème comme une rectification de celui de Löwenheim, une 'précision', écrit-il (1931c, page 4, ou 1968, page 225), faisant usage d'un germanisme. Cette correction consistait à éliminer les notions ensemblistes, mais ainsi le sens même du théorème se trouve évidemment changé.

C'est d'ailleurs à une conclusion analogue que semble aboutir Herbrand lui-même, car, après avoir parlé de proposition n'ayant 'aucun sens précis' et de démonstration 'totalement insuffisante' ayant de 'graves lacunes', il écrit (1930, page 118, ou 1968, page 144) : 'Nous pouvons dire que la démonstration de Löwenheim était suffisante en mathématiques ; mais il nous a fallu, dans ce travail, la rendre "métamathématique" (voir l'Introduction) pour qu'elle nous soit de quelque utilité'. Dans cette introduction à laquelle il nous réfère, Herbrand avait essayé de délimiter les notions et les arguments utilisés en métamathématique de ceux qu'acceptent les mathématiques en général. Alors qu'il rejette la notion d'ensemble infini en métamathématique, il admet la théorie classique des ensembles en mathématiques.

Les critiques que Herbrand fait du théorème de Löwenheim et de sa démonstration s'étendent à la solution que Löwenheim avait donnée du problème de l'élimination, en théorie de la quantification, de symboles de prédicat ayant plus de deux arguments. Tout ce que nous venons de dire s'applique aussi aux reproches faits à ce sujet. Là aussi, Löwenheim et Herbrand établissent des résultats différents. Pour le premier, la transformée d'une formule valide est valide. Pour le second, il s'agit de démonstrabilité dans un système donné. Quant aux méthodes, Löwenheim se sert d'arguments sémantiques, alors que Herbrand applique son théorème et peut ainsi donner à la réduction un sens constructif.

Herbrand mentionne plusieurs fois dans ses écrits Skolem 1920 et certaines de ses remarques peuvent nous faire penser qu'il lut l'article, sans que nous puissions en avoir l'assurance complète (il ne dit rien qui s'appuie sur autre chose que ce que Hilbert et Ackermann (1928) mentionnent). Pour le cas d'une seule formule, la version du théorème de Löwenheim dont Skolem donne, dans cet article de 1920, une démonstration amendée est celle qui extrait un sous-modèle dénombrable du modèle qui est supposé exister. Mais, pour le cas d'un ensemble dénombrable de formules, la démonstration que Skolem esquisse tend à établir l'autre version du théorème, celle dans laquelle on obtient le modèle dénombrable en le construisant à l'aide d'approximations finies successives.

C'est cette esquisse de démonstration que Skolem développe dans la section 3 de son article de 1922. Là, nous avons, dans le cas particulier d'une formule prénexe pour laquelle tous les quantificateurs universels précèdent les quantificateurs existentiels, la construction, pour tout nombre naturel n, de ce que Skolem appelle les solutions de niveau n, qui vont donner les approximations finies du modèle dénombrable. Ces solutions forment, en fait, une suite de champs finis, au sens de Herbrand.

L'article de Skolem de 1922 ne semble avoir été guère lu. Il ne suscita, autant que je sache, que deux réactions, un compte rendu de Fraenkel (1927) et une mention par von Neumann (1925, page 232) ; mais, dans l'un et l'autre cas, ce furent les problèmes de la théorie des ensembles qui retinrent l'attention, et non la méthode employée pour établir le théorème de Löwenheim. Dans son compte rendu, Fraenkel, passant en revue l'article de Skolem, ne mentionne même pas l'existence de la section 3. Il semble bien que Herbrand ne connut pas l'article.

Il ne connut pas non plus, on peut l'affirmer sans crainte, Skolem 1928. Cet article révèle, de la part de Skolem, des préoccupations tout à fait analogues à celles de Herbrand. Aussi est-il important d'en examiner les similitudes et les différences avec les travaux de Herbrand. A la différence de celui-ci, Skolem ne nous présente pas un système défini. Après nous avoir dit ce qu'est une expression bien formée de la théorie de la quantification, il écrit que l'on peut 'représenter les démonstrations mathématiques comme des transformations de telles expressions logiques selon certaines règles'. Ces règles, il ne nous en donne pas une liste exhaustive, et il mêle arguments sémantiques et arguments syntactiques. Les règles qu'il énonce explicitement lui permettent de passer d'une formule close quelconque F à l'une de ses formes prénexes ; ce sont les règles de passage de Herbrand, avec la différence triviale que la conjonction est, chez Skolem, primitive alors qu'elle est, chez Herbrand, définie.

Skolem affirme ensuite l'équivalence sémantique de cette formule prénexe et de sa forme fonctionnelle stricte pour satisfaisabilité, F_{ffss}, les deux formules 'voulant dire' la même chose. Considérons les termes obtenus à partir d'une constante initiale 0 par composition avec les symboles fonctionnels de F_{ffss}, le niveau d'un terme étant k si et seulement si ses arguments sont au plus de niveau $k-1$ et l'un au moins est de ce niveau (le niveau de 0 est 0). Soit $C(F,k)$ la conjonction des cas particuliers de F_{ffss} obtenus lorsque ses variables (qui sont les variables universelles de F) sont remplacées, de toutes les manières possibles, par des termes de niveau k au plus.

Nous avons maintenant l'alternative suivante : Ou bien, pour un certain k, $C(F,k)$ n'est pas satisfaisable (par les tables de vérité), ou bien, pour tout k, $C(F,k)$ est satisfaisable. Dans le premier cas, Skolem déclare, F est réfutable. Il ne donne aucun argument. Il n'est peut-être pas impossible d'imaginer ce qu'était son raisonnement : $C(F,k)$ n'étant pas satisfaisable, $\sim C(F,k)$ est valide, donc démontrable (dans le fragment propositionnel du système ébauché plus haut) ; puis on peut passer d'une démonstration de $\sim C(F,k)$ à $\sim F_{ffss}$ en appliquant certaines règles qui sont fiables, c'est-à-dire qui conduisent d'une formule démontrable à une formule démontrable, et qui sont les règles de substitution et la règle de simplification. Le raisonnement suit de près ce que l'intuition nous dicte et peut-être Skolem avait-il une vue assez précise de l'argument, mais il ne nous en dit rien.

Dans le second cas présenté par l'alternative, Skolem cherche à montrer que, si F_{ffss} est réfutable, alors, pour un certain nombre k_o, $C(F,k_o)$ n'est pas satisfaisable. Selon l'ébauche de raisonnement qu'il nous présente, une démonstration conditionnelle de p & $\sim p$ en prenant F_{ffss} comme hypothèse peut être transformée, si l'on remplace les variables libres par 0, en une démonstration conditionnelle de p & $\sim p$ en prenant comme hypothèse un certain nombre fini de cas particuliers de F_{ffss}, c'est-à-dire un certain nombres de termes de la conjonction $C(F,k_o)$, où k_o est un nombre suffisamment grand.

Il est curieux de noter que les reproches que Herbrand fait à Löwenheim, il aurait pu, avec beaucoup plus de justesse, les adresser à Skolem, s'il avait connu les travaux de celui-ci. Car Skolem se donne le même but que Herbrand. Après avoir, tout comme Löwenheim, considéré les expansions et les champs comme des moyens techniques pour obtenir des résultats sémantiques, Skolem en est arrivé en 1928 à voir là une méthode qui doit supplanter les dérivations axiomatiques telles que les avaient conçues Frege, Russell ou Hilbert. C'est ainsi qu'il déclare : 'Je crois qu'il est possible d'aborder les problèmes de déduction d'une autre manière, une manière plus commode' (1928, page 130, ou van Heijenoort 1967, page 516). Tout comme Herbrand, il montre par des exemples que cette nouvelle méthode permet de traiter certains problèmes de décision. Les deux avantages de Herbrand, ce sont une plus grande précision et une plus grande généralité. Il spécifie exactement le système qu'il considère, il entreprend de donner des démonstrations complètes de ses résultats (même s'il y a une erreur involontaire dans une de ses démonstrations), il considère des formules quelconques, alors que Skolem laisse son système dans le vague, y mêle des considérations sémantiques, ébauche à peine ses raisonnements et se borne à considérer des formules prénexes. Tous les reproches que Herbrand fait à Löwenheim s'appliquent exactement à Skolem.

Il faut noter ici que Skolem avait, envers les fondements des mathématiques, une attitude différente de celle de Herbrand. Celui-ci, à la suite de Hilbert, sépare nettement, quant à leur objet et leurs méthodes, mathématiques et métamathématique. Skolem ne fait pas cette distinction et pour lui les mathématiques, ce sont les mathématiques constructives. Il se place, quant à cette question, à un point de vue assez semblable à celui de Brouwer. Ceci nous conduit à examiner l'attitude de Herbrand envers l'intuitionnisme.

Les constructivistes en mathématiques ont toujours mis en avant la conception que la valeur d'une fonction, pour chaque suite d'arguments donnée, doit être effectivement calculable. Au début des années vingt, les fonctions récursives primitives, introduites par Dedekind dès 1888, étaient devenues le paradigme même de fonction calculable et en 1923 Skolem avait publié une version de l'arithmétique basée sur l'emploi de ces fonctions. Mais il était vite devenu clair, ne fût-ce que par un argument diagonal, qu'il est des fonctions calculables qui ne sont pas récursives primitives ; en 1925 Hilbert mentionnait déjà une telle fonction et en 1928 Ackermann en publiait une étude détaillée.

En 1931 Herbrand proposa par trois fois d'introduire une classe de fonctions calculables qui fussent plus générales que les fonctions récursives primitives.

La première fois, au début de 1931, il écrit que, selon l''intuitionnisme' (et dans ce passage il entendait par ce mot le finitisme de Hilbert, qu'il avait adopté), 'toutes les fonctions introduites devront être effectivement calculables pour toutes les valeurs de leurs arguments, par des opérations décrites entièrement d'avance' (Herbrand 1931a, page 187, ou 1968, page 210).

La deuxième fois, ce fut lorsque, à peu près au même moment, il envoya à Gödel une lettre dans laquelle il proposait une définition de la notion de fonction récursive (générale), définition qui, dans le texte des conférences que Gödel donna à Princeton en 1934, est reproduite comme suit : 'Si φ est une fonction inconnue et $\psi_1,\ldots,\psi_k$ sont des fonctions connues, et si les fonctions ψ et φ sont substituées les unes aux autres de toutes les manières possibles et certaines paires des expressions ainsi obtenues sont égalées, alors, si l'ensemble ainsi obtenu d'équations fonctionnelles a une solution et une seule pour φ, φ est une fonction récursive' (Gödel 1934, page 26, ou Davis 1965, page 70). A la définition de Herbrand, Gödel ajouta deux clauses, que nous examinerons dans un instant.

La troisième fois que Herbrand proposa une définition des fonctions récursives (générales), ce fut dans l'article qu'il termina quelques jours avant sa mort (Herbrand 1931c, page 5, ou 1968, pages 226-227). Là, il écrivait : 'On pourra aussi introduire un nombre quelconque de fonctions $f_i\, x_1\, x_2 \ldots x_{n_i}$ avec des hypothèses telles que :

(a) *Elles ne contiennent pas de variables apparentes* ;

(b) *Considérées intuitionnistiquement, elles permettent de faire effectivement le calcul de* $f_i\, x_1 x_2 \ldots x_{n_i}$ *pour tout système particulier de nombres ; et l'on puisse démontrer intuitionnistiquement que l'on obtient un résultat bien déterminé*.'

Et à la première occurrence du mot 'intuitionnistiquement' Herbrand avait attaché une note, la note 5, qui disait : 'Cette expression signifie : traduites en langage ordinaire, considérées comme une propriété des entiers, et non comme un pur symbole'.

Nous avons là trois suggestions, car il faut parler ici de suggestions plutôt que de définitions, et elles diffèrent. Le calcul effectif de la valeur de la fonction est mentionné dans la première et la troisième, mais non dans la deuxième. Dans la première ce calcul doit se faire 'par des opérations décrites entièrement d'avance'. Dans la troisième le calcul se base sur les propriétés intuitives des entiers, propriétés indépendantes, comme semblent l'indiquer les derniers mots de la note 5, de toute définition des entiers dans un système formel.

Des deux clauses dont Gödel jugea nécessaire de compléter la deuxième suggestion de Herbrand, la première donnait une forme canonique au côté gauche des équations et la seconde, plus importante, présentait la liste finie des opérations admises dans le calcul de la valeur d'une fonction pour des arguments donnés. Ainsi la définition de Gödel combinait la deuxième suggestion de Herbrand (communiquée par Herbrand à Gödel dans une lettre) avec la première (alors inconnue de Gödel), en outre réalisant pour la première ce qui, chez Herbrand, n'était encore qu'un programme ; il avait bien réclamé 'des opérations décrites entièrement d'avance', mais il n'en avait pas donné de liste.

Dans une lettre datée du 23 avril 1963 à van Heijenoort, Gödel faisait sur la manière dont la notion de fonction récursive avait été acquise les commentaires suivants : 'Je n'ai jamais rencontré Herbrand. Sa suggestion fut faite par lettre en 1931, et elle était formulée *exactement* comme ce l'est à la page 26 du texte de mes conférences [*Gödel 1934*, ou *Davis 1965*, page 70], c'est-à-dire sans qu'il fût fait mention de la calculabilité. Mais, comme Herbrand était un intuitionniste, pour lui cette définition signifiait évidemment qu'il existe une démonstration *constructive* de l'existence et de l'unicité de φ. Il croyait probablement qu'on ne pouvait donner une telle démonstration qu'en présentant un procédé de calcul. (Notez que, si la thèse de Church est correcte, il est *vrai* que, au cas où $\exists ! \varphi \underline{A}(\varphi)$ est acceptable intuitionnistiquement, alors la fonction $\iota\varphi \underline{A}(\varphi)$ est récursive générale, bien que, pour obtenir le procédé de calcul de φ, il puisse être nécessaire d'ajouter certaines équations à celles qui sont déjà contenues dans $\underline{A}(\varphi)$. Donc je ne pense pas qu'il y ait un désaccord quelconque entre ses deux définitions [la deuxième et la troisième, Gödel ignorait alors la première] telles qu'il les entendait. Ce qu'il n'a pas vu (ou n'a pas clairement exprimé), c'est que le calcul, pour *toutes* les fonctions calculables, se fait selon *des règles qui restent exactement les mêmes*. C'est là le fait qui rend possible une définition précise de la récursivité générale. Je n'ai malheureusement pas retrouvé la lettre de Herbrand dans mes papiers. Elle a été probablement perdue à Vienne pendant la Seconde Guerre mondiale, comme tant d'autres choses. Mais mon souvenir est bien net et il était encore tout frais en 1934.'

On a parfois pensé que cette deuxième suggestion de Herbrand, telle qu'elle est formulée par Gödel, visait une classe de fonctions plus générales que les fonctions récursives, la classe des fonctions hyperarithmétiques. Telle n'était pas, comme on le voit, l'opinion de Gödel et, à la lumière des deux autres suggestions de Herbrand, l'hypothèse semble peu fondée.

Dans une lettre datée du 14 août 1964, Gödel écrivait à van Heijenoort que c'était une exagération que de dire que Herbrand avait 'introduit' la notion de fonction récursive ; il fallait, selon lui, plutôt parler d''ébauche', 'car c'est précisément en *spécifiant* les règles de calcul qu'un concept mathématiquement maniable et fécond avait été obtenu. Herbrand, de son côté, *exclut explicitement* la spécification de règles formelles de calcul par la locution "considérées intuitionnistiquement" (et l'explication qu'il en donne dans sa note 5). La question de savoir si la conception de Herbrand [dans *1931c*, la troisième] est équivalente à la récursivité générale est considérée par Heyting, moi-même et d'autres comme non résolue. C'est mon opinion que la thèse de Church est incontestablement correcte pour la calculabilité mécanique, mais peut-être incorrecte pour la calculabilité intuitionniste (comme je l'ai clairement dit dans la note ajoutée au texte de mes conférences de 1934 [*Davis 1965*, pages 71-73]).'

Nous voici donc amenés à essayer de comprendre ce que Herbrand entendait par 'intuitionnisme', un mot qu'il utilise à partir de 1930. Le plus souvent, dans les écrits de Herbrand, le mot dénote simplement les méthodes adoptées par Hilbert en métamathématique et qualifiées par celui-ci de 'finit', un mot allemand peu usité et aujourd'hui traduit par 'finitiste' ou 'finitaire'. Nous lisons par exemple dans Herbrand 1931b (Herbrand 1968, page 216) : 'Hilbert a de plus exigé, pour échapper à la critique destructive de Brouwer, que tout raisonnement fait en métamathématique soit du type dit "intuitionniste".' On pourrait multiplier les citations de ce genre.

Herbrand, c'est bien clair, ne s'est pas plongé dans la lecture des écrits de Brouwer, et ce qu'il dit de celui-ci, c'est ce qu'il a pu apprendre en lisant les écrits de Hilbert ou en conversant avec von Neumann ou Bernays. Cette connaissance de seconde main prend un tour caricatural lorsqu'il en vient à attribuer à Brouwer des idées de Hilbert explicitement rejetées par Brouwer : 'On a le droit de se servir de ces notions interdites [c'est-à-dire des notions non-finitistes] puisque tout résultat démontré en les utilisant comme intermédiaires ne peut être faux. Seulement ces notions devront être considérées par Brouwer comme des éléments sans signification réelle, des éléments idéaux, comme dit Hilbert' (Herbrand 1930a, page 252, ou 1968, pages 163-164).

Il est à noter qu'à la fin des années vingt et au début des années trente il n'était pas rare d'entendre dire que la métamathématique de Hilbert se proposait de se limiter aux méthodes 'intuitionnistes' (voir, par exemple, von Neumann 1927, page 3, lignes 1-5). Cela même incita Bernays à souligner la distinction entre finitisme et intuitionnisme (Hilbert et Bernays 1934, pages 34 et 43, Bernays 1934, page 69, 1934a, pages 89-90, 1935, page 212, 1938, page 146, et 1967, page 502).

Comme nous l'avons vu plus haut, Gödel, dans les années soixante, considère encore Herbrand comme un intuitionniste. Le seul passage dans les écrits de Herbrand qui indique que celui-ci ait pu avoir de l'intuitionnisme une conception allant au-delà du finitisme de Hilbert est la note 5 de 1931c, que je vais citer encore une fois. Parlant des conditions imposées aux fonctions récursives, Herbrand les dit 'considérées intuitionnistiquement' et il ajoute en note : 'Cette expression signifie : traduites en langage ordinaire, considérées comme une propriété des entiers, et non comme un pur symbole'. La remarque est si brève qu'elle reste ambigüe. Je vois deux interprétations possibles :

(a) La remarque n'est rien d'autre qu'une allusion au caractère intuitif de la métamathématique (qui, comme l'écrivait von Neumann (1927, page 3), doit être 'un enchaînement d'aperceptions intuitives immédiatement évidentes') ; c'est simplement une glose sur le mot 'inhaltlich', si usité par Hilbert, et nous sommes là sur un terrain commun à Hilbert et à Brouwer ;

(b) La remarque vise des procédés de calcul non formalisés et peut-être non formalisables, et elle laisse la porte ouverte à des méthodes intuitionnistes, non-mécaniques.

La seconde interprétation est celle qu'adopte Gödel, qui voit dans la note de Herbrand un refus de se cantonner dans des règles formelles de calcul. Mais Herbrand n'avait-il pas écrit quelques mois plus tôt que les calculs devaient se faire 'par des opérations décrites entièrement d'avance'? Il n'avait pas donné une liste de ces opérations, mais peut-on dire que la note 5 exclut catégoriquement la possibilité d'une telle liste ?

Cet examen de l'intuitionnisme de Herbrand se complique encore du fait qu'il introduit lui-même une nouvelle distinction lorsqu'il parle de l'intuitionnisme 'dans sa forme extrême' (1931a, page 187, ou 1968, page 210). Tout ce que dit Herbrand de cette forme extrême s'accorde très bien avec le finitisme de Hilbert. Le malheur est que Herbrand ne parle jamais d'une forme non extrême de l'intuitionnisme, et la seule suggestion dans ce sens, c'est, peut-être, la note 5, que nous avons déjà discutée.

Dans 1931c (page 3, ou 1968, page 225) Herbrand énonce son théorème et ajoute : 'La démonstration et l'énoncé de ce théorème sont intuitionnistes'. La phrase se comprend très bien si l'on prend 'intuitionniste' dans le sens de 'finitiste'. Elle prend cependant, dans le contexte des affirmations répétées de Herbrand que son théorème supplée à l'imprécision du résultat ensembliste de Löwenheim, une résonance plus profonde. Herbrand remplace la notion de satisfaisabilité d'une formule F dans un ensemble par celle-ci : pour tout nombre naturel n, la n-ième expansion conjonctive de F a la valeur logique v pour une certaine interprétation, c'est-à-dire pour certaines valeurs logiques attribuées aux formules atomiques. Chaque interprétation qui vérifie la ℓ-ième expansion conjonctive de F est une extension d'une interprétation qui vérifie la k-ième expansion, pour $k<\ell$. Ces interprétations se situent donc aux noeuds d'un arbre et, comme pour tout nombre n il y a un nombre fini d'interprétations de la n-ième expansion, cet arbre est finitaire. Si la formule $\sim F$ est démontrable en théorie de la quantification, l'arbre est fini. Si elle ne l'est pas, l'arbre a au moins une branche infinie, laquelle donne précisément l'interprétation de la formule F dans un ensemble dénombrable. Pour chaque interprétation qui se trouve à l'un des noeuds de l'arbre, un symbole de prédicat, P, a la valeur v pour un certain nombre fini de termes, et la valeur f pour un nombre fini d'autres termes. Lorsqu'on s'avance le long d'une branche infinie de l'arbre, ces deux ensembles finis de termes croissent (ou du moins ne décroissent pas) et peuvent être considérés comme des approximations finies de plus en plus fines des deux ensembles associés l'un à P, l'autre à ∿P, dont l'union forme le domaine de l'interprétation-limite. Nous rejoignons ici la notion de déploiement telle que Brouwer l'avait introduite pour fonder l'analyse intuitionniste et qu'il considérait comme devant se substituer à la notion d'ensemble, trop suspecte. Herbrand, on peut en être certain, n'avait pas lu les articles de Brouwer sur ce sujet, mais nous voyons son constructivisme l'amener par une pente naturelle à remplacer, tout comme Brouwer, la notion d'ensemble classique par celle de déploiement. Certes, pour Herbrand, cette substitution se réalise uniquement en métamathématique. Il accepte la théorie des ensembles classiques dans les mathématiques. Aussi cette manière de considérer l'arbre des interprétations d'une expansion conjonctive de Herbrand ne sort-elle pas du cadre du finitisme. On ne trouve chez Herbrand aucune trace de l'analyse intuitionniste lorsqu'on passe aux mathématiques. Le rapprochement avec Brouwer sur ce point précis n'en est pas moins suggestif.

L'attitude négative de Herbrand envers la théorie des ensembles l'amène à prendre sur certaines questions une attitude plus stricte que celle de Hilbert et de ses collaborateurs. Il est plus royaliste que le roi. La métamathématique de Hilbert se proposait avant tout d'établir la non-contradiction de certaines branches des mathématiques et, ainsi, de les justifier ; là, il lui fallait se limiter aux méthodes finitistes. Mais, pour l'étude de la logique en dehors du problème de la non-contradiction de théories mathématiques, l'école de Hilbert acceptait volontiers les notions ensemblistes. Hilbert et Ackermann avaient, dans leur 1928, posé le problème de la complétude sémantique de la théorie de la quantification. Bernays (voir Bernays et Schönfinkel 1928) et Ackermann (1928) s'étaient occupés de certains cas du problème de la décision posés en termes de validité (ou de satisfaisabilité). Herbrand, lui, veut se limiter aux méthodes finitistes dans toute étude de la logique. Quand il aborde les problèmes de décision ou de réduction, la première chose qu'il fait, c'est de les reformuler en termes de démonstrabilité. Puis il se sert de son théorème pour éliminer, dans le traitement de ces problèmes, tout recours à des ensembles infinis. Dans son 1930 (page 119, ou 1968, page 144), il examine la solution apportée par Ackermann (1928) au problème de la décision dans le cas du préfixe $\forall\ldots\forall\,\exists\forall\ldots\forall$ et il écrit que la notion de satisfaisabilité ('Erfüllbarkeit'), utilisée par Ackermann, 'ne nous paraît pas définie d'une manière suffisamment complète (on ne nous donne pas de moyen précis permettant d'affirmer qu'une proposition est "erfüllbar")'. Cette accusation de manque de précision manque elle-même de précision. S'il s'agit simplement de réclamer de chaque auteur des définitions explicites des notions sémantiques utilisées, c'était là un point qui n'était peut-être pas superflu au moment où Herbrand écrivait, mais qui reste néanmoins secondaire. S'il s'agit d'exiger une notion décidable, c'est utopique. Ce que Herbrand avait en vue, c'était en fait l'utilisation de son théorème. Il reprend l'argument d'Ackermann en le modifiant et peut, grâce à son théorème, éliminer de la démonstration toute considération d'ensembles infinis.

C'est également parce qu'il a banni les notions ensemblistes de toute étude de la logique que Herbrand ne se soucie guère du problème de la complétude sémantique de la théorie de la quantification. Il n'ignorait nullement que la suite infinie de champs finis attachés à une formule non réfutable nous permet d'obtenir une interprétation de cette formule dans un domaine dénombrable (voir 1930, page 109, ou 1968, page 136, les deux première lignes), mais il laisse de côté cette conclusion. (Sur Herbrand et le problème de la complétude sémantique de la théorie de la quantification voir Note H, écrite par Dreben, dans van Heijenoort 1967, pages 578-580, et Note N, écrite par Dreben et Goldfarb, dans Herbrand 1971, pages 265-271).

Lorsqu'il s'attaqua au problème de la complétude, Gödel, bien que ne connaissant pas encore les écrits de Herbrand, jugea nécessaire de souligner que l'emploi des notions non-finitistes était légitime en logique dans les questions autres que le problème de la non-contradiction et que la complétude sémantique de la théorie de la quantification constituait un problème parfaitement respectable. Il écrit (1930, page 5) : 'Enfin, il faut encore ne pas oublier que le problème traité ici n'est nullement apparu à cause de la querelle sur les fondements (alors que ce fut sans doute le cas pour le problème de la non-contradiction des mathématiques), mais que, même si l'on n'avait jamais douté que la mathématique "naïve" fût valable quant à son contenu, ce problème pouvait être posé d'une manière nullement dénuée de sens à l'intérieur de cette mathématique (contrairement, par exemple, au problème de la non-contradiction), car une limitation des moyens de démonstration ne paraît pas plus s'imposer ici que pour n'importe quel autre problème mathématique.' Ces dernières lignes semblent avoir été écrites comme pour répondre directement à Herbrand.

Je viens de parler des origines et du contexte historique de l'oeuvre de Herbrand. Je n'ai pas la possibilité de parler ici de son impact, de ses applications ou des développements ultérieurs de la théorie de la démonstration. Je ne peux pas, cependant, ne pas dire quelques mots sur les rapports que les résultats de Gerhard Gentzen ont avec ceux de Herbrand. Gentzen publia en 1934 deux théorèmes, son Hauptsatz et son verschärfter Hauptsatz, qui s'appliquent à un calcul de séquences, introduit par lui pour formaliser la théorie de la quantification. Le Hauptsatz dit que toute démonstration peut, d'une manière effective, être transformée en une démonstration, dite démonstration normale, qui, tout en ayant la même séquence finale, ne fait pas usage de la règle de coupure ; par suite, chaque formule qui entre dans une démonstration normale est une sous-formule de la séquence finale de cette démonstration. 'On n'y introduit [dans une démonstration normale] aucun concept qui ne soit contenu dans son résultat final' (Gentzen 1934, page 177, 1955, pages 4-5). C'est là, évidemment, un résultat analogue à l'instauration, par Herbrand, du système $\underline{Q}_H$, mentionné plus haut, qui se passe de la règle de détachement et pour chaque règle duquel la conclusion ne contient aucun connecteur propositionnel, aucun symbole de prédicat, aucun symbole de fonction qui ne soit contenu dans la prémisse. Et lorsque Gentzen dit qu'une démonstration normale 'ne fait pas de détours' (1934, page 177, ou 1955, page 4), ces mots sont l'écho de ceux de Herbrand qui, pour $\underline{Q}_H$, parlait de 'démonstration ne faisant appel à aucun artifice' (Herbrand 1930, page 120, ou 1968, page 145).

Le *verschärfter Hauptsatz* de Gentzen (1934, page 408, ou 1955, pages 114-115) énonce que toute démonstration d'une séquence constituée de formules prénexes peut, d'une manière effective, être transformée en une démonstration qui, premièrement, est sans coupure et, ensuite, possède une séquence moyenne ou intermédiaire ('*Mittelsequenz*') telle que

(1) La démonstration de la séquence moyenne à partir des axiomes ne contient aucun quantificateur ;

(2) Les seules règles employées dans la démonstration de la séquence finale à partir de la séquence moyenne sont les règles qui permettent l'introduction d'un quantificateur (universel ou existentiel) dans (l'antécédent ou le subséquent d')une séquence et les règles structurales (la dilution, la contraction et l'interchange).

La parenté de ce résultat avec celui de Herbrand est frappante. Une *Mittelsequenz* correspond à une expansion de Herbrand qui est (propositionnellement) valide et à partir de laquelle on obtient la formule finale par les règles généralisées d'universalisation, d'existentialisation et de simplification. Dans les deux cas, le travail quantificationel vient après le travail propositionnel. Mais Gentzen obtient la *Mittelsequenz* à partir d'axiomes au moyen de règles qui détaillent le jeu de chaque connecteur propositionnel, alors que Herbrand part de son expansion, jugée valide par n'importe quel moyen (tables de vérité ou système axiomatique) ; il se désintéresse du travail propositionnel.

Quant aux méthodes employées pour établir les résultats, elles sont différentes. Gentzen se sert de manipulations syntactiques qui dépendent étroitement de la forme spécifique de ses axiomes et de ses règles. Grâce aux suites de champs finis, les arguments de Herbrand ont une fragrance sémantique qui les rend, dans une certaine mesure, indépendants de la forme particulière de son système initial. Mais, d'un autre côté, cet aspect quasi-sémantique des arguments et le délaissement du travail propositionnel font que les résultats de Herbrand ne s'appliquent qu'à la logique classique, alors que le Hauptsatz de Gentzen (mais non, évidemment, son verschärfter Hauptsatz) s'étend de façon naturelle à la logique intuitionniste. Toutes les tentatives de généraliser le théorème de Herbrand dans cette direction là n'ont conduit qu'à des résultats partiels et peu maniables (voir van Heijenoort 1971).

La démonstration que donne Herbrand de son théorème fondamental contient, comme on sait, une lacune. Déjà dans Hilbert et Bernays 1939, page 158, note 1, Bernays écrivait que cette démonstration était 'difficile à suivre', mais il n'indiquait aucune erreur. En 1943 Gödel était arrivé à la conclusion que l'argument de Herbrand était défectueux et il écrivit sur ce point une petite note à son usage personnel, sans rien rendre public. En 1963 Dreben, Andrews et Aanderaa publièrent (1963, 1963a) des contre-exemples refutant certains lemmes sur lesquels reposait la démonstration de Herbrand. La lacune fut comblée par Dreben et Denton (1966).

Disons qu'une formule F est d'ordre p si p est le plus petit nombre naturel tel que l'expansion de F dans le champ C_p soit valide. Herbrand pensait que les règles qui permettent de passer d'une formule à l'une de ses formes prénexes laissent invariant l'ordre et il tenta de le démontrer. En réalité, la prénexisation peut faire croître l'ordre.

Une partie du théorème que Herbrand veut établir dit que, si une formule F est démontrable en théorie de la quantification, il existe un nombre naturel r tel que la r-ième expansion de F est (propositionnellement) valide. Herbrand entreprend de démontrer ceci par récursion sur la démonstration de F. Il vérifie donc, premièrement, que, pour chaque axiome A du système considéré, il existe un nombre naturel r tel que la r-ième expansion de A est valide ; il doit ensuite établir que, pour chaque règle, la conclusion est d'un certain ordre, s, si l'on suppose que la ou les prémisses sont de certains ordres donnés. Pour la règle de détachement, par exemple, nous supposons que la formule G est d'ordre p, la formule $G \supset H$ d'ordre q et il nous faut trouver une borne supérieure pour l'ordre s de H. Herbrand pensa erronément que $\max(p,q)$ était une telle borne. Soient j le nombre des quantificateurs de G et k celui des quantificateurs de H. Dreben et Aanderaa (1964) établissent, au moyen de contre-exemples, qu'une borne supérieure pour l'ordre de H ne peut être fournie ni par une fonction de p et q, ni par une fonction de j, p et q, ni par une fonction de k, p et q ; ils présentent une fonction récursive primitive, des quatre arguments, j, k, p et q, qui donne une borne supérieure. Appelons une telle fonction une fonction analysante ('analysing function') pour la règle de détachement. Comme l'ordre d'une formule reste encore aujourd'hui la mesure la moins artificielle que nous ayons de sa complexité logique, l'étude des fonctions analysantes constitue un domaine dans lequel certaines questions mériteraient d'être éclaircies.

Le présent article a bénéficié d'un certain nombre de remarques que m'ont faites B. Dreben, W. Goldfarb et G. Kreisel.

REFERENCES :

Aanderaa, Stål, Voir Dreben, Burton, et Stål Aanderaa ; voir aussi Dreben, Burton, Peter Andrews et Stål Aanderaa.

Ackermann, Wilhelm,

1924 Begründung des 'tertium non datur' mittels der Hilbertschen Theorie der Widerspruchsfreiheit, Mathematische Annalen 93, 1-36.

1928 Über die Erfüllbarkeit gewisser Zählausdrücke, ibid. 100, 638-649.

1940 Zur Widerspruchsfreiheit der Zahlentheorie, ibid. 117, 162-194.

Voir Hilbert, David et Wilhelm Ackermann.

Andrews, Peter, Voir Dreben, Burton, Peter Andrews et Stål Aanderaa.

Bernays, Paul,

1934 Sur le platonisme dans les mathématiques, L'enseignement mathématique 34 (1935), 52-69.

1934a Quelques points essentiels de la métamathématique, ibid., 70-95.

1935 Hilberts Untersuchungen über die Grundlagen der Arithmetik, dans Hilbert 1935, 196-216.

1936 Logical calculus, miméographé, The Institute for Advanced Study, Princeton.

1938 Sur les questions méthodologiques actuelles de la théorie hilbertienne de la démonstration, dans Gonseth 1938, 144-152 ; Discussion, 153-161.

1967 Hilbert, David, dans Edwards 1967, 496-504.

Voir Hilbert, David, et Paul Bernays.

Bernays, Paul, et Moses Schönfinkel

1928 Zum Entscheidungsproblem der mathematischen Logik, Mathematische Annalen 99, 342-372.

Chevalley, Claude

1934 Sur la pensée de J. Herbrand, L'enseignement mathématique 34 (1935), 97-102 ; réimprimé dans Herbrand 1968, 17-23.

Chevalley, Claude, et Albert Lautman

1931 Notice biographique sur Jacques Herbrand, Annuaire de l'Association amicale de secours des anciens élèves de l'Ecole normale supérieure, 66-68 ; réimprimé dans Herbrand 1968, 13-15.

Davis, Martin

1965 *The undecidable*, New York.

Denton, John, Voir Dreben, Burton, et John Denton.

Dreben, Burton, et Stål Aanderaa

1964 Herbrand analysing functions, *Bulletin of the American Mathematical Society 70*, 697-698.

Dreben, Burton, Peter Andrews et Stål Anderaa

1963 Errors in Herbrand, *American Mathematical Society, Notices 10*, 285.

1963a False lemmas in Herbrand, *Bulletin of the American Mathematical Society 69*, 699-706.

Dreben, Burton, et John Denton

1966 A supplement to Herbrand, *The journal of symbolic logic 31*, 393-398.

Edwards, Paul

1967 *The encyclopedia of philosophy*, vol. 3.

Fraenkel, Abraham

1927 Compte rendu de *Skolem 1922*, *Jahrbuch über die Fortschritte der Mathematik 49* (pour 1923, publié en 1927), 138-139.

Frege, Gottlob

1879 *Begriffsschrift, eine der arithmetischen nachgebildete Formelsprache des reinen Denkens*, Halle.

Gentzen, Gerhard

1934 Untersuchungen über das logische Schliessen, *Mathematische Zeitschrift 39*, 176-210, 405-431.

1955 *Recherches sur la déduction logique*, traduction et commentaire par Robert Feys et Jean Ladrière, Paris.

Gödel, Kurt

1930 *Über die Vollständigkeit des Logikkalküls*, texte dactylographié, Thèse à l'Université de Vienne.

1934 *On undecidable propositions of formal mathematical systems*, texte de conférences rédigé par Stephen Cole Kleene et John Barkley Rosser, The Institute for Advanced Study, Princeton, New Jersey ; réimprimé avec corrections, amendements et une postface dans *Davis 1965*, 39-74.

Goldfarb, Warren D.

1971 Introduction, in *Herbrand 1971*, 1-20.

Gonseth, Ferdinand

1938 *Les entretiens de Zurich sur les fondements et la méthode des sciences mathématiques, 6-9 décembre 1938*, Zurich (1941).

Herbrand, Jacques

1928 Sur la théorie de la démonstration, Comptes rendus hebdomadaires des séances de l'Académie des sciences (Paris) 186, 1274-1276 ; réimprimé dans Herbrand 1968, 21-23.

1929 Non-contradiction des axiomes arithmétiques, ibid. 188, 303-304 ; réimprimé dans Herbrand 1968, 25-26.

1929a Sur quelques propriétés des propositions vraies et leurs applications, ibid., 1076-1078 ; réimprimé dans Herbrand 1968, 27-29.

1929b Sur le problème fondamental des mathématiques, ibid. 189, 554-556 ; réimprimé dans Herbrand 1968, 31-33.

1930 Recherches sur la théorie de la démonstration, Thèse à l'Université de Paris, publiée dans Prace Towarzystwa Naukowego Warszawkiego, Wydział III, n° 33 ; réimprimé dans Herbrand 1968, 35-153.

1930a Les bases de la logique hilbertienne, Revue de métaphysique et de morale 37, 243-255 ; réimprimé dans Herbrand 1968, 155-166.

1931 Sur le problème fondamental de la logique mathématique, Sprawozdania z posiedzeń Towarzystwa Naukowego Warszawskiego, Wydział III, 24, 12-56 ; réimprimé dans Herbrand 1968, 167-207.

1931a Note non signée sur Herbrand 1930, Annales de l'Université de Paris 6, 186-189 ; réimprimé dans Herbrand 1968, 209-214.

1931b Notice pour Jacques Hadamard, dans Herbrand 1968, 215-219.

1931c Sur la non-contradiction de l'arithmétique, Journal für die reine und angewandte Mathematik 166, 1-8 ; réimprimé dans Herbrand 1968, 221-232.

1968 Ecrits logiques, Paris.

1971 Logical writings, Dordrecht.

Hilbert, David

1925 Über das Unendliche, Mathematische Annalen 95 (1926), 161-190 ; traduction française d'André Weil, Acta mathematica 48 (1926), 91-122.

1929 Probleme der Grundlegung der Mathematik, Mathematische Annalen 102, 1-9.

Hilbert, David, et Wilhelm Ackermann

1928 Grundzüge der theoretischen Logik, Berlin.

Hilbert, David, et Paul Bernays

1939 Grundlagen der Mathematik, vol. 2, Berlin.

1970 ———————————, 2-ième édition.

Löwenheim, Leopold

1915 Über Möglichkeiten im Relativkalkül, Mathematische Annalen 76, 447-470 ; traduction anglaise de Stefan Bauer-Mengelberg dans van Heijenoort 1967, 228-251.

Russell, Bertrand, Voir Whitehead, Alfred North, et Bertrand Russell.

Scanlon, Thomas Michael, Jr.

1973 The consistency of number theory via Herbrand's theorem, *The journal of symbolic logic* *38*, 29-58.

Skolem, Thoralf

1920 Logisch-kombinatorische Untersuchungen über die Erfüllbarkeit oder Beweisbarkeit mathematischer Sätze nebst einem Theoreme über dichte Mengen, *Videnskapsselskapets skrifter*, I., *Matematisk-naturvidenskabelig klasse*, n°4 ; réimprimé dans *Skolem 1970*, 103-136 ; traduction anglaise partielle de Stefan Bauer-Mengelberg dans *van Heijenoort 1967*, 252-263.

1922 Einige Bemerkungen zur axiomatischen Begründung der Mengenlehre, *Matematikerkongressen i Helsingfors den 4-7 Juli 1922*, *Den femte skandinaviska matematikerkongressen*, *Redogörelse*, Helsinki (1923), 217-232 ; réimprimé dans *Skolem 1970*, 137-152 ; traduction anglaise de Stefan Bauer-Mengelberg dans *van Heijenoort 1967*, 290-301.

1928 Über die mathematische Logik, *Norsk matematisk tidsskrift* *10*, 125-142 ; réimprimé dans *Skolem 1970*, 189-206 ; traduction anglaise de Stefan Bauer-Mengelberg et Dagfinn Føllesdal dans *van Heijenoort 1967*, 508-524.

1970 *Selected works in logic*, edited by Jens Erik Fenstad, Oslo.

van Heijenoort, Jean

1967 *From Frege to Gödel, A source book in mathematical logic, 1879-1931*, Cambridge, Massachusetts.

1971 Comptes rendus d'articles de G.E. Mints, *The journal of symbolic logic* *36*, 524-528.

von Neumann, John

1925 Eine Axiomatisierung der Mengenlehre, *Journal für die reine und angewandte Mathematik* *154*, 219-240 ; Berichtigung *ibid*. *155*, 128 ; réimprimé dans *von Neumann 1961*, 34-56. Traduction anglaise de Stefan Bauer-Mengelberg et Dagfinn Føllesdal dans *van Heijenoort 1967*, 393-413.

1927 Zur Hilbertschen Beweistheorie, *Mathematische Zeitschrift* *26*, 1-46 ; réimprimé dans *von Neumann* 1961, 256-300.

1961 *Collected works*, vol. 1, New York.

Whitehead, Alfred North, et Bertrand Russell

1910 *Principia mathematica*, vol. 1, Cambridge, Grande-Bretagne.

1925 ————————, 2-ième édition.

Zaremba, Stanislas

1926 *La logique des mathématiques*, Mémorial des sciences mathématiques, fasc. 15, Paris.

PROCEEDINGS OF THE HERBRAND SYMPOSIUM
LOGIC COLLOQUIUM '81, J. Stern (editor)

CASE DISTINCTIONS ARE NECESSARY FOR REPRESENTING POLYNOMIALS AS SUMS OF SQUARES

Charles N. Delzell
Department of Mathematics
Louisiana State University
Baton Rouge, Louisiana 70803
USA

Suppose $f \in \mathbb{Z}[C;X]$ is the general form in $n + 1$ variables $X = (X_0,\ldots, X_n)$ of even degree d with coefficients $C = (C_j | 1 < j \leqslant \binom{n+d}{n})$. Let P_{nd} be the set of coefficients from the real closed ground field R for which f is positive semidefinite in X. We give a simple geometric proof of a conjecture raised in the early sixties by Kreisel: there do not exist rational functions $g_j \in \mathbb{Q}(C)$ and $r_j \in \mathbb{Q}(C;X)$ such that $f = \Sigma_j g_j r_j^2$ and $\forall c \in P_{nd}$, all $g_j(c) \geqslant 0$. The only exception to the conjecture is $d \leqslant 2$. This negative result complements the construction by Kreisel and Daykin of a disjunction of polynomial nonnegatively-weighted sum-of-squares representations of f, by means of Herbrand's Theorem.

Contents:

1. History of Logical Aspects of Hilbert's 17th Problem

(a) Geometric Origin of the Problem[1]

In his book [1899] on the foundations of geometry, Hilbert showed that those problems in plane geometrical construction which can be solved by means of only his five groups of axioms, can always be carried out by the use of straightedge and gauge (an example of a gauge is a compass whose use is restricted to the laying off of distances on a straight line). He gave two algebraic characterizations (Theorems 41 and 44) of the set of points so constructible, in terms of their Cartesian coordinates $(f_1(x), f_2(x))$, where the given points are expressed as rational functions of the parameters $x = (x_0,$

Research supported in part by NSF grant No. MCS8102744.
1980 Mathematics Subject Classification: Primary 03F55, 10C04, 10C10; Secondary 01A65, 14G30.

[1] Prestel [1978] has also given a synopsis of the geometric origin of the problem.

$\ldots, x_n) \in \mathbb{R}^{n+1}$. The second of his two characterizations was a necessary and sufficient condition, namely that $f_i(x)$ be a totally real algebraic number for all $x \in \mathbb{Q}^{n+1}$. (In the posthumous Seventh Edition of his book [1971], a minor error in his formulation of this criterion was corrected, in Supplement IV.) The proof required the fact that

(1.1) if a rational function $f \in \mathbb{Q}(X)$, where $X = (X_0, \ldots, X_n)$ are indeterminates, is positive semidefinite ("psd"),[2] then f equals a sum of squares of rational functions in $\mathbb{Q}(X)$.

We shall abbreviate "sum(s) of squares" as "SOS." In the first edition [1899] of the book, Hilbert gave the proof of 1.1 only for $n = 0$, which was enough for some applications, such as showing that those regular polygons constructible by means of a compass and straightedge can also be constructed using only a gauge and a straightedge (here not even 1 parameter is involved).

Hilbert left the case $n > 0$ as his 17th problem; in later editions of the book, e.g. [1971], Hilbert mentioned Artin's solution [1927] (see (b) below). There is an equivalent formulation of the problem in terms of homogeneous rational functions (i.e. quotients of psd forms); also we can convert back and forth between a quotient of SOS and a SOS of quotients. "At the same time, it is desirable . . . to know whether the coefficients of the forms to be used in the expression may always be taken from the realm of rationality given by the form represented." [Hilbert 1900] Thus, to solve the problem in its full generality, we must allow the coefficients to come from any ordered field K: K is called <u>ordered</u> once we have specified an <u>ordering</u>, i.e. a set $P \subset K$ (the "positive" elements), such that $P + P \subseteq P$, $P \cdot P \subseteq P$, $P \cup (-P) = K$, and $P \cap (-P) = \{0\}$. K will always denote an ordered field, and we shall write K^+ for P.

(b) Classical Solutions

Besides the case $n = 0$ mentioned above, one other case had already been solved, namely the well-known case of <u>quadratic</u> forms: given a psd form $f(X) = \Sigma a_{ij} X_i X_j$ $(0 \leq i,j \leq n,\ a_{ij} = a_{ji} \in K)$, then

$$f(X) = \sum_{i=0}^{n} p_i \Big(\sum_{j=0}^{n} b_{ij} X_j \Big)^2,$$

some p_i, $b_{ij} \in K$ with $p_i \geq 0$. For the special case where $K = \mathbb{R}$, the problem had been solved for three additional classes of forms, namely (1) psd binary forms, which are SOS of (two) forms, by the 2-square identity and the factorization of binary forms over $\mathbb{R}$ (see, e.g., [Landau 1903]); (2) psd ternary quartic forms, which are SOS of (quadratic) forms ([Hilbert 1888]; see also [Choi and Lam 1977b] for a more elementary proof, using "extremal forms"); and (3) ternary forms, which are SOS of (four, homogeneous) rational functions [Hilbert 1893]. While for the binary, the quadratic, and the ternary quartic forms, one could choose the square summands to be <u>forms</u> over $\mathbb{R}$ (without using denominators), Hilbert had found examples [1888] of psd ternary sextic and quaternary quartic forms which were not SOS of forms. Work of Ellison (1968 unpublished), Motzkin [1967, p. 217], and R. M. Robinson [1973] toward simpler and/or more explicit such examples, culminated in Choi and Lam's [1977(a), (b)] construction of the psd forms

$$X^2Y^4 + Y^2Z^4 + Z^2X^4 - 3X^2Y^2Z^2 \quad \text{and}$$

---

[2] I.e. for all $x \in \mathbb{R}^{n+1}$ at which f is defined, $f(x) \geq 0$.

$$W^4 + X^2Y^2 + Y^2Z^2 + Z^2X^2 - 4XYZW,$$

which they easily showed to be not representable as SOS of (real) forms. It is not known if the result (2) above on ternary quartic forms still holds with $\mathbb{Q}$ in place of $\mathbb{R}$, but it is known that the result (1) on binary forms does (though more than two summands are required): an algorithm of Landau [1906] transforms the rational functions given by Hilbert for $n = 0$ into polynomials, still with rational coefficients; it only remains to homogenize these polynomials. However by then, Landau had already [1903] obtained directly this improvement of the classical result on binary forms. In passing from $\mathbb{R}$ to $\mathbb{Q}$, the number of required squares in Landau's representation of [1903] increased from 2 to $2d+2$, where d is the degree of the form. In [1904] he lowered this number of squares to 5 for quadratics (smallest possible) and ≤ 6 for quartics; Fleck [1906] lowered the 6 to 5. Using the 8-square identity, Landau [1906] finally proved that, regardless of the degree, 8 squares are enough. Pourchet [1971] extended Landau's result by replacing $\mathbb{Q}$ with any algebraic number field, and simultaneously reduced the number of required squares to 5, the smallest possible.

The main step in the history of the 17th problem was Artin's [1927] nonconstructive proof of 1.1, and not only for the ground field $\mathbb{Q}$, but even if $\mathbb{Q}$ is replaced by any uniquely orderable subfield K of $\mathbb{R}$ (e.g. $\mathbb{R}$ or the real algebraic numbers); dropping the unique orderability hypothesis, Artin represented psd (rational) functions f as $f = \Sigma p_i r_i^2$ where $p_i \in K^+$ and $r_i \in K(X)$. Artin proved this using his result that in any field F of characteristic $\neq 2$, an element f is "totally positive with respect to the ordered subfield k"[3] if an only if $f = \Sigma p_i r_i^2$, where $p_i \in k^+$ and $r_i \in F$. Thus it remained to show that a psd function $f \in K(X)$ is totally positive in $K(X)$ with respect to K. For this he used a series of "specialization lemmas" using Sturm's Theorem.

In solving the problem, Artin had recognized that it had more to do with the algebraic than the arithmetic properties of $\mathbb{Q}$ and $\mathbb{R}$. Specifically, he was led to introduce the axioms for a <u>real closed field</u> R: (i) R is a formally real field (i.e. -1 is not a SOS in R), (ii) for all $a \in R$, either a or $-a$ is a square in R, and (iii) any odd degree polynomial in $R[T]$ has a root in R. For future reference, we mention the fact that every ordered field is contained in a(n essentially unique) smallest real closed field called its <u>real closure</u>. This axiomatization not only led to greater generality, but it actually made the problem easier; thus his solution was perhaps the first spectacular use of the axiomatic method for mathematical as opposed to metamathematical purposes, such as independence results.

A more recent effort to generalize Artin's theorem focused on his hypothesis that $K \subseteq \mathbb{R}$, which is essentially the hypothesis that K be Archimedean: an ordered field K is <u>Archimedean</u> over the subfield k if $\forall c \in K$ $\exists d \in k$ such that $c < d$; if k is $\mathbb{Q}$, we omit "over k." It was well-known that the Archimedean hypothesis played a role in his original formulation of his theorem. For example,[4] over the non-Archimedean ordered field $\mathbb{Q}(X)$ (where $X^{-1} > \mathbb{Q}$, i.e the indeterminate X is infinitesimally small compared to $\mathbb{Q}$, and positive), the (quartic) polynomial $f(Y) = (Y^2-X)^2 - X^3 \in \mathbb{Q}(X)[Y]$ (Y an indeterminate) is psd over $\mathbb{Q}(X)$ but not a SOS even in $R(Y)$ (where R is the real closure of $\mathbb{Q}(X)$). Indeed, upon factoring over R, we see that

---

[3] I.e., nonnegative in every ordering of F extending the order on k; of course, F need not have any ordering. When $k = \mathbb{Q}$, we say simply "totally positive"

[4] P. 99 of [Artin and Schreier 1927].

$f < 0$ precisely on the two intervals I and $-I$, where $I = (\sqrt{X}(1-\sqrt{X}), \sqrt{X}(1+\sqrt{X}))$, which contain no point of $\mathbb{Q}(X)$.[5]

However, it was not known whether the Archimedean hypothesis could be dropped from Artin's theorem provided the unique orderability was retained. An incorrect proof by Lang [1965] was followed by a counterexample from Dubois [1967]: Let F be the "Euclidean closure" of $\mathbb{Q}(t)$ (t an indeterminate, $t^{-1} > \mathbb{Q}$) i.e. the smallest extension closed under extraction of square roots of positive elements. Then F, being Euclidean, has a unique order, relative to which Dubois showed $f(X) = (X^3-t)^2 - t^3 \in F[X]$ to be (strictly) definite; on the other hand, $f(t^{1/3}) < 0$, so f cannot be a SOS in $F(X)$.

(c) Constructivizations and Bounds

Artin wondered if a constructive version of his solution could be given, and he considered this question in a seminar which he led between the wars. In particular, he wished to eliminate his appeal to an infinite tower of field extensions, and he desired a bound on the number and degree of summands in his representation.

Habicht [1940] gave an elementary, explicit construction of a SOS-representation of forms f strictly definite over $\mathbb{R}$. In fact, the denominator he gives is $(X_0^2 + \cdots + X_n^2)^m$, some $m \in \mathbb{N}$, and the numerator contains only rational coefficients if f does. He derived his representation by combining the "Rabinowitch trick" (i.e. adding a new indeterminate X_{n+1}) with a theorem of Pólya on the representation of forms which are positive when all $X_i \geq 0$ (except when all $X_i = 0$).[6] Habicht's algorithm is fully constructive: it can easily be made to produce a representation correct to any desired accuracy in an estimable amount of time.

A. Robinson used lower predicate calculus and the model completeness of $\mathbb{R}$ to prove a number of overlapping results. To describe them, we first introduce the following notation. Let the ordered field K be contained in the real closed order-extension field R. Then for any finite subset $\{g_i\} \subset K[X]$, write[7]

$$Z\{g_i\} = \{x \in R^{n+1} \mid \bigwedge_i g_i(x) = 0\},$$

$$U\{g_i\} = \{x \in R^{n+1} \mid \bigwedge_i g_i(x) > 0\}, \text{ and}$$

$$W\{g_i\} = \{x \in R^{n+1} \mid \bigwedge_i g_i(x) \geq 0\}.$$

We shall call a set of the form $W\{g_i\}$ a basic closed semi-algebraic set, or simply a "W," and similarly with "U" and "open" in place of "W" and "closed." We shall call $S \subseteq R^{n+1}$ a basic semi-algebraic set if it is the intersection of a U and a W. A set $S \subseteq R^{n+1}$ is called semi-algebraic (s.a. for short) if it is a finite union of basic s.a. sets. Thus a basic s.a. set is one which can be defined by an elementary formula of the language of ordered

---

[5] A. Robinson [1955] gave a similar example.

[6] See the second edition [1952] of [Hardy, Littlewood, and Pólya 1934] for an enjoyable English version of both results.

[7] $\bigwedge_i$ [resp. $\bigvee_i$] means iterated conjunction [resp. disjunction], indexed by i.

fields, with $n + 1$ free variables, which has no quantifiers, negations, or disjunctions, while an arbitrary s.a. set is one which can be defined by any elementary (quantifier-free[8]) formula of the language of ordered fields, with $n + 1$ free variables. See [Brumfiel 1979] for an extensive development of semi-algebraic geometry.

Robinson's first result [1955] is that if K is either real closed or Archimedean, then if $f(x) \geq 0 \ \forall x \in K^{n+1} \cap U\{g_i\}$ (where $\{f, g_i\} \subset K[X]$) then $f = \Sigma c_I g_I r_I^2$, where $c_I \in K^+$, where the g_I are (not necessarily distinct) products of the g_i, and $\{r_I\} \subset K(X)$; further, if the ordering on K is unique, then the c_I are totally positive, hence SOS in K, so that the c_I may be absorbed into the the r_I; better still, for K real closed, he proved the existence of a bound on the number and degrees of the summands which depends on $\{g_i\}$ and $\deg f$ but not on the coefficients of f (or of course, on R). In [1956] Robinson extended the real closed case as follows: if $V \neq \emptyset$ is an irreducible algebraic variety in R^{n+1} with prime ideal P, $\{f, g_i\} \subset R[X]$, and $f(x) \geq 0 \quad \forall x \in V \cap U\{g_i\}$, then $h^2 f = \Sigma g_I h_I^2 \pmod P$ for some $\{h, h_I\} \subset R[X]$, where the g_I are products of the g_i; we still have a bound on the number and degrees of the $\{h, h_I\}$, which depends only on the $\{g_i\}$ and $\deg f$, not on the coefficients of f.

In October 1955 Artin asked Kreisel if explicit bounds could be found. In Nov. 1955, somewhat before the appearance of Robinson's result, Kreisel succeeded in obtaining, by two proof theoretic methods, a primitive recursive bound (Robinson's was only general recursive). The first method ([1957], pp. 165-6 of [1958], and [1960]) used proof theoretical results: Hilbert's first and second ε-Theorems (or Herbrand's Theorem). The second method [1960] consisted of extracting the constructive content of Artin's original argument, by replacing Artin's use of a real closed extension of an ordered field with a specific finite extension sufficient for the result; in this replacement some elegance and clarity is lost, but some explicitness is gained; here the ideas but no theorem of proof theory for first order logic are used. In [1957(I)] Kreisel gave a rough estimate (for $n = 2$) of this primitive resursive bound. A sharper estimate is

$$2^{2^{\cdot^{\cdot^{\cdot^{2^{cd}}}}}},$$

where there are n 2's, where $d = \deg f$, and where c is a positive constant.

Stimulated by these results, Henkin [1960] used model theoretic methods similar to Robinson's to prove what is now accepted as the most natural formulation of the answer to Hilbert's question: if $f \in K[X]$ is psd (over R) and if $\deg f \leq d$, then $f = \Sigma c_i r_i^2$, where $r_i \in K(X)$ and $c \in K^+$ (Artin had obtained this representation under the hypothesis that $K \subseteq \mathbb{R}$ and that f be psd over K; for $K \subseteq \mathbb{R}$, psd over K is equivalent to psd over $\mathbb{R}$ and to psd over R, since K is then dense in $\mathbb{R}$). Henkin also showed that the (bounded number of) c_i and the (bounded number of) coefficients of the r_i can be taken to be functions of the coefficients of f which are "piecewise-ratio-

[8] In classical mathematics we need not exclude quantifiers, since they can be eliminated if necessary by the Tarski-Seidenberg Theorem. In intuitionistic mathematics, however, quantifier-elimination is not generally valid, unless both K and R are recursive (see footnote 9), so here we do exclude quantifiers.

nal" over $\mathbb{Z}$, where the finitely many "pieces" are s.a. subsets of $R^{\binom{n+d}{n}}$, the space of coefficients of f; the coefficients of these rational functions and the polynomials defining their domains are recursive but not necessarily primitive recursive functions of n and d. L. van den Dries [1977] generalized Henkin's results in a certain direction, to polynomials which are "psd over good preordered regular rings;" case distinctions were formally avoided, but at the cost of an artificial definition of rational function.

Robinson gave a correspondingly improved formulation of his results. In §5 of [1957] he proved for $f,g \in K[X]$, that if $f(x) \geq 0$ $\forall x \in Z\{g\}$, then $h^2 f = \Sigma c_i h_i^2 + kg$ for some $\{h,h_i,k\} \subset K[X]$, where $c_i \in K^+$; this time the bound is on the number and degrees of h, k, and the h_i, and it depends on $\deg f$ and $\deg g$, but not on K or the coefficients of f and g. In §8.5 of [1963] he replaced $Z\{g\}$ above with $Z\{g\} \cap U\{g_i\}$ (any $\{g_i\} \subset K[X]$) provided that g generates the ideal of $Z\{g\}$ and that $g \nmid g_i$; the conclusion then is $h^2 f = \Sigma_I c_I g_I h_I^2 + kg$, where the g_I are products of the g_i. The bound no longer applies to $\deg k$, and now the bound depends also on the degree of the g_i.

It is no accident that in the logical treatments of the 17th problem, the Archimedean property was replaced by the condition that the given polynomial be psd over the <u>real closure</u> of the ordered field of coefficients, because the Archimedean property <u>cannot</u> be expressed by an elementary statement. Since Archimedean ordered fields are isomorphic to subfields of $\mathbb{R}$, and are therefore dense in their real closures, "psd" over an Archimedean ordered field already implies "psd" over its real closure.

Robinson further proved [1957] that if p is totally positive in a finite, formally real extension F of K, then $p = \Sigma_{i=1}^{r} c_i r_i^2$, where $c_i \in K^+$ and $r_i \in F$; what was new was that r depends only on $[F:K]$, not on F, K, or p. Thus if all the positive elements of K are SOS, and if the number of required squares is bounded, then we may absorb the c_i into the r_i in the above representation, but make r dependent also on this bound; this overlaps an important theorem stated by Hilbert (first proved by Siegel [1921]) that if $K = \mathbb{Q}$, then $r = 4$, independent even of $[F:K]$. Here we have an interesting historical twist: while work on Hilbert's 17th problem led to a result much like Siegel's theorem, Siegel's theorem helped lay the foundation for the 17th problem; indeed, one of the first uses of Siegel's theorem (even before anyone had published a proof!) was by Hilbert in his solution [1899] of the case $n = 0$ of the problem.

Daykin [1960] constructed a primitive recursive, piecewise-rational solution which was superior to the Henkin-Robinson solutions, by working out Kreisel's [1960] sketch of the constructivization of Artin's original proof. A little more notation at this point will help us describe Daykin's representation (and eventually others as well). Let $\alpha = (\alpha_0,\ldots,\alpha_n) \in \mathbb{N}^{n+1}$ be a multi-index, let $|\alpha| = \Sigma\alpha_i$, fix an even $d \in \mathbb{N}$, let $C = (C_\alpha)_{|\alpha|=d}$ be $\binom{n+d}{n}$ indeterminates (in some fixed order), let $c = (c_\alpha)_{|\alpha|=d}$ be an element of $R^{\binom{n+d}{n}}$, let $f \in \mathbb{Z}[C;X]$ be the general form of degree d in X with coefficients C (i.e. $f(C;X) = \Sigma_{|\alpha|=d} C_\alpha X^\alpha$, where $X^\alpha = X_0^{\alpha_0}\cdots X_n^{\alpha_n}$), and let

$$P_{nd} = \{c \in R^{\binom{n+d}{n}} \mid f(c;X) \text{ is psd (over } R) \text{ in } X\}.$$

Daykin showed how to compute effectively, from n and d alone, finitely many $p_{ij} \in \mathbb{Z}[C]$ and $r_{ij} \in \mathbb{Q}(C;X)$ (homogeneous in X) such that

(1.2) $$\bigwedge_i \quad f = \Sigma_j\, p_{ij} r_{ij}^2 \quad \text{and}$$

(1.3) $$\forall c \in P_{nd}, \quad \bigvee_i \; \bigwedge_j \left[\begin{array}{l} p_{ij}(c) \geq 0, \text{ and the denominator of } r_{ij}(c;X) \\ \text{does not vanish identically in } X. \end{array}\right]$$

Thus, as in most applications of Herbrand's theorem, the answer is expressed as a disjunction. The superiority of this representation consists not only in the explicitness of the bound but also in the choice of pieces on which the rational functions are defined: the earlier pieces were s.a., but Daykin's are basic closed s.a., namely, $W_i = W\{p_{ij}\}$. Daykin's proof was long and difficult; in §2 we give a quick proof and refinement of his representation, using powerful results in s.a. geometry.

The main result in the sixties was Pfister's elegant "2^n bound" [1967] on the number of square summands required to represent a homogeneous, psd $f \in R(X)$, where R is real closed. The bound is independent of $\deg f$. (Hilbert had proved this for $n = 2$ in [1893].) More precisely, Pfister has shown [1974]: if

$$f = \sum_{i=1}^{2^n+m} f_i^2$$

with $f_i \in R(X)$ homogeneous of degree d, then there is a representation

$$f = \sum_{i=1}^{2^n} g_i^2$$

with $g_i \in R(X)$ homogeneous of degree $\leq C(n)^{\frac{n^m-1}{n-1}} d^{n^m}$: the constant $C(n)$ depends only on n, and could be determined explicitly; it probably grows quickly with n.

Pfister's proof uses (1) a special case of the Tsen-Lang Theorem: if C is an algebraically closed field and F is a field of transcendence degree n over C, then every quadratic form with coefficients in F, of dimension $> 2^n$, has a non-trivial zero in F: and (2) his theorem that the non-zero elements of a field F of characteristic $\neq 2$ represented by (what is now called) a "Pfister form," form a subgroup of F^*. (An independent, unpublished study by Ax in 1966, showed that 8 squares suffice when $n = 3$.) It is not known whether Pfister's bound applies in the case of ordered coefficient fields K, in particular $\mathbb{Q}$; again we should allow positive constant weights on the squares in the ordered field case, since positive elements of K need not be sums of (even an unbounded number of) squares.

For real closed fields it is not known whether 2^n is best possible, except for $n \leq 2$; Cassels, Ellison, and Pfister [1971] showed that the (psd) Motzkin polynomial $1 + X^2Y^4 + X^4Y^2 - 3X^2Y^2$ is not a sum of three squares in $\mathbb{R}(X,Y)$: but their method uses the theory of elliptic curves, and does not extend to $n > 2$. The only known lower bound is $n + 1$: Cassels [1964] showed that $1 + X_1^2 + \cdots + X_n^2$ is not a sum of n squares in $\mathbb{R}(X_1,\ldots,X_n)$, by sharpening Landau's [1906] result that a SOS of rational functions can be reduced to a SOS of rational functions in which any one variable, say X_1, does not occur in the (common) denominator; Cassels did this without increasing the number of summands. Hsia and Johnson [1974] have conjectured that a homogeneous, psd $f \in \mathbb{Q}(X)$ must be a sum of $2^n + 3$ squares in $\mathbb{Q}(X)$ (this is Lagrange's Theorem (1770) for $n = 0$ and Fourchet's Theorem [1971] for $n = 1$, but it is not known whether <u>any</u> bound, independent of degree, exists for $n > 1$).

(d) Continuous Solutions

The improvements found in the fifties to Artin's solution brought only temporary satisfaction, and by the early sixties Kreisel wondered if one could not do better. In particular, the piecewise character of the representations meant that when computing a representation from given coefficients of f, one first had to determine in which piece of P_{nd} the coefficients lay. This amounts to testing various polynomial inequalities in the coefficients. For recursive ordered fields[9] such as $\mathbb{Q}$ or the real algebraic numbers, the test is effective. But such testing is precisely what we cannot do in, say, $\mathbb{R}$, an element of which must be presented as, say, a decimal, or an oscillating decimal used in computer science, or a pair $((r_n),\mu)$ of some kind of Cauchy sequence of rationals and a "modulus of convergence" function μ satisfying $\forall k > 0$, $\forall n,m \geq \mu(k)$ $[|r_n - r_m| < 1/k]$. Thus it is at the <u>discontinuities</u> that we are unable to compute the representation. While continuity is not required by classical algebraists (who implicitly use the discrete topology when doing algebra, as if real numbers were presented with infinite precision), the lack of it is enough of a problem to leave Hilbert's 17th problem still unsolved from a constructivist, or, for that matter, a topological, point of view. Thus, by the early sixties, the following two questions were open.

First Question: Are the case distinctions of 1.2 and 1.3 unnecessary? I.e., do there exist $p_j \in \mathbb{Q}(C)$ and $r_j \in \mathbb{Q}(C;X)$ such that $f = \Sigma p_j r_j^2$ and $\forall c \in P_{nd}$, each $p_j(c) \geq 0$?

Second Question: Can a topological or "continuous" version of Artin's theorem be given? This question has two parts: (a) can we choose representing rational functions which are continuous in R^{n+1} (where by a "continuous rational function" we mean a continuously-extendible rational function; e.g., the representation

$$1 = \frac{X^2}{X^2 + Y^2} + \frac{Y^2}{X^2 + Y^2}$$

is discontinuous at the origin in R^2), and (b) can the weights, and the coefficients of the numerator and denominator in each rational function, be chosen to be continuous functions of the given coefficients (in P_{nd})? When we refer to continuity, we are mainly interested in the usual interval topology on K or R; when $K = R = \mathbb{R}$, we should also consider various "computational" topologies on "enrichments" of $\mathbb{R}$ by specific representations, say Cauchy sequences of rationals with the topology inherited from the product topology on $\mathbb{Q}^{\omega}$. These questions appeared in print, e.g., on p. 102 of Kreisel [1969], on pp. 115-6 of [1977a], in footnote 1 of [1977b], and in [1978].

In view of the geometric origin of Hilbert's 17th problem, stressed in his own presentation, it seems natural enough to impose topological conditions. Kreisel's interest is logical: To determine the extent to which current mathematical notions express adequately or better the aims usually stated in terms of so-called constructive, in particular, of intuitionistic foundations.

---

[9] By a <u>recursive</u> ordered field we mean a numbering of its underlying set for which the field operations and the order (hence also equality) relation are recursive.

The main result of this paper is a simple, geometric proof (4.1) of the negative answer to Kreisel's first question, except when $d \leq 2$. The proof amounts to showing (4.2) that P_{nd} is not a basic s.a. set, except when $d \leq 2$, when it is a basic closed s.a. set. When $d = 2$, not only is the answer affirmative, but the r_j can be found in $Q(C)[X]$, that is, they can be chosen to be linear forms in X (see also next page). The case $d > 2$ reduces to the case $n = 1$, $d = 4$, and it is interesting to compare what 4.2 says about $P_{1,4}$ to what classical algebra texts say about it. The standard approach is to reduce to monic polynomials with no X^3 term: $a(X) = X^4 + qX^2 + rX + s$. Write $L = 8qs - 2q^3 - 9r^2$ and let D be the discriminant of a, namely

$$D = 4\left(4s + \frac{q^2}{3}\right)^3 - 27\left(\frac{8}{3}qs - r^2 - \frac{2}{27}q^3\right)^2.$$

Write $\overline{P}'_{1,4} \subset R^3$ for the set of (coefficients (q,r,s) of) reduced, monic quartic polynomials a with no real or multiple roots. By Sturm's theorem, $\overline{P}'_{1,4} = (U\{D\} \cap W\{q\}) \cup (U\{D\} \cap W\{-L\})$ (cf. [Jacobson 1974], pp. 299-300). The closure in R^3 of $\overline{P}'_{1,4}$ equals the set $P'_{1,4}$ of reduced, monic quartic psd polynomials. Thus we get a description of $P_{1,4}$ in terms of two basic s.a. sets. 4.2 now says that this description is simplest possible, in the sense that no single basic s.a. set will suffice.

For part (a) of the second question, continuity in the variables of the rational functions, Kreisel had found [1978] a simple proof of the affirmative answer, using Stengle's [1974] "Positivstellensatz:" for $\{f, g_i\} \subset K[X]$, if $\forall x \in W\{g_i\}$, $f(x) > 0$, then

$$f = \frac{\Sigma_I\, p_{1,I}\, g_I h_{1,I}^2}{f^{2s} + \Sigma_I\, p_{2,I}\, g_I h_{2,I}^2}, \tag{1.4}$$

some $s \in N$, $p_{jI} \in K^+$, and $h_{jI} \in K[X]$ $(j = 1,2)$, where the g_I are products of the g_i. (In [1979], Stengle arranged for the h_{jI} to be homogeneous if f is.) If we want to transform this into a (nonnegatively-weighted) SOS of rational functions, we just multiply the numerator and denominator of 1.4 by the denominator. Since $W\{g_i\} \cap Z\{\text{denominator}\} \subseteq Z\{f\} \subseteq R^{n+1}$, the Squeeze Theorem easily implies that each of the resulting rational functions (and not merely the sum of their squares, of course) extends--by 0--to a function of X which is continuous throughout $W\{g_i\}$.

In forthcoming papers we shall prove the affirmative answer to both parts of Kreisel's second question, at least when the ground field K is real closed [Delzell, in preparation (a)], and when it is a countable subfield of R [Delzell, in preparation (b)]. Precisely, we construct a finite set of functions $p_i\colon P_{nd} \to R^+$ and $r_i\colon P_{nd} \times R^{n+1} \to R$ taking points with coordinates in K to points with coordinates in K, satisfying, $\forall c \in P_{nd}$, $f(c;x) = \Sigma p_i(c) r_i(c;x)^2$, with each summand $p_i r_i^2$ a function which is continuous, relative to the usual interval topology on R, hence also relative to the computational topologies in case $R = \mathbb{R}$, (and "s.a." in the case when K is real closed) simultaneously in c and x for $(c;x) \in P_{nd} \times R^{n+1}$, and with r_i homogeneous and rational in X. For the case $R = \mathbb{R}$, this last sentence provides the first constructive, in particular, intuitionistic, solution to

Hilbert's 17th problem, since (1) the s.a. descriptions of the p_i and r_i (and also of the X-coefficients of the r_i) are recursive in n and d, and (2) while elements of $\mathbb{R}$ can be given only as approximations, we can aproximate p_i and r_i by approximating c, by continuity.

In [1980] and [to appear], we also give continuity results, positive and negative, for representing psd quadratic and psd ternary quartic forms as SOS of forms: the former can be represented over K by continuous summands, at least if the number of summands is increased greatly beyond n+1, while if the latter are represented as SOS of quadratic forms (even over R), certain coefficients must jump at $(X^2 + Y^2)^2$.

Our history has dealt mainly with the logical aspects of Hilbert's original problem; we have tried to include especially those results which appear to be less well-known. However, a large literature on other aspects of the problem has developed, and we now mention enough references to guide the interested reader into these areas. Bochnak and Efroymson [1980] cover the current knowledge of SOS of C^∞ functions, Nash functions (real algebraic analytic functions), and real analytic functions. They generalize both Stengle's and Pfister's Theorems to certain subrings of the ring of Nash functions on open s.a. subsets of irreducible nonsingular algebraic sets in $\mathbb{R}^n$. They consider both global functions and germs of functions. They similarly generalize Procesi's [1978] representations of symmetric psd functions to Nash functions invariant under a Lie group action on $\mathbb{R}^n$. (The study of SOS of real algebraic functions was initiated by Artin [1927].)

Lam [1980] gives a bibliography on the 17th problem, including references to a non-commutative generalization of the problem, a p-adic analog, and a generalization to psd symmetric matrices over polynomial rings. Pfister [1976] also gives historical references. We shall not try to duplicate here these three main bibliographies, but instead conclude with some references not included in these bibliographies.

Berg, Christenson, and Ressel [1976] studied positive definite functions on Abelian semigroups, and approximated definite polynomials by SOS of polynomials, of not necessarily bounded degrees. Bose [1976] gave algorithms to test polynomials for psd-ness. Ellison [1969] considered a "Waring's problem" for forms. Dickmann [1980] characterized definite polynomials over "real closed rings."

2. NEW PROOF OF DAYKIN'S DISJUNCTIVE SUM-OF-SQUARES REPRESENTATION

As in §1, let $f \in \mathbb{Z}[C;X]$ be the general form in X.

Theorem 2.1: There exist finitely many $g_{iJ} \in \mathbb{Z}[C]$, $h_{kiJ} \in \mathbb{Q}[C;X]$, and $s_i \in \mathbb{N}$ $(k = 1,2)$ such that

(2.1.1) $$\bigwedge_i \quad f = \frac{\Sigma_J\, g_{iJ} h_{1iJ}^2}{f^{2s_i} + \Sigma_J\, g_{iJ} h_{2iJ}^2} \quad \text{and}$$

(2.1.2) $$\forall c \in P_{nd}, \quad \bigvee_i \bigwedge_J \quad g_{iJ}(c) \geq 0.$$

Remark: 2.1 refines Daykin's representation (1.2), since we can transform 2.1.1 into a nonnegatively-weighted SOS of rational functions by multiplying the numerator and denominator of 2.1.1 by the denominator. The resulting rational functions need not be homogeneous in X, but since f is, we can extract the lowest homogeneous components from the numerator and denominator and still have an identity, though its denominator may not have the special structure shown in 2.1.1.

The proof of 2.1 will use the Positivstellensatz (1.4) and

Theorem 2.2 (The Finiteness Theorem, [Delzell 1980 and 1981]): _If_ $S \subseteq R^{n+1}$ _is s.a., then_

(a) S _is open if and only if_ $S = \bigcup_i U\{g_{ij}\}$, _some finite set_ $\{g_{ij}\} \subset K[X]$; _equivalently,_

(b) S _is closed if and only if_ $S = \bigcup_i W\{g_{ij}\}$, _some finite set_ $\{g_{ij}\} \subset K[X]$.

_If_ K _is recursive, the_ g_{ij} _are computable from the presentation of_ S _as a s.a. set._

The "if" directions are trivial. The equivalence of (a) and (b) follows by taking complements and distributing. 2.2 is deceptively simple for $n = 0$, and deceptively difficult for $n > 0$: for $n = 0$, we combine Rolle's Theorem with induction on the degree (over K) of the endpoints of the intervals comprising S; for $n > 0$ we combine the "Good Direction Lemma" with a parametrized version of the case $n = 0$.

Proof of 2.1: P_{nd} is a s.a. set in $R^{\binom{n+d}{n}}$, since $c \in P_{nd}$ if and only if the elementary formula $\forall x_0 \cdots \forall x_n\ f(c;x_0,\ldots,x_n) \geq 0$ holds, and the Tarski-Seidenberg Theorem produces an equivalent quantifier-free formula. Also, P_{nd} is closed, since a limit of psd forms is still psd. (Furthermore, it is easy to see that P_{nd} is even a convex cone.) Therefore we may apply the Finiteness Theorem (b) to P_{nd}: there exist finitely many $g_{ij} \in \mathbb{Z}[C]$ such that, writing $W_i = W\{g_{ij}\}$, $P_{nd} = \bigcup_i W_i$. For each i we apply the Positivstellensatz (1.4) to f, which is nonnegative on $W_i \subset R^{\binom{n+d}{n}+n+1}$ (we are now viewing $\{g_{ij}\}$ as being in the larger ring $\mathbb{Z}[C;X]$). 2.1 follows immediately, taking the g_{iJ} to be products of the g_{ij}. Q. E. D.

3. TWO LOGICAL POINTS CONCERNING THE FINITENESS THEOREM (2.2)

(a) Intuitionistic Considerations

The first point is addressed to readers interested in intuitionistic mathematics. Since we do not assume in 2.2 that K and R are recursive, several steps in the proof of 2.2 are, intuitionistically, problematic. Specifically, even if S is defined by a quantifier-free formula (and not by an arbitrary first-order formula) in the language of ordered fields, and the g_{ij} are presented by terms in that language, the equivalence $(\forall x \in R^{n+1})[x \in S \leftrightarrow x \in \bigcup_i W\{g_{ij}\}]$ may be valid classically but not intuitionistically. In broad terms: to each $x \in S$ we are to find i such that $x \in W\{g_{ij}\}$, and so, if S happens to be connected the map $x \mapsto i$ must be constant, since intuitionistic functions are continuous. A concrete example refuting 2.2(b) will be given after the proof of 4.2 below. Evidently, the situation changes if we replace, in 2.2(b), the union by its classically equivalent form $R^{n+1} - \bigcap_i \bigcup_j U\{-g_{ij}\}$.

These specifically intuitionistic or, in a related context, sheaf-theoretic requirements should be distinguished from questions about the (recursion-theoretic) complexity of the terms (defining) g_{ij} as functions of the presentation of S, either by a first order formula or more specifically a quantifier-free one. These questions simply concern the validity of 2.2(a) and 2.2(b) in the classical theory. Reasonably good bounds on the complexity can be extracted from our proof in [1980] and [1981]. But it is worth noting that mere recursiveness of some suitable g_{ij}, as functions of (the presentation of) S follows trivially from the logical form of 2.2 by the (recursiveness of the) Tarski-Seidenberg algorithm. We begin by recursively enumerating all presentations of s.a. sets S, and all presentations of, that is, formulae defining basic open, resp. closed, sets, denoted above using g_{ij}. The latter enumeration induces enumerations of finite unions $\bigcup_i U\{g_{ij}\}$, resp. $\bigcup_i W\{g_{ij}\}$, of such basic sets. Now, for any given presentation of S (with parameters, interpreted to denote elements ε K) and any given presentation of a union of basic sets, the validity of $S = \bigcup_i U\{g_{ij}\}$, resp. $S = \bigcup_i W\{g_{ij}\}$, is recursively decidable by the Tarski-Seidenberg algorithm. In other words, the relation, say $\mathcal{R}$, between presentations of S and g_{ij} satisfying 2.2(a), resp. 2.2(b), is recursive. (For the conclusion below it would be sufficient if that relation were recursively enumerable.) Thus $\forall S \exists \{g_{ij}\} \mathcal{R}(S,\{g_{ij}\})$ is true and hence there is a recursive map $\gamma : S \to \{g_{ij}\}$ satisfying $\forall S \mathcal{R}(S,\gamma(S))$. Evidently, corresponding results apply to 1.4 and 2.1.

Remark: The same considerations are used in the work of Robinson and Henkin, discussed in §1, providing recursive bounds, and more generally, in applications of first-order model theory to theorems with a resursively enumerable set of axioms. This brings us to our second logical point:

(b) The Terminology "Finiteness"

Can general (classical) model theory be used to give a simple, or at least new, proof of 2.2? The question arose--if for no other reason--because of a pun: the compactness theorem of logic was originally called the "Finiteness Theorem" just like our 2.2. The latter is called "Finiteness Theorem" because, at least superficially (that is under quite general conditions), it is trivial that S is open if and only if it is a possibly infinite union of basic open sets. It is worth pausing a moment to look at the assumptions behind the question.

First of all, the "trivial" representation of any (not only: s.a.) open set S by an infinite union of basic open sets defined by use of g_{ij}--tacitly, with coefficients in K--assumes that K is dense in R. In that case $x \varepsilon S$ can be surrounded by an (n+1)-ball in S, and the latter by a basic open K-R-s.a. (n+1)-ball, also lying in S. (The center of the (n+1)-ball need not be x itself, for example if $x \notin K^{n+1}$.) But, in general, that is if K is not Archimedean, K need not be dense in R at all, with consequences illustrated by the penultimate paragraph of §1(b). More specifically, the representation of S as an infinite union of basic open sets defined by use of parameters for elements ε K only, is by no means trivial.

What can be done--and this is implicit in a suggestion made by R. L. Vaught at the conference--is this. Suppose that the gap mentioned in the last paragraph is filled, but possibly in a non-uniform way. Specifically, suppose it is shown that, for each open s.a. set S, there is some infinite family of

$U_i = U\{g_{ij}\}$ such that $S = \bigcup_i U_i$, but possibly a different family for different R (and the same description of S). Then the compactness theorem of logic ensures also a uniform representation as follows. We first expand the language by adding a constant a, and consider the conjunction, for all U_i (built up from the parameters in S, in other words, in the field generated by them), of $(a \in S \wedge a \notin U_i) \vee (a \in U_i \wedge a \notin S)$. This is inconsistent for each real closed field, by our hypothesis that there are non-uniform representations. The compactness theorem then ensures that a finite subset is inconsistent, and hence uniformity.

It is evident that the Tarski-Seidenberg Theorem by itself does not provide the elimination of negations and $=$ and $\leqslant$ relations from quantifier-free presentations of open sets S.

4. THE NEGATIVE ANSWER TO KREISEL'S FIRST QUESTION

Theorem 4.1: For $d \leqslant 2$, there are $s \in \mathbb{N}$, $\{g_J\} \subset \mathbb{Z}[C]$, and $\{h_{kJ}\} \subset \mathbb{Q}[C;X]$ $(k = 1,2)$, such that

$$f = \frac{\Sigma_J\, g_J h_{1J}^2}{f^{2s} + \Sigma_J\, g_J h_{2J}^2} \quad \text{and} \quad \forall c \in P_{nd},\ \text{each}\ g_J(c) \geqslant 0;$$

for $d > 2$, there do not exist $p_j \in \mathbb{Q}(C)$, and $r_j \in \mathbb{Q}(C;X)$ such that

$$(4.1.1) \quad f = \Sigma_j\, p_j r_j^2 \quad \text{and} \quad (4.1.2) \quad \forall c \in P_{nd},\ \text{each}\ p_j(c) \geqslant 0.$$

The proof is based on a finer analysis of P_{nd} than that given by the Finiteness Theorem:

Theorem 4.2: P_{nd} is not a basic s.a. set, except when $d \leqslant 2$, when it is a basic closed s.a. set.

(4.2 is to be expected: basic s.a. sets are rather special among all s.a. sets.)

Proof of 4.2: For the case $d = 2$ we use induction on n. For $n = 1$, $P_{1,2} = W\{A, C, AC - B^2\}$ (writing $f(A,B,C;X,Y) = AX^2 + 2BXY + CY^2$). To prove P_{n2} is a single W for $n > 1$, we may suppose, inductively, that the condition for a quadratic form in $X_0,\ldots,X_{n-1}$ to be psd is a conjunction of non-strict inequalities in C. Write

$$f(X_0,\ldots,X_n) = f_2(X_0,\ldots,X_{n-1}) + 2f_1(X_0,\ldots,X_{n-1})X_n + f_0 X_n^2,$$

where $\deg f_i = i$ $(i = 0,1,2)$. Then f is psd if and only if f_2, f_0, and $f_0 f_2 - f_1^2$ are all psd in $X_0,\ldots,X_{n-1}$; this is just a conjunction of three conjunctions, since these three forms are quadratic (except the constant form f_0, for which the psd property is an "improper" conjunction, namely, with only one conjunct).

For the case $d > 2$, note that if P_{nd} were a basic s.a. set for $d \geqslant 4$, then by setting some coefficients equal 0, we would have that $P_{1,4}$ is a basic s.a. set. To derive a contradiction from this, let Y be a single indeterminate and consider $f_{a,b}(Y) = (Y^2 + a)^2 + b \in R[Y]$ for $(a,b) \in R^2$. On the one hand, the set $A = \{(a,b) \in R^2 \mid f_{a,b}(Y)$ is psd in Y over $R\}$ must be of the form $W\{g_i(a,b)\} \cap U\{h_i(a,b)\}$ for some $g_i, h_i \in \mathbb{Z}[a,b]$. On the other hand, $A = \{(a,b) \mid b \geq 0 \vee \sqrt{-b} - a \leq 0\} = \{(a,b) \mid b \geq 0 \vee (a \geq 0 \wedge b \geq -a^2)\}$ (striped in the figure below). Then some g_i or h_i would have to change sign across a Zariski-dense subset of the negative a-axis, hence the Sign-Changing Theorem in [Dubois-Efroymson 1970] would then imply that this g_i or h_i would have to be divisible by (precisely) an odd power of b, which would make it change sign across even the positive a-axis, thereby excluding part of A. Q. E. D.

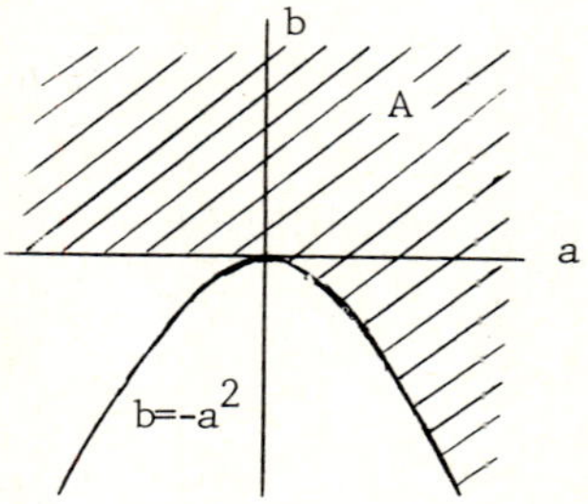

Note that the closed set A is a counterexample to the Finiteness Theorem (b), when interpreted intuitionistically. The above proof showed, in effect, that no basic s.a. subset of A can contain a set of the form $B \cap A$, where B is any open disk about the origin in R^2. Therefore if a Cauchy sequence of pairs of rationals $(r_n, s_n) \to (0,0) \in A$ is given, we will not always be able to compute i such that $(r_n, s_n) \in W_i$.

Proof of 4.1: For $d = 2$ we just combine 2.2 and 1.4, as in the proof of 2.1, with a single W_i, so that we may drop the i.

For $d > 2$, suppose that 4.1 were false, and write each $p_j = r_j/s_j$, with r_j, s_j relatively prime in $\mathbb{Q}[C]$; then we could conclude that $P_{nd} = W\{r_j s_j\} \cap U\{s_j\}$ ($\supseteq$ by 4.1.1 and $\subseteq$ by 4.1.2), contradicting 4.2. Q. E. D.

The results of this paper appeared in my dissertation [1980]. I am grateful to Professor Gregory Brumfiel, my thesis advisor, and to Professor Georg Kreisel, for many helpful conversations on this subject.

BIBLIOGRAPHY

Artin, E. Uber die Zerlegung definiter Funktionen in Quadrate, Abhandl. Math. Sem. Hamburg **5** (1927), 100-15.

Artin, E. and Schreier, O., Algebraische Konstruktionen reeller Korper, Abhandl. Math. Sem. Hamburg **5** (1927), 85-99.

Berg, C., Christenson, J.P.R., and Ressel, P., Positive definite functions on Abelian semigroups, Math. Ann. **223**(3), (1976), 253-74.

Bochnak, J. and Efroymson, G., Real algebraic geometry and the 17th Hilbert problem, Math. Ann. (1980).

Bose, N.K., New techniques and results in multi-dimensional problems, J. Franklin Inst. **301**(1-2), (1976), 83-101.

Brumfiel G., Partially Ordered Rings and Semi-Algebraic Geometry. Lecture Note Series of the London Math. Soc. (Cambridge Univ. Press, Cambridge, 1979).

Cassels, J.W.S., On the representation of rational functions as sums of squares, Acta Arith. **9** (1964), 79-82.

Cassels, J.W.S., Ellison, W.J., and Pfister, A., On sums of squares and on elliptic curves over function fields, J. Number Theory **3**(2), (1971), 125-49.

Choi, M.D., and Lam, T.-Y., An old question of Hilbert, Proc. Conf. on Quadratic Forms--1976, (Queen's Papers on Pure and Applied Math., No. 46 (G. Orzech, ed.), Queen's Univ., Kingston, Ontario, 1977), 385-405.

Extremal positive semidefinite forms, Math. Ann. **231** (1977), 1-18.

Daykin, Thesis, Univ. of Reading, 1960 (unpublished); cited by Kreisel, A survey of proof theory, J. Symb. Logic **33** (1968), 321-88.

Delzell, C., A constructive, continuous solution to Hilbert's 17th problem, and other results in semi-algebraic geometry, Ph.D. dissertation, Stanford Univ., 1980 (Univ. Microfilms International, Order No. 8024640). Cf. also Dissertation Abstracts International **41**, no. 5, 1980.

A finiteness theorem for open semi-algebraic sets, with applications to Hilbert's 17th problem, Proc. AMS Special Session on Ordered Fields and Real Algebraic Geometry, Jan. 7-8, 1981, San Francisco, D.W. Dubois, ed., Contemporary Math. Series, AMS.

Continuous sums of squares of forms, to appear in Proc. L.E.J. Brouwer Centenary Symposium, June 8-13, 1981, Nordwijkerhout, Holland, A.S. Troelstra and D. van Dalen, eds., North Holland.

A continuous, constructive solution to Hilbert's 17th problem, in preparation (a). See also a preliminary abstract in AMS Abstracts, Jan. 1981.

On Hilbert's 17th problem over countable subfields of $\mathbb{R}$, in preparation (b).

Dickmann, M.A., Sur les anneaux de polynôme à coefficients dans un anneaux réel clos, Comptes Rendus (Paris), Serie A **290** (1980), 57.

Dubois, D.W., Note on Artin's solution of Hilbert's 17th problem, Bull. Amer. Math. Soc. **73** (1967), 540-1.

Dubois, D.W., and Efroymson, G., Algebraic theory of real varieties, I, Studies and Essays Presented to Yu-Why Chen on his 60th Birthday, pp. 107-135, Taiwan University, 1970.

Ellison, W.J., A "Waring's problem" for homogeneous forms, Proc. Cambridge Philos. Soc. **65** (1969), 663-72.

Fleck, A., Zur Darstellung definiter binäre Formen als Summen von Quadraten ganzer Rationalzahligen Formen, Arkiv der M. & Physik 3d Ser. **10** (1906), 23-38, and 3d Ser. **16** (1910), 275-6.

Habicht, W., Über die Losbarkeit gewissen algebraischer Gleichungssysteme, Comm. M. Helv. **12** (1940), 317-22. (An English version is in the second [1952] edition of [Hardy, Littlewood, and Pólya, 1934], below.)

Hardy, G.H., Littlewood, J.E., and Pólya, G., Inequalities (Cambridge Univ. Press, Cambridge 1934); second edition, (1952).

Henkin, L., Sums of squares, Summaries of Talks Presented at the Summer Institute of Symbolic Logic in 1957 at Cornell University (Institute Defense Analyses, Princeton, 1960), 284-91.

Hilbert, D., Über die Darstellung definiter Formen als Summe von Formenquadraten, Math. Ann. **32** (1888), 342-50; see also Ges. Abh. **2**, (Springer, Berlin, 1933), 154-61.

Über ternäre definite Formen, Acta Math. **17** (1893), 169-97; see also Ges. Abh. **2**, (Springer, Berlin, 1933), 345-66.

Grundlagen der Geometrie (Teubner, 1899); transl. by E.J. Townsend (Open Court Publishing Co., La Salle, IL, 1902); transl. by L. Unger from the tenth German edition (Open Court, 1971).

Mathematische Probleme, Göttinger Nachrichten (1900), 253-97, and Archiv der Mathematik und Physik 3d ser. **1** (1901), 44-53, 213-37. Transl. by M.W. Newson, Bull. Amer. Math. Soc. 8 (1902), 437-79; reprinted in Mathematical Developments Arising from Hilbert Problems (F. Browder, ed.), Proc. Symp. in Pure Math. **28**, (Amer. Math. Soc., Providence, 1976,) 1-34.

Hsia, J.S., and Johnson, R.P., On the representation in sums of squares for definite functions in one variable over an algebraic number field, Amer. J. Math. **96**(3), (1974), 448-53.

Jacobson, N., Basic Algebra **1** (Freeman, San Francisco, 1974).

Kreisel, G., Hilbert's 17th problem, I and II, Bull. Amer. Math Soc. **63** (1957), 99 and 100.

Mathematical significance of consistency proofs, J. Symb. Logic **23** (1958), 155-82 (reviewed by A. Robinson, J. Symb. Logic **31**, 128).

Sums of squares, Summaries of Talks Presented at the Summer Institute for Symbolic Logic in 1957 at Cornell University, (Institute for Defense Analyses, Princeton, 1960), 313-20.

Two notes on the foundations of set theory, Dialectica **23** (1969), 93-114.

On the kind of data needed for a theory of proofs, Logic Colloquium 1976 (Gandy, R.O. and Hyland, J.M.E., eds.), North-Holland Publishing Co. Amsterdam, 1977), 111-28. (MR**58**#21397)

Review of L.E.J. Brouwer, Collected Works, Vol. I, Philosophy and Foundations of Mathematics (A. Heyting, ed.), Bull. Amer. Math. Soc. **83** (1977), 86-93.

Review of Ershov, Zentralblatt **374** (1978), 18-9, #02027.

Lam, T.-Y., The theory of ordered fields, Ring Theory and Algebra III, B. McDonald, ed., (Marcel Bekker, New York, 1980).

Landau, E., Über die Darstellung definiter binärer Formen durch Quadrate, Math. Ann. **57** (1903), 53-65.

Über die Zerlegung definiter Funktionen durch Quadraten, Ark. fur Math. und Physic, 3d Ser. **7** (1904), 271-7.

Über die Darstellung definiter Functionen in Quadraten, Math. Ann. **62** (1906), 272-85.

Lang, S., Algebra (Addison-Wesley, Reading, MA, first edition, 1965).

Motzkin, T.S., The arithmetic-geometric inequality, Inequalities **1**, Shisa, O., ed., (Academic Press, New York, 1967), 204-24.

Pfister, A., Zur Darstellung definiter Functionen als Summe von Quadraten, Invent. Math. **4** (1967), 229-37. (An English version is in [1976] below.)

Letter to Kreisel, The Kreisel Papers (Stanford Univ. Archives, 1974).

Hilbert's 17th problem and related problems on definite forms, Mathematical Developments Arising from Hilbert Problems (F. Browder, ed.), Proc. Symp. in Pure Math. **28**, (Amer. Math. Soc., 1976), 483-9.

Pourchet, Y., Sur la représentation en somme de carrés des polynômes à une indeterminée sur un corps de nombres algébriques, Acta Arith. **19** (1971), 89-104.

Prestel, A. Sums of squares over fields, Atas da 5ª Escola de Algebra (Soc. Brasileira de Matemática, Rio de Janeiro, 1978).

Procesi, C., Positive symmetric functions, Advances in Math., **29** (1978), 219-25.

Robinson, A., On ordered fields and definite forms, Math. Ann. **130** (1955), 257-71.

Further remarks on ordered fields and definite forms, Math. Ann. **130** (1956), 405-9.

Some problems of definability in the lower predicate calculus, Funda. Math. **44** (1957), 309-29.

Introduction to Model Theory and to the Metamathematics of Algebra (North-Holland Publishing Co., Amsterdam, 1963).

Robinson, R.M., Some definite polynomials which are not sums of squares of real polynomials, Notices Amer. Math. Soc. **16** (1969), 554; Selected Questions in Algebra and Logic (Vol. dedicated to the memory of A.I. Mal'cev), Izdat. "Nauka." Sibirsk Otdel Novosibirsk (1973), 264-82; or Acad. Sci. USSR. (MR**49**#2647).

Siegel, C.L., Darstellung total positiver Zahlen durch Quadrate, Math. Zeit. **11** (1921), 246-75; Ges. Abh. **1** (1966), 47-76.

Stengle, G., A Nullstellensatz and a Positivstellensatz for semi-algebraic geometry, Math Ann. **207** (1974), 87-97.

Integral solution of Hilbert's 17th problem, Math. Ann. **246** (1979), 33-9.

Van den Dries, L., Artin-Schreier theory for commutative regular rings, Ann. Math. Logic **12** (1977), 113-50.

PROCEEDINGS OF THE HERBRAND SYMPOSIUM
LOGIC COLLOQUIUM '81, J. Stern (editor)

ON LOCAL AND NON-LOCAL PROPERTIES

Haim Gaifman
The Hebrew University Jerusalem
and University Paris VI

§0. INTRODUCTION

The result to be presented here was motivated by questions of first-order definability within the class of finite relational structures. These questions arose in the research of suitable languages for databases (cf [AHU], [AU] and [CH]). A standard example is the following : Can we express in first-order language the property that a graph (i.e. a binary symmetric relation) is connected : The negative answer is easily proved either by a compactness argument, or by forming the ultrapower of a sequence of connected graphe, M_i, $i<\omega$, such that for each n almost all M_i have diameter $> n$. If, however, we pose the question in the domain of finite graphs : is there a first-order sentence φ such that for all finite M, $M \models \varphi$ iff M is connected ? then these easy arguments do not carry over. The negative answer needs another kind of proof (cf [AU], [CH]) which, though not difficult, involves a finer analysis of the situation.

Since detabases are, essentially, finite relational structures, their investigation leads naturally to questions concerning finite models. Some questions which have easy solutions for infinite models become not so easy and sometimes quite difficult when transferred to the finite domain.

The method to be presented is an analysis of first-order formulas in terms of local properties. We use a natural simple metric in the model and define the concept of a k-local formula where k is any natural number. Roughly speaking, a k-local formula is one which asserts something about some k-neighborhood around a point, x, i.e. about the model $(V^{(k)}(x), x)$, where $V^{(k)}(x)$ is the set of all points of distance $\leq k$ from x ; this means that all quantifiers are relativized to $V^{(k)}(x)$ and x is the free variable of the formula. The main theorem asserts that every first order sentence, φ, is logically equivalent to a Boolean combination of sentences that assert, each, something of the following form :

There exist s disjoint r-neighborhoods, each satisfying the r-local formula ψ.

If φ is a formula, one has to add to the combination r-local formulas in the free variables of φ.

The theorem is proved by quantifier elimination. The proof yields an effective translation of the formula φ into the Boolean combination, as well as upper bounds for the neighborhood radius, r, and the number, s, of neighborhoods, in terms of the quantifier depth of φ. If the quantifier depth is n then, roughly, speaking, the upper bound for r is 7^{n-1}. By considering particular examples, one can establish a lower bound which is 2^{n-2}. (For the exact details see §1). This poses the problem of narrowing the gap. The reader not interested in these details or in the technicalities of the proof can go on directly to §2, once the theorem is understood. The point is to establish a precise sense in which first order sentences are local and to use it in order to show that such and such properties are not characterized by first-order sentences because they are not local. Thus, to cite one of the examples given in §2, within the class of finite graphs one cannot define by a first-order sentence those that are planar.

In general, let C_o and C_1 be two classes of models. Assume that for every n one can find $M_o \in C_o$, $M_1 \in C_1$ such that for all $r, s \leqslant n$, every combination of s disjoint r-neighboods in one model is isomorphic to a similar combination in the other. Then C_o and C_1 cannot be seperated by means of a first-order sentence. Moreover, as proved in §2, such classes cannot be seperated in any richer logic obtained by introducing predicates which denote local properties - i.e., properties of k-neighborhoods, where k ranges over the natural numbers. There is no restriction on the properties allowed, provided only that they are preserved under isomorphisms and that they are local.

The examples given in §2 are all from graph theory, which seems to be the most natural domain for applying the method. The proofs are very easy - for they consist in drawing prototypes of the M_o and M_1 mentioned above. A glance suffices to shows that the same combinations of a "small" number of "small" neighborhoods are realized in both.

In §3 a different kind of application is given - a much shorter proof of a previous result of the author, answering a set-theoretical question of Levy, [G]. As stated at the end, this leads to the existence of a certain curious transitive set.
As it is, the method does not apply to models possessing a more regular mathematical structure - such as a linear ordering, for then the whole model is equal to some small neighborhood. Any extension of the method to such models will have to use a much more sophisticated distance function.

We should mention that a result of L. Marcus [M] which is an immediate corollary of the theorem, can be regarded as a forerunner of it. There seems to be no direct proof of the theorem, or the applications, from the corollary.

§1. THE THEOREM

Let L be a first-order language with finitely many predicates including equality and no function symbols (and no individual constants). The restriction on function symbols is not essential, since they can be eliminated in the usual way by using predicates. Our results can be formulated in general, but it is more convenient to formulate and prove them for a language that has only predicates.

We shall use "x", "y", "z" both for individual variables in L as well as for members of models. We shall also use the members of models as names of themselves and occasionally substitute them for free variables in formulas. $\bar{a},\bar{b},\ldots,\bar{x},\bar{y},\bar{z}$ are used for tuples. If $\bar{x} = x_0,\ldots,x_{k-1}$ and $\bar{y} = y_0,\ldots,y_{n-1}$ then $\bar{x},\bar{y} = x_0,\ldots,x_{k-1}, y_0,\ldots,y_{n-1}$. By "$\bar{x} \in X$" we mean that all the elements of $\bar{x}$ are in X. By "$\bar{x} \subset \bar{y}$" we mean that every element in $\bar{x}$ is also in $\bar{y}$.

Let M be a model for L, say $M = (M, R_0,\ldots)$. Define a metric $d = d_M$ over M as follows :

$$d(x,x) =_{Df} 0$$

$d(x,y) \leqslant 1 \iff_{DF}$ for some predicate R (including equality) and some tuple $\bar{x} \in M$ containing both x and y, $M \models R(\bar{x})$.

$$d(x,y) \leqslant n+1 \iff_{DF} \text{ for some } z \ d(x,z) \leqslant n \text{ and } d(y,z) \leqslant 1$$

$$d(x,y) = n \iff_{DF} d(x,y) \leqslant n \text{ and } d(x,y) \nleqslant n-1$$

$$d(x,y) = \infty \iff_{DF} \text{ for all } n, \ d(x,y) \nleqslant n.$$

If L has one binary predicate besides equality then M can be regarded as a directed graph and d_M is the usual distance : $d_M(x,y)$ = length of the smallest path connecting x and y, disregarding the direction of the edges.

If all predicates of L are monadic or binary then $d(x,y) = 1$ is definable by the formula

$$W_i(R_i(x,y) \vee R_i(y,x))$$

where the R_i's are all the binary predicates in L. In general, $d(x,y) = 1$ is definable by a formula that involves quantifiers. Here the finiteness of the number of predicates is used. Since the forthcoming result concerns single formulas and each single formula involves finitely many predicates, the result holds for any language provided that we define the distance using only the predicates that occur in the formula in question.

$d(x,y) \leq n$ is definable by $\exists v_0,\ldots,v_n (x = v_0 \wedge y = v_n \wedge \bigwedge_{i<n} d(v_i,v_{i+1}) \leq 1)$.

If $M' \subset M$ then evidently $d_{M'}(x,y) \geq d_M(x,y)$ for all $x,y \in M'$. If L has only monadic or binary predicates then $d_M(x,y) = 1 \iff d_{M'}(x,y) = 1$. But in general the additional members of M which cause $d_M(x,y)$ to be 1 need not be in M'.

$$V^{(k)}(x) = \{y : d(x,y) \leq k\}$$

$V^{(k)}(x)$ is the <u>k-neighborhood of x</u>.

Since $V^{(k)}(x)$ is definable in L we can express inside L the bounded quantifiers $\exists v \in V^{(k)}(x)$ and $\forall v \in V^{(k)}(x)$.

<u>Definition</u> : A <u>local k-formula around x</u> is a formula with a single free variable x, in which all quantifiers are of the form $Q u \in V^{(k)}(x)$.

The satisfaction of a k-local formula around x depends only on the model $(V^{(k)}(x), x)$.

These notions are generalized in the obvious way for tuples. If $\bar{k} = k_0,\ldots,k_{m-1}$ and $\bar{x} = x_0,\ldots,x_{m-1}$ then define the $\bar{k}$-neighborhood around $\bar{x}$ as :

$$V^{(\bar{k})}(\bar{x}) = \bigcup_{i<m} V^{(k_i)}(x_i)$$

and define a $\bar{k}$-local formula around $\bar{x}$ as a formula with free variables $\bar{x}$ in which all quantifiers are bounded by $V^{(\bar{k})}(\bar{x})$.

We use "$V^{(k)}(\bar{x})$" for $V^{(\bar{k})}(\bar{x})$ for the case where $k_i = k$ for all $i < m$ and speak accordingly of k-neighborhoods of $\bar{x}$ and of k-local formulas around $\bar{x}$. We put $d(\bar{x},y) =_{Df} \mathrm{Min}_{i<m}\, d(x_i,y)$ and then $V^k(\bar{x}) = \{y : d(\bar{x},y) \leq k\}$.

A <u>k-local formula</u> is a formula which is k-local around some $\bar{x}$ and a <u>simple k-local formula</u> is one which is k local around some x. Similarly k-neighborhoods around a single x are referred to as simple k-neighborhoods. We use $\varphi^{(k)}(\bar{x})$, $\psi^{(k)}(\bar{x}),\ldots,\varphi^{(k)}(\bar{x})$, $\psi^{(k)}(\bar{x})$, ... to range over $\bar{k}$-local (k-local) formulas around $\bar{x}$. Before the main theorem we bring some helpful trivial observations.

If $x_1, x_2 \in V^{(k)}(x)$ and $\mathrm{Min}\,(d(x,x_1)\ d(x,x_2)) < k$ then $d(x_1, x_2) = 1 \iff V^{(k)}(x) \models (d(x_1, x_2) = 1)$. For if $M \models R(\bar{x})$ and x_1,x_2 occur in $\bar{x}$ then $\mathrm{Min}(d(x,x_1), d(x,x_2)) < k$ implies that all members of $\bar{x}$ are in $V^{(k)}(x)$. (Evidently if L has only monadic or binary predicates then $x_1, x_2 \in V^k(x)$ is sufficient.) Now, if $d(x,x_1) = m$, $d(x_1,x_2) = n$ and $m+n \leqslant k$ then all the members of the path of length n that connects x_1 and x_2 except, possibly, the last are in $V^{(k-1)}(x)$. Hence $V^{(k)}(x) \models (d(x_1,x_2) = n)$ and this also holds with $V^{(k)}(x)$ replaced by any submodel M' such that $M' \supset V^{(k)}(x)$. From this we get :

(1) For $m+n \leqslant k$, $d(\bar{x},y) \leqslant m$ implies

$$\exists u \in V^{(n)}(y)\ \varphi \leftrightarrow \exists u \in V^{(k)}(\bar{x})\ [(d(u,y) \leqslant n)^{V^{(k)}(\bar{x})} \wedge \varphi]$$

where $\psi^{V^{(k)}(\bar{x})}$ is obtained from ψ by bounding all quantifiers in it by $V^{(k)}(\bar{x})$. This enables one to use, instead of quantifiers bounded by $V^{(n)}(y)$, quantifiers bounded by $V^{(k)}(\bar{x})$. In particular, if $k \leqslant k'$ then every $\varphi^{(k)}(\bar{x})$ can be rewritten as a $\psi^{(k')}(\bar{x})$.

If $v_0,\ldots,v_n$ are such that $d(v_i, v_{i+1}) \leqslant 1$, $x_1 = v_0$ and $x_2 = v_n$, then $v_0,\ldots,v_k \in V^{(k)}(x_1)$ and $v_k,\ldots,v_n \in V^{(n-k)}(x_2)$. From this we get

(2) If $k_1+k_2 \geqslant n$ then

$$d(x_1,x_2) \leqslant n \leftrightarrow (d(x_1,x_2) \leqslant n)^{V^{(k_1,k_2)}(x_1,x_2)}.$$

Main Theorem. Every first-order formula $\alpha(\bar{u})$ with the free variables $\bar{u} = u_0,\ldots,u_{m-1}$ is logically equivalent to a Boolean combination of

(I) Sentences of the form

$$\exists v_0,\ldots,v_{s-1}\ [\bigwedge_{i<s} \psi^{(r)}(v_i) \wedge \bigwedge_{i<j<s} d(v_i,v_j) > 2r]$$

where $\psi^{(r)}(v_i)$ is obtained from $\psi^{(r)}(v_j)$ by substituting v_i for v_j.

(II) Local formulas $\varphi^{(t)}(\bar{w})$ where $\bar{w} \subset \bar{u}$.

If α has no free variables then only the sentences of (I) occur in the Boolean combination.

If n = quantifier depth of α then the following inequalities can be guaranteed

$$r \leq 7^{n-1}, \quad s \leq m+n, \quad t \leq \frac{1}{2}(7^n-1).$$

Explanation. From (2) it follows that $d(v_i,v_j) > 2r$ can be written in the form $\sigma^{(2r)}(v_i,v_j)$. Hence, by (1), sentences of form (I) are also of form

(I*) $\exists\bar{v}\lambda^{(k)}(\bar{v})$

and the bound on r becomes : $k \leq 2.7^{n-1}$. (I*) asserts the existence of a k-neighborhood satisfying a certain first-order property (where a "k-neighborhood" is to be understood here as a submodel of the form $(V^k(v_0,\ldots,v_{s-1}),v_0,\ldots,v_{s-1})$). (I) is of course a very special case of (I*). As is easily seen $d(x_1,x_2) > k_1 + k_2$ iff $V^{(k_1)}(x_1) \cap V^{(k_2)}(x_2) = \phi$. Hence (I) asserts :

There are at least s pairwise disjoint simple r-neighborhood, each satisfying the first-order property expressed by ψ.

In the case where $d(x_1,x_2) \leq k$, for all $x_1,x_2 \in M$, the whole model coincides with some simple k-neighborhood and for models for which this k is small the theorem is of no interest. For example, in models for set theory the distance between any two points is ≤ 2 (take z such that $x \in z$ and $y \in z$). Or if we have a ternary relation R such that $\forall x,y\exists z\, R(x,y,z)$ holds, the distance is always ≤ 1. On the other hand many families of graphs considered by graph theorists give rise to models with non-trivial distance functions. That is, roughly speaking, to say, that for k which is not very large the simple k-neighborhoods are considerably simpler than the whole model. In such cases the theorem is very helpful in describing what can and what cannot be expressed in first-order logic. Also models with trivial distances may sometimes be described in terms of a different collection of basic relations which give rise to non-trivial distances. For example, if for some binary predicate R, our models satisfy $\forall x,y R(x,y)$ then the distance is alway ≤ 1, but then one can omit this R without reducing the expressive power of the language (i.e., formulas involving R are translatable into formulas not involving it). In general, we can consider any change of basic relations provided that we do not reduce the expressive power of our language. We shall later use this procedure in one of our applications.

Note that if all our predicates are monadic then $x \neq y$ implies $d(x,y) = \infty$. Consequently the sentence in (I) becomes $\exists \bar{v}(\bigwedge_{i<j<s} v_i \neq v_j)$ and we get the elementary well-known result on quantifier elimination in monadic first-order logic. The theorem can be viewed as the generalizing of this phenomenon.

(An easy corollary of the theorem is that for any given model M and any formula $\varphi(\bar{v})$ there exist k and a formula $\psi(\bar{v})$ such that $M \models \varphi(\bar{x}) \Longleftrightarrow V^{(k)}(\bar{x}) \models \psi(\bar{x})$, for all $\bar{x} \in M$. This was first proved syntactically by Leo Marcus [M] for models of one monadic function. In [G] it was generalized and used by the author who gave it a shorter semantic but non-constructive proof. Some years later the author found an easy semantic non-constructive proof for a weaker version of the theorem where (I) is replaced by (I^*). Then, using this as a guideline he arrived at the present proof which amounts to a quantifier elimination).

We shall use the following fact :

(3) For any formula $\beta(\bar{u},\bar{v})$ there is a Boolean combination $\sigma(\ldots\gamma_i(\bar{u}) \ldots \delta_j(\bar{v}) \ldots)$ of formulas $\gamma_i(\bar{u})$, $i \in I$, and $\delta_j(\bar{v})$, $j \in J$, such that γ_i and δ_j have no common free variables and

$$\beta(\bar{x},\bar{y}) \longleftrightarrow \sigma(\ldots\gamma_i^{M_1}(\bar{x}) \ldots \beta_j^{M_2}(\bar{y}) \ldots)$$

holds in any model M which is the disjoint sum of the two models M_1, M_2 and for any $\bar{x} \in M_1$, $\bar{y} \in M_2$. Here λ^{M_i} is the relativization of λ to M_i (obtained by replacing every Qu by $Qu \in M_i$). By a "disjoint sum" we mean that $M = M_1 \cup M_2$, $M_1 \cap M_2 = \phi$ and no relation holds between members of M_1 and members of M_2.

(3) is a special case covered by the results of Fefferman-Vaught, [FV]. It is also provable directly by induction on β : If β is quantifier-free, replace each atomic formula which contains variables both from $\bar{u}$ and from $\bar{v}$ by $u_0 \neq u_0$ (This is the only place where the disjointness of M_1 and M_2 is used). For the induction step use the fact that $\exists v\gamma$ is equivalent in M to $(\exists v \in M_1)\gamma \vee (\exists v \in M_2)\gamma$.

<u>Proof of the Main theorem</u> :

By induction on α. The main step uses the following lemma.

<u>Lemma</u> : <u>Let $\beta = \beta^{(k)}(\bar{v},z)$ be a k-local formula around $\bar{v},z$, where $\bar{v} = v_0,\ldots,v_{\ell-1}$. Then $\exists z\beta$ is logically equivalent to a Boolean combination of sentences of the form</u> $\exists \bar{x}[\bigwedge_{i<s} \psi^{(r)}(x_i) \wedge \bigwedge_{i<j<s} d(x_i,x_j) > 2r]$ <u>with</u>

$\overline{x} = x_0,\dots,x_{s-1}$, $r = 2k+1$, $s \leq \ell + 1$, <u>and of local formulas $\varphi^{(t)}(\overline{v})$, with $t \leq 7k+3$.</u>

<u>Proof</u> : $\exists z\beta$ is equivalent to

$$\exists z[(d(\overline{v},z) \leq 2k+1) \wedge \beta] \vee \exists z[(d(\overline{v},z) > 2k+1) \wedge \beta].$$

Call the first disjunct β_1 and the second β_2. We shall transform each into the required form ; first - β_1.
Assuming $d(\overline{v},z) \leq 2k+1$ it follows that $W_{i<\ell}(V^{(k)}(z) \subset V^{(3k+1)}(v_i))$ and, by (1), instead of using quantifiers bounded by $V^{(k)}(z)$ we can use quantifiers bounded by $V^{(3k+1)}(\overline{v})$. Also $\exists y \in V^{(k)}(\overline{v},z)\varphi$ is equivalent to $\exists y \in V^{(k)}(\overline{v})\varphi \vee \exists y \in V^{(k)}(z)\varphi$ and $\exists y \in V^{(k)}(\overline{v})$ can be eliminated in favour of quantifiers bounded by $V^{(3k+1)}(\overline{v})$. Hence β_1 can be rewritten as

$$\exists z \in V^{(2k+1)}(\overline{v})\ \beta_1'(\overline{v},z)$$

where all quantifiers in β_1' are bounded by $V^{(3k+1)}(\overline{v})$. This, again can be rewritten as $\varphi^{(3k+1)}(\overline{v})$.

Now consider β_2. If $d(\overline{v},z) > 2k+1$ then for every $x \in V^{(k)}(\overline{v})$ and every $y \in V^{(k)}(z)$, $d(x,y) > 1$. This is easily seen to imply that the submodel $V^{(k)}(\overline{v},z)$ is the disjoint sum of $V^{(k)}(\overline{v})$ and $V^{(k)}(z)$. By (3), β is in this case equivalent to a Boolean combination of formulas of the forms $\gamma^{(k)}(\overline{v})$ and $\delta^{(k)}(z)$. Putting this combination in disjunctive normal form and using the fact that the k-local formulas around the same string of variables are closed under Boolean combinations, β_2 can be rewritten as

$$\exists z[(d(\overline{v},z) > 2k+1) \wedge W_{i\in I}(\gamma_i^{(k)}(\overline{v}) \wedge \delta_i^{(k)}(z))]$$

which is equivalent to

$$W_{i\in I}\exists z[(d(\overline{v},z) > 2k+1) \wedge \gamma_i^{(k)}(\overline{v}) \wedge \delta_i^{(k)}(z)].$$

It remains to transform the formula of the form

$$\exists z[(d(\overline{v},z) > 2k+1) \wedge \gamma^{(k)}(\overline{v}) \wedge \delta^{(k)}(z)].$$

Since z does not occur in $\gamma^{(k)}$, this becomes :

$$\gamma^{(k)}(\overline{v}) \wedge \exists z[(d(\overline{v},z) > 2k+1) \wedge \delta^{(k)}(z)].$$

Thus, we have reduced our problem to

$$\exists z[(d(\bar{v},z) > 2k+1) \wedge \delta^{(k)}(z)].$$

Call this formula $\eta(\bar{v})$. It asserts the existence of z satisfying $\delta^{(k)}(z)$, whose distance from each v_i, $i = 0,\dots,\ell-1$ is $>2k+1$. Let B_m be the sentence asserting the existence of at least m z's satisfying $\delta^{(k)}(z)$ whose mutual distances are $>2(2k+1)$. $B_m = \exists z_0,\dots,z_{m-1} A_m(z_0,\dots,z_{m-1})$, where $A_m(z_0,\dots,z_{m-1})$ is

$$\bigwedge_{i<m} \delta^{(k)}(z_i) \wedge \bigwedge_{i<j<m} (d(z_i,z_j) > 2(2k+1)).$$

(For $m = 1$ the second conjunct disappears.) Since $\delta^{(k)}(z_i)$ is also of the form $\psi^{(2k+1)}(z_i)$, the sentences B_m, for $m = 1,\dots,\ell+1$ are as required in the lemma. Evidently $\eta(\bar{v})$ is equivalent to $\eta(\bar{v}) \wedge \exists z\delta^{(k)}(z)$ and $\exists z\delta^{(k)}(z)$ is equivalent to :

$$(B_1 \wedge \neg B_2) \vee (B_2 \wedge \neg B_3) \vee \dots \vee (B_\ell \wedge \neg B_{\ell+1}) \vee B_{\ell+1}.$$

Hence it suffices to consider the formulas $\eta(\bar{v}) \wedge B_j \wedge \neg B_{j+1}$ for $1 \leqslant j \leqslant \ell+1$, and $\eta(\bar{v}) \wedge B_{\ell+1}$.

First, assume $B_{\ell+1}$ and let $z_0,\dots,z_\ell$ satisfy $A_{\ell+1}(z_0,\dots,z_\ell)$. No two z_i's can belong to a simple $(2k+1)$-neighborhood, because, for $i \neq j$, $d(z_i, z_j) > 2(2k+1)$; hence some z_i does not belong to any of the $V^{(2k+1)}(v_j)$, $j < \ell$. This implies $\eta(\bar{v})$. Thus, $\eta(\bar{v}) \wedge B_{\ell+1}$ is equivalent to $B_{\ell+1}$.

For $1 \leqslant j \leqslant \ell$ rewrite $\eta \wedge B_j \wedge \neg B_{j+1}$ as :

$$(\eta(\bar{v}) \wedge B_j \wedge \neg B_{j+1} \wedge C_j(\bar{v})) \vee (\eta(\bar{v}) \wedge B_j \wedge \neg B_{j+1} \wedge \neg C_j(\bar{v}))$$

where $C_j(\bar{v})$ asserts the existence of j z_i's in $V^{(2k+1)}(\bar{v})$ such that $A_j(z_0,\dots,z_{j-1})$:

$$\exists z_0 \in V^{(2k+1)}(\bar{v}) \dots \exists z_{j-1} \in V^{(2k+1)}(\bar{v})\, A_j(z_0,\dots,z_{j-1}).$$

Note that $A_j(z_0,\dots,z_{j-1})$ can be written as a $(2k+1)$-local formula around $z_0,\dots,z_{j-1}$ (because $d(z_i,z_j) > 2(2k+1)$ is expressible by a $(2k+1)$-local formula around z_i, z_j). If $z_0,\dots,z_{j-1} \in V^{(2k+1)}(\bar{v})$, then quantifiers bounded by $V^{(2k+1)}(z_0,\dots,z_{j-1})$ can be eliminated in favour of quantifiers bounded by $V^{(4k+2)}(\bar{v})$.

This implies that C_j is equivalent to a $(4k+2)$-local formula around $\bar{v}$. Now

$B_j \wedge \neg C_j(\bar{v})$ implies that, for some z not in $V^{(2k+1)}(\bar{v})$, $\delta(z)$ holds. Hence $B_j \wedge \neg C_j(\bar{v})$ implies $\eta(\bar{v})$ and the second disjunct can be replaced by $B_j \wedge \neg B_{j+1} \wedge \neg C_j(\bar{v})$ which is of the required form.

$\neg B_{j+1} \wedge C_j(\bar{v})$ implies the existence of j z_i's in $V^{(2k+1)}(\bar{v})$ such that every x satisfying $\delta^{(k)}(x)$ has distance $\leqslant 2(2k+1)$ from some z_i. In this case, all x's satisfying $\delta^{(k)}(x)$ must be found in $V^{3(2k+1)}(\bar{v})$ and $\eta(\bar{v})$ becomes equivalent to :

$$\exists x \in V^{(6k+3)}(\bar{v})\ [\delta^{(k)}(x) \wedge d(\bar{v},x) > 2k+1].$$

If $x \in V^{(6k+3)}$ the quantifiers in $\delta^{(k)}(x)$, being bounded by $V^{(k)}(x)$, are eliminable in favour of quantifiers bounded by $V^{(6k+3+k)}(\bar{v})$. Also $d(\bar{v},x) > 2k+1$ is expressible by a formula in which all quantifiers are bounded by $V^{(2k+1)}(\bar{v})$. Consequently $\neg B_{j+1} \wedge C_j(\bar{v})$ implies $\eta(\bar{v}) \leftrightarrow \varphi^{(7k+3)}(\bar{v})$ for some $\varphi^{(7k+3)}(\bar{v})$, and the first disjunct can be rewritten as $B_j \wedge \neg B_{j+1} \wedge C_j(\bar{v}) \wedge \varphi^{(7k+3)}(\bar{v})$.

q.e.d. lemma.

<u>End of the proof</u> : Since every quantifier-free formula is of the form $\varphi^{(0)}(\bar{u})$, the claim holds for n = quantifier depth = 0. The claim carries over, trivially, to Boolean combinations. It suffices to prove the claim for $\alpha = \exists u \alpha'(\bar{v},u)$ that has quantifier depth n, $n \geqslant 1$, assuming it for all formulas of smaller quantifier depth. Let $\bar{v} = v_0,\dots,v_{m-1}$. Apply the claim to $\alpha'(\bar{v},u)$, write the resulting Boolean combination of formulas of forms (I) and (II) in disjunctive normal form and distribute $\exists u$ over this disjunction. Using the fact that k-local formulas around $\bar{v},u$ are closed under Boolean combinations we get a disjunction of formulas, each of the form

$$\exists u[\beta^{(k)}(\bar{v},u) \wedge \bigwedge_{i \in I} \lambda_i]$$

where the λ_i's are sentences of form (I) or their negations. This is equivalent to

$$(\exists u\ \beta^{(k)}(\bar{v},u)) \wedge \bigwedge_{i \in I} \lambda_i .$$

By the induction hypothesis each λ_i is either

$\exists z_0,\dots,z_{s'-1}[\bigwedge_{i<s'} \psi^{(r')}(z_i) \wedge \bigwedge_{i<j<s'} d(z_i, z_j) > 2r']$ or a negation of such a sentence, where $r' \leqslant 7^{n-2}$ and $s' \leqslant (m+1) + (n-1) = m+n$ (the number of the free variables of α' is more by 1, its quantifier depth-less by 1). Also

$k \leq \frac{1}{2}(7^{n-1}-1)$. Applying the lemma we can rewrite $\exists u\, \beta^{(k)}(\bar{v},u)$ as a Boolean combination of the desired form, such that in the sentences of form (I) we have $r = 2k+1 \leq 2\,\frac{1}{2}(7^{n-1}-1) + 1 = 7^{n-1}$, $s \leq m+1 \leq m+n$ and in the formulas $\varphi^t(\bar{v})$ we have $t \leq 7k+3 \leq 7.\ \frac{1}{2}(7^{n-1}-1) + 3 = \frac{1}{2}(7^n-1)$.

q.e.d.

The question that naturally arises is : Can the bounds on r,s,t, be improved ? Let us introduce the following classification of sentences, where "LS" stands for "Local Sentence".

Definition : $LS(r,s)$ is the set of all sentences of the form

$$\exists v_0,\ldots,v_{\ell-1} \left(\bigwedge_{i<\ell} \psi^{(k)}(v_i) \wedge \bigwedge_{i<j<\ell} d(v_i,v_j) > 2k \right)$$

where $k \leq r$ and $\ell \leq s$. This is defined for $r \geq 0$, $s \geq 1$.

The theorem asserts that if φ is a sentence of quantifier depth n then it is equivalent to a Boolean combination of sentences from $LS(7^{n-1},n)$ and if it is a formula with m free variables then it is equivalent to a Boolean combination of members of $LS(7^{n-1}, n+m)$ and of $\frac{1}{2}(7^n-1)$-local formulas. A careful checking of the end of the proof shows, however, that we actually get, in the case of sentences an equivalent Boolean combination of members of

$$\bigcup_{i<n} LS(7^{n-1-i}, i+1)$$

In the case of formulas this is to be replaced by $\bigcup_{i<n} LS(7^{n-1-i}, m+i+1)$. Note that the construction of the proof uses members of each $LS(7^{n-1-i}, i+1)$ for $i = 0,\ldots,n-1$. Now the following fact holds :

There exist sentences $\eta_{r,s} \in LS(r,s)$, such that for every sequence $(r_0,s_0),\ldots,(r_{k-1},s_{k-1})$ in which $r_0 > r_1 > \ldots > r_{k-1} \geq 0$ and $1 \leq s_0 < s_1 < \ldots < s_{k-1}$ the conjunction $\eta_{r_0,s_0} \wedge \ldots \wedge \eta_{r_{k-1},s_{k-1}}$ is not equivalent to a Boolean combination of members of $\bigcup_{0\leq i} \bigcup_{1\leq j} LS(i,j)$ unless the combination contains members from each $LS(r_i,s_i)$ $i = 0,\ldots,k-1$. In particular $\eta_{r,s}$ is not equivalent to any Boolean combination of members of

$$\bigcup_{0\leq i<r} \bigcup_{1\leq j} LS(i,j) \cup \bigcup_{0\leq i} \bigcup_{1\leq j<s} LS(i,j).$$

This means that the LS(r,s) form a doubly indexed proper hierarchy in a strong sense. There are many ways of constructing such $\eta_{r,s}$. Having them we can get lower bounds on r and s. The lower the quantifier depth that is needed for the $\eta_{r,s}$ the higher (and therefore the better) the lower bound. Our $\eta_{r,s}$ will be in a language consisting of equality, one bindary predicate R and one monadic predicate P (which is used only for $\eta_{0,s}$). For $r > 0$, $\eta_{r,s}$ asserts the existence of s disjoint neighborhoods $V^{(r)}(x_i)$, $i<s$, each containing a point of distance $r-1$ from its centre, x_i, but no point of distance r from x_i :

$$\exists x_0,\ldots,x_{s-1} \left[\bigwedge_{i<s} \xi_r(x_i) \wedge \bigwedge_{i<j<s} d(x_i,x_j) > 2r \right]$$

where $\xi_r(x) = \exists u(d(x,u) = r-1) \wedge \neg \exists u(d(x,u) = r)$; evidently $\xi_r(x)$ can be put in the form $\varphi^{(r)}(x)$. As $\eta_{0,s}$ take $\exists \bar{x}(\bigwedge_{i<s} P(x_i) \wedge \bigwedge_{i<j<s} d(x_i,x_j) > 0)$.

To prove the desired properties let C_ℓ, for $\ell \geq 1$, be the model consisting of 2ℓ points : $0,\ldots,2\ell-1$, with R interpreted as $\{<i,i+1> : i>\ell\} \cup \{<2\ell-1,0>\}$ and P interpreted as ϕ. Let C_0 consist of one point, 0, with R and P interpreted as empty and let C_{-1} consist of, 0, with P interpreted as $\{0\}$ and R as empty. Finally, for $\ell \geq 0$ let C^*_ℓ be obtained from the cycle C_ℓ by tacking on it a "tail" of length $\ell+1$; i.e., add new members $a_0,\ldots,a_\ell$ and add to the binary relation the pairs $<0,a_0>$, $<a_0,a_1>$, ..., $<a_{\ell-1},a_\ell>$. Now consider models which are disjoint unions of isomorphic copies of the C_i's, $i \geq -1$ and the C^*_j's, $j \geq 0$ (where any number of isomorphic copies of each of these models may be used). Note that for x in a copy of C_ℓ the ℓ-neighborhood coincides with the copy of C_ℓ and as there is no point of distance $\ell+1$ from x, this is also its $(\ell+1)$-neighborhood. Note also that the ℓ-neighborhood of the point that corresponds to ℓ in the copy of C^*_ℓ is isomorphic to C_ℓ but that for all x in the copy of C^*_ℓ there are points of distance $\ell+1$ from x (because of the tail). Now fix $r \geq 0$ and let M contain infinitely many disjoint copies of each C^*_ℓ, $\ell = 0,1,\ldots$, exactly s copies of C_{r-1}, where $0 < s < \infty$, and arbitrary numbers of copies of the other basic models. Let M' be obtained from M by deleting one copy of C_{r-1}. It is not difficult to see that, for all $0 \leq \ell < r$ and for each collection $V^{(\ell)}(x_0),\ldots,V^{(\ell)}(x_{j-1})$ of j disjoint simple ℓ-neighborhoods in M there is a similar collection $V^{(\ell)}(x'_0),\ldots,V^{(\ell)}(x'_{j-1})$ in M' such that the corresponding neighborhoods $V^{(\ell)}(x_i)$ and $V^{(\ell)}(x'_i)$ are isomorphic. This is also true for $\ell \geq r$, provided that $j < s$ (for in that case each of the copies of C_{r-1} that occur in the collection in M can be matched by a copy of C_{r-1} in M'). Consequently Boolean combinations of sentences from

$\bigcup_{0\leq i<r}\bigcup_{1\leq j} LS(i,j) \cup \bigcup_{0\leq i}\bigcup_{1\leq j<s} LS(i,j)$ have the same truth-values in M and M'. But $\eta_{r,s}$ is seen to hold in M and to fail in M'. Our claim can be now derived by considering a model which is the disjoint union of s_0 copies of C_{r_0-1}, s_1 copies of $C_{r_1-1},\ldots,s_{k-1}$ copies of C_{r_k-1} and of infinitely many copies of C^*_ℓ, $\ell = 0,1,\ldots$.

Now $d(v_1,v_2) \leq n$ (for $n>0$) can be expressed using quantifier depth $\lceil \log n \rceil$. Where $\log = \log_2$ and $\lceil x \rceil$ = smallest integer $\geq x$. This is easily established, by induction, using the equivalences :

$$d(u,v) \leq 1 \iff R(u,v), \qquad d(u,v) \leq 2k \iff \exists w[d(u,w) \leq k \wedge d(v,w) \leq k] \text{ and}$$

$$d(u,v) \leq 2k+1 \iff \exists w[d(u,w) \leq k \wedge d(v,w) \leq k+1].$$

Consequently $d(u,v) > n$, which is equivalent to $\neg(d(u,v) \leq n)$, needs $\lceil \log_2 n \rceil$ quantifier depth and this is also true for $d(u,v) = n$ (which is equivalent to $(d(u,v) \leq n \wedge d(u,v) > n-1)$). Hence, for $r > 0$, $\xi_r(x)$ can be written with $\lceil \log r \rceil + 1$ quantifier depth. From this it follows easily that, for $r > 0$, $\eta_{r,s}$ can be expressed in $\lceil \log r \rceil + 1 + s$ quantifier depth. Evidently, $\eta_{0,s}$ needs quantifier depth s. From this we get the following lower bound.

Theorem : For each $n \geq 2$ there exists a sentence, σ, of quantifier depth n such that any equivalent Boolean combination of sentences of $\bigcup_{0\leq i}\bigcup_{1\leq j} LS(i,j)$ must contain members form each of $LS(2^{n-2-i}, i+1)$, $i = 0,\ldots,n-1$ where, for $i = n-1$, $LS(2^{-1}, n)$ is to be reread as $LS(0,n)$.

This simply means that we get the lower bound by replacing in the upper bound everywhere 7^{n-1} by 2^{n-2}. (and rereading 2^{-1} as 0).

Problem : Narrow down the gap between the upper bound 7^{n-1} and the lower bound 2^{n-2}.

In the case of formulas a similar construction yields the same kind of lower bound where, in addition, $\frac{1}{2}(7^n-1)$ is to be replaced by 2^{n-2}.

As far as the applications of the next section are concerned there is no need for estimates on the values of r and s. Here is an outline of a shorter semantic proof of the theorem as stated :

For $\bar{a} \in M$ <u>let the local type of $\bar{a}$</u> be the set of all local formulas $\varphi^{(k)}(\bar{v})$, $k = 0,1,\ldots$ such that $M \models \varphi^{(k)}(\bar{a})$. Now assume that for all $i \geq 0$, $j \geq 1$ and all $\sigma \in LS(i,j)$, $M_1 \models \sigma \Leftrightarrow M_2 \models \sigma$. Assume furthermore that $\bar{a} = a_0,\ldots,a_{m-1} \in M_1$, $\bar{b} = b_0,\ldots,b_{m-1} \in M_2$ and $\bar{a}$ and $\bar{b}$ have, in their respective models, the same local type. Finally, assume that M_1 and M_2 are ω-saturated. Then for every $a_m \in M_1$ there exists $b_m \in M_2$ such that $\bar{a},a_m$ and $\bar{b},b_m$ have the same local type. (This holds also for $m = 0$). Playing a Fraissé-Ehrenfeucht game we deduce that $(M_1,\bar{a}) \equiv (M_2,\bar{b})$. If M_1 and M_2 are not ω-saturated we can get elementary extensions M_1^* and M_2^* that are ω-saturated and from $(M_1^*,\bar{a}) \equiv (M_2,\bar{b})$ we deduce again $(M_1,\bar{a}) \equiv (M_2,b)$. Now use the theorem that if every two models (of a theory T) that satisfy the same sentences out of some class S are elementary equivalent, then every sentence is equivalent (in T) to a Boolean combination of members of S.

§2. LOCAL INSEPERABILITY

<u>Definition</u> : Let C_0 and C_1 be two classes of models for L and let φ be a sentence. Say that <u>φ seperates C_0 and C_1</u> if φ is true in all members of one class and false in all members of the other. C_0 and C_1 are <u>first-order seperable</u> if some sentence φ in L seperates them. They are <u>first-order inseperable</u> if no sentence in L seperates them.

If P is some property of models then we say that P is <u>first-order within the class C</u> if $\{M \in C : M \text{ satisfies } P\}$ and $\{M \in C : M \text{ does not satisfy } P\}$ are first-order seperable. This means that for some φ we have : M satisfies P $\Longleftrightarrow$ $M \models \varphi$, for all $M \in C$.

If Φ is any set of formulas, then <u>$(M,a_0,\ldots,a_{k-1}) \equiv_\Phi (M',b_0,\ldots,b_{k-1})$</u> means that $M \models \varphi(a_0,\ldots,a_{k-1}) \Longleftrightarrow M' \models \varphi(b_0,\ldots,b_{k-1})$ for all $\varphi(v_0,\ldots,v_{k-1}) \in \Phi$.

<u>Definition</u> : (I) Say that C_0 and C_1 are <u>locally inseperable</u> if for every natural number n there exist a pair of models $M_0 \in C_0$, $M_1 \in C_1$ such that, for all $r,s \leq n$, the same (up to isomorphism) collections, of s disjoint simple r-neighborhoods are realized in M_0 and M_1. By this we means that for every collection $V^{(r)}(x_0),\ldots,V^{(r)}(x_{s-1})$ of disjoint neighborhoods in M_i there is a corresponding collection $V^{(r)}(y_0),\ldots,V^{(r)}(y_{s-1})$ in M_{1-i} (also disjoint) such that $(V^{(r)}(x_j), x_j) \cong (V^{(r)}(y_j), y_j)$, for all $j<s$.

(II) Say that C_0 and C_1 are <u>locally first-order inseperable</u> if for every n and every finite set of formulas Φ there exist $M_0 \in C_0$, $M_1 \in C_1$ such that for all $r,s \leq n$ and every collection of disjoint neighborhoods $V^{(r)}(x_0),\ldots,V^{(r)}(x_{s-1})$ in M_i, $(i = 0,1)$ there is a corresponding collection $V^{(r)}(y_0),\ldots V^{(r)}(y_{s-1})$ in M_{1-i} such that $(V^{(r)}(x_j),x_j) \equiv_{\Phi} (V^{(r)}(y_j),y_j)$, for all $j<s$.

Evidently, if C_0 and C_1 are locally inseperable they are also locally first-order inseperable, but the converse need not hold. As a corollary of the Main Theorem we have :

Theorem 2.1. : If C_0 and C_1 are locally first-order inseperable then they are first-order inseperable.

Proof : Let φ be any first-order sentence. By the Main Theorem φ is equivalent to a Boolean combination of sentences of the form $\exists\bar{v}(\bigwedge_{i<s} \psi^{(r)}(v_i) \wedge \bigwedge_{i<j<s} d(v_i,v_j)>2r)$. Let n be greater than all the r and s that are involved and let Φ consist of all the ψ's such that $\psi^{(r)}(v_i)$ occurs in this combination. Let $M_0 \in C_0$ $M_1 \in C_1$ be the two models satisfying the requirements of the definition with respect to n and Φ. Then $M_o \models \varphi \iff M_1 \models \varphi$, hence φ does not seperate C_0 and C_1. q.e.d.

In order to show that a certain property P is not first-order within C it suffices to show that $\{M \in C : M \text{ satisfies } P\}$ and $\{M \in C : M \text{ does not satisfy } P\}$ are locally first-order inseperable. In most examples that we have thought of these two classes satisfy the stronger requirement of local inseperability and it is this that one proves directly. Yet there are cases in which only the first property holds and for this reason we have introduced this weaker version.

If two classes C_0 and C_1 are locally inseperable then they are inseperable in the somewhat stronger logic obtained as follows :

Let $\{P_i\}_{i\in I}$ be any family of properties of models of the form (M,a) where M is a model for L and $a\in M$. There is no restriction on the P_i except that they be preserved under isomorphisms : If (M,a) satisfies P_i and $(M',a')\cong(M,a)$ then (M',a') satisfies P_i. For each P_i and each $k\in\omega$ add a monadic predicate $P_i^{(k)}$ to L and let L^* be the language thus obtained. If M is a model for L, enlarge it to a model M^* for L^* by interpreting each $P_i^{(k)}$ as the set of $x\in M$ such that $(V^{(k)}(x),x)$ satisfies P_i. If $\varphi(\bar{v})$ is a formula of L^* put : $M \models \varphi(\bar{a}) \iff_{Df} M^* \models \varphi(\bar{a})$.

We shall call the logic thus obtained a monadic arbitrary local logic (of L), or for short, a MAL-logic. The word "arbitrary" is used to indicate that there is no restriction on the properties P_i of the neighborhoods. We could define this logic, equivalently, by adding, instead of monadic standardly interpreted predicates, new quantifiers $\exists_i^{(k)}$, such that "$\exists_i^{(k)}x\ldots$" is interpreted as "there exists x such that $(V^{(k)}(x),x)$ satisfies P_i and ...".

Theorem 2.2. If C_0 and C_1 are locally inseperable then they are inseperable in the MAL-logic (i.e. no formula of L^* seperates them).

In order to prove it we first show :

Lemma 2.3. If C_0 and C_1 are locally inseperable then for every $m \geq 1$ there are $M_0 \in C_0$, $M_1 \in C_1$ which realize the same m-neighborhoods of m-tuples. That is to say, for each $\bar{x} = x_0,\ldots,x_{m-1} \in M_i$, $i = 0,1$, there exists $\bar{y} = y_0,\ldots,y_{m-1} \in M_{1-i}$ satisfying :

$$(V^{(m)}(\bar{x}),\bar{x}) \simeq (V^{(m)}(\bar{y}),\bar{y})$$

It is easily seen that any isomorphisms of $(V^{(m)}(\bar{x}),\bar{x})$ onto $(V^{(m)}(\bar{y}),\bar{y})$ carries $V^{(r)}(x_{j_0},\ldots,x_{j_{s-1}})$ isomorphically onto $V^{(r)}(y_{j_0},\ldots,y_{j_{s-1}})$ for all $r,s \leq m$. Hence the condition in the lemma is a natural generalization of the condition defining local inseperability. The lemma asserts that this generalization is already implied by local inseperability.

Proof of 2.3. : We show the following :

Let $t \geq 0$, $k \geq 1$ and assume that for all $s \leq n$, and all $r \leq 3^{k-1}t + \frac{1}{2}(3^k-3)$ the models M_0,M_1 realize the same collections of s disjoint r-neighborhoods. Then M_0 and M_1 realize the same t-neighborhoods of k-tuples.

This implies the lemma. For, given m, take $M_0 \in C_0$, $M_1 \in C_1$ which satisfy the requirement in the definition of local inseperability for $n = 3^{m-1}.m + \frac{1}{2}(3^m-3)$.

The proof is by induction on k. If $k = 1$, then $3^{k-1}.t + \frac{1}{2}(3^k-3) = t$, hence the condition is that both models realize the same simple t-neighborhoods, i.e. the same t-neighborhoods of 1-tuples.

Assume the conditions to hold for t and k+1 and let $\bar{x} = x_0,\ldots,x_k$ be a (k+1)-tuple in one of the models say M_0. We have to find the corresponding y_j's in M_1. First consider the case where $d(x_i,x_j) > 2(t+1)$ for all $i<j\leqslant k$. Then $V^{(t+1)}(x_j)$, $j = 0,\dots,n$ form a collection of $k+1$ disjoint (t+1)-neighborhoods. Since $t+1 \leqslant 3^{k+1}.\ t + \frac{1}{3}(3^{k+1}-3)$ the condition implies the existence of $y_0,\dots,y_k \in M_1$ and of isomorphisms $f_j : (V^{(t+1)}(x_j),x_j) \simeq (V^{(t+1)}(y_j)y_j)$, where the $V^{(t+1)}(y_j)$, $j = 0,\dots,k$, are disjoint. If $z\in V^{(t)}(x_j)$, $z'\in V^{(t)}(x_{j'})$ and $j \neq j'$ then $d(z,z') > 2$. Hence $V^{(t)}(\bar{x})$ is the disjoint sum of the models $V^{(t)}(x_j)$, $j = 0,\dots,k$. The same argument applies to the $V^{(t)}(y_j)$'s. Hence $\bigcup_{j\leqslant k} f_j$ is an isomorphism of $(V^{(t)}(\bar{x}),\bar{x})$ onto $(V^{(t)}(\bar{y}),\bar{y})$.

Now assume that for some $i<j\leqslant k$ $d(x_i,x_j) \leqslant 2(t+1)$; say, without loss of generality, $d(x_{k-1}, x_k) \leqslant 2(t+1)$. Replace $k+1$ by k and t by $3(t+1)$; since $3^{k-1}(3(t+1)) + \frac{1}{2}(3^k-3) = 3^k.t + \frac{1}{2}(3^{k+1} - 3)$ the condition for $k+1$ and t implies the condition for k and $3(t+1)$. By the induction hypothesis for k (with $3(t+1)$ instead of t) there is a k-tuple $y_0,\dots,y_{k-1} \in M_1$ and an isomorphism :

$$f : (V^{(3(t+1))}(x_0,\dots,x_{k-1}),x_0,\dots,x_{k-1}) \simeq (V^{(3(t+1))}(y_0,\dots,y_{k-1}),y_0,\dots,y_{k-1}).$$

Since $x_k \in V^{(2(t+1))}(x_{k-1})$, we have $V^{(t)}(x_k) \subset V^{(3t+2)}(x_{k-1})$. Put $y_k =_{Df} f(x_k)$. It is easily seen that $y_k \in V^{(2(t+1))}(y_{k-1})$ and, consequently, that $V^{(t)}(y_k) \subset V^{(3t+2)}(y_{k-1})$. Moreover, for members of $V^{(t)}(x_k)$ the distance function inside $V^{(3(t+1))}(x_0,\dots,x_{k-1})$ is the same as the distance function in M_0 and similarly for $V^{(t)}(y_k)$ and M_1. (This follows from the argument for (1) in §1). Hence f maps $V^{(t)}(x_k)$ onto $V^{(t)}(y_k)$. It follows that f maps $V^{(t)}(\bar{x})$ onto $V^{(t)}(\bar{y})$. q.e.d.

<u>Proof of Theorem 2.2</u> : Given φ in L^* we have to show that φ does not seperate C_0 and C_1. There are alltogether finitely many $P_j^{(k)}$ in φ. Let $\tilde{L}$ be obtained from L by adding these predicates and for every model M for L let $\tilde{M}$ the enriched model for $\tilde{L}$. Let $\tilde{C}_i = \{\tilde{M} : M\in C_i\}$, $i = 0,1$. Let $\tilde{k}$ be the maximum of all k such that $P_j^{(k)}$ is in $\tilde{L}$. For each n, there are by lemma 2.3 models $M_0\in C_0$ and $M_1\in C_1$ that realize the same $(n+\tilde{k}+1)$ - neighborhoods of n-tuples. Let $\bar{x} = x_0,\dots,x_{n-1} \in M_0$, $\bar{y} = y_0,\dots,y_{n-1} \in M_1$ and let $f : V^{(n+\tilde{k}+1)}(\bar{x}) \simeq V^{(n+\tilde{k}+1)}(\bar{y})$. If $z\in V^{(n)}(\bar{x})$ then for each $k\leqslant\tilde{k}$ f maps $V^{(k)}(z)$ onto $V^{(k)}(f(z))$. Whether $P_j^{(k)}(a)$ holds in M_i $(i = 0,1)$ depends only

on the isomorphism type of $(V^{(k)}(a),a)$. Hence for $z \in V^{(n)}(\bar{x})$ we have $\tilde{M}_0 \models P_j^{(k)}(z) \Longleftrightarrow \tilde{M}_1 \models P_j^{(k)}(fz)$. Thus f maps $V^{(n)}(\bar{x})$ isomorphically onto $V^{(n)}(\bar{y})$, where these are considered as neighborhoods in $\tilde{M}_0$ and $\tilde{M}_1$. Consequently $\tilde{C}_0$ and $\tilde{C}_1$ are locally inseperable. By Theorem 2.1, φ does not seperate C_0 and C_1.

q.e.d.

By analogy to monadic local properties we can consider n-ary local properties ; that is to say, n-ary relations, R, such that the holding of $R(\bar{a})$ is determined by the isomorphism type of $(V^{(k)}(\bar{a}),\bar{a})$, where $k = k_R$ is an arbitrary, but fixed, natural number. Let the AL-logic (i.e, arbitrary local logic) be obtained Ly adding predicates denoting arbitrary local properties and let the n-AL-logic be the one in which only predicates of arity $\leqslant n$ are added.

Does Theorem 2.2. remain true if we replace "MAL-logic" by "AL-logic" ?

The argument for the monadic case does not carry over ; for, by adding relations of arity $\geqslant 2$ one usually changes the distance and a k-neighborhood in the enriched model may contain members which are not in the k-neighborhood in the original model.

We do have however :

Theorem 2.4. If C_0 and C_1 are locally inseperable and for each k, only finitely many isomorphism types of k-neighborhoods of (n-1)-tuples are realized in the models of $C_0 \cup C_1$, then C_0 and C_1 are inseperable in the n-AL-logic. Actually, it suffices to have a locally inseperable pair of subclasses $C_0' \subset C_0$, $C_1' \subset C_1$ satisfying the condition of finitely many realizable isomorphism types.

The second claim is trivially implied by the first. For if C'_0 and C'_1 are inseperable in some logic so are any pair of classes C_0,C_1 that include them.

If, for some m, all simple 1-neighborhoods in the models of $C_0' \cup C_1'$ have $\leqslant m$ members, then also, for each k,n, the k-neighborhoods around n-tuples are uniformly bounded in size. (This situation obtains in the forthcoming examples). Then, evidently, the condition of theorem 2.4. holds ; but the theorem becomes superfluous, because within $C_0' \cup C_1'$, every formula in the AL-logic is then equivalent to a first-order formula. Theorems 2.2. and 2.4. are of interest only when the sizes of 1-neighborhoods realized in $C_0' \cup C_1'$ are not uniformly bounded by some integer.

The proof of 2.4. is given in the appendix

Examples

In all cases the property in question, P, is not first-order definable within the indicated class, C. The subclasses {M∈C : M satisfies P } and {M∈C : M does not satisfy P } are locally inseperable (except for the last example where they are only first-order inseperable). This is shown by pointing out a pair M_0, $M_1 \in C$ of prototypes such that M_0 satisfies P, M_1 does not and, for all r,s sufficiently small with respect to the size of the prototype, the same collections of s disjoint simple r-neighborhoods are realized in both. The models can be "blown up" in the obvious way to any desired size, while retaining these properties.

1. P = Connectedness, C = class of finite graphs.

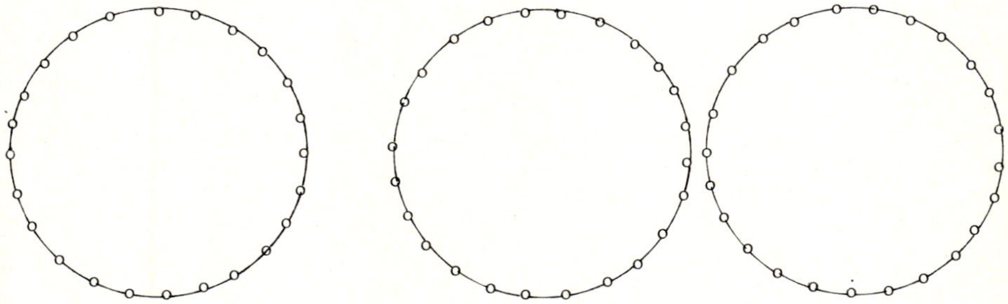

M_0 M_1

Fig 1

If on each circle we have > 2r+1 points then all simple r-neighborhoods are of the form :

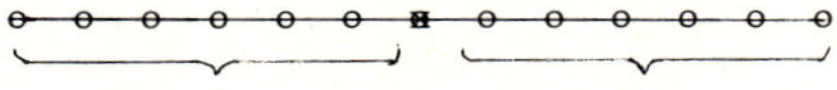

and if, on each circle, we have > (2r+1).s points then s disjoint simple r-neighborhoods can be realized.

In [CH] the method of Fraissé-Ehrenfeucht games is applied to this prototype pair in order to show that connectedness is not first order definable in C. In principle the method of games is applicable to the other examples. However, if the models are not as homogoneous, a description of the 2nd player's winning strategy for n moves can be quite involved, whereas a glance may suffice to see that the neighborhoods are the same. What is more important is that the method indicates the way in which the prototype models should be constructed.

In the following examples we let the drawings speak for themselves.

2. P = ℓ - connectedness C = finite (ℓ-1)-connected graphs.

(A graph is k-connected if the removal of any k-1 edges does not disconnect it).

Fig 2.2. corresponds to the case $\ell = 4$. In general, $\ell - 1$ is the number of bridges between the two components of M_1. For ℓ-even, ℓ is the nomber of edges issuing from any ordinary vertex (i.e. vertex not connected by a bridge). The case of an odd ℓ is obtained from that of $\ell+1$ by removing an edge between two ordinary vertices in M_0 and the corresponding edge in one of the components in M_1.

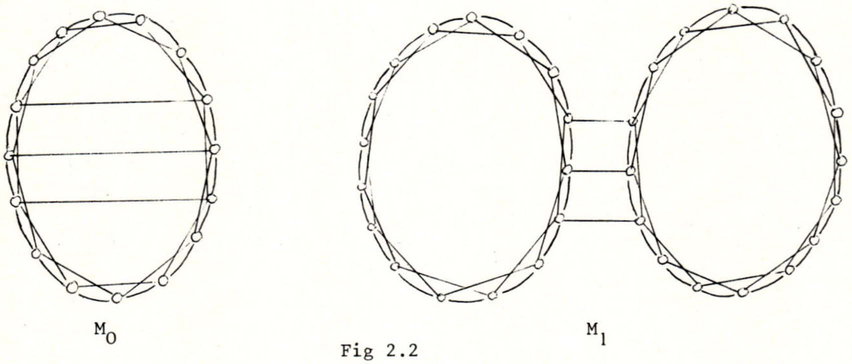

Fig 2.2

3. P = being planar C = class of finite graphs

(A graph is planar if it is representable in the plane so that each edge is an arc and the arcs do not intersect except at their common end points. In Fig. 2.3. M_0 is planar, but M_1 is not).

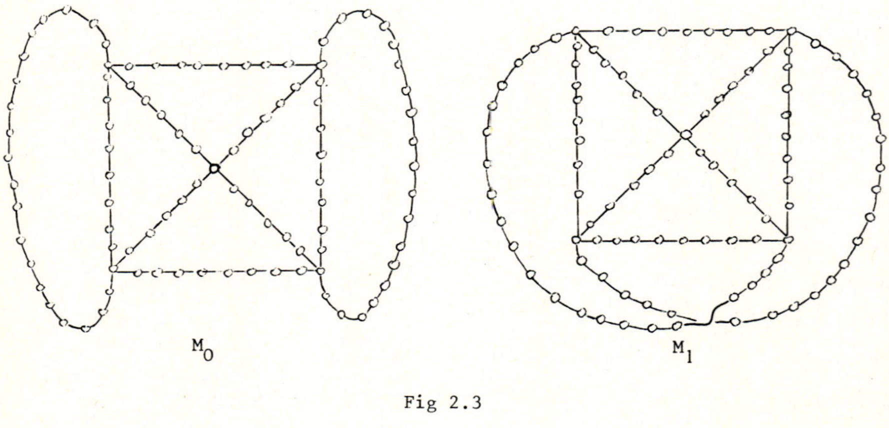

Fig 2.3

4. P = Hamiltonian (i.e., having a Hamilton cycle)

C = class of finite k-regular graphs with a Hamilton path. Here k should be $\geqslant 3$. For $k = 3,4,5$ we can restrict C further by adding the requirement that the graphs be planar.

(A Hamilton path is a path in the graph passing through each vertex exactly once. If such a path is also a cycle then it is a Hamilton cycle. A graph is k-regular if every vertex has degree k, where degree(x) = number of edges containing x).

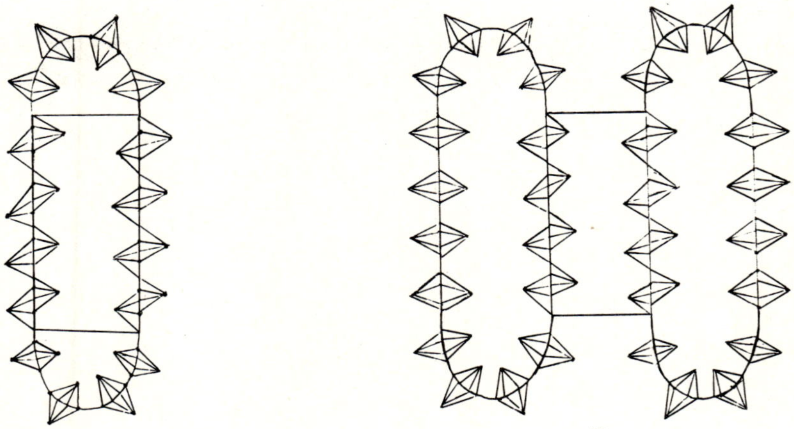

M_0 M_1

Fig 2.4

Figure 2.4. is the construction for C = class of 4-regular planar graphs having a Hamilton path. The construction for $k>4$ is obtained by replacing each vertex in Fig. 2.4. by a graph G as follows. Let G^* be a k-regular Hamiltonian graph. Let (a,a') and (a,b) be two edges in G^* such that there is an Hamilton cycle containing (a,a') and not containing (a,b) . Get G by removing these two edges (without removing vertices). Now replace each vertex of 2.4. by a copy of G and use edges issuing from a, a', b in order to connect the different copies. Figure 2.5. indicates how this is to be done. For $k = 5$ take G^* to be the icosahedron. Since this is planar the construction can be carried out so as to yield a planar graph. (For $k>5$ the graph cannot be planar, since each planar graph has a vertex of degree $\leqslant 5$). We leave the case $k = 3$ for the reader.

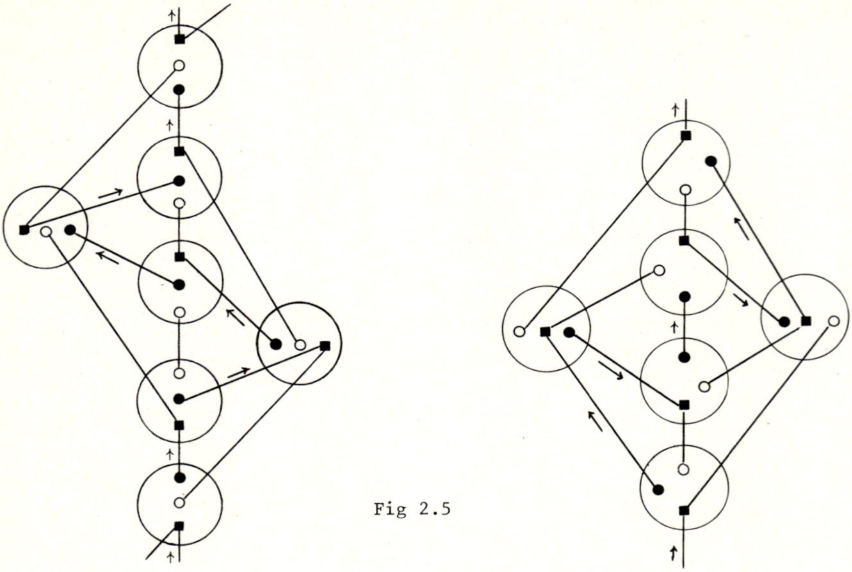

Fig 2.5

(■ = a, ● = a', ○ = b. arrows show parts of a Hamilton path)

4. P = Eulerian C = connected finite graphs

(A graph is Eulerian if it contains an Euler cycle, i.e. a cycle passing through each edge exactly once).

As is well known, a (finite) connected graph is Eulerian iff each vertex has an even degree. This is obviously a local property. Consequently, within the class of connected graphs those that are Eulerian are not locally inseperable from those that are not ; in fact - they are definable in the MAL - logic. Yet they are first-order locally inseperable. This is seen by letting C^* consist of all graphs of the form :

Fig 3.7

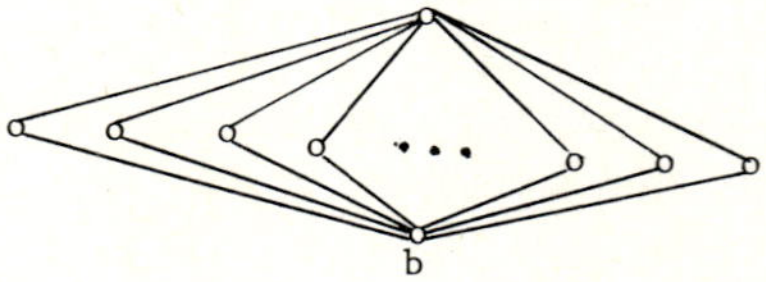

Let $C_0 \subset C^*$ consist of those having an even number of vertices and let $C_1 = C^* - C_0$. Then, for $G \in C^*$, G is Eulerian or not according as $G \in C_0$ or $G \in C_1$. Now replace the binary predicate R of our language, L, by a monadic predicate, P. For each $M \in C^*$ let M' be the model for the new language, L', obtained by interpreting

P as $\{x \in M : x \neq a$ and $x \neq b\}$. Then :

$$M \models R(x,y) \iff M' \models (P(x) \vee P(y)) \wedge \neg(P(x) \wedge P(y)).$$

This implies that every sentence $\varphi \in L$ is translatable into some $\varphi' \in L'$ such that, for all $M \in C^*$,

$$M \models \varphi \iff M' \models \varphi'.$$

If φ were to seperate C_0 from C_1 then φ' would have seperated $C'_0 = \{M' : M \in C_0\}$ from $C'_1 = \{M' : M \in C_1\}$. But this is impossible, because for models of L' we have $d(x,y) = \infty$ for all $x \neq y$, implying that each simple neighborhood consist of one point and consequently, that C'_0 and C'_1 are locally inseperable.

§3. CERTAIN TRANSITIVE MODELS

Levy's hierarchy classifies formulas in the language of set theory as follows : Σ_0 - formulas are those in which all quantifiers are bounded, i.e., of the forms $\exists x \in y$, $\forall x \in y$. The higher levels are obtained in the usual way by tacking on alternating blocks of unbounded quantifiers. We have :

$x = 0 \iff \forall y \in x\ (y \neq y)$

x is transitive $\iff \forall y \in x\ \forall z \in y\ (z \in x)$

x is an ordinal $\iff$ x is transitive $\wedge \forall u \in x\ \forall v \in x\ [u=v \vee u \in v \vee v \in u]$

x is zero or a successor $\iff$ x is an ordinal $\wedge (x=0 \vee \exists u \in x\ \forall v \in x (v=u \vee v \in u))$

x is a natural number $\iff$ x is zero or a successor $\wedge\ \forall y \in x$ (y is zero or a successor).

Consequently all these notions are Σ_0 (i.e. expressible in ZFC by Σ_0-formulas). Now in ZFC finiteness can be defined in either of the two ways :

x is finite $\iff \exists z \exists y$ [y is a natural number $\wedge$ z maps y onto x]

x is finite $\iff \forall z \forall y \in x$ [z is not a mapping of $x - \{y\}$ onto x]

The first is easily seen to be Σ_1 , the second - Π_1. Thus finiteness is, in ZFC, a Δ_1 - notion. The natural question is whether it is Σ_0. The answer is negative in the strongest possible sense :

There is a transitive infinite set A, of rank ω, such that for any Σ_0 - formula

$\varphi(v)$, there exists a finite transitive subset $A' \subset A$, such that $\varphi(v)$ is true for A iff it is true for A'. Furthermore A is primitive recursive. Thus being finite is not characterizable by a Σ_0 - formula in the real world.

Note that for any $\varphi(v) \in \Sigma_0$, with v as its only free variable, there exists $\varphi'(v) \in \Sigma_0$, such that in φ' all quantifiers are of the form $Qx \in v$ and we have, in pure logic :

$$v \text{ is transitive } \rightarrow (\varphi(v) \leftrightarrow \varphi'(v))$$

$\varphi'(v)$ is obtained by replacing every "$\exists x \in u \ldots$" by "$\exists x \in v\ (x \in u \wedge \ldots)$" and every "$\forall x \in u \ldots$" by "$\forall x \in v\ (x \in u \rightarrow \ldots)$". Furthermore, we can assume that in $\varphi'(v)$, v occures only as a bound for quantifiers. For if x is any variable different from v then its occurences in any quantifier-free part are within the scope of some $Qx \in v$, hence the occurences (in a quantifier - free part) of $v \in x$ and $v = x$ can be replaced by $x \neq x$, and the occurences of $x \in v$ - by $x = x$. Similarly $v = v$ and $v \in v$ are replaceable by, say $\forall y \in v\ (y=y)$ and $\exists y \in v (y \neq y)$, respectively. Now let φ'' be the sentence obtained by removing the bound on the quantifiers, i.e., replacing each $Qx \in v$ by Qx. It is easily seen that for a transitive A we have :

$$\varphi(A) \iff (A, \in \restriction A) \models \varphi''.$$

Vice versa the satisfaction of any sentence in $(A, \in \restriction A)$ is equivalent to the satisfaction by A of some Σ_0 - formula $\varphi(v)$. Hence we have to construct a transitive infinite model M such that for any sentence φ there exists a finite transitive $M' \subset M$ such that

$$M \models \varphi \iff M' \models \varphi.$$

The construction to be given here is similar to the one used in [G]. To show that M has the desired property a rather involved argument was used there, which relied on the generalization of Marcus's result. Here we get it at a glance by realizing that the classes $\{M\}$ and $\{M' \subset M : M' \text{ is finite and transitive}\}$ are locally inseperable.

<u>Construction of M.</u>

Let $1 < n_0 < n_1 < \ldots < n_i < n_{i+1} < \ldots$ be such that $n_{i+1} - n_i$ is strictly increasing with i. Let $\{0,1\}^*$ = set of all finite binary sequences. For $w \in \{0,1\}^*$, let $\ell(w)$=length of w, Λ = empty sequence. For $X \subset \{0,1\}^*$, put $|X|$ = cardinality of X, $Xi = \{wi : w \in X\}$ (where $i \in \{0,1\}$). We construct, level by level, a tree

$T = \bigcup_{i\in\omega} L_i \subset \{0,1\}^*$, where $L_i = i^{\underline{tn}}$ level of $T = \{w\in T : \ell(w) = i\}$.

$L_0 = \{\Lambda\}$

For j not of the form n_i, $L_{j+1} = L_j 0$

For $j = n_{2k}$, $L_{j+1} = L_j 0 \cup L_j 1$

For $j = n_{2k+1}$, choose any $L_j^* \subset L_j$ such that $|L_j^*| = \frac{1}{2}|L_j|$ and put $L_{j+1} = L_j^* 0 \cup L_j^* 1$.

Thus, branching occures only at the n_i - levels ; but for i odd half of the branches reaching this level are cut off, namely the branches whose ends are not in L_j^*. It follows that for $j = n_{2k}$ $|L_{j+1}| = 2|L_j|$ and for j not of this form $|L_{j+1}| = |L_j|$. Consequently, for $j \leq n_0$, $|L_j| = 1$ and, for $n_{2k} < j \leq n_{2(k+1)}$, $|L_j| = 2^{k+1}$. For $j = n_{2k+1}$, the $j^{\underline{th}}$ level contains 2^{k+1} leaves. Also, every leaf of T is in some n_{2k+1} - level. The choice of L_j^* can be made effectively, giving rise, say to a primitive recursive T. The model, M, constructed from T will be as effective as T.

With each $w\in T$ we associate a set $h(w)$ as follows :

$h(\Lambda) = 0$, $h(w0) = \{h(w)\}$. If $w1\in T$, then, as is easily seen, w must be of the form $w_0 0$ and we put $h(w1) = \{h(w), h(w_0)\}$.

Let M be the model whose domain is $\{h(w) : w\in T\}$. Evidently M is transitive.

Fig 3.1. describes a portion of M. Members $a,b \in M$ such that $a\in b$ are exactly those connected by an edge going upwards from a to b. Filled circles correspond to leaves of T. It is clear that the shortest path connecting $h(w_1)$ and $h(w_2)$ in M corresponds to the path connecting w_1 and w_2 in T, except that at branching points we can use a small short cut : instead of going from w to $w0$ and then to $w01$ we can go directly from $h(w)$ to $h(w01)$.

Let M_m = the m - level trancation of $M = \{h(w)\in M : \ell(w) \leq m\}$. It is not difficult to see (by actual looking) that, for given r,s, if m is of the form n_k and sufficiently large then M and M_m realize the same collections of s disjoint simple r-neighborhoods. Here is a more formal proof :

Suppose that $n_{j+1} < \ell(w) \leq n_{j+2}$ and $2r < n_{j+1} - n_j$. If $\ell(w) - n_{j+1} \leq r$ then there exists exactly one $h(w')\in V^{(r)}(h(w))$ such that $\ell(w') \in \{n_0, n_1, \ldots\}$ and this is a branching point at level n_{j+1}. If $n_{j+2} - \ell(w) \leqslant r$ then, again there is one such w'. It is at level n_{j+2} and is either a branching point or a leaf.

Fig 3.1

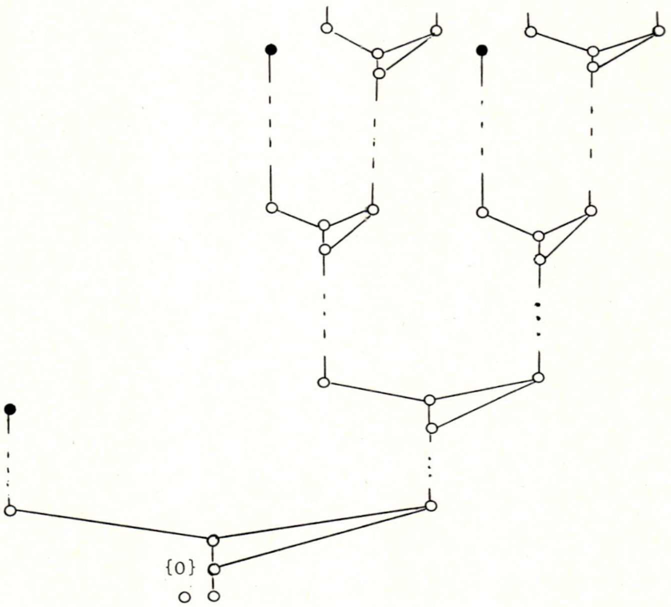

In all other cases there is no such w'. Hence we can divide the r-neighborhoods of h(w) into the following 3 types.

(I) A simple chain with h(w) not in the centre. This is the case where, for some leaf w', $d(h(w),h(w')) < r$.

(II) A simple chain with h(w) at the centre. This is the case where the neighborhood contains no leaf of distance < r and no branching point.

(III) There is a branching point of distance $\leqslant r$ from w. In this case the neighborhood is either of the king of Fig 3.2. (III) or a substructure obtained by omitting one of the 3 branches. $h(w)$ may be either the branching point or on one of the branches. It is not indicated and does not matter in the argument.

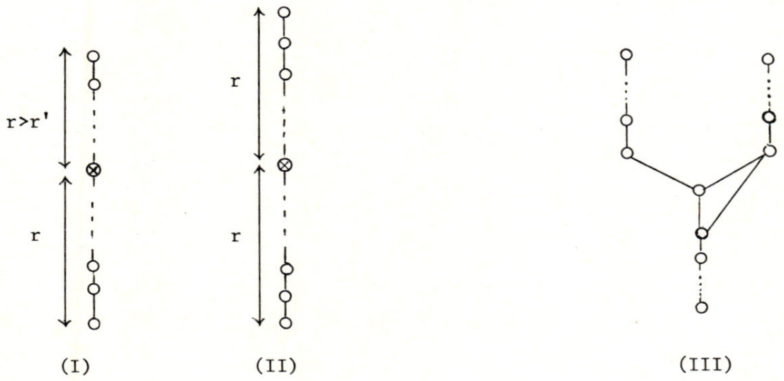

Fig 3.2

Assume that j is odd and $n_{j+1} - n_j > 2r+1$. Let $\ell(v) > n_{j+1}$, say $n_k < \ell(v) \leqslant n_{k+1}$, where $k \geqslant j+1$. Then there exists w, satisfying $n_{j+1} < \ell(w) \leqslant n_{j+2}$, such that the r-neighborhoods of $h(w)$ and $h(v)$ are isomorphic. For if $\ell(v) - n_k \leqslant r$ choose w satisfying $\ell(w) - n_{j+1} = \ell(v) - n_k$. And if $n_{k+1} - \ell(v) \leqslant r$ choose w satisfying $n_{j+2} - \ell(w) = n_{k+1} - \ell(v)$ and such that the n_{j+2}- level descendant of w is of the same type (leaf or branching point) as the n_{k+1}- level descendant of v. In all other cases choose w such that $\ell(w) - n_{j+1}$ and $n_{j+2} - \ell(w)$ are both $> r$.

Now consider M_m, for $m = n_k$ where $k \geqslant j+3$ and $2r+1 < n_{j+1} - n_j$. The r-neighborhoods around points whose level is in (n_{j+1}, n_{j+2}) are the same in M and M_m. If furthermore, j is odd, then every simple r-neighborhood in M_m around some $h(v)$ at level $> n_{j+1}$ is isomorphic to the r-neighborhood of some $h(w)$ at level $\in (n_{j+1}, n_{j+2}]$. The argument given for M works also for M_m. The essential point is that by truncating M at level n_k one converts all branching points at that level into leaves and consequently neighborhoods may change from type (III) to types (II) and (I). But these can be simulated in the lower region, using the leaves at level n_{j+2}. (It is for this reason that T should contain many leaves).

Given r and s choose $j = 2j_0+1$ such that $n_{j+1} - n_j > 2r+1$ and $2^{j_0} > 2s$. If $m = n_k$ and $k \geq j+3$ then any collection of s disjoint simpler r-neighborhoods in M or in M_m can be matched by an isomorphic collection in the other model as follows : If $\ell(v) \leq n_{j+1}$, the r-neighborhood of $h(v)$ is matched by itself. The rest can be realized around $h(w)$'s such that $n_{j+1} < \ell(w) \leq n_{j+2}$. There are 2^{j_0+1} disjoint branches going from level $n_{j+1}+1$ to n_{j+2}, of which 2^{j_0} terminate in leaves and 2^{j_0} in branching points ; also each neighborhood around a point al level $\leq n_{j+1}$ intersects at most r of these. Hence there are enough disjoint branches to realize the rest. Q.E.D.

Note that, for M_m in which $m \notin \{n_0, n_1, \ldots\}$, the claim need not be true, even if m is arbitrarily large. For M_m may have r-neighborhoods of type III whose upper end nodes are of distance $< r$ from its distinguished member. Such neighborhoods are not realized in M. However, using a somewhat more complicated tree we can construct a transitive M of rank ω having the following curious property :

<u>For every first-order sentence φ and every ascending infinite chain $M_0 \subset M_1 \subset \ldots \subset M_n \subset \ldots$ of transitive models such that $\cup M_i = M$, $M \models \varphi$ iff $M_i \models \varphi$ for all but finitely many i's.</u>

APPENDIX

Proof of Theorem 2.4

Consider a predicate $P^{(k)}(\bar{v})$ which denotes a property of the k-neighborhood of $\bar{v}$, where $\bar{v} = v_o,\dots,v_{n-1}$. We show that, as far as models of $C'_0 \cup C'_1$ are concerned, $P^{(k)}(\bar{v})$ is equivalent to a first-order formula in a language $L \cup \{R_i^{(k+1)} : i<s\}$, such that each $R_i^{(k+1)}$ is of arity $\leqslant n$, $R_i^{(k+1)}(\bar{u})$ denotes a property of the $(k+1)$-neighborhood of $\bar{u}$ and the following is valid :

$$R_i^{(k+1)}(u_o,\dots,u_{m-1}) \longrightarrow \bigwedge_{i<j<m} (d(u_i,u_j) \leqslant (2k+1)(n-1)).$$

If φ is a sentence in the n-AL-logic we can get $\tilde{\varphi}$, such that $M \vdash \varphi <\!\!-\!\!> \tilde{\varphi}$ for all $M \in C'_0 \cup C'_1$ and $\tilde{\varphi}$ is in some $\tilde{L} = L \cup \{R_i^{(k+1)} : i \in I\}$, I - finite, such that, that for some constant ℓ, we have :

$$R_i^{(k+1)}(u_o,\dots,u_{m-1}) \longrightarrow (\bigwedge_{j<j'<m} d(u_j,u_{j'}) \leqslant \ell)$$

It follows that for every model M for L, if d is the distance in M and $\tilde{d}$ is the distance in the enriched model $\tilde{M}$ (for $\tilde{L}$) we have :

$$d(a,b) \leqslant \ell.\tilde{d}(a,b).$$

Hence if "V" and "$\tilde{V}$" denote, respectively, neighborhoods in M and $\tilde{M}$ we have : $\tilde{V}^{(j)}(\bar{a}) \subset V^{(\ell.j)}(\bar{a})$. The technique of the proof of Theorem 2.2. can be now used to show that no sentence of $\tilde{L}$ seperates C'_0 and C'_1.

The predicate $R^{(k+1)}$ and the translation of $P^{(k)}(\bar{v})$ are obtained as follows : Let $\{\varphi_j(\bar{v}) : j < p\}$. be the set of all conjunctions of the form $\bigwedge_{i<j<n} \pm (d(v_i,v_j) \leqslant 2k+1)$, where ' + ψ' denotes the formula ψ, '-ψ'- its negation and for each (i,j) either + or - is chosen. Here $p = 2^{\frac{1}{2}n(n-1)}$. Note that each $\varphi_j(\bar{v})$ is a first-order formula of the form $\psi^{(k+1)}(\bar{v})$. Evidently

$$\vdash P^{(k)}(\bar{v}) <\!\!-\!\!> \bigvee_{j<p} (P^{(k)}(\bar{v}) \wedge \varphi_j(\bar{v})).$$

Each $\varphi_j(\bar{v})$ determines a graph $G_j(\bar{v})$ in which the edjes consist of all (v_i,v_j) for which the formula $d(v_i,v_j) \leqslant 2k+1$ is a conjunct. It is easily seen that if v_i and v_j are in the same connected component of $G_j(\bar{v})$ then $\vdash \varphi_j(\bar{v}) \rightarrow d(v_i,v_j) \leqslant (n-1)\ (2k+1)$. Hence for those j for which $G_j(\bar{v})$ is connected we can take $P^{(k)}(\bar{v}) \wedge \varphi_j(\bar{v})$ as one of the predicates $R_i^{(k+1)}(\bar{v})$. Consider $\varphi(\bar{v}) = \varphi_q(\bar{v})$ for which $G_q(\bar{v})$ is not connected and let $\bar{v}_0,\ldots,\bar{v}_{t-1}$ be its components. Then each $\bar{v}_i$ has $< n$ variables and for each v_i, v_j in the same component, $\vdash \varphi(\bar{v}) \rightarrow d(v_i,v_j) \leqslant (2k+1)(n-1)$. If $M \models \varphi(\bar{a})$ then $V^{(k)}(\bar{a})$ is the disjoint sum of the $V^{(k)}(\bar{a}_i)$, $i<t$. Hence the isomorphism type of $(V^{(k)}(\bar{a}),\bar{a})$ is completely determined by the isomorphism types of $(V(\bar{a}_i),\bar{a}_i)$, $i<t$. Consequently an equivalence of the following form is valid

$$\varphi(\bar{v}) \wedge P^{(k)}(\bar{v}) \longleftrightarrow \varphi(\bar{v}) \wedge \bigvee_{j \in J} (S^{(k)}_{0,j}(\bar{v}_0) \wedge \ldots \wedge S^{(k)}_{t-1,j}(\bar{v}_{t-1})$$

where $S^{(k)}_{i,j}(\bar{v}_i)$ is a new predicate describing an isomorphism type of $(V^{(k)}(\bar{v}_i),\bar{v}_i)$ and the disjunction ranges over all combinations which, given $\varphi(\bar{v})$, imply $P^{(k)}(\bar{v})$. This disjunction can be infinite. But for $n'<n$ only finitely many isomorphism types of k-neighborhoods of n'-tuples are realized in the models of $C'_0 \cup C'_1$. Hence for some finite $J' \subset J$, the equivalence, with J replaced by J', holds in all models of $C'_0 \cup C'_1$. Let $\psi_i^{(k+1)}(\bar{v}_i)$ be the conjunction of all $d(v_j,v_{j'}) \leqslant 2k+1$ which occur as conjuncts in $\varphi(\bar{v})$, where $v_j, v_{j'} \in \bar{v}_i$, and let $R^{(k+1)}_{i,j}(\bar{v}_i)$ be a new predicate equivalent to $S^{(k)}_{i,j}(\bar{v}_i) \wedge \psi_i^{(k+1)}(\bar{v}_i)$. Then the $R^{(k+1)}_{i,j}$ have the desired property and

$$\varphi(\bar{v}) \wedge P^{(k)}(\bar{v}) \longleftrightarrow \varphi(\bar{v}) \wedge \bigvee_{j \in J'} (\bigwedge_{i<t} R^{(k+1)}_{i,j}(\bar{v}_i)).$$

q.e.d.

REFERENCES

[AHU] Aho, A.V.,Y. Sagiv and J.D. Ullman, Equivalences among relational expressions. SIAM J. Computing (1978).

[AU] Aho, A.V. and J.D. Ullman, Universality of data retreival languages. Proceeding 6th ACM Symp. on Principles of Programming languages. San-Antonio, Texas (Jan 1979) pp. 110-117.

[CH] Chandra A.K. and D. Harel, Computable queries for relational data bases, Proc. 21st FOCS, Syracuse, New York (Oct 1980) pp. 333-347.

[FV] Feferman, S. and R.L. Vaught, The first order properties of algebraic systems. Fund. Math. 47 (1959) pp. 57-103.

[G] Gaifman, H. Finiteness is not a Σ_o-property. Israel J. of Math. 19 (1974) pp. 359-368.

[M] Marcus, L. Minimal models of theories of one function symbol. Israel J. of Math. 18 (1974) pp. 117-131 (also Doctoral Thesis, The Hebrew University, Jerusalem 1975).

PROCEEDINGS OF THE HERBRAND SYMPOSIUM
LOGIC COLLOQUIUM '81, J. Stern (editor)

ITERATING ADMISSIBILITY IN PROOF THEORY

GERHARD JÄGER [1)]
Mathematisches Institut
Universität München
BRD

Several subsystems of set theory are introduced in order to provide a natural framework for iterating admissibility. These theories are compared with well-known subsystems of analysis and systems for explicit mathematics.

INTRODUCTION

It is the aim of the present paper to continue the systematic study of the hyperjump operation in a proof-theoretic context. From work of Feferman and Friedman we know that certain subsystems of classical analysis can be reduced to systems of iterated hyperjumps or iterated inductive definitions. So we have for example

$$(\Delta^1_2\text{-CA}) \equiv (\Pi^1_1\text{-CA})_{<\varepsilon_0} \equiv ID_{<\varepsilon_0} \quad .$$

Whilst in [3] the investigation of iterations of the hyperjump along arbitrary well-orderings $<\omega_1^{ck}$ is completely laid out, in this paper we introduce a framework for iterations of the hyperjump along 'virtual well-orderings' of order type $\geq\omega_1^{ck}$. Since there are close connections between definability theory in the constructible hierarchy L and proof theory, it is natural and useful to work with subsystems of set theory rather than subsystems of analysis. By the Spector-Gandy theorem it makes no difference whether you iterate the hyperjump or admissibility.

We shall also mention how this work is related to Feferman's theories for explicit mathematics. In this direction our main result is

$$T_0 \equiv (\Delta^1_2\text{-CA})+(BI) \equiv KPi$$

where KPi is a theory of iterated admissible sets which formalizes a recursively inaccessible universe.

§1. Let L be the language of number theory with constants for all primitive recursive relations and functions. All the theories considered are formulated in the first order language $L^* = L(\in,=,N,M)$ where we augment L by $\in$, $=$, a set constant N for the set of the natural numbers and a relation constant M for sets. We define Δ_0-, Σ_n-, Π_n-, Σ- and Π-formulas as usual and write $\underline{u}$ for a finite string $u_1,\ldots u_n$. The notation $A[\underline{u}]$ is used to indicate that all free variables of A come from the list $\underline{u}$; $A(\underline{u})$ may contain other free variables besides $\underline{u}$.

All theories in L^* will be assumed to contain axioms which state that no element of N is a set and that N is precisely the set of urelements; urelements contain no elements. Elementary set theory ES is given by the additional axioms of number theory with the scheme of complete induction on N and the following set-theoretical axioms:

(Extensionality) $M(a) \,\&\, M(b) \,\&\, \forall x(x\in a \leftrightarrow x\in b) \rightarrow a=b$,
(Pair) $\exists x(a\in x \,\&\, b\in x)$,
(Union) $\exists x(\cup a \subset x)$,
(Δ_0-Separation) $\exists z(M(z) \,\&\, \forall x(x\in z \leftrightarrow x\in a \,\&\, A(x)))$ for all Δ_0-formulas A ,
(Foundation scheme) $\forall x(\forall y\in x A(y) \rightarrow A(x)) \rightarrow \forall x A(x)$ for all formulas A .

For arbitrary sets a, r and formulas F we define

$Prog(a,r,F) := r \subset a\times a \,\&\, \forall x\in a(\forall y(\langle y,x\rangle\in r \rightarrow F(y)) \rightarrow F(x))$,
$TI(a,r,F) := Prog(a,r,F) \rightarrow \forall x\in a F(x)$,
$Wf(a,r) := \forall x TI(a,r,x)$.

The Bar Rule (BR) is the rule of inference

$$\frac{Wf(N,\prec)}{TI(N,\prec,F)}$$

for every 2-place primitive recursive relation $\prec$ and for every formula F. By Axiom β we mean the universal closure of

$$Wf(a,r) \rightarrow \exists f(Fun(f) \,\&\, dom(f)=a \,\&\, \forall x\in a(f(x) = \{f(y):\langle y,x\rangle\in r\}))\ .$$

An ordinal α is said to be provable in the theory T if there exists a primitive recursive well-ordering $\prec$ of order type α such that $T\vdash Wf(N,\prec)$. The proof-theoretic ordinal of T is the least ordinal α that is not provable in T; it is denoted by $|T|$. We write $S \equiv T$ to express the proof-theoretic equivalence of the theories S and T.

Feferman's constructive theory T_0 for explicit mathematics and its subsystems are presented in [6] and [7]. There you will also find detailed information about the philosophical background of these theories and their relation to mathematical practice. For the ordinal notations we refer to [2] and the forthcoming [17].

We obtain a natural translation F^* of every sentence F in the language L_2 of second order arithmetic by replacing all numerical quantifiers $\forall x(\ldots x\ldots)$, $\exists x(\ldots x\ldots)$ in F by $\forall x\in N(\ldots x\ldots)$, $\exists x\in N(\ldots x\ldots)$ and all set quantifiers $\forall Y(\ldots Y\ldots)$, $\exists Y(\ldots Y\ldots)$ by $\forall y(y\subset N \rightarrow \ldots y\ldots)$, $\exists y(y\subset N \ \& \ \ldots y\ldots)$.

In what follows we shall be mainly concerned with three different types of theories:

(1) $L_\alpha \models T$ implies α admissible ,

(2) $L_\alpha \models T$ implies α limit of admissibles ,

(3) $L_\alpha \models T$ implies α admissible limit of admissibles ,

where L_α is the αth level of the constructible hierarchy relativized to the standard structure N of the natural numbers in the sense of [1].

If T is a theory that contains ES then W-T ($T\restriction$) denotes the theory which is obtained from T by restricting the scheme of foundation (the scheme of foundation and complete induction on N) to sets.

§2. The situation of the universe being an admissible set is appropriately described by Kripke-Platek set theory, which is discussed in [1] at full length. KPu denotes Barwise's KPU^+ with the natural numbers as urelements (cf. [1]). The axioms of KPu consist of the axioms of elementary set theory ES and the scheme of Δ_0-collection

(Δ_0-Coll) $\quad \forall x\in a\exists y A(x,y) \rightarrow \exists z\forall x\in a\exists y\in z A(x,y)$

for arbitrary Δ_0-formulas A. Fundamental for KPu are the Σ-reflection theorem, the Δ-separation theorem and the Σ-recursion theorem.

Let $L^*(X)$ be the language L^* extended by the 1-place relation constant X. If $A(X,u)$ is an X-positive Σ-formula then there exists a Σ-function symbol I_A such that KPu proves for all ordinals α

$$I_A^\alpha = \{x\in N : A(I_A^{<\alpha},x)\}\ , \qquad \text{where} \quad I_A^{<\alpha} := \cup\{I_A^\xi : \xi<\alpha\}$$

We define $P_A(u) := \exists\xi(u\in I_A^\xi)$ and prove

($ID_A.1$) $\quad \forall x\in N(A(P_A,x) \rightarrow P_A(x))$

by Σ-reflection and for arbitrary formulas F

($ID_A.2$) $\quad \forall x\in N(A(F,x) \rightarrow F(x)) \rightarrow \forall x\in N(P_A(x) \rightarrow F(x))$

by transfinite induction on the ordinals. Therefore the theory ID_1 of one inductive definition (see for example [5]) is contained in KPu. Since the Δ-separation theorem can already be proved in W-KPu, we obtain that W-KPu is an extension of the theory $(\Delta^1_1\text{-CA})$ of second order arithmetic. In both cases we have proof-theoretic equivalence as follows from [10] and [12].

Theorem 1.
(a) $\text{W-KPu} \equiv (\Delta^1_1\text{-CA})$ and $|\text{W-KPu}| = \overline{\theta}\varepsilon_0 0$.
(b) $\text{KPu} \equiv \text{ID}_1$ and $|\text{KPu}| = \overline{\theta}\varepsilon_{\Omega_1+1}0$.

§3. In the following two sections we shall consider theories whose models are limits of admissible sets, but not necessarily admissible. Let Ad(a) be an abbreviation for $a \models \ulcorner \text{KPu} \urcorner \,\&\, \text{Tran}(a)$. KPl is the theory ES with the additional axioms

(Lim) $\forall x \exists y (\text{Ad}(y) \,\&\, x \in y)$,
(Linear ordering) $\text{Ad}(a) \,\&\, \text{Ad}(b) \rightarrow a \in b \vee a = b \vee b \in a$,

which allow the reduction of Σ^1_2-properties of the analytical hierarchy to Σ_1-predicates of the set-theoretic hierarchy (cf. quantifier theorem below). It is clear that all theorems of KPu can be relativized to admissible sets in $\text{KPl}^{\upharpoonright}$. Barwise shows in [1] that Axiom β is provable in $\text{KPU}+(\Sigma_1\text{-Separation})$. He needs $(\Sigma_1\text{-Separation})$ in order to show that the well-founded part of the 2-place relation r on the set b forms a set. This is already provable in $\text{KPl}^{\upharpoonright}$.

Theorem 2. $\text{KPl}^{\upharpoonright} \vdash \text{Axiom } \beta$.

Proof. Suppose b and r are arbitrary sets with Wf(b,r). By (Lim) we choose an admissible set a that contains b and r. By the relativized Σ-recursion theorem we define a function F which is Σ on a such that $\text{KPl}^{\upharpoonright}$ proves for all ordinals α of a:

$F(\alpha) = \{x \in b : \forall y \in b(\langle y,x\rangle \in r \rightarrow \exists \beta < \alpha (y \in F(\beta)))\}$.

The predicate $P(u) := \exists \xi \in a (u \in F(\xi))$ is Σ on a. Let a^+ denote the least admissible set which contains a. It follows by Δ-separation in a^+ that $\{x \in b : P(x)\}$ is a set in a^+. The rest of the proof is exactly as in [1]. ./.

Corollary. $\text{KPl} \vdash \text{Wf}(b,r) \rightarrow \text{TI}(b,r,F)$ for every formula F .

Let T be an arbitrary extension of ES. We call a formula $A[\underline{u}]$ a $\Pi^1(T)$-formula, if there exists a Δ_0-formula $B[\underline{u},v,w]$ such that

$T \vdash A[\underline{u}] \leftrightarrow \text{Wf}(N, \prec_{\underline{u}})$

for the relation $\prec_{\underline{u}} := \{\langle x,y\rangle \in N^2 : B[\underline{u},x,y]\}$. The following quantifier theorem is extremely useful when studying the connections between set theories and theories of second order arithmetic.

Theorem 3. For every $\Pi^1(\mathrm{KPl}^r)$-formula $A[\underline{u}]$ there exists a Δ_0-formula $B[\underline{u},v]$ such that $\mathrm{KPl}^r \vdash \mathrm{Ad}(a) \ \& \ \underline{u}\in a \rightarrow (A[\underline{u}] \leftrightarrow B[\underline{u},a])$.

Proof. The proof is similar to the proof of Axiom β. Again we use the fact that $\mathrm{Wf}(N, \prec_{\underline{u}})$ is provably equivalent to a Σ-formula on every admissible set a that contains $\underline{u}$. The details can be found in [11]. ./.

The quantifier theorem is something like a formalized version of the Spector-Gandy theorem. It is called quantifier theorem, since it allows us to reduce every Σ^1_{n+1}- or Π^1_{n+1}-formula of second order arithmetic to a Σ_n- or Π_n- formula of set theory, using the fact that every Π^1_1-formula is a $\Pi^1(\mathrm{KPl}^r)$-formula.

Theorem 4.

(a) $\mathrm{KPl}^r \equiv (\Pi^1_1\text{-CA})^r \equiv \mathrm{ID}_{<\omega} \equiv (\mathrm{EM}_0)^r + (\mathrm{IG})^r$ and $|\mathrm{KPl}^r| = \overline{\theta}\Omega_\omega 0$.

(b) $\text{W-KPl} \equiv (\Pi^1_1\text{-CA}) \equiv \text{W-ID}_\omega \equiv (\mathrm{EM}_0)+(\mathrm{IG})^r$ and $|\text{W-KPl}| = \overline{\theta}(\Omega_\omega\cdot\varepsilon_0)0$.

(c) $\mathrm{KPl} \equiv (\Pi^1_1\text{-CA})+(\mathrm{BI}) \equiv \mathrm{ID}_\omega \equiv (\mathrm{EM}_0)+(\mathrm{IG})$ and $|\mathrm{KPl}| = \overline{\theta}\varepsilon_{\Omega_\omega+1}0$.

Proof. The previous considerations show that the set theories mentioned contain the corresponding systems of second order arithmetic. The proof-theoretic equivalence of the second order systems with the theories of inductive definition and explicit mathematics follows from [5], [9], [19], [18] and [3]. Buchholz and Pohlers proved in [4] that $\overline{\theta}\Omega_\omega 0 \le |\mathrm{ID}_{<\omega}|$ and $\overline{\theta}\varepsilon_{\Omega_\omega+1}0 \le |\mathrm{ID}_\omega|$. Buchholz proved in [3] that $\overline{\theta}(\Omega_\omega\cdot\varepsilon_0)0 \le |\text{W-ID}_\omega|$ holds. In [11] we showed $|\mathrm{KPl}^r| \le \overline{\theta}\Omega_\omega 0$, $|\text{W-KPl}| \le \overline{\theta}(\Omega_\omega\cdot\varepsilon_0)0$ and $|\mathrm{KPl}| \le \overline{\theta}\varepsilon_{\Omega_\omega+1}0$. ./.

§4. Given a well-ordering $\prec$ on a set b we define the hierarchy of admissible sets along $\langle b,\prec\rangle$ starting with the set c by the following transfinite recursion (for $x\in b$):

(1) $f(0) := c$,

(2) $f(x) := f(y)^+$, if x is the successor of y in $\langle b,\prec\rangle$,

(3) $f(x) :=$ disjoint union of $f(y)$ $(y\prec x)$, if x is a limit in $\langle b,\prec\rangle$.

We write $H(c,f,b,\prec)$ to express that the function f satisfies (1)-(3); we assume that 0 is the least element of $\langle b,\prec\rangle$ and a^+ is the least admissible set containing a.

We shall introduce systems Aut(Ad) and $\mathrm{Aut}^E(\mathrm{Ad})$ which can both be understood, in a way, as autonomous closures of admissibility. Besides the axioms of KPl^r, Aut(Ad) contains the Bar Rule (BR) and the following rule

$$\frac{Wf(N,\prec)}{\forall x\exists f H(x,f,N,\prec)}$$

where $\prec$ is a primitive recursive well-ordering. It is easy to check that Aut(Ad) is proof-theoretically equivalent to the theory Aut(Π^1_1) of autonomously iterated hyperjumps, introduced in [8], and to the theory Aut(ID) of [3]. Therefore we have

Theorem 5. $|\mathrm{Aut(Ad)}| = |\mathrm{Aut}(\Pi^1_1)| = |\mathrm{Aut(ID)}| = \overline{\theta}\Omega_{\Omega_1}0$.

Aut(Ad) describes the autonomous closure of admissibility from outside. Whenever a primitive recursive well-ordering on N has been proved to be well-founded, an iteration of admissibility along this relation is permitted. So all iterations of admissibility carried through in Aut(Ad) are iterations along well-founded relations of order type $<\omega_1^{ck}$. Let Ω be the function which enumerates the admissible ordinals and their limits. Then the least standard model of Aut(Ad) is $L_{\Omega_{\Omega_1}}$. The concept of admissibility (of the hyperjump), however, is not exhausted by Aut(Ad). In KPl$\upharpoonright$ we can prove the existence of $\omega_1^{ck} = \Omega_1$ and therefore iterations along Ω_1 should be possible. Doing this we obtain Ω_{Ω_1}. Now iterate along Ω_{Ω_1} and so on.

To carry out these considerations, we have to extend the Bar Rule and the rule which produces hierarchies. By the Extended Bar Rule (E-BR) we mean the rule of inference

$$\frac{\exists!\alpha A[\alpha]}{\forall\alpha(A[\alpha]\rightarrow TI(\alpha,F))}$$

for every Σ-formula A and every formula F. Aut^E(Ad) consists of KPl$\upharpoonright$, the Extended Bar Rule and the following rule

$$\frac{\exists!\alpha A[\alpha]}{\forall\alpha(A[\alpha]\rightarrow\forall x\exists f H(x,f,\alpha))}$$

for Σ-formulas A. Since Axiom β is provable in KPl$\upharpoonright$, we see immediately that Aut(Ad) is contained in Aut^E(Ad). Let g10 denote the first fixed point of Ω. Then L_{g10} is the least standard model of Aut^E(Ad). We know $\overline{\theta}(g10)0 \leq |\mathrm{Aut}^E(\mathrm{Ad})|$ and believe that the converse direction also holds.

Conjecture. $|\mathrm{Aut}^E(\mathrm{Ad})| = \overline{\theta}(g10)0$.

Let us now consider the fundamental difference between iterations along ordinal $<\omega_1^{ck}$ and $\geq\omega_1^{ck}$ by comparing Aut(Ad) and Aut^E(Ad):

1. Whenever Aut(Ad) allows a hierarchy of length α, then it allows the construction of a hierarchy of length β for every $\beta<\alpha$. So in Aut(Ad) the hierarchies become

longer and longer 'in the course of time'.

2. In $Aut^E(Ad)$ we have an impredicativity in the construction of hierarchies. It may happen, that $Aut^E(Ad)$ allows a hierarchy of comparatively large length α and that this hierarchy is used for the construction of a hierarchy of length $\beta<\alpha$.

Remark. Feferman's theory $ID_{\prec^*}$ (cf. [5]) is intermediate between Aut(Ad) and $Aut^E(Ad)$. It allows iterations of inductive definitions along the accessible part $Acc(\prec)$ of the primitive recursive relation $\prec$. Since the provable part of $Acc(\prec)$ depends on the whole theory, this is an iteration from inside.

§5. Up to this point we have studied theories which have minimal models L_α with α admissible or limit of admissibles. Now we go a step further and consider a universe which is recursively inaccessible, i.e. admissible and limit of admissibles. The theory KPi is obtained by combining the theories KPu and KPl and describes exactly this situation. Recall that W-KPi ($KPi\upharpoonright$) is obtained from KPi by restricting the scheme of foundation (the scheme of foundation and complete induction on N) to sets. Then we have the following characterizations

Theorem 6.

(a) $KPi\upharpoonright \equiv (\Delta^1_2\text{-CA})\upharpoonright \equiv ID_{<\omega} \equiv (EM_0)\upharpoonright+(IG)\upharpoonright+(J)$ and $|KPi\upharpoonright| = \bar{\theta}\Omega_\omega 0$.

(b) $W\text{-}KPi \equiv (\Delta^1_2\text{-CA}) \equiv ID_{<\varepsilon_0} \equiv (EM_0)+(IG)\upharpoonright+(J)$ and $|W\text{-}KPi| = \bar{\theta}\Omega_{\varepsilon_0} 0$.

Proof. By Theorem 3 and Δ-separation we obtain an embedding of $(\Delta^1_2\text{-CA})\upharpoonright$ into $KPi\upharpoonright$ and of $(\Delta^1_2\text{-CA})$ into W-KPi. From [5], [9], [18] and [3] we know the proof-theoretic equivalence of the second order systems with the theories of inductive definitions and explicit mathematics. In [4] it is shown that $\bar{\theta}\Omega_\omega 0 \leq |ID_{<\omega}|$ and $\bar{\theta}\Omega_{\varepsilon_0} 0 \leq |ID_{<\varepsilon_0}|$. In [11] we proved $|KPi\upharpoonright| \leq \bar{\theta}\Omega_\omega 0$ and $|W\text{-}KPi| \leq \bar{\theta}\Omega_{\varepsilon_0} 0$. ./.

These results show that the weak versions of KPi can be reduced to iterations of admissibility up to all ordinals $\nu<\omega$ or $\nu<\varepsilon_0$. The situation changes dramatically in the presence of the full scheme of foundation. It is easy to check that even $Aut^E(Ad)$ is a proper subtheory of KPi. KPi goes beyond $Aut^E(Ad)$ as ID_1 goes beyond predicative mathematics. In treating KPi proof-theoretically we need to extend Buchholz' notation system so that it reflects the idea of the first recursively inaccessible ordinal I_0. In [17] Pohlers develops higher notation systems which are suited for the purpose of analysing KPi. The main result is

Theorem 7. $KPi \equiv (\Delta^1_2\text{-CA})+(BI) \equiv T_0$ and $|KPi| = \bar{\theta}^0(\bar{\theta}^1\varepsilon_{I_0+1}0)0$.

Proof. From [7] we know that T_0 is interpretable in $(\Delta^1_2\text{-CA})+(BI)$. By Theorem 3, the Corollary to Theorem 2 and Δ-separation we obtain an embedding of $(\Delta^1_2\text{-CA})+(BI)$

into KPi. [14] proves $|KPi| \leq \bar{\theta}^0(\bar{\theta}^1\varepsilon_{I_0+1}0)0$, and the recent [13] gives $\bar{\theta}^0(\bar{\theta}^1\varepsilon_{I_0+1}0)0 \leq |T_0|$.

Conjecture. $W\text{-}KPi + (E\text{-}BR) \equiv Aut^E(Ad)$.

§6. The considerations of the last sections should have sufficiently motivated the fact that for the proof-theoretic analysis of an impredicative theory T not only the ordinals below ω_1^{ck} are important but also the ordinals of the higher constructive number classes. With good Σ_1-definitions in mind (cf. [1]) we introduce the following notion.

Definition. Let σ be the least ordinal with $L_\sigma \models T$ and let κ be an admissible ordinal less than σ or equal to σ.
1. An element $a \in L_\kappa$ is κ-provable in T if there exists a Σ-formula $A[\underline{u}]$ such that $T \vdash \exists! x A[x]$ and $L_\kappa \models A[a]$.
2. $Core_T(\kappa)$ is the set of all elements of L_κ which are κ-provable in T.

For every T and κ as above we have that $Core_T(\kappa)$ is a proper subset of L_κ. This is a typical 'inside-outside-effect' in proof theory. If you look, say, at the ordinal ω_1^{ck} from inside the theory T it has the meaning of the first admissible which contains ω. From outside, however, its meaning is given by $Core_T(\omega_1^{ck}) \cap \omega_1^{ck}$. $Core_T(\omega_1^{ck})$ in general does not contain the provable ordinals of T, but we have $|T| \subset Core_T(\omega_1^{ck})$ as soon as Axiom β is provable in T.

The notion $Core_T(\kappa)$ of a theory T has only recently been introduced and will have to be studied carefully. It appears that it will be of use in reaching a better understanding of the connections between proof theory and definability theory. There should be some relation to Pohlers' notion of Spectrum(T) of a theory T. For more information about this concept and about the literature cf. [15] and [16]. A survey of admissibility in proof theory is given in [16].

REFERENCES

1. J. Barwise: Admissible Sets and Structures. Springer, Berlin, Heidelberg, New York (1975).

2. W. Buchholz: Normalfunktionen und konstruktive Systeme von Ordinalzahlen. Proof Theory Symposion Kiel 1974. Springer Lecture Notes in Math. 500 (1975).

3. W. Buchholz & S. Feferman & W. Pohlers & W. Sieg: Iterated inductive definitions and subsystems of analysis: recent proof-theoretical studies. To appear in Springer Lecture Notes in Math.

4. W. Buchholz & W. Pohlers: Provable well-orderings of formal theories for transfinitely iterated inductive definitions. J. Symb. Logic 43 (1978).

5. S. Feferman: Formal theories for transfinite iterations of generalized inductive definitions and some subsystems of analysis. Intuitionism and Proof Theory. North Holland, Amsterdam (1970).

6. S. Feferman: A language and axioms for explicit mathematics. Algebra and Logic. Springer Lecture Notes in Math 450 (1975).

7. S. Feferman: Constructive theories of functions and classes. Logic Colloquium 78. North Holland, Amsterdam (1979).

8. S. Feferman & G. Jäger: Choice principles, the bar rule and autonomously iterated comprehension schemes in analysis. To appear in J. Symb. Logic.

9. H. Friedman: Iterated inductive definitions and Σ^1_2-AC. Intuitionism and Proof Theory. North Holland, Amsterdam (1970).

10. G. Jäger: Beweistheorie von KPN. Archiv f. Math. Logik u. Grundl. 20 (1980).

11. G. Jäger: Die konstruktible Hierarchie als Hilfsmittel zur beweistheoretischen Untersuchung von Teilsystemen der Mengenlehre und Analysis. Dissertation, München (1979).

12. G. Jäger: Zur Beweistheorie der Kripke-Platek-Mengenlehre über den natürlichen Zahlen. To appear in Archiv f. Math. Logik u. Grundl.

13. G. Jäger: A well-ordering proof for Feferman's theory T_0. To appear in Archiv f. Math. Logik und Grundl.

14. G. Jäger & W. Pohlers: Eine beweistheoretische Untersuchung von $(\Delta^1_2\text{-CA})+(\text{BI})$ und verwandter Systeme der Mengenlehre. To appear in Sitzungsberichte der Bayerischen Akademie der Wissenschaften.

15. W. Pohlers: Beweistheorie der iterierten induktiven Definitionen. Habilitation, München (1977).

16. W. Pohlers: Admissibility in Proof Theory; a Survey. To appear in the Proceedings of the VI. International Congress for Logic, Methodology and Philosophy of Science, 1979. North Holland, Amsterdam.

17. W. Pohlers: Higher notation systems. To appear in Archiv f. Math. Logik und Grundl.

18. W. Sieg: Trees in metamathematics (Theories of inductive definitions and subsystems of analysis). Ph.D. Thesis, Stanford (1977).

19. J.I. Zucker: Iterated inductive definitions, trees and ordinals. Springer Lecture Notes in Math. 344 (1973).

FOOTNOTE

1) This work is part of a project supported by the Deutsche Forschungsgemeinschaft.

PROCEEDINGS OF THE HERBRAND SYMPOSIUM
LOGIC COLLOQUIUM '81, J. Stern (editor)

INTRODUCING HOMOGENEOUS TREES

Herman R. Jervell
University of Tromsø
Tromsø
Norway

Homogeneous trees are introduced. They generalize both hierarchies of numbertheoretic functions and ordinalnotations. We give generalizations of common proof-theoretic operations and show how to define the new and powerful recursion over homogeneous trees.

1. INTRODUCTION

This paper gives an introduction to the homogeneous trees which we introduced in [2] to give a geometrical interpretation of Jean-Yves Girards Π^1_2-logic [1]. Here we shall emphasize ideas and examples. For the proofs we refer mostly to [2] or the dissertation of Udo Lenz [3]. Lenz has not only written out the proofs of [2] carefully, but there is an improvement in the definition, corrections of some mistakes, and the details for how to get the Howard ordinal. In [4] Marcel Masseron has given functorial connections between the homogeneous trees and Jean-Yves Girards rungs. For further development of the theory we refer to the revised version of [1].

2. NOTATIONS AND DEFINITIONS

2.1 <u>Ordinals and the extra symbol ∞</u>

ON is the class of all ordinals.

As variables for ordinals we use $\alpha, \beta, \gamma, \delta, \ldots$

As variables for natural numbers we use $n_1, n_2, m_1, m_2, \ldots$

An ordinal is the set of its predecessors $\alpha = \{ \beta \mid \beta < \alpha \}$

$0, 1, \omega$ are the usual ordinals

Ω is the first uncountable ordinal

The extra symbol ∞ is the name for an object larger than any ordinal. For any $\alpha \in ON$ we have $\alpha < \infty$

2.2 <u>Increasing functions</u>

Let $\alpha \leq \beta \leq \gamma$. We have the following properties

F1. $f: \beta \cup \{\infty\} \to \gamma \cup \{\infty\}$

F2. $f(0) = 0$

F3. $f(\infty) = \infty$

F4. $b < c \Rightarrow f(b) < f(c) \quad \forall b,c \in \beta \cup \{\infty\}$

F5. $b < \alpha \Rightarrow f(b) = b$

F6. $\sup \{ f(b) \mid b < \beta \} = \gamma$

We introduce the following sets of functions

$f \in I(\beta,\gamma) := F1 \wedge F2 \wedge F3 \wedge F4$

$f \in (\alpha)I(\beta,\gamma) := F1 \wedge F2 \wedge F3 \wedge F4 \wedge F5$

$f \in J(\beta,\gamma) := F1 \wedge F2 \wedge F3 \wedge F4 \wedge F6$

$f \in (\alpha)J(\beta,\gamma) := F1 \wedge F2 \wedge F3 \wedge F4 \wedge F5 \wedge F6$

2.3 <u>Finite sequences</u>

α-SEQ is the set of all finite sequences of elements from $\alpha \cup \{\infty\}$

As variables for sequences we use $\mu, \nu, \sigma, \tau, \ldots$.

A typical sequence is $\sigma = < s_1, s_2, \ldots, s_n >$

$< >$ is the empty sequence.

$\sigma * \tau$ is the concatenation of the two sequences σ and τ.

2.4 <u>Trees</u>

$S \in \alpha$-TREE if

TREE 1. S is a non-empty subset of α-SEQ.

TREE 2. $\sigma * \tau \in S \Rightarrow \sigma \in S$

TREE 3. $\sigma * <\beta> \in S \wedge \gamma < \beta \Rightarrow \sigma * <\gamma> \in S$

A typical picture of a tree is:

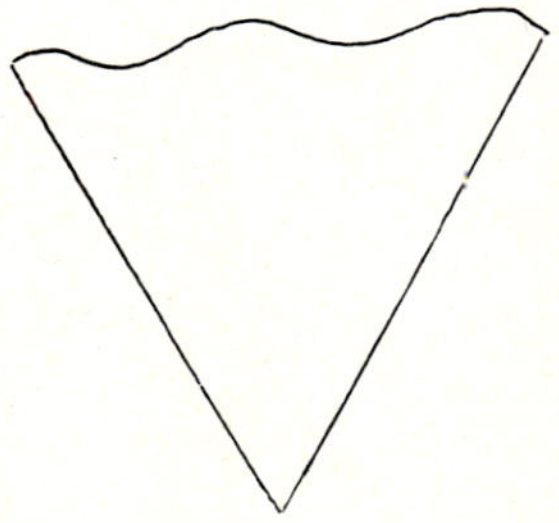

The downmost node is < >.
Immediately above node σ we may have

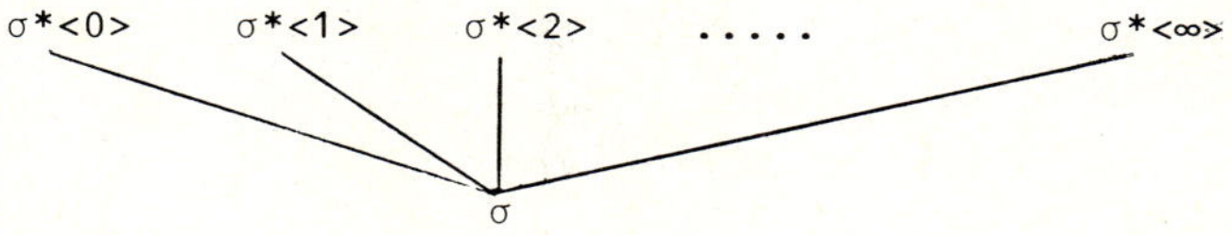

Another useful picture of node $\sigma = <s_1, s_2, \ldots\ldots, s_n>$ in tree S is:

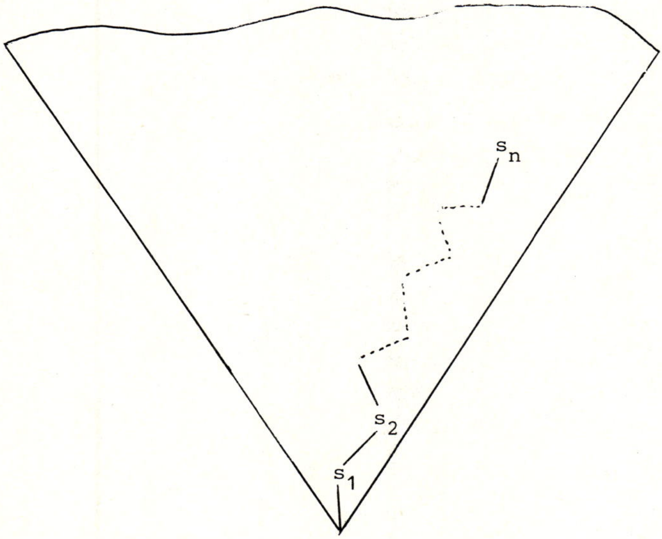

Here we have written the names of the elements $s_1, s_2, \ldots, s_n$ of σ on the nodes leading up from < > to σ .

2.5 <u>The generated tree</u>

If $S \subseteq \alpha$-SEQ , we write Tree S for the smallest α-TREE which contains S . To be more precise we should have an α as an index but we omit it where it is clear from the context which ordinal we should have as index.

3. THE ELEMENTARY THEORY OF TREES

3.1 Some examples of trees

ONE := Tree$\{<0>\}$

ID := Tree$\{<n> | n \in \omega\}$

We can picture these trees as:

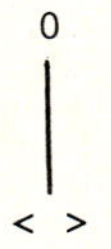

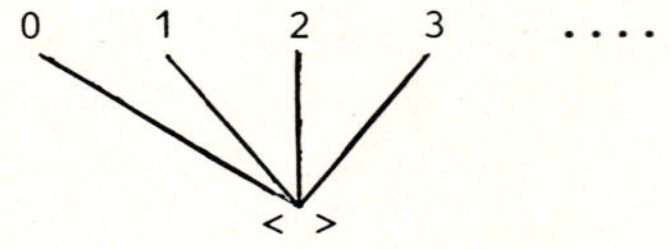

SQUARE := Tree$\{ <m,n> \mid m,n \in \omega \}$ is pictured as:

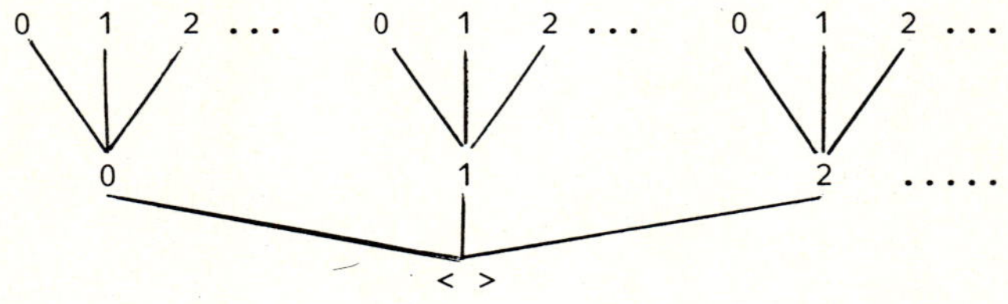

2-EXP := $\{ <n_1,n_2,\ldots\ldots,n_k> \mid k \in \omega \wedge \omega > n_1 > n_2 > \ldots\ldots > n_k \}$

EXP := $\{ <n_1,m_1,n_2,m_2,\ldots\ldots,n_k,m_k> \mid k \in \omega \wedge \forall i \leqq k \; m_i \in \omega \wedge \omega > n_1 > n_2 > \ldots\ldots > n_k \}$

3.2 Orderings of the nodes

The Brouwer-Kleene ordering of the nodes in a tree is given as the linear ordering satisfying:

BK 1. $\tau \neq < > \;\Rightarrow\; \sigma < \sigma * \tau$

BK 2. $\sigma * <\alpha> * \tau_0 < \sigma * <\beta> * \tau_1 \;\Leftrightarrow\; \alpha < \beta$

3.3 Well-founded trees

A tree is well-founded if the Brouwer-Kleene ordering of the nodes in the tree is a well-ordering. This is equivalent to that there is no infinite ascending sequence of nodes in the tree.

3.4 Topmost nodes

τ is a topmost node in the tree T iff there is no $\sigma \neq \langle\ \rangle$ $\tau * \sigma \in T$. We let $\hat{T}$ be the set of the topmost nodes of T .

3.5 Ordertypes

Let T be a well-founded tree. The length of T is

$|T|$:= the ordertype of the topmost nodes $\hat{T}$ of T under the Brouwer-Kleene ordering.

If $\tau \in \hat{T}$ we define

$|\tau|_T$:= the ordertype of the topmost nodes preceding τ .

3.6 Examples

The ordertypes of the examples in 3.1 are

$|ONE| = 1$

$|ID| = \omega$

$|SQUARE| = \omega^2$

$|2\text{-}EXP| = 2^{\omega} = \omega$

$|EXP| = \omega^{\omega}$

3.7 Sum

Let S and T be two trees - none of them equal to the trivial tree $\{\langle\ \rangle\}$. We define :

$$S + T := \{\langle\ \rangle\} \cup \{\langle 0\rangle * \sigma \mid \sigma \in S\} \cup \{\langle\infty\rangle * \tau \mid \tau \in T\}$$

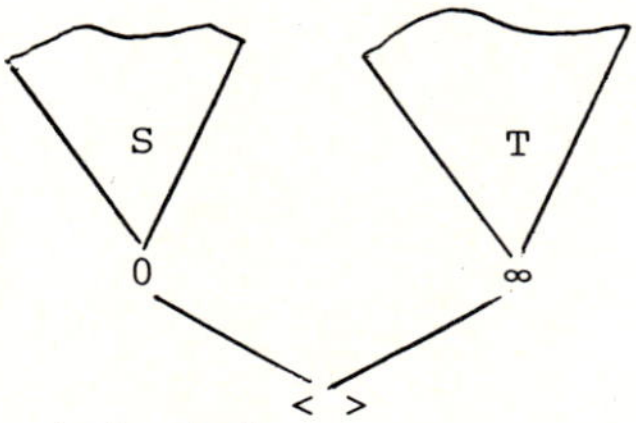

We have $|S+T| = |S| + |T|$

Two new trees can be defined by

DOUBLE := ID + ID

SUC := ID + ONE

4. HOMOGENEOUS TREES

4.1 Mappings of sequences

Let $f \in I(\alpha,\beta)$

We extend f to $\bar{f} : \alpha\text{-SEQ} \to \beta\text{-SEQ}$ by

M1. $\bar{f}(\langle s\rangle) = \langle f(s)\rangle$

M2. $\bar{f}(\sigma * \tau) = \bar{f}(\sigma) * \bar{f}(\tau)$

4.2 Diagrams of sequences

Let $\alpha \leqq \beta$ and $\sigma \in \beta\text{-SEQ}$.

Let $\{ s \mid s \geqq \alpha$ and s is an element of $\sigma \} = S$ and the number of elements of S be n .

There is a unique $f \in (\alpha)I(\alpha+n,\beta)$ such that all elements of σ is in the range of f .

The α-diagram of σ is defined as $(\sigma)_\alpha = \bar{f}^{-1}(\sigma)$

So for example for $\langle 6,2,4,0,2\rangle \in 7\text{-SEQ}$ we have

$\langle 6,2,4,0,2\rangle_1 = \langle 3,1,2,0,1\rangle$

$\langle 6,2,4,0,2\rangle_5 = \langle 5,2,4,0,2\rangle$

4.3 Collapsing of trees

Let $f \in I(\alpha,\beta)$

We define $f^{-1}: \beta\text{-TREE} \to \alpha\text{-TREE}$ by

$f^{-1}(T) = \{ \sigma \in \alpha\text{-SEQ} \mid \bar{f}(\sigma) \in T \}$

Our picture of this collapsing operation is as follows

1. Start with T
2. Erase all nodes in T containing elements not from the range of f
3. Then collapse to get the tree $f^{-1}(T)$

4.4 Homogeneous trees

Let $\alpha \leqq \beta$. The β-TREE T is $[\alpha,\beta[$-homogeneous if

(1) $\sigma \in T \wedge (\sigma)_\alpha = (\tau)_\alpha \Rightarrow \tau \in T$

Either of the following is equivalent to (1) :

(2) $\sigma \in T \Leftrightarrow (\sigma)_\alpha \in T$

(3) $\forall\gamma \in [\alpha,\beta[\ \forall f,g \in (\alpha)I(\gamma,\beta)\quad f^{-1}(T) = g^{-1}(T)$

(4) $\forall\gamma \in [\alpha,\beta[\ \forall f \in (\alpha)I(\gamma,\beta)\quad f^{-1}(T) = T \cap \gamma\text{-SEQ}$

The trees ONE, ID, SUC, DOUBLE, SQUARE, 2-EXP, EXP are all $[0,\omega[$-homogeneous.

4.5 Extensions of homogeneous trees

Assume that T is $[\alpha,\beta[$-homogeneous and $\gamma \geqq \alpha$ and $\gamma > 0$.
We define $T[\gamma] = \{ \sigma \in \gamma\text{-SEQ} \mid (\sigma)_\alpha \in T \}$

It is easy to see that $T[\gamma]$ is $[\alpha,\gamma[$-homogeneous .
If $\alpha + \omega \leqq \beta \leqq \gamma$ then $T[\gamma]$ is the unique $[\alpha,\gamma[$-homogeneous tree extending T .
If $\gamma \in [\alpha,\beta[$, then $T[\gamma] = T \cap \gamma\text{-SEQ}$.

The ordertypes of some extensions of trees are :

$|\ \text{ONE}[\alpha]\ | = 1$
$|\ \text{ID}[\alpha]\ | = \alpha$
$|\ \text{SUC}[\alpha]\ | = \alpha+1$
$|\ \text{DOUBLE}[\alpha]\ | = \alpha\cdot 2$
$|\ \text{SQUARE}[\alpha]\ | = \alpha^2$
$|\ \text{2-EXP}[\alpha]\ | = 2^\alpha$
$|\ \text{EXP}[\alpha]\ | = \alpha^\alpha$

For finite α we get numbertheoretic functions. So for example the homogeneous tree SQUARE gives the squaring-function. For ω we get a way of building up ordinals from sequences of length ω . For an arbitrary α we get ordinals built up from sequences of length α , i.e. something like ordinalnotations.

4.6 Strongly well-founded homogeneous trees

Let T be $[\alpha,\beta[$-homogeneous. T is strongly well-founded if $T[\gamma]$ is well-founded for all $\gamma \geqq \alpha$.

It is sufficient to consider only all $\gamma < \alpha+\Omega$ where Ω is the first uncountable ordinal, or even to consider whether only $T[\alpha+\Omega]$ is well-founded. The following example shows that it is necessary to consider all $\gamma < \alpha+\Omega$ in the definition.

Let δ be an infinite countable ordinal and $d_0,d_1,d_2,d_3,\ldots\ldots\ldots$ an enumeration of all ordinals $< \delta$. We define

$$D = \{ \langle n_0,n_1,n_2,\ldots,n_{k-1},n_k\rangle \mid \langle n_0,\ldots,n_{k-1}\rangle_0 = \langle d_0,\ldots,d_{k-1}\rangle_0 \}$$

D is a $[0,\omega[$-homogeneous tree. $D[\gamma]$ is well-founded if and only if $\gamma < \Omega$.

5. COMPOSITION

5.1 Composition

Assume T is a well-founded $[\alpha,\beta[$-homogeneous tree.

Let $\varphi_T : |T| \to \hat{T}$ be the function which enumerates the topmost nodes of T .

We extend φ_T to $\overline{\varphi}_T : |T|\text{-SEQ} \to \beta\text{-SEQ}$ by:

$$\overline{\varphi}_T(<s_1,s_2,\dots,s_n>) = \varphi_T(s_1)*\varphi_T(s_2)*\cdots*\varphi_T(s_n)$$

Assume S is $[\alpha,\beta[$-homogeneous. We define the composition by

$$S \mathbin{\square} T = \text{Tree } \overline{\varphi}_T(\ S[|T|]\ )$$

We picture the composition as follows

1. Start with $[\alpha,\beta[$-homogeneous tree S
2. Extend S to $[\alpha,\gamma[$-homogeneous tree $S[\gamma]$ where $\gamma = |T|$
3. Replace each γ-branching in $S[\gamma]$ with the tree T to get the $[\alpha,\beta[$-homogeneous tree $S \mathbin{\square} T$.

5.2 Simple properties

If S and T are $[\alpha,\beta[$-homogeneous trees and T is well-founded, then

$$|\ S\square T\ | = |\ S[\,|T|\,]\ |$$

We have for example

$$|\ \text{EXP} \mathbin{\square} \text{EXP}\ | = (\omega^\omega)^{(\omega^\omega)} = \omega^{(\omega^\omega)}$$

$$|\ \text{SUC} \mathbin{\square} \text{SUC}\ | = \omega+2$$

5.3 Some auxiliary definitions

We define α^k ($k < \omega$) by

$$\alpha^0 = <\ >$$

$$\alpha^{k+1} = \alpha^k*<\alpha>$$

A γ-TREE T starts with identity if for some k

$$0^k*<\beta> \in T \text{ for all } \beta < \gamma .$$

5.4 The treeclasses $H\alpha[\beta]$ $\alpha < \beta$ β is a limitordinal

T is in $H\alpha[\beta]$ if

H1. T is $[\alpha,\beta[$-homogeneous tree

H2. T is strongly well-founded

H3. T starts with identity

The following trees are in $H0[\omega]$

ID, SUC, DOUBLE, SQUARE, ID + 2-EXP, EXP

while 2-EXP and ONE are not.

5.5 <u>α-commuting operations</u>

An n-ary operation π is α-commuting if

O1. $\pi : H\alpha[\beta]^n \to H\alpha[\beta]$ for any limitordinal β

For any $T_1, T_2, \ldots, T_n \in H\alpha[\beta]$

O2. $f \in (\alpha)J(\gamma,\beta) \Rightarrow f^{-1}(\pi(T_1,\ldots,T_n)) = \pi(f^{-1}(T_1),\ldots,f^{-1}(T_n))$.

O3. $\delta > \beta \Rightarrow \pi(T_1[\delta],\ldots,T_n[\delta]) = \pi(T_1,\ldots,T_n)[\delta]$ (δ limit)

One of the programs in the theory of homogeneous trees is to show that the usual ordinalnotations can be given by members of $Hn[\omega]$ and that operations on ordinalnotations can be given by n-commuting operations. We have for example that $+$ and $\square$ are both 2-ary 0-commuting operations.

The use of the functionclass $J(\gamma,\beta)$ in O2 above comes from the dissertation of Lenz [3] . In my München-lectures there is used the functionclass $I(\gamma,\beta)$. See [2] . In that case it is not true that $\square$ is commuting.

5.5 <u>Iteration</u>

Two basic operations in proof-theory are

- the Grzegorczyk-operation in hierarchies of number-theoretic functions
- the derivative of normal functions of ordinals

The 2-ary 0-commuting operation iteration generalizes both. It is defined as follows:

1. Start with the well-founded tree S.
2. For each topmost node σ of S : if $\sigma = \langle s_1, s_2, \ldots, s_n \rangle$, then above σ tack on T^n . (Here $T^1 = T$, $T^{k+1} = T^k \square T$)
3. The result is the tree $It(S,T)$.

Using the theory so far we can give ordinalnotations for all predicative ordinals. We have for example:

$$|\, It(\mathrm{EXP},\mathrm{EXP}) \,| = \varepsilon_0$$

6. RECURSION

6.1 Restriction

Let τ be a node in tree T . $\tau = <t_1, t_2, \ldots, t_n>$.
The restriction of τ to T is given by:

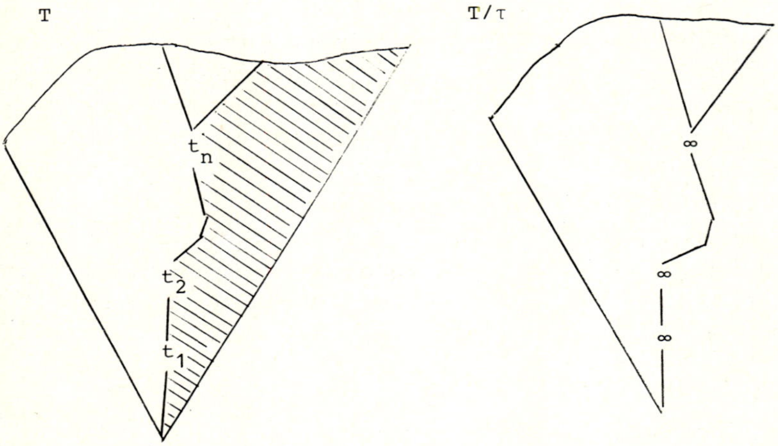

Here we have cut off everything to the right of $<t_1, t_2, \ldots, t_n>$ and replaced t_i with ∞ $1 \leq i \leq n$ to get T/τ .

It is straightforward that

1. T/τ is a tree
2. If T is $[\alpha, \beta[$-homogeneous and $\tau \in \gamma$-SEQ with $\gamma \in [\alpha, \beta[$, then $f^{-1}(T/f(\tau)) = f^{-1}(T)/\tau$ for $f \in I(\delta, \beta)$.
3. If T is $[\alpha, \beta[$-homogeneous and $\tau \in \gamma$-SEQ with $\gamma \in [\alpha \ \beta[$, then T/τ is $[\gamma, \beta[$-homogeneous.

6.2 Critical nodes

The critical node of the tree T is the ∞^k such that

1. $\infty^k * <0> \in T$
2. $\infty^{k+1} * <0> \notin T$

6.3 Recursion

Recursion is a powerful way of producing new homogeneous trees. It was first used by Jean-Yves Girard [1] to get $\Lambda(T)$ from T. The other operations defined above are all predicative, while the recursion is not.

Assume S,T trees, T well-founded and π a unary operation on trees. We define $Rec(\pi,T/\tau)S$ by recursion over the Brouwer-Kleene ordering of nodes τ of T.

Recursionstart:

If T/τ does not contain any critical node, then $Rec(\pi,T/\tau)S = S$. In this case T/τ contains only nodes of the form ∞^n , and τ is among the first few nodes in the Brouwer-Kleene ordering.

Recursionstep:

If ∞^k is the critical node of T/τ , then $Rec(\pi,T/\tau)S$ is the tree built up from

1. $\delta * \pi Rec(\pi,(T/\tau)/\infty^k * \delta)S$ where $\infty^k * \delta$ is a topmost node of T/τ .
2. $\delta * Rec(\pi,(T/\tau)/\infty^k * \delta)S$ where $\infty^k * \delta$ is a not-topmost node of T/τ .

We then define $Rec(\pi,T)S \quad := \quad Rec(\pi,T/\langle\,\rangle)S$

6.4 The recursiontheorem

If π is a unary α-commuting operation, then $\lambda S,T\ Rec(\pi,T)S$ is a 2-ary α-commuting operation.

For the proof we refer to our München-lectures [2] or Lenz [3] .

6.5 Some applications

We give a few applications to indicate the power of recursion.

Define $\quad$ INDUCTIVE-0 $:=$ $It(\text{EXP},\text{EXP})$

$\qquad$ INDUCTIVE-n+1 $:=$ $Rec(It,\text{INDUCTIVE-n})\text{EXP}$

We then have that $|$ INDUCTIVE-n $|$ is the proof-theoretic ordinal of n times iterated inductive definitions, and in particular INDUCTIVE-1 gives an ordinalnotation for the Howard ordinal. It is interesting to note that these ordinalnotations are here defined without using higher number classes. These definitions can be extended by using the unary commuting operation Ind^n defined by

$$Ind^1(T) \quad := \quad Rec(It,T)\text{EXP}$$
$$Ind^{n+1}(T) \quad := \quad Rec(Ind^n,T)\text{EXP}$$

We have not worked out the details for how these match up with the usual proof-theoretic ordinals.

7. REFERENCES

[1] Girard, J.Y., Π_2^1-logic, part 1. Manuscript 1979. To appear in Annals of Mathematical Logic. (Revised version 1981)

[2] Jervell, H.R., Homogenous trees. Lectures given at the University of München, Summer 1979. Manuscript. Tromsø.

[3] Lenz, U., Homogene Bäume und die Howardzahl. Dissertation. University of München.

[4] Masseron, M., Rungs and Trees. Manuscript 1980. To appear in Journal of Symbolic Logic. (Revised version 1981)

PROCEEDINGS OF THE HERBRAND SYMPOSIUM
LOGIC COLLOQUIUM '81, J. Stern (editor)

EXPONENTIAL DIOPHANTINE REPRESENTATION OF RECURSIVELY ENUMERABLE SETS

J.P. Jones and Ju. V. Matijasevič

University of Calgary and Steklov Mathematical Institute of the Academy of Science, Leningrad

En 1961 M. Davis, H. Putnam et J. Robinson ont démontré que tout ensemble récursivement énumérable est défini par une équation diophantienne pouvant comporter la fonction exponentielle. De façon plus précise, tout ensemble récursivement énumérable, A peut se définir par $n \in A$ si et seulement si $\exists x_1, x_2, \ldots, x_\nu \, (P(n, x_1, \ldots, x_\nu) = Q(n, x_1, \ldots, x_\nu))$, dans laquelle P et Q sont des fonctions d'entiers naturels obtenues à partir des opérations d'addition de multiplication et d'exponentiation. En 1979, ce résultat fut amélioré par Ju. V. Matijasevič qui a prouvé que trois inconnues x_1, x_2 et x_3 suffisent (i.e. $\nu = 3$) et que la representation est univoque (uni-pli) : c'est-à-dire, la solution x_1, x_2, x_3 est unique pour $n \in A$. En anglais cette représentation est appelé 'singlefold', en russe 'odno-kratno'.

Dans le présent travail, ce résultat est encore amélioré. Nous démontrons que tout ensemble énumérable peut être représenté par une inégalité $\exists z, y[P(n,z,y) \leqslant Q(n,z,y)]$, dans laquelle z et y sont des entiers uniques (représentation univoque), P et Q sont des fonctions sur les entiers naturels définies à partir des opérations d'addition, de multiplication et de l'exponentielle 2^x (en base 2). Ceci implique une forme forte du théorème initial des trois inconnues ci-dessus, car ici l'exponentiation x^y sous sa forme générale n'apparait que sous la forme 2^x. De plus, on montre que deux itérations de l'exponentielle suffisent, i.e. 2^{2^x}.

Des résultats connexes et variés sont démontrés : un ensemble énumérable a une représentation sous la forme $\exists z \forall y[R(n,z,y) \leqslant S(n,z,y)]$, dans laquelle R et S sont obtenus à partir de l'addition, de la multiplication et de l'exponentielle en base 2.

On montre aussi qu'une relation élémentaire de Kalmar a aussi une représentation, avec une variable (que l'on peut supposer bornée), i.e. sous les formes $\exists y[P(n,y) \leq Q(n,y)]$ et $\forall y[R(n,y) \leq S(n,y)]$; et donc sous les formes $\exists y,x[P(n,y,x) = Q(n,y,x)]$ et $\forall y \exists x[R(n,y,x) = S(n,y,x)]$ (où les quantificateurs sont bornés). Ces énoncés caractérisent les relations élémentaires de Kalmar. Comme l'ensemble des nombres premiers est élémentaire de Kalmar, il s'ensuit que l'ensemble des nombres premiers possède une représentation sous la forme ci-dessus.

In 1961 it was shown that every recursively enumerable set is exponential diophantine. M. Davis, H. Putnam and Julia Robinson [1961] proved that for each r.e. set W there exists an exponential diophantine equation

$$P(a, z_1, \ldots, z_\nu) = Q(a, z_1, \ldots, z_\nu) \tag{1}$$

solvable in the unknowns $z_1, \ldots, z_\nu$, if, and only if a belongs to the set W. Here P and Q are functions built up from natural numbers and variables $a, z_1, \ldots, z_\nu$ by addition, multiplication and exponentiation. The unknowns $z_1, \ldots, z_\nu$ are understood to range over natural numbers.

In the papers of Matijasevič [1974] and [1976] this theorem was improved to the effect that each r.e. set has a singlefold exponential diophantine representation, i.e. one in which $x_1, \ldots, x_\nu$, when they exist, are unique. Thus for each value of the parameter a such that $a \in W$, there is one and only one solution $z_1, \ldots, z_\nu$.

In the paper of Matijasevič [1979] this result was further improved to the effect that there always exists a singlefold exponential diophantine representation in three unknowns, i.e. every r.e. set can be represented in the form

$$\exists x,y,z \quad [P(x,y,z) = Q(x,y,z)]. \tag{2}$$

Here we have suppressed mention of the parameter a. P and Q are functions obtained from natural numbers and variables a,x,y,z by addition, multiplication and iterated exponentiations of type u^ν, in two variables.

In the paper of Jones and Matijasevič [1981] this result was further improved to unary singlefold, three unknown, exponential diophantine representation, i.e. one based only on powers of two. We obtained, a representation of type (2) in which P and Q are functions built up from natural numbers and variables a,x,y,z using only the operations of addition, multiplication and the raising of 2 to a power, i.e. exponentiations of type 2^v. Since u^v is a two-place function and 2^v is a one-place function, we call our new base 2 representation unary. Unlimited iteration of exponentiation is to be understood here but actually two levels of exponentiation are sufficient for Theorems 1 and 2. (Terms of the type $2^{2^{f(z)}}$ appear, where $f(z)$ is a polynomial in z.)

The results of the present paper are essentially the same as those of Jones - Matijasevič [1981]. The only difference being one of completeness. The proof here is self contained. Necessary results of Matijasevič [1979] are here included where needed so that it is not necessary to refer to this earlier paper to understand the proofs. We prove the following theorems.

THEOREM 1. Every recursively enumerable set can be represented in the form

$$\exists z \exists y \; [P(z,y) \leqslant Q(s,y)]$$

where P and Q are unary exponential diophantine expressions. Furthermore the representation is singlefold and the second quantifier, $\exists y$ may be bounded.

This is actually a stronger result than the three unknown theorem. Taking $P(x,y,z) = Q(z,y) + x$ we have as an immediate corollary the three unknown theorem, every r.e. set can be singlefold represented in the form (2). Also P and Q are here unary exponential diophantine functions, built up only of powers of 2. Further, x and y may be supposed to be less than some bound, where the bound takes the form of an iterated exponential function of the first unknown, z. We will also prove

THEOREM 2. Every r.e. set can be represented in the form $\exists z \forall y \; [R(z,y) \leqslant S(z,y)]$ where R and S are unary exponential expressions. Here the universal quantifier, $\forall y$ may be bounded if we wish.

The representations of Theorems 1 and 2 are best possible both in terms of the number of quantifiers and the bounds. One cannot delete any quantifiers or bound the first quantifier, $\exists z$. (However we do not know if the representation $\exists z,y,z \; [P(z,y,x) = Q(z,y,x)]$ is best possible in terms of the number of quantifiers.)

If we replace "r.e." by "recursive" in Theorems 1 and 2, then one might expect to be able to bound all the quantifiers. However a simple diagonal argument shows that this is not the case. (See Davis, Matijasevič, Robinson [1975].) The same argument shows also that even in this case one cannot delete any quantifiers.

In the case of certain particular recursive sets we may be able to delete a quantifier. We will show that this is the case for primes, Mersenne primes, perfect numbers and certain other recursive sets occurring in classical number theory. These sets are all particular examples of Kalmar Elementary Relations. For this type of set we can delete a quantifier from Theorems 1 and 2.

THEOREM 3. Every Kalmar elementary relation is representable in both of the forms

$$\exists y[P(y) \leqslant Q(y)], \quad \text{and} \quad \forall y[R(y) \leqslant S(y)].$$

Here P,Q,R and S are unary exponential diophantine expressions. Furthermore y can be bounded in both cases.

Theorem 3 implies immediately the following

THEOREM 4. Every Kalmar elementary relation is representable in both the forms

$$\exists y,\exists x[P(y,x) = Q(y,x)] \quad \text{and} \quad \forall y,\exists x[R(y,x) = S(y,x)]$$

where P,Q,R and S are unary exponential diophantine expressions. Furthermore all of the quantifiers can be bounded.

With the understantding that the quantifiers are bounded, the converses of Theorems 3 and 4 also hold. Hence the two theorems provide (four) new characterizations of the Kalmar elementary sets. The quantifier bounds take the form of unary iterated exponential functions of the parameters. In the terminology of Davis, Matijasevič, Robinson [1975] such ∃ quantifier bounds are referred to as "iterated exponential test functions". Further information about Kalmar elementary sets and functions is given in §2. There we give several new examples of bases for these sets and a slight improvement of a theorem of S. S. Marchenkov [1980].

As a corollary of Theorem 4 we find that the set of prime numbers, the set of Mersenne primes and the set of perfect numbers can each be represented in the above forms. We do not know if these forms are best possible.

In at least one interesting case it is possible to do better. The set of Fermat primes can be exponentially defined in only one unknown. It is possible to construct exponential expressions P(n,x) and Q(n,x) such that for all natural numbers

$$n \text{ is a Fermat prime} \iff \exists x[P(n,x) = Q(n,x)].$$

For the details of how to construct P and Q cf. Jones [1979], (Lemma 4.3). Of course it is not known whether this set is finite or infinite. If there are only finitely many Fermat primes, then we can delete another quantifier, ∃x.

§1. All presently known methods of constructing a diophantine representation for an r.e. predicate exploits the exponential diophantine representation in the capacity of an intermediate step. But in the present work it will be the other way around, namely the diophantine representation (cf. Matijasevič [1971]) will be used to construct an exponential diophantine representation. More precisely the following will be proved.

LEMMA 1. For any polynomial equation

$$D(w_1, w_2, \ldots, w_\nu) = 0, \qquad (3)$$

there exist unary exponential diophantine functions $P(z,y)$ and $Q(z,y)$ such that (3) has a solution if and only if there exist (unique) natural z and y such that

$$P(z,y) \leqslant Q(z,y). \qquad (4)$$

Here lower case latin letters represent variables for natural numbers (non-negative integers). All unknowns are assumed to range over natural numbers.

The polynomial $D(w_1, \ldots, w_\nu)$ may be written in the form

$$D(w_1, \ldots, w_\nu) = \sum_{i_1 + \ldots + i_\nu \leqslant \delta} D_{i_1, \ldots, i_\nu} w_1^{i_1}, \ldots, w_\nu^{i_\nu} \qquad (5)$$

where the coefficients, $D_{i_1, \ldots, i_\nu}$ are integer valued functions of the parameters (these functions of parameters will correspondingly appear in the inequality (4).

We will not place any ceiling on the solutions of equation (3), but rather we may suppose that any solution to (3) has the property that $w_1 > w_2, \ldots, w_1 > w_\nu$ and that each of $w_2, w_3, \ldots, w_\nu$ is uniquely determined by w_1. (If that is not the case, then we introduce a new variable w_0 and consider instead of (3) the equation

$$(w_0 - 1 - w_1 - (w_1 + w_2)^2 - (w_1 + w_2 + w_3)^3 \ldots$$

$$\ldots - (w_1 + \ldots + w_\nu)^\nu)^2 + D^2(w_1, \ldots, w_\nu) = 0,$$

which has the same solutions as (3) and which possesses the required property with w_0 playing the role of w_1.)

Now it will be necessary for us to use a number-theoretic theorem of E. Kummer [1852], afterwards restated and reproved in other papers, for example Singmaster [1974].

KUMMER'S THEOREM. Let $\kappa(t)$ denote the multiplicity to which 2 divides t and let $\sigma(t)$ denote the sum of the digits in the binary expansion of t. Then

$$\kappa\left(\binom{2t}{t}\right) = \sigma(t). \tag{6}$$

We note two obvious properties of the function σ : if $t_2 < 2^s$, then

$$\sigma(t_1 2^s + t_2) = \sigma(t_1) + \sigma(t_2). \tag{7}$$

Also if T is an integer such that $|T| < 2^s \leqslant 2^r$, then

$$\sigma(2^r + T) \in \begin{cases} [1,\ s+1] & \text{if } T \geqslant 0, \\ [r-s+1, r], & \text{if } T < 0. \end{cases} \tag{8}$$

We introduce two functions

$$F^-(\alpha,\beta) = (2^\alpha + \beta)\cdot 2^{\alpha+1} + (2^\alpha - \beta)\ ; \tag{9}$$

$$F^+(\alpha,\beta) = (2^\alpha + \beta - 1)\cdot 2^{\alpha+1} + (2^\alpha - \beta - 1). \tag{10}$$

On the basis of (7) and (8) it is easy to show the following :

If T is any integer such that $|T| < 2^s \leqslant 2^r$, then

$$\sigma(F^*(r,T)) \in \begin{cases} [r-s+2*(r-s),\ r+s+2*(r-s)], & \text{if } T = 0 \\ [r-s+2,\ r+s+2], & \text{if } T \neq 0. \end{cases} \tag{11}$$

Here and in what follows $*$ denotes either one of the two signs + or − .

The first step in the construction of the inequality (4) consists of finding a natural number valued function A such that $\sigma(A(z))$ is relatively large, if z is the least possible value of w in a solution of equation (3), but small in the opposite case. (The words "great" and "small" will be made clear in what follows.)

First we choose a function S, taking only positive integer values, and such that the inequality

$$w_1 \leqslant z,\ \ w_2 \leqslant z, \ldots,\ w_\nu \leqslant z, \tag{12}$$

implies the inequality

$$|D(w_1,\ w_2, \ldots, w_\nu)| < 2^{S(z)}. \tag{13}$$

For example we can put

$$S(z) = \left[\sum_{i+\ldots+i_\nu \leqslant \delta} (D_{i_1,\ldots,i_\nu})^2\right](z+1)^\delta. \tag{14}$$

(If $D_{i_1,\ldots,i_\nu}$ is a polynomial in the parameters $a_1,\ldots,a_\mu$, then in $D_{i_1,\ldots,i_\nu}$ in (14), it is necessary to substitute polynomials in which all the minus signs, - have been replaced by + signs, so that the subtraction operation does not enter into and therefore appear in the end product polynomials P and Q of (4).)

Further we put

$$R(z) = (2z^\nu + 3)S(z), \tag{15}$$

$$T(z) = T(z) - S(z) + 2. \tag{16}$$

From (11) - (16) it follows that, for $D = D(w_1, w_2, \ldots, w_\nu)$, we have

$$\sigma(F^*(R(z)),D)) \in \begin{cases} [T(z)*(R(z) - S(z)),\ T(z) + 2S(z)*(R(z) - S(z))], & \text{if } D = 0 \\ [T(z),\ T(z) + 2S(z)], & \text{in the opposite case.} \end{cases} \tag{17}$$

Put

$$B^*(z,w) = \sum_{w_2=0}^{z-1} \ldots \sum_{w_\nu=0}^{z-1} F^*(R(z),\ D(w,w_2,\ldots,w_\nu) \times$$
$$\times\ 2^{(2R(z)+3)(w_2+zw_3+\ldots+z^{\nu-2}w_\nu)}. \tag{18}$$

According to (7), for $w \leqslant z$,

$$\sigma(B^*(z,w)) = \sum_{w_2=0}^{z-1} \ldots \sum_{w_\nu=0}^{z-1} \sigma(F^*(R(z),D(w,w_2,\ldots,w_\nu))). \tag{19}$$

Taking into account that, according to our assumption about equation (3) (each solution $w_1,\ldots,w_\nu$ is uniquely determined by and also smaller than, w_1), it follows from (17) and (19) that for $w \leqslant z$ we have

$$\sigma(B^*(z,w)) \in \begin{cases} [z^{\nu-1}T(z)*(R(z)-S(z)), \\ \ z^{\nu-1}(T(z)+2S(z))*(R(z)-S(z))] & \text{or} \\ [z^{\nu-1}T(z),\ z^{\nu-1}(T(z)+2S(z))] \end{cases} \tag{20}$$

according as $\exists w_2, w_3,\ldots,w_\nu\ D(w, w_2,\ldots,w_\nu) = 0$ or not. Now put

$$A(z) = B^{+}(z,z) + \sum_{w_1=0}^{z-1} B^{-}(z,w_1)\cdot 2^{(2R(z)+3)z^{\nu-1}(w_1+1)}. \tag{21}$$

According to (7)

$$\sigma(A(z)) = \sigma(B^{+}(z,z)) + \sum_{w_1=0}^{z-1} \sigma(B^{-}(z,w_1)). \tag{22}$$

From this and (20) it follows that

$$\sigma(A(z)) \begin{cases} \geqslant (z^{\nu}+z^{\nu-1})T(z) + (R(z) - S(z)), \\ \text{if } (\exists w_2,\dots,w_\nu)(D(z,w_2,\dots,w_\nu) = 0)) \ \& \\ \& \ (\neg\exists w_1,\dots,w_\nu)(w_1 < z \ \& \ D(w_1,\dots,w_\nu) = 0)), \\ \leqslant (z^{\nu}+z^{\nu-1})(T(z)+2S(z)) \ \text{ in the other case.} \end{cases} \tag{23}$$

Put

$$M(z) = (z^{\nu} + z^{\nu-1})T(z) + (R(z) - S(z)). \tag{24}$$

According to (15) and (24)

$$(z^{\nu} + z^{\nu-1})(T(z) + 2S(z)) < M(z). \tag{25}$$

From this and (23) we find that

$$\sigma(A(z)) \geqslant M(z) \tag{26}$$

holds if and only if z is the least value of w_1 in any solution to equation (3). (This will imply the uniqueness of z in condition (4). If this singlefold aspect of the representation is unimportant to the reader, then the construction can be simplified somewhat. It is enough to put $A(z) = B^{+}(z,z)$ and $M(z) = z^{\nu-1}T + R - S$).

Now we make use of the identity (6) to rewrite (26) in the form

$$2^{M(z)} \,\Big|\, \binom{2A(z)}{A(z)} \tag{27}$$

Thus (27) holds if and only if z is the least value of w_1 in any solution to equation (3).

In order to give a unary definition of the condition (27), we will use the following generating function for the symmetric binomial coefficient, $\binom{2n}{n}$:

$$\frac{1}{\sqrt{1-4x}} = \sum_{n=0}^{\infty} \binom{2n}{n} x^{n} \qquad (\text{for } |x| < 1/4) \tag{28}$$

This series (28) can be derived from the binomial series

$$(1-4x)^{-1/2} = \sum_{n=0}^{\infty} \binom{-1/2}{n} (-4x)^n, \qquad (29)$$

by using the identity

$$\binom{-1/2}{n}(-4)^n = \frac{(-1/2)(-3/2)\ \dots\ (-1/2-n+1)}{n!}(-2)^n\,2^n =$$

$$(30)$$

$$= \frac{1.3.5\ \dots\ (2n-1)}{n!} = \frac{1.2.3.4\ \dots\ (2n-1).(2n)}{n!\ n!} = \binom{2n}{n}$$

Now replace x by $1/u$ in (28) and multiply by u^n to obtain the following series (valid for $|u| > 4$)

$$\frac{u^n}{\sqrt{1-4/u}} = u^n+2u^{n-1}+6u^{n-2}+\dots+\binom{2n-2}{n-1}u+\binom{2n}{n}+\sum_{k=1}^{\infty}\binom{2n+2k}{n+k}u^{-k}. \qquad (31)$$

We can use a geometric series to estimate the size of the fractional part of (31).

$$\sum_{k=1}^{\infty}\begin{bmatrix}2n+2k\\ n+k\end{bmatrix}u^{-k} < \sum_{k=1}^{\infty} 2^{2n+2k}\,u^{-k} = \frac{4\cdot 4^n}{u-4}\ . \qquad (32)$$

The series (31) also gives a direct formula for the symmetric binomial coefficient in terms of the remainder function and the integer part function, [ ].

$$\binom{2n}{n} = \mathrm{rem}\left(\left[\frac{u^n}{\sqrt{1-4/u}}\right], u\right) \qquad (\text{for } 4\cdot 4^n+4<u). \qquad (33)$$

We will use this in §2. To return to condition (27), we see that if 2^M divides u, then (27) can be rewritten in the form

$$2^M \,\Big|\, \left[\frac{u^A}{\sqrt{1-4/u}}\right]. \qquad (34)$$

For any $\varepsilon \leqslant 1/2$, if $(4/\varepsilon)4^A + 4 < u$, then by (32) we can rewrite condition (34) in the form that, for some (unique) natural number y

$$\left|\frac{u^A}{\sqrt{1-4/u}} - 2^M y\right| \leqslant \varepsilon \qquad (35)$$

Taking $\varepsilon = 1/8$, we see that since $\varepsilon^2 < 1/8$, condition (35) implies

$$\left| \frac{\frac{u^{2A}}{1-4/u} - 2^{2M}y^2}{2^{M+1}y} \right| \leqslant 1/4. \tag{36}$$

Similarly, condition (36) implies condition (35) with $\varepsilon = 3/8$. Hence for $32 \cdot 4^A + 4 < u$, condition (34) is equivalent to condition (36). Condition (36) is in turn equivalent to

$$2\left| \frac{u^{2A}}{1-4/u} - 2^{2M}y^2 \right| \leqslant 2^M y. \tag{37}$$

which is in turn equivalent to the condition

$$4\left[\frac{u^{2A+1}}{u-4} - 2^{2M}y^2 \right]^2 \leqslant 2^{2M}y^2, \tag{38}$$

which is in turn equivalent to the condition

$$4\left[u^{2A+1} - 2^{2M}y^2(u-4)\right]^2 \leqslant 2^{2M}y^2(u-4)^2. \tag{39}$$

Finally (39) can be rewritten in the form

$$\begin{aligned} &4\left[u^{2A+1} + 4\cdot 2^{2M}y^2\right]^2 + 4\cdot 2^{4M}y^4u^2 + 4\cdot 2^{2M}y^2 \\ &\leqslant u\, 2^{2M}y^2 \left[32\cdot 2^{2M}y^2 + 8u^{2A+1} + 1\right]. \end{aligned} \tag{40}$$

This is the inequality (4). The functions $M(z)$ and $A(z)$ appear in (40). The function $M(z)$ is already a polynomial function of z. We wish to express the function $A(z)$ in a unary exponential form. Using (9), (10), (18) and (21) we can first express $A(z)$ in terms of iterated nested summations (with variables as upper summation limits). Then, by changing the order of these summations, it is possible to represent $A(z)$ in the form of a finite linear combination (with coefficients depending on z) of sums of the form

$$\sum_{w_1=0}^{z-1} \cdots \sum_{w_\nu=0}^{z-1} w_1^{i_1} \cdots w_\nu^{i_\nu} \cdot 2^{Q_1(z)w_1 + \ldots + Q_\nu(z)w_\nu} \tag{41}$$

and analogous sums without external summation on w_1.

The above sum can be rewritten in the form

$$\prod_{k=1}^{\nu} \left(\sum_{w_k=0}^{z-1} w_k^{i_k} \cdot 2^{Q_k(z)w_k} \right). \tag{42}$$

For sums of the form

$$\sum_{w=0}^{z-1} w^i q^w, \tag{43}$$

generalizations of geometric series, it is not difficult to derive formulas of the type

$$\sum_{w=0}^{z-1} w^i q^w = \frac{G_i(z,q)q^z + H_i(z,q)}{(q-1)^{i+1}} \tag{44}$$

where G_i and H_i are polynomials with integer coefficients.

Using these, it is possible to represent the function A in the form

$$2A(z) + 1 = \frac{E_1 - E_2}{E_3 - E_4} . \tag{45}$$

Here each E_i is a unary exponential function obtained from the variable z and natural numbers by application of the operations of addition, multiplication and raising to a power of 2. In addition, it is easy to see that the inequality $E_3 > E_4$ holds for all values of z.

If we put

$$u = 2^{(2E_1 + 4M)(E_3 - E_4)}, \tag{46}$$

then u satisfies the conditions

$$2^M | u \quad \text{and} \quad 32 \cdot 4^A + 4 < u. \tag{47}$$

Also u and u^{2A+1} may be written in the form

$$u = 2^{E_5 - E_6} \quad \text{and} \quad u^{2A+1} = 2^{E_7 - E_8} \tag{48}$$

where E_5, E_6, E_7 and E_8 are unary exponential functions of z.

Hence inequality (40) may be replaced by

$$4\left(2^{E_7 - E_8} + 4 \cdot 4^M y^2\right) + 4 \cdot 16^M y^4 4^{E_5 - E_6} + 4 \cdot 4^M y^2 \leqslant$$
$$\leqslant 2^{E_5 - E_6} 4^M y^2 \left[32 \cdot 4^M y^2 + 8 \cdot 2^{E_7 - E_8} + 1\right]. \tag{49}$$

Now multiply both sides of (49) by $4^{E_6 + E_8}$ to obtain

$$4\left[2^{E_7} + 4\cdot 4^{M}y^2 2^{E_8}\right]^2 \cdot 4^{E_6} + 4\cdot (16)^{M}y^2 4^{E_5+E_8} + 4\cdot 4^{M}y^2 4^{E_6+E_8}$$

$$\leqslant 2^{E_5+E_6+E_8}\, 4^{M}y^2 \left[32\cdot 4^{M}y^2 2^{E_8} + 8\cdot 2^{E_7} + 2^{E_8}\right]. \tag{50}$$

This completes the proof of Lemma 1. The promised unary exponential diophantine expressions $P(z,y)$ and $Q(z,y)$ are given in (50). Note that in the expressions $P(z,y)$ and $Q(z,y)$ the unknown y does not appear among the exponents, only z.

When the initial diophantine equation (3) has a solution in $w_1,\ldots,w_\nu$, then the inequality (50) has a unique solution $\langle z,y\rangle$. When (3) has no solution in $w_1,\ldots,w_\nu$, then the inequality (50) is also unsatisfiable.

Next we proceed to the proof of Theorem 2. This will follow from

LEMMA 2. For any polynomial $D(w_1,\ldots,w_\nu)$, we can construct unary exponential diophantine functions $R(z,y)$ and $S(z,y)$ such that $D = 0$ has a solution in $w_1,\ldots,w_\nu$ if and only if

$$\exists z \forall y\ (R(z,y) \leqslant S(z,y)). \tag{51}$$

PROOF OF LEMMA 2. If we redefine the functions $A(z)$ and $M(z)$ by putting

$$A'(z) = B^{-}(z,z) \tag{52}$$

and

$$M'(z) = z^{\nu-1}(T(z) + 2S(z)) - R(z) + S(z) + 1, \tag{53}$$

then we may replace condition (26) by its negation

$$\sigma(A'(z)) < M'(z). \tag{54}$$

Hence (3) has a solution if and only if there exists a z satisfying

$$2^{M'(z)}\binom{2A'(z)}{A'(z)}. \tag{55}$$

If as before $2^{M'}$ divides u, then using (31) we may see that (55) may be rewritten in the form

$$2^{M'}\left[\frac{u^{A'}}{\sqrt{1-4/u}}\right]. \tag{56}$$

If u is defined as in (46), then (47) holds. Hence (56) may be rewritten in the form that, for all y

$$\left| \frac{u^{A'}}{\sqrt{1-4/u}} - 2^{M'}y \right| > \frac{1}{2}. \tag{57}$$

Now inequality (57) is exactly inequality (35) with the sign reversed. Since inequality (35) was equivalent to inequality (36), inequality (57) is equivalent to (36) with sign reversed. Continuing in this way through (37), (38), (39), (40), (49), (50), we find that for $R(z,y)$ and $S(z,y)$ in (51) we may take $S = P'(z,y)$ and $R(z,y) = Q'(z,y) + 1$ where $P'(z,y)$ and $Q'(z,y)$ are essentially the same as $P(z,y)$ and $Q(z,y)$ in (50) but with different E_i's due to the changes in $M(z)$ and $A(z)$. (In this case $M(z)$ is not unary exponential diophantine. However it becomes so after replacing z by $z+1$). This completes the proof of Lemma 2. Note that the universal, $\forall y$ quantifier in (51) may be taken to be bounded. By (31), we may suppose that $y < (1+u)^{2E_1'} <$ some unary exponential diophantine function of z.

§2. APPLICATION TO THE THEORY OF KALMAR ELEMENTARY FUNCTIONS.

The class K of Kalmar elementary functions is a proper subclass of the class of primitive recursive functions. In fact $K = \mathcal{E}^3$, the class of functions at the third level of the Grzegorczyk [1953] hierarchy. One possible definition of the class K is to define it as the smallest class of functions containing $x+1$, $x+y$ and $x \dot{-} y$ and closed under composition, bounded summation and bounded products. In 1953 A. Grzegorczyk asked the following question : Is it possible to give a finite bases for the set of Kalmar elementary functions ? i.e., Can one find a finite set of initial functions from which all others may be obtained by composition ? Under composition we permit also closely related operations such as permutation and identification of variables.

Grzegorczyk's question was answered affirmatively by D. Rodding [1964]. Subsequently various other authors made further progress, e.g. Ch. Parsons [1968] and S. S. Marchenkov [1980]. Marchenkov considered also the closely related class of Kalmar elementary relations, the relations whose characteristic functions belong to the class K, and proved that

$$M = \{x+1,\ x \dot{-} y,\ [x/y],\ x^y\} \tag{58}$$

is a basis for the Kalmar elementary relations. We will show here essentially that x^y can be replaced by $[\sqrt{x}]$ and 2^x. We will also give some other possible bases for the Kalmar elementary relations. (It should be mentioned that Marchenkov defines $[x/0] = 0$.)

THEOREM 5. Every Kalmar elementary relation belongs to the closure of the following class

$$M_1 = \{x+y,\ x \dot{-} y,\ [x/y],\ [\sqrt{x}],\ 2^x\}. \tag{59}$$

Here Marchenkov's exponential function x^y, of two variables, has been replaced by two functions, each of one variable. Unfortunately it was also necessary to replace $x+1$ by $x+y$ (since the closure of M_1 would otherwise contain no function $f(x,y)$ with the property that $x \leq f(x,y)$ and $y \leq f(x,y)$ and hence it would not even contain the function $x+y$).

First we prove a lemma, then Theorem 5 and then we give further examples of bases for Kalmar elementary relations.

LEMMA 3. Each of the functions x^2, $x.y$, $\left[\frac{\sqrt{x}}{\sqrt{y}}\right]$, rem (y,x) and $\binom{2x}{x}$ belongs to the compositional closure of the class M_1.

PROOF.

$$x^2 = \left[\frac{2^{2^{2^x}}}{\left[\frac{\left[\frac{2^{2^{2^x}}}{x}\right]}{x}\right]}\right]. \tag{60}$$

$$y.x = \left[\frac{((y+x)^2 \dot{-} y^2) \dot{-} x^2}{2}\right]. \tag{61}$$

$$\left[\frac{\sqrt{x}}{\sqrt{y}}\right] = \left[\frac{\sqrt{x}\sqrt{y}}{y}\right] = \left[\frac{\sqrt{xy}}{y}\right] = \left[\frac{[\sqrt{xy}]}{y}\right]. \tag{62}$$

$$\text{rem}\ (x,y) = y \dot{-} (x[y/x]). \tag{63}$$

$$\binom{2x}{x} = \text{rem}\left(\left[\frac{2^{8x^2+4x}}{\sqrt{2^{8x} \dot{-} 4}}\right],\ 2^{8x}\right), \qquad (0<x). \tag{64}$$

Formula (64) is obtained from (33) by taking $u = 2^{8x}$.

This completes the proof of Lemma 3 (which is analogous to Marchenkov's Lemma 1). Now the idea of the proof of Theorem 5 is to replace Marchenkov's use of x^y by use of 2^x and his use of $\binom{x}{y}$ by $\binom{2x}{x}$. This could be done by following

Marchenkov's [1980] argument and using Lemma 3.5 of Matijasevič [1977] in place of Matijasevič Lemma 3.1 [1977]. However we shall give here a different proof which is much shorter.

PROOF OF THEOREM 5. Let $R(x_1,\dots,x_n)$ be an arbitrary Kalmar elementary relation. By Davis, Putnam & Robinson [1961] and Matijasevič [1971] (cf. also L. Adelman and K. Manders [1975]) it is possible to find a polynomial $D(w_1,\dots,w_\nu)$ and a Kalmar function $g(x_1,\dots,x_n)$ such that

$$R(x_1,\dots,x_n) \Leftrightarrow (\exists w_1,\dots,w_\nu)_{< g(x_1,\dots,x_n)} [D(x_1,\dots,x_n, w_1,\dots,w_\nu) = 0]. \tag{65}$$

We may suppose that g is an iterated exponential function of the type

$$g(x_1,\dots,x_n) = 2^{2^{\cdot^{\cdot^{2^{x_1+\dots+x_n}}}}} \tag{66}$$

so that $g(x_1,\dots,x_n)$ belongs to the closure of M_1.

Now let $T(z)$, $R(z)$, $S(z)$ and $B^+(z,w)$ be as in (14), (15), (16) and (18) but redefine the functions $M(z)$ and $A(z)$ as follows :

$$A(z) = \sum_{w_1=0}^{z-1} B^+(z,w_1).2^{(2R(z)+3)z^{\nu-1}(w_1+1)} \tag{67}$$

and

$$M(z) = z^\nu T(z) + R(z) - S(z). \tag{68}$$

The using (20) it is not difficult to see that for any value of z

$$(\exists w_1,\dots,w_\nu)_{<z} [D(x_1,\dots,x_n, w_1,\dots,w_\nu) = 0] \tag{69}$$

if and only if

$$M(z) \leqslant \sigma(A(z)). \tag{70}$$

Now since the equivalence of (69) and (70) holds for any value of z, we may put $z = g(x_1,\dots,x_n)$. By (65) we see that $R(x_1,\dots,x_n)$ is then equivalent to

$$M(g(x_1,\dots,x_n)) \leqslant \sigma(A(g(x_1,\dots,x_n))). \tag{71}$$

Now the function $M(z)$ is unary exponential diophantine. So the function $B = M(g(x_1,\dots,x_n))$ belongs to the closure of M_1. Also, as we have seen, there are unary exponential functions E_1, E_2, E_3 and E_4 such that $E_3 > E_4$ and $E_1 > E_2$

and $A = [E_1 \dot{-} E_2/E_3 \dot{-} E_4]$. Hence every Kalmar elementary relation, $R(x_1,\ldots,x_n)$ is represented in the form

$$B(x_1,\ldots,x_n) \leq \sigma(A(x_1,\ldots,x_n)) \tag{72}$$

where $A(x_1,\ldots,x_n)$ and $B(x_1,\ldots,x_n)$ belong to the closure of M_1. As usual (72) can be replaced by

$$2^B \;\Big|\; \binom{2A}{A}. \tag{73}$$

Then (73) can be replaced by

$$\text{rem}\left[\binom{2A}{A},\ 2^B\right] = 0. \tag{74}$$

Hence the characteristic function of the relation $R(x_1,\ldots,x_n)$ can be represented in the form

$$1 \dot{-} \text{rem}\left[\binom{2A}{A},\ 2^B\right] \tag{75}$$

where $A = A(x_1,\ldots,x_n)$ and $B = B(x_1,\ldots,x_n)$ are functions in the closure of M_1. By Lemma 3, (75) is a function in the closure of M_1. Hence Theorem 5 is proved.

Since the functions of M_1 belong to the closure of each of the following sets, (76) and (77), these may also be taken as a basis for the Kalmar elementary relations :

$$\{\ x \dot{-} y,\quad [x/y],\quad [\sqrt{x}],\quad 2^{x+y}\ \} \tag{76}$$

$$\{\ x \dot{-} y,\quad [\sqrt{x}/y],\quad x^2,\quad 2^{x+y}\ \}. \tag{77}$$

Whether or not the following set, (78) is a basis, is an open problem :

$$\{\ x \dot{-} y,\quad [\sqrt{x}/y],\ 2^{x+y}\}. \tag{78}$$

It may be possible to do better here by considering other functions such as $\sigma(x)$, $\binom{2x}{x}$ or $x!$. For example it is easy to prove that the following sets (79), (80) and (81) are bases :

$$\{\ x+y\ ,\quad x \dot{-} y\ ,\quad [x/y],\quad \sigma(x),\quad 2^x\ \} \tag{79}$$

$$\{\ x+y\ ,\quad x \dot{-} y\ ,\quad [x/y],\quad \binom{2x}{x},\quad 2^x\ \} \tag{80}$$

$$\{\ x+y\ ,\quad x \dot{-} y\ ,\quad [x/y],\quad x!,\quad 2^x\ \}. \tag{81}$$

REMARK. As regards functions (as opposed to relations), Theorem 5 implies also that finite valued Kalmar elementary functions belong to the compositional closure of each of our bases. This is because a function taking only finitely many values is a finite sum of characteristic functions.

PROOF OF THEOREM 3. Let $R(x_1,\dots,x_n)$ be an arbitrary Kalmar elementary relation. Using (71) and proceeding as in (26), (27), (34), (35), (36), (37), (38), (39), (40), (49), (50) one obtains for $R(x_1,\dots,x_n)$ the representation

$$(\exists y)\ [P(y) \leqslant Q(y)] \tag{82}$$

where as before the quantifier $\exists y$ can be bounded. The complement of a Kalmar relation is also Kalmar and the negation of a predicate of type (82) can be written in the form

$$(\forall y)\ [R(y) \leqslant S(y)] \tag{83}$$

by putting $S(y) = P(y)$ and $R(y) = Q(y) + 1$. This proves Theorem 3. Observe that in the unary expressions $P(y)$, $Q(y)$, $R(y)$ and $S(y)$ only the parameters, $x_1,\dots,x_n$ appear in the exponents, not the variable y.

REFERENCES

[1975] L. Adelman and K. Manders, Computational complexity of decision procedures for polynomials, Proc. 16 th Annual Symposium on Foundations of Comp. Science, 1975 New York, N.Y., pp. 169-177.

[1961] M. Davis, H. Putnam and J. Robinson, The decision problem for exponential diophantine equations, Ann. of Math. 74 (1961), pp. 425-436 = Matematika 8.5 (1964), pp. 69-79.

[1976] M. Davis, Ju. Matijasevič and Julia Robinson, Hilbert's tenth problem. Diophantine equations : Positive aspects of a negative solution, Proc. Symp. on the Hilbert Problems, Dekalb, Illinois, May 1974, Amer. Math. Soc. 28 (1976), pp. 323-378.

[1953] A. Grzegorczyk, Some classes of recursive functions, Rozprawy matematyczne no. 4, Instityt Matematyczny Polskiej Akademie Nauk, Warsaw, 1953.

[1975] N.K. Kossovski, A hierarchy of diophantine representations for primitive recursive functions (Russian), Computation techniques and questions of Cybernetics, Leningrad State University, vol. 12, (1975), pp. 99-107.

[1979] J.P. Jones, Diophantine representation of Mersenne and Fermat primes, Acta Arithmetica 35 (1979), pp. 209-221.

[1981] J.P. Jones and Ju. V. Matijasevič, A new representation for the symmetric binomial coefficient and its applications, Annales des Sciences Math. du Quebec (to appear).

[1852] E. E. Kummer, Über die Erganzungssätze zu den algemeinen Reciprocitätsgesetzen. - J. reine und angew. Math., 1852, 44, S. pp. 93-146.

[1980] S.S. Marchenkov, On a new basis whose closure under composition gives the Kalmar elementary functions (Russian), Math.Zam.27(1980), no.3, pp. 321-331. MR 81e 03039.

[1971] Ju. V. Matijasevič, Diophantine representation of enumerable predicates (Russian), Izvestija Akademii Nauk SSSR, Serija Matematika, Vol. 35 (1971), pp. 3-30. English translation : Mathematics of the USSR - Izvestija, Vol. 5 (1971), pp. 1-28.

[1974] Ju. V. Matijasevič, The existence of noneffectizable estimates in the theory of exponential diophantine equations, Zapiski Naučnyh Seminarov Leningradskogo Otdelenija Matematičeskogo Instituta im. V.A. Steklova, Akad. Nauk. SSSR, Vol. 40 (1974), pp. 77-93. English translation : Jour. Soviet Math. (Plenum Publishers), Vol. 8 (1977).

[1976] Ju. V. Matijasevič, A new proof of the theorem on exponential diophantine representation of enumerable sets (Russian), Zapiski Naučnyh Seminarov Leningradskogo Otdelenija Matematičeskogo Instituta im. V.A. Steklova, Akad. Nauk. SSSR 60 (1976), pp. 75-89. English translation Jour. Soviet Math. 14 (1980) 1475-1486, MR 81 f 03055.

[1977] Ju. V. Matijasevič, A class of primality criteria formulated in terms of the divisibility of binomial coefficients (Russian), Zapisky Naučnyh Seminarov Leningradskogo Otdelenija Matematičezkogo Instituta im. V.A. Steklova, Akad. Nauk. SSSR Vol. 67 (1977), pp. 167-183. English translation Jour. Soviet Math. 16 (1981), 874-885.

[1979] Ju. V. Matijasevič, Algorithmic unsolvability of exponential diophantine equations in three unknowns (Russian), Studies in the Theory of Algorithms and Mathematical Logic, Moscow, Nauka, 1979, pp. 69-78. English translation available from University of Calgary, Dept. of Mathematics and Statistics.

[1968] Ch. Parsons, Hierarchies of primitive recursive functions, Zeitschrift fur Math. Logik und Grundlagen der Mathematik, Vol. 14, No. 4 (1968), pp. 357-376.

[1964] D. Rödding, Über die Eliminierbarkeit von Definitionschemata in der Theorie der Rekursiven Funktionen, Zeit. fur Math. Logik und Grundlagen der Mathematik, Vol. 10, No. 4 (1964), pp. 315-330.

[1974] D. Singmaster, Notes on binomial coefficients I, II, III. - J. London Math. Soc. (1974), 8, No 3, pp. 545-548, 549-554, 555-560.

PROCEEDINGS OF THE HERBRAND SYMPOSIUM
LOGIC COLLOQUIUM '81, J. Stern (editor)

EFFECTIVE RAMSEY THEOREMS IN THE PROJECTIVE HIERARCHY

Alexander S. Kechris[1]

Department of Mathematics
California Institute of Technology
Pasadena, California 91125
U.S.A.

Let $\omega = \{0,1,2,...\}$ be the set of natural numbers, and $[\omega]^{\omega}$ the collection of all infinite subsets of ω. More generally, for each infinite set $H \subseteq \omega$, let $[H]^{\omega}$ denote the set of all infinite subsets of H. Given any set $A \subseteq [\omega]^{\omega}$ recall that A is said to be *Ramsey* if there is $H \in [\omega]^{\omega}$ such that $[H]^{\omega} \subseteq A$ or $[H]^{\omega} \subseteq \sim A = [\omega]^{\omega} - A$. Such an H is called *homogeneous for* (the partition of $[\omega]^{\omega}$ determined by) A. If $[H]^{\omega} \subseteq A$ we say that H is *homogeneous landing in* A, while if $[H]^{\omega} \subseteq \sim A$ we say that H *is homogeneous avoiding A*.

We are interested here in questions of the following type:

Given a level of complexity of A, what is the level of complexity of a homogeneous set H for A?

A prototype result along these lines is the following theorem of Solovay:

Theorem (Solovay [7]). If $A \in \Sigma^0_1$, then either there is a homogeneous set H avoiding A (which can be chosen to be recursive in the Kleene $\mathcal{O}$), or else there is a Δ^1_1 homogeneous set H landing in A. Every Δ^1_1 set A has a homogeneous set $H \in L_{\sigma}$, where σ is the first recursively inaccessible ordinal.

On the other hand (see also [7]) there is a Σ^0_1 set A which has no Δ^1_1 homogeneous set H.

Our main result in this paper is the following:

Theorem 1. Assume $n > 0$ and Projective Determinacy holds. If $A \in \Pi^1_{2n+1}$, either there is a homogeneous set H avoiding A (which can be chosen to be Δ^1_{2n+1} in the first non-trivial Π^1_{2n+1}-singleton y^0_{2n+1}), or else there is a Δ^1_{2n+1} homogeneous set H landing in A. Moreover, every Δ^1_{2n+1} set A has a Δ^1_{2n+1} homogeneous set.

Thus we see a sharp contrast between the $n = 0$ and the $n > 0$ case. This points out to a new discrepancy between the structure theory of the first and the higher odd levels of the analytical hierarchy.

The proof of this theorem makes substantial use of Q-theory, although its statement clearly has a priori nothing to do with it. An exposition of Q-theory can be found in [3], and the reader is assumed to be more or less familiar with it, although we will summarize below the relevart parts that will be needed here. Our application of Q-theory in this paper is quite elementary, so it is hoped that it will bring out clearly some of its essential ideas and methods, that can be basic ingredients in more elaborate uses and applications of this theory.

This paper is organized as follows: In §1 we prove the main theorem. In §2 we discuss some related results on Δ^1_{2n+1} - and Q_{2n+1} - encodability, and in §3 we mention some open problems.

In conclusion, we would like to thank Dr. Ilias Kastanas, for many stimulating discussions on the subject matter of this paper.

§1. Proof of the main theorem. We will first give the proof for the case n = 1 in 1.1. without worrying about the amount of determinacy used, in order to make the key ideas involved more transparent. Then we will discuss the generalization to all $n \geq 1$ in 1.2., and finally in 1.3. the technical modifications needed to bring down the level of determinacy used.

1.1. In this subsection we shall assume full AD (beyond the basic theory ZF + DC) and prove Theorem 1 for n = 1. (Actually $AD^{L[\mathbb{R}]}$ is enough, since the theorem is absolute between the real world and $L[\mathbb{R}]$, so we can work entirely within $L[\mathbb{R}]$ for this proof).

The proof will use Mathias forcing (see [5]) over an appropriate inner model of ZFC. This model is an "analog" of L for the third level of the analytical hierarchy and is defined as follows:

For each real $\alpha \in \omega^\omega$ let $L[\alpha]$ be the relativized to α constructible universe, and consider $HOD^{L[\alpha]}$, the inner model of all HOD within $L[\alpha]$ sets. For each constructibility degree $d = [\alpha]_c$ let

$$L[d] = L[\alpha],$$

and consider the ultrapower

$$M_3 = \Pi_d \; HOD^{L[d]}/_\mu,$$

where μ is the Martin measure on constructibility degrees, i.e. the one generated by cones. We will need the following facts about M_3 (proved in [3] and due to Kechris, Martin, Solovay and Woodin -- see [3] for relevant references).

(1) The set of reals in M_3 is Q_3. (By reals we always mean members of ω^ω, i.e. functions from ω into ω.) A quick definition of Q_3 is the following: A real α *belongs to* Q_3 iff α is Δ^1_3 in a countable ordinal, i.e. there is $\xi < \omega_1$ such that $\alpha \in \Delta^1_3(w)$ for all (real) codes w of ξ.

More than that, for each real α, if $M_3[\alpha]$ is the smallest inner model of ZFC containing M_3 and α then

$$\omega^\omega \cap M_3[\alpha] = Q_3(\alpha),$$

where $Q_3(\alpha)$ is the relativized to α Q_3.

(2) M_3 satisfies a "dual Shoenfield absoluteness" theorem at the third level of the analytical hierarchy, i.e. for each Σ^1_3 formula $\varphi(\alpha)$ there is a Π^1_3 formula $\varphi^*(\alpha)$ (effectively constructed from φ) such that

$$\varphi(\alpha) \Leftrightarrow M[\alpha] \models \varphi^*(\alpha),$$

and similarly interchanging the roles of Σ^1_3 and Π^1_3.

It can be shown that M_3 is not Σ^1_3-correct (i.e. Σ^1_3 formulas are not absolute between M_3 and the universe).

This strange "dual absoluteness" will turn out to be crucial to the argument below.

(3) $M_3 \models \exists$ measurable cardinals.

(4) Q_3 has the following reflection property: If $P(\alpha)$ is Π^1_3 then

$$\exists\alpha \in Q_3 P(\alpha) \Leftrightarrow \exists\alpha \in \Delta^1_3 P(\alpha).$$

(5) $(\omega^\omega)^{M_3}$ and $(\text{power}(\omega^\omega))^{M_3}$ are countable (in the universe).

Finally we will need the following definability estimate, strengthening (5).

(6) Let y^0_3 be the first non-trivial Π^1_3-singleton (i.e. a Π^1_3-singleton which is not Δ^1_3 and it is Δ^1_3 in any other such Π^1_3-singleton -- this is a good analog of the Kleene $\mathcal{O}$ at the third level). Then there is a real $x^0 \in \Delta^1_3(y^0_3)$ such that x^0 enumerates Q_3, i.e. $Q_3 = \{(x^0)_n : n \in \omega\}$, $(x^0)_n \neq (x^0)_m$ if $m \neq n$ and if

$$\alpha \in P \Leftrightarrow \alpha \in 2^\omega \wedge \{(x^0)_n : \alpha(n) = 0\} \in M_3$$

then there is $x^1 \in \Delta^1_3(y^0_3)$ enumerating P (P is countable by (5)). This just says that there is an enumeration of $(\text{power}(\omega^\omega))^{M_3}$ which is "Δ^1_3 in y^0_3".

We proceed now to prove our theorem.

Let $\mathbb{P}$ be Mathias' notion of forcing in M_3. Thus $\mathbb{P}$ consists of all pairs $(s,S) \in M_3$ where s is a finite subset of ω, S an infinite subset of ω and $\max(s) < \min(S)$. Order these conditions by

$$(s,S) \leq (t,T) \Leftrightarrow t \subseteq s \wedge S \subseteq T \wedge s - t \subseteq T.$$

Let $A \subseteq [\omega]^\omega$ be a given Π^1_3 set. Let φ be a Π^1_3 formula defining A and let φ^* be the Σ^1_3 formula associated to φ by (2), so that for any $X \in [\omega]^\omega$:

$$X \in A \Leftrightarrow \varphi(X) \Leftrightarrow M_3[X] \models \varphi^*(X).$$

By a basic fact on Mathias' forcing (see [5]) there is a Mathias condition $\langle\emptyset,S\rangle \in M_3$ deciding the sentence of the forcing language $\varphi^*(\dot{H})$, where $\dot{H}$ is a name for the Mathias generic real, i.e.

$$\langle\emptyset,S\rangle \Vdash_{\mathbb{P}} \varphi^*(\dot{H}) \quad \text{or} \quad \langle\emptyset,S\rangle \Vdash_{\mathbb{P}} \neg\varphi^*(\dot{H}).$$

Consider now separately the two cases.

Case I. $\langle\emptyset,S\rangle\Vdash_{\mathbb{P}}\neg\varphi^*(\dot{H})$.

Let then H be a Mathias generic real over M_3 satisfying the condition $\langle\emptyset,S\rangle$, i.e. $H\subseteq S$. Such a real exists and can be found in a $\Delta^1_3(y^0_3)$ fashion by (6), which implies that there is a $\Delta^1_3(y^0_3)$ enumeration of all dense for $\mathbb{P}$ sets belonging to M_3. Then by another basic property of Mathias' forcing every infinite $X\subseteq H$ is also Mathias generic (see [5]), and since it satisfies $\langle\emptyset,S\rangle$ as well, we must have

$$M_3[X]\models\neg\varphi^*(X),$$

thus $X\notin A$. So H is a $\Delta^1_3(y^0_3)$ homogeneous set avoiding A, so our first alternative holds.

Case II. $\langle\emptyset,S\rangle\Vdash_{\mathbb{P}}\varphi^*(\dot{H})$.

Let again H_0 be Mathias generic over M_3. Exactly as before H_0 is homogeneous landing in A, i.e.

$$\forall X\subseteq H_0\varphi(X).$$

Since φ is Π^1_3 there is a Π^1_3 formula ψ such that

$$\psi(H)\Leftrightarrow\forall X\subseteq H\varphi(X).$$

Let ψ^* be the Σ^1_3 formula associated with ψ by (2), so that

$$\psi(H)\Leftrightarrow M_3[H]\models\psi^*(H).$$

In particular, $M_3[H_0]\models\psi^*(H_0)$, so $M_3[H_0]\models\exists H\psi^*(H)$. Now $\sigma\Leftrightarrow\exists H\psi^*(H)$ is a Σ^1_3 sentence. But $M_3[H_0]$ is a mild generic extension of M_3, which is a model containing a measurable cardinal by (3), so by Martin-Solovay [4], M_3 is Σ^1_3-correct in $M_3[H_0]$, thus $M_3\models\sigma$, i.e. $M_3\models\exists H\psi^*(H)$. So pick $H_1\in M_3$ with $M_3\models\psi^*(H_1)$; thus since $M_3[H_1]=M_3$, we have $M_3[H_1]\models\psi^*(H_1)$, so $\psi(H_1)$ holds. Thus $\exists H\in Q_3\psi(H)$ and so by the reflection property (4), $\exists H\in\Delta^1_3\psi(H)$. Thus finally let $\overline{H}\in\Delta^1_3$ be such that $\psi(\overline{H})$, i.e. $\forall X\subseteq\overline{H}(X\in A)$. So we see that there is a Δ^1_3 homogeneous set landing in A, which means that our second alternative holds, and our proof is complete. (It is rather obvious from the above argument that in case $A\in\Delta^1_3$, A has a Δ^1_3 homogeneous set).

1.2. We now indicate the modifications needed to carry out this proof for each odd level $2n+1\geq 3$, still under the assumption of full AD. Assume $n\geq 1$ below.

Let T_{2n-1} be the free associated to a Π^1_{2n-1}-scale on a complete Π^1_{2n-1} set (see [6]). By replacing $L[\alpha]$ by $L[T_{2n-1},\alpha]$ ($L[T_1,\alpha]=L[\alpha]$) and HOD by $\mathrm{HOD}_{T_{2n-1}}$ (the HOD from T_{2n-1} sets) in the definition of the model M_3 in 1.1, we can construct a model M_{2n+1} satisfying all the conditions (1)-(6) with 3 replaced by $2n+1$ everywhere (see [3]).

To be precise, for each real α let $d=[\alpha]_{\mathcal{C}_{2n}}$ be its $\mathcal{C}_{2n}$-degree defined by

$d = \{\beta : \alpha \in \mathbb{C}_{2n}(\beta) \wedge \beta \in \mathbb{C}_{2n}(\alpha)\}$, where $\mathbb{C}_{2n}(\alpha)$ is the largest countable $\Sigma^1_{2n}(\alpha)$ set. As (see [2])

$$\alpha \in \mathbb{C}_{2n}(\beta) \Leftrightarrow \alpha \in L[T_{2n-1},\beta]$$

this notion of degree coincides with constructibility degree modulo T_{2n-1}. Let μ_{2n} be Martin's measure on the $\mathbb{C}_{2n}$-degrees (i.e. again the one generated by cones) and let

$$M_{2n+1} = \Pi_d \, \mathrm{HOD}_{T_{2n-1}}^{L[T_{2n-1},d]} / \mu_{2n}$$

be the associated ultrapower.

However there is one more thing to worry about. Condition (3) was enough to guarantee the Σ^1_3-correctness of M_3 within any $M_3[H]$, where H is Mathias generic over M_3, but it is not enough to guarantee the Σ^1_{2n+1}-absoluteness of M_{2n+1} within $M_{2n+1}[H]$, when $n \geq 2$. So at this point in the proof we need a different argument to handle the general case.

First let us recall a further property of M_{2n+1} and its extensions (see [3]) which is automatic in the case n = 1, and this explains why it was not mentioned explicitly before.

(7) For every real α, $M_{2n+1}[\alpha]$ is Σ^1_{2n}-correct, i.e. Σ^1_{2n} formulas are absolute between $M_{2n+1}[\alpha]$ and the real world.

Now let us go back to the argument in 1.1, Case II. (Case I can be handled in the general case exactly as before). It is clear that given a Mathias condition $\langle\emptyset,S\rangle \in M_{2n+1}$ it is enough to find a model $N = M_{2n+1}[\alpha]$, such that there is in N a Mathias generic over M_{2n+1} real H, and M_{2n+1} is Σ^1_{2n+1}-correct in N. (Fact (7) is used here). As $(\mathrm{power}(\omega^\omega))^{M_{2n+1}}$ is countable in the universe, $\kappa = (2^{2^{\aleph_0}})^{M_{2n+1}} < \omega_1$, so if $g : \omega \twoheadrightarrow \kappa$ is a generic over M_{2n+1} collapse, and we take $N = M_{2n+1}[g] = M_{2n+1}[\alpha]$ where α is a real coding the prewellordering on ω induced by g, then there is such an H in N, so it is enough to show M_{2n+1} is Σ^1_{2n+1}-correct in N. Since N is a homogeneous generic extension of M_{2n+1} this will follow if we can show that *in N* every Π^1_{2n} set in $\omega^\omega \times \omega^\omega$ has a definable uniformization.

For each integer N > 0 let ZFC_N be the first N axioms of ZFC. If $T_{N,n} = ZFC_N + \utilde{\Delta}^1_{2n}$-DETERMINACY, let $\mathrm{Prov}_{T_{N,n}}\utilde{\Delta}^1_{2n}$-DETERMINACY be the assertion that all provable in $T_{N,n}$ $\utilde{\Delta}^1_{2n}$ games are determined. Then we have the following further fact (see [3])

(8) For each real α, $M_{2n+1}[\alpha] \models \neg\utilde{\Delta}^1_{2n}$-DETERMINACY, but for each N,

$$M_{2n+1}[\alpha] \models \mathrm{Prov}_{T_{N,n}}\utilde{\Delta}^1_{2n}\text{-DETERMINACY}.$$

Now from the proof of the Transfer Theorem for Scales under the game quantifier (see [6], 6E.15) it follows that assuming $\mathrm{Prov}_{T_{N,n}} \underset{\sim}{\Delta}^1_{2n}$-DETERMINACY, we can show that every Π^1_{2n} set carries a definable scale, so can be definably uniformized (here N is a large enough integer). Thus our proof is complete.

1.3. Finally, we discuss the technical changes needed to make this proof work using PD only.

The only place where full AD was really used in the preceding proof was in constructing the ultrapower $M_{2n+1} = \Pi_d \, \mathrm{HOD}_{T_{2n-1}}^{L[T_{2n-1},d]}/\mu_{2n}$. But this can be easily avoided (at the expense of some loss of clarity, which explains our presentation in terms of the model M_{2n+1} first) by working with $\mathrm{HOD}_{T_{2n-1}}^{L[T_{2n-1},\alpha]}$, for large enough $\mathcal{C}_{2n}$-degree of α.

Instead of results (1)-(8) about M_{2n+1} that we quoted in 1.1. and 1.2. we can now use the following facts, all proved in [3] and *using only PD*, from which by exactly the same proof as before we can establish our main theorem.

(1′) There is a real z_0 such that

$$z_0 \in L[T_{2n-1},\beta] \Rightarrow \omega^\omega \cap \mathrm{HOD}_{T_{2n-1}}^{L[T_{2n-1},\beta]} = Q_{2n+1}.$$

Moreover for each real α there is a real z (depending on α), such that

$$\alpha, z \in L[T_{2n-1},\beta] \Rightarrow \omega^\omega \cap \mathrm{HOD}_{T_{2n-1}}[\alpha]^{L[T_{2n-1},\beta]} = Q_{2n+1}(\alpha).$$

(2′) Given a Π^1_{2n+1} formula $\varphi(\alpha)$ there is a Σ^1_{2n+1} formula $\varphi^*(\alpha)$ (effectively constructed from φ) such that for any real α and any model N of ZFC with $N \cap \omega^\omega = Q_{2n+1}(\alpha)$ we have

$$\varphi(\alpha) \Leftrightarrow N \models \varphi^*(\alpha),$$

and similarly interchanging the roles of Π^1_{2n+1} and Σ^1_{2n+1}.

(3′) If $P(\alpha)$ is Π^1_{2n+1} then

$$\exists\alpha \in Q_{2n+1} P(\alpha) \Leftrightarrow \exists\alpha \in \Delta^1_{2n+1} P(\alpha).$$

(4′) There is a real z_1 such that

$$z_1 \in L[T_{2n-1},\beta] \Rightarrow (\mathrm{power}\ (\omega^\omega))^{\mathrm{HOD}_{T_{2n-1}}^{L[T_{2n-1},\beta]}} = R,$$

where R is a countable set independent of β.

(5′) Let y^0_{2n+1} be the first non-trivial Π^1_{2n+1}-singleton. Then there is a real $x^0 \in \Delta^1_{2n+1}(y^0_{2n+1})$ such that x^0 enumerates Q_{2n+1}, and if

$$P(\alpha) \Leftrightarrow \alpha \in 2^\omega \wedge \{(x^0)_n : \alpha(n) = 0\} \in R$$

(where R is as in (4′)), then there is $x^1 \in \Delta^1_{2n+1}(y^0_{2n+1})$ enumerating P.

(6′) If N is a model of ZFC with $N \cap \omega^\omega = Q_{2n+1}(\alpha)$, then N is Σ^1_{2n}-correct and

$$N \models \mathrm{Prov}_{T_{N,n}}\ \underset{\sim}{\Delta}^1_{2n}\text{-DETERMINACY},$$

where as in (8) of 1.2., $T_{N,n} = ZFC_N + \underset{\sim}{\Delta}^1_{2n}$-DETERMINACY.

§2. *Δ^1_{2n+1} - and Q_{2n+1} - encodability.* Consider any notion of reducibility $\leq_r$ among reals, like for instance $\leq_T$ (Turing reducibility), $\leq_n$ (Δ^1_n-reducibility; $\alpha \leq_n \beta \Leftrightarrow \alpha \in \Delta^1_n(\beta)$), or $\leq^Q_{2n+1}$ (Q_{2n+1}-reducibility;

$$\alpha \leq^Q_{2n+1} \beta \Leftrightarrow \alpha \in Q_{2n+1}(\beta)).$$

A real α is called $\leq_r$-*encodable* iff $\forall X \in [\omega]^\omega\ \exists Y \subseteq X[\alpha \leq_r Y]$. When $\leq_r\ =\ \leq_T$ we talk about *recursively encodable* reals, when $\leq_r\ =\ \leq_n$ about Δ^1_n-*encodable* reals and when $\leq_r\ =\ \leq^Q_{2n+1}$ about Q_{2n+1}-*encodable* reals.

The following result of Solovay computes the recursively and Δ^1_1-encodable reals.

Theorem (Solovay [7]). The recursively encodable reals are exactly the Δ^1_1 reals. The Δ^1_1-encodable reals are exactly the reals in $L_{\sigma^1_1}$, where σ^1_1 is the first Σ^1_1-reflecting ordinal.

Since for n = 0, $\leq^Q_1\ =\ \leq_1$ the Δ^1_1-encodable and Q_1-encodable reals are the same. We can use the techniques of Q-theory to compute now the Q_{2n+1}-encodable reals for n > 0. Again the situation is different from the case n = 0.

Theorem 2. Let n > 0 and assume Projective Determinacy. Then the Q_{2n+1}-encodable reals are exactly the Q_{2n+1}-reals.

Proof. We will need first the following lemma.

Lemma 1. There is a model M of ZFC such that

i) M = L[A], A a bounded subset of ω_1,

ii) If H is Mathias generic over M, then $M[H] \models \underset{\sim}{\Delta}^1_{2n}$-DETERMINACY, and M[H] is Σ^1_{2n}-correct.

(Since $(\mathrm{power}(\omega^\omega))^M$ is countable such reals exist)

Proof (Following ideas of Solovay). Let T_{2n+1} be the tree associated with a Π^1_{2n+1}-scale on a complete Π^1_{2n+1} set.

Every extension of $L[T_{2n+1}]$ is Σ^1_{2n+2} absolute, so it also satisfies $\underset{\sim}{\Delta}^1_{2n}$-DETERMINACY. Let κ be the cardinality of ω^ω in $L[T_{2n+1}]$ and λ the cardinality of power(ω^ω) in $L[T_{2n+1}]$, so that $\kappa < \lambda < \omega_1$. Let f, g $\in L[T_{2n+1}]$ be such that

$$L[T_{2n+1}] \models f: \kappa \xrightarrow[\text{onto}]{1-1} \omega^\omega$$

$$L[T_{2n+1}] \models g: \lambda \xrightarrow[\text{onto}]{1-1} \mathrm{power}(\omega^\omega).$$

Then there is a bounded subset A of ω_1, $A \in L[T_{2n+1}]$, encoding f, g, thus L[A] contains the same reals and sets of reals as $L[T_{2n+1}]$. Hence L[A], $L[T_{2n+1}]$ have the same Mathias conditions, and for every Mathias generic real H over L[A], H is Mathias generic over $L[T_{2n+1}]$ and $\omega^\omega \cap L[A,H] = \omega^\omega \cap L[T_{2n+1}, H]$, so L[A,H] is also Σ^1_{2n+2}-correct and $L[A,H] \models \underset{\sim}{\Delta}^1_{2n}$-DETERMINACY. ⊣

Call a real α M-*encodable*, where M is a model of ZFC with $(\text{power}(\omega^\omega))^M$ countable, if

$$\forall H[H \text{ Mathias generic over } M \Rightarrow \exists X \subseteq H(\alpha \in M[X])].$$

The following result is a special instance of a more general theorem of Solovay about M-encodability (its proof is implicit in [7] and explicit in [1]).

Theorem (Solovay). Let M = L[A], where A is a bounded subset of ω_1. Then every M-encodable real is in M.

Finally we shall need the following fact from [3].

Lemma 2. If M is a Σ^1_{2n}-correct model of ZFC + $\underset{\sim}{\Delta}^1_{2n}$-DETERMINACY, then M is downward closed under $\leq^Q_{2n+1}$, i.e. $\beta \in M \wedge \alpha \in Q_{2n+1}(\beta) \Rightarrow \alpha \in M$.

Fix now some simple way of encoding models L[A], where A is a bounded subset of ω_1, by reals. If this is done in any reasonable fashion, the set

S = {α: α codes a model L[A], where A is a bounded subset of ω_1, and every Mathias generic extension of L[A] is Σ^1_{2n}-correct and satisfies $\underset{\sim}{\Delta}^1_{2n}$-DETERMINACY}

will be Σ^1_{2n+1}. By Lemma 1, $S \neq \phi$.

As usual, the 2n+1-*hull* of S is defined by

$$\text{Hull}_{2n+1}(S) = \{\alpha : \forall\beta \in S(\alpha \in \Delta^1_{2n+1}(\beta))\}$$

(see [3]) and a basic fact of Q-theory is that $\text{Hull}_{2n+1}(S) \subseteq Q_{2n+1}$, since Q_{2n+1} is the largest 2n + 1-hull of a nonempty Σ^1_{2n+1} set. So it is enough to show that if α is Q_{2n+1}-encodable, then $\alpha \in \text{Hull}_{2n+1}(S)$.

Let $\beta \in S$ and let L[A] be the model it encodes. We have to show that $\alpha \in \Delta^1_{2n+1}(\beta)$. It is clearly enough to show $\alpha \in L[A]$. By the theorem of Solovay previously mentioned, it is enough to show that α is M-encodable, where M = L[A]. So fix H Mathias generic over M. We have to find $X \subseteq H$ such that $\alpha \in M[X]$. But by the Q_{2n+1}-encodability of α, there is $X \subseteq H$ with $\alpha \in Q_{2n+1}(X)$. Now X is also Mathias generic over M, thus by the properties of M = L[A], $M[X] \models \underset{\sim}{\Delta}^1_{2n}$-DETERMINACY, and M[X] is Σ^1_{2n}-correct, so by Lemma 2 $\alpha \in M[X]$, and the proof is complete. ⊣

As every Δ^1_{2n+1}-encodable real is Q_{2n+1}-encodable, it follows that every Δ^1_{2n+1}-encodable real is in Q_{2n+1}. However we believe that actually the following much stronger statement is true.

Conjecture. Assume Projective Determinacy. Then for $n > 0$ the Δ^1_{2n+1}-encodable reals are exactly the Δ^1_{2n+1} reals.

§3. Some open problems. The following questions related to the topics discussed in this paper are, to the best of our knowledge, still open.

3.1. There is an ordinal $\tau < \omega_1$ such that: If $A \subseteq [\omega]^\omega$ is Π^1_1 then either there is a homogeneous set avoiding A or else there is a homogeneous set in L_τ landing in A. What is the least such τ?

3.2. What are best possible estimates for homogeneous sets for Σ^1_{2n} and Δ^1_{2n} subsets of $[\omega]^\omega$, assuming PD? (Added in Proof: This problem is solved in H. Woodin's paper in this volume).

3.3. What is the set of Δ^1_n-encodable reals for $n \geq 2$? (cf. the conjecture in §2).

REFERENCES

[1] Ennis, G., Mathias forcing and encodability, preprint, Caltech (1978).

[2] Harrington, L. A. and Kechris, A. S., On the determinacy of games on ordinals, Ann. Math. Logic, 20(1981), 109-154.

[3] Kechris, A. S., Martin, D. A. and Solovay, R. M., Introduction to Q-theory, Cabal Seminar (1979-81), to appear.

[4] Martin, D. A. and Solovay, R. M., A basis theorem for Σ^1_3 sets of reals, Ann. of Math. 89 (1969), 138-160.

[5] Mathias, A.R.D., Happy families, Ann. Math. Logic, 12 (1977), 59-111.

[6] Moschovakis, Y. N., Descriptive Set Theory, North Holland (1980).

[7] Solovay, R. M., Hyperarithmetically encodable sets, Trans. Amer. Math. Soc. 239 (1978), 99-122.

[1]The author is an A. P. Sloan Foundation Fellow. Research partially supported by NSF Grant MCS79-20465.

PROCEEDINGS OF THE HERBRAND SYMPOSIUM
LOGIC COLLOQUIUM '81, J. Stern (editor)

FINITE HOMOGENEOUS SIMPLE DIGRAPHS

A. H. Lachlan

Department of Mathematics,
Simon Fraser University
Burnaby, British Columbia,
Canada V5A 1S6

Simple directed graphs are regarded as structures $G = \langle |G|, E_G \rangle$ where E_G is an irreflexive binary relation on a nonempty set of vertices $|G|$. A classification is given of all finite simple digraphs which are homogeneous in the sense of Fraissé. The classification extends that given by Gardiner of homogeneous graphs.

By a <u>digraph</u> $G = \langle |G|, E_G \rangle$ we mean an ordered pair whose first member is a nonempty set and whose second member is an irreflexive binary relation on that set. Thus, in the language of graph theory, we confine consideration to digraphs without loops and without multiple edges. We call H a <u>subdigraph</u> of G if $|H| \subset |G|$ and $E_H = E_G \cap |H|^2$; our subdigraphs are what are usually called induced subdigraphs. The <u>power</u> or <u>order</u> of G is defined to be the cardinality of $|G|$ and is denoted $\|G\|$. Following Fraissé [1] G is called <u>homogeneous</u> if any isomorphism between subdigraphs of smaller power can be extended to an automorphism of G. In this paper we classify all finite homogeneous digraphs.

Let I_G denote the identity relation on $|G|$. The <u>complement</u> of G denoted $\overline{G}$ is $\langle |G|, |G|^2 - (E_G \cup I_G) \rangle$. G is <u>symmetric</u> if $(x, y) \in E_G$ implies $(y, x) \in E_G$ and <u>antisymmetric</u> if $(x, y) \in E_G$ implies $(y, x) \notin E_G$. Two examples of symmetric digraphs are given in Figure 1.

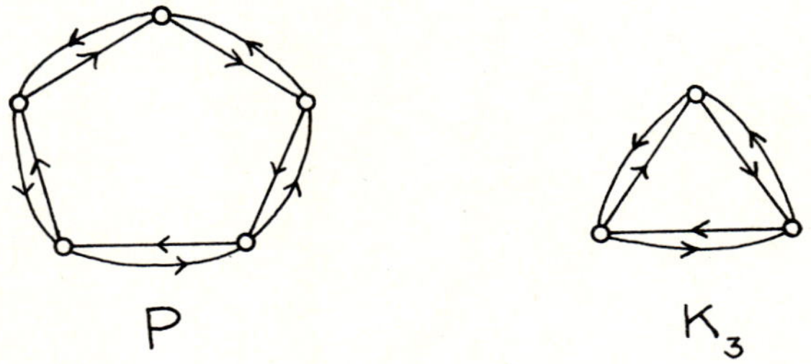

Figure 1

In general K_n denotes the complete symmetric digraph of order n. If G, H are digraphs then $G \times H$ denotes their product and $G[H]$ denotes the composition of G and H. In [2] Gardiner showed that G is a finite symmetric homogeneous digraph if and only if G or $\overline{G}$ is isomorphic to one of:

$$P,\ K_3 \times K_3,\ K_m[\overline{K}_n] \quad (1 \leqslant m,n < \omega)\ .$$

We shall need to refer to three well known asymmetric digraphs T, C, and D depicted in Figure 2.

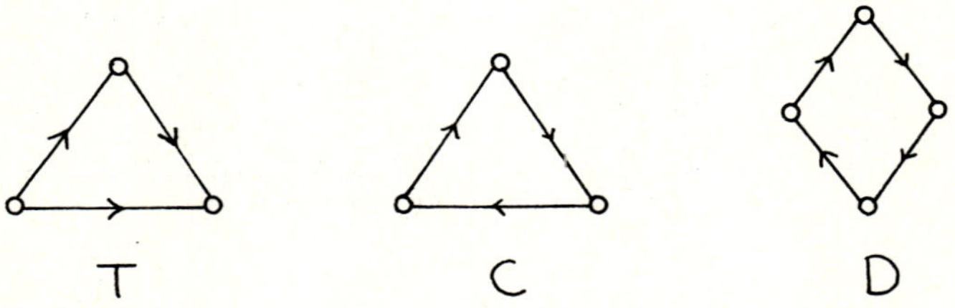

Figure 2

Note that C and D are homogeneous but T is not. Three more homogeneous digraphs H_0, H_1, H_2 are required for the statement of our results. These are shown in Figures 3, 4, and 5 respectively. I am indebted to M. Dubiel for improving my original diagrams.

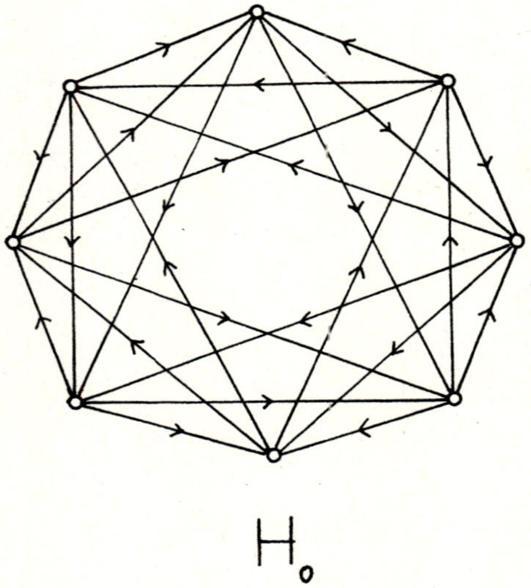

Figure 3

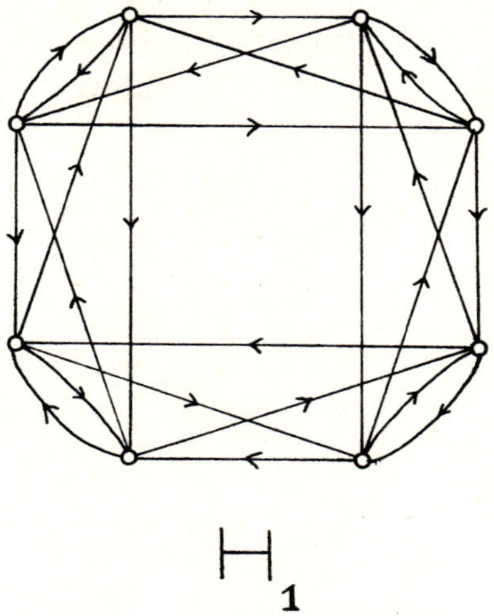

Figure 4

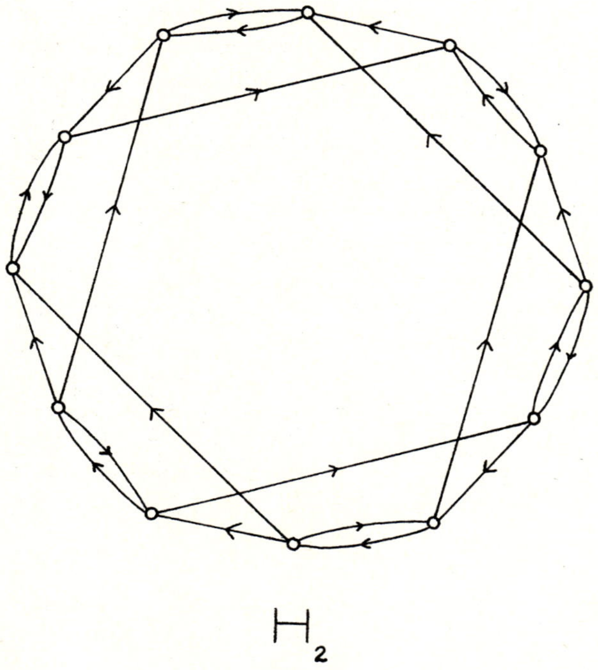

Figure 5

In the diagram of H_2 we have omitted most of the edges in the hope that this will make the picture clearer. The following rule is applied to obtain the remaining edges. Call vertices v, w double-linked if there are edges both from v to w and from w to v, and say that v dominates w if there is an edge from v to w but not from w to v. In H_2 each vertex has a unique mate to which it is double-linked. If v dominates (is dominated by) w then v is dominated by (dominates) the mate of w. This leads to the insertion of another 36 edges.

We can now present our findings. Let $\mathcal{H}$, $\mathcal{A}$, $\mathcal{S}$ denote the classes of all finite digraphs which are homogeneous, homogeneous and antisymmetric, homogeneous and symmetric, respectively.

THEOREM 1. $G \in \mathcal{A}$ iff G is isomorphic to one of D, $\overline{K}_n$, $\overline{K}_n[C]$, $C[\overline{K}_n]$, H_0 for some n, $1 \leq n < \omega$.

THEOREM 2. $G \in \mathcal{H}$ iff either G or $\overline{G}$ is isomorphic to a digraph with one of the following forms: $K_n[A]$, $A[K_n]$, S, C[S], S[C], H_1, H_2 where $1 \leq n < \omega$, $A \in \mathcal{A}$, and $S \in \mathcal{S}$.

These results may be viewed as a contribution to the study of countable homogeneous structures in general. Given a finite relational language L (in the case studied here L consists of one binary relation) we can ask: Is there a nice classification of the countable homogeneous L-structures satisfying specified axioms? A more precise question along the same lines asks whether the first-order theory of countable homogeneous L-structures satisfying specified axioms is decidable. The same questions may be posed with "finite" instead of "countable." The only evidence available indicates that, at least when we are dealing with a single binary relation, there is likely to be a simple classification for the countable homogeneous L-structures. Schmerl [12] has made the classification for strict partial orderings. Woodrow and the author [9] have made one for graphs, and the author [8] has found all countable homogeneous tournaments. In all cases that have been worked out the classification of homogeneous structures is quite simple although the calçulations needed to justify the classification are very complicated in some cases.

The plan of the paper is as follows. After some preliminaries in §1, the two theorems stated in the Introduction are broken down into thirteen special cases in §2. In §3 through §8 we treat the special cases one by one.

1. PRELIMINARIES. As mentioned above a digraph G is a structure $\langle |G|, E_G \rangle$ such that $G \models \neg E(x, x)$. For convenience, we require that $|G| \subset \omega = \{0, 1, 2, \ldots\}$. There is no loss of generality because we shall only be concerned with the case in which G is finite. Members of $|G|$ are called vertices. We shall often abuse notation by writing $v \in G$ instead of $v \in |G|$. By $v_0(G)$, or v_0 if it is clear which digraph we are talking about, we mean the least vertex of G. Vertices v, w are said to be linked if one of them dominates the other or they are double-linked, and to be unlinked otherwise.

Given a digraph G, we single out subdigraphs G', $'G$, G^u, G^d by:

$$|G'| = \{v \in G\colon v_0 \text{ dominates } v\}$$

$$|'G| = \{v \in G\colon v \text{ dominates } v_0\}$$

$$|G^u| = \{v \in G\colon v \text{ and } v_0 \text{ are unlinked}\}$$

$$|G^d| = \{v \in G\colon v \text{ and } v_0 \text{ are double-linked}\}.$$

When G is homogeneous so are G', $'G$, G^u, and G^d and their isomorphism types are dependent only on that of G. We call G', $'G$, G^u, and G^d the derived digraphs of G.

Our results are based on the following kind of induction which Gardiner [2] used in classifying finite homogeneous graphs and which Woodrow [14] used in showing there are no finite homogeneous tournaments of order > 3. A finite digraph G is said to have rank 0 if $|G'| = \phi$ and to have rank $1 + \text{rank}(G')$ otherwise. Our proofs proceed by induction on rank and it turns out that no finite homogeneous digraph has rank > 2.

Let G be a digraph and $1 \leq n < \omega$. An equivalence relation $\overset{n}{\sim}$ on $|G|^n$ is defined by:

$$(u_1, \ldots, u_n) \overset{n}{\sim} (v_1, \ldots, v_n) \equiv \begin{array}{l}\text{there exists an automorphism } f \text{ of } G \\ \text{such that } f(u_1) = v_1, \ldots, f(u_n) = v_n.\end{array}$$

The equivalence classes with respect to $\overset{n}{\sim}$ are called n-types and the n-type of $(v_1, \ldots, v_n)$ is denoted $tp(v_1, \ldots, v_n)$. If G is homogeneous and $v_1 \neq v_2$ then there are at most four possibilities for $tp(v_1, v_2)$. Also, by homogeneity $tp(v_1, \ldots, v_n)$ is fixed by the 2-types $tp(v_i, v_j)$, $1 \leq i < j \leq n$. The converse of $tp(v_1, v_2)$ is $tp(v_2, v_1)$, and $tp(v_1, v_2)$ is called symmetric if $tp(v_1, v_2) = tp(v_2, v_1)$.

Let G be a given digraph. An n-ary relation on $|G|$ is called <u>definable</u> if it is invariant under all automorphisms of G. Let $\tilde{E}_G$ denote the converse of the relation E_G.

1.1. LEMMA. Let G be a homogeneous digraph.

(1) If a binary relation on $|G|$ is definable, it is a Boolean combination of I_G, E_G, and $\tilde{E}_G$.

(2) If there are three symmetric pairwise incompatible binary relations on $|G|$, which are definable and none of which includes I_G, then in some order they are:

x and y are unlinked

x and y are double-linked

x dominates y or y dominates x.

PROOF. This is straightforward so we leave it as an exercise.

For any set A, $|A|$ denotes the cardinality of A. For any digraph G, $\tilde{G}$ the <u>converse of</u> G is the digraph $\langle |G|, \tilde{E}_G \rangle$. Clearly, whenever we prove something about digraphs we can draw immediate corollaries by replacing G by $\bar{G}$ and $\tilde{G}$. This principle will be used without special mention. A <u>congruence relation</u> on a digraph G is an equivalence relation R on $|G|$ such that

$$x_0 R x_1 \wedge y_0 R y_1 \wedge x_0 \not{R} y_0 \rightarrow tp(x_0, y_0) = tp(x_1, y_1)$$

for all $x_0, x_1, y_0, y_1 \in G$.

Since there are no loops any homogeneous digraph G is vertex transitive. Thus if G is also finite we have $\|G'\| = \|{}'G\|$.

For $1 \leq n < \omega$ let $[G]^n$ denote the set of subsets of $|G|$ of power n; alternatively, we may think of $[G]^n$ as the set of subdigraphs of G of power n. Let G^0, G^1 be two of the derived digraphs of G. A mapping of $|G^0|$ into $[G^1]^n$ or $|G^1|^n$ is called <u>definable</u> if it is invariant under every v_0-automorphism of G. Such definable mappings will be important below because they allow us to draw conclusions about the structure of G^0 from that of G^1. For example, if there is a definable one-one correspondence between $|G^0|$ and $|G^1|$ and G^1 is symmetric, then G^0 is also symmetric.

The next lemma collects together some basic observations about finite homogeneous digraphs which will be useful later. We write $G^0 \approx G^1$ to mean G^0 and G^1 are isomorphic.

1.2. LEMMA. Let G be a finite homogeneous digraph.

(1) If $G' \approx \overline{K}_n$, $n \geq 3$, $|G'| = \{v'_1, \ldots, v'_n\}$ and v'_1, v'_2 dominate v, then at most one element of G' does not dominate v. Similarly, for $'G$ instead of G'.

(2) If $G' \approx K_m[\overline{K}_n]$, where $m \geq 3$ and $n \geq 2$, or $m = 2$ and $n \geq 4$, and if v is dominated by two linked vertices of G' then v is dominated by at least two of the m disjoint copies of $\overline{K}_n$ in G'.

(3) If $G' \approx {}'G$ and G' contains both a pair of unlinked vertices and a pair of double-linked vertices, then G' is symmetric and $G \approx C[G']$.

(4) $'(G') \approx ('G)'$.

PROOF. (1) Let v fall in the derived digraph G^0 and for proof by contradiction suppose the conclusion fails. Then $n \geq 4$. Suppose v is dominated by precisely m vertices of G', $2 \leq m \leq n - 2$. For simplicity suppose $n = 4$ so that $m = 2$; other cases may be handled similarly or a little more easily.

Letting $F(x) = \{u \in G' : u \text{ dominates } x\}$, $x \in G^0$, we see that F is a definable mapping of $|G^0|$ into $[G']^2$. If $F(x) \cap F(y) \neq \emptyset$ then $tp(x, y)$ is symmetric. Otherwise, G embeds T contradicting $G' \approx \overline{K}_n$.

Consider $v_1, v_2, v_3 \in G^0$ such that $|F(v_i) \cap F(v_j)| = 1$ for all i, j $(1 \leq i < j \leq 3)$. Suppose there exist $x, y \in G^0$ such that $x \neq y$ and $F(x) = F(y)$ then $t_0 = tp(x, y)$ is symmetric. Let t_1 be the other symmetric 2-type, then $tp(v_i, v_j) = t_1$ for all i, j $(1 \leq i < j \leq 3)$. Thus $tp(v_1, v_2, v_3)$ is fixed. However, v_1, v_2, v_3 may be chosen to make $F(v_1) \cap F(v_2) \cap F(v_3)$ empty or not, contradiction. Thus F is one-one. Given any two copies of $\overline{K}_2$ in G' there is a v_0-automorphism switching them. Hence G^0 is symmetric and $tp(v_i, v_j)$ does not depend on i, j $(1 \leq i < j \leq 3)$. Now we get the same contradiction as before.

(2) We shall treat two cases which illustrate the principles required in general. Firstly, suppose $m = 2$ and $n = 4$ and that $|G'| = \{v'_{i,j} : 1 \leq i \leq 2, 1 \leq j \leq 4\}$ where $v'_{i,j}$ and $v'_{k,\ell}$ are linked iff $i \neq k$. Let v fall in the derived digraph G^0. Define $f(w) = \{u \in G' : u \text{ dominates } w\}$, $w \in G^0$. Suppose v is dominated by $v'_{1,1}$ and $v'_{2,1}$ and no other vertex of G', then F is a definable mapping of $|G^u|$ into $[G']^2$.

Suppose F is one-one and consider the 4-tuples

$$(F^{-1}\{v'_{1,1}, v'_{2,1}\}, F^{-1}\{v'_{1,1}, v'_{2,2}\}, F^{-1}\{v'_{1,3}, v'_{2,3}\}, F^{-1}\{v'_{1,3}, v'_{2,4}\})$$

$$(F^{-1}\{v'_{1,1}, v'_{2,1}\}, F^{-1}\{v'_{1,1}, v'_{2,2}\}, F^{-1}\{v'_{1,3}, v'_{2,3}\}, F^{-1}\{v'_{1,4}, v'_{2,3}\}) .$$

Any ordered pair from the first 4-tuple realizes the same 2-type as the corresponding ordered pair from the second 4-tuple, because there exist corresponding automorphisms of G'. Thus there is a v_0-automorphism of G taking the first 4-tuple into the second. Such an automorphism fixes $v'_{1,1}$ and takes $v'_{1,3}$ onto $v'_{2,3}$, contradiction.

Now suppose F is not one-one. If $x, y \in G^0$ are distinct and $F(x) \cap F(y) \neq \emptyset$ then $tp(x, y)$ is symmetric, because G does not embed T. Now $|F(x) \cap F(y)| = i \wedge x \neq y$ for $i = 0, 1, 2$ are three mutually exclusive symmetric definable relations on G^0 none of which contains I_G. From the previous remark and 1.1

$$|F(x) \cap F(y)| = 0 \equiv x \text{ dominates } y \vee y \text{ dominates } x .$$

Choose vertices $v_i \in G^0$ $(1 \leqslant i \leqslant 4)$ such that $F(v_i) = \{v'_{1,i}, v'_{2,i}\}$. Then for all i, j $(1 \leqslant i < j \leqslant 4)$ one of v_i, v_j dominates the other. Hence G embeds T, contradiction.

If v is dominated by three out of four vertices in each copy of $\overline{K}_4$ in G' rather than by just one in each copy as supposed above, define $F(w) = \{u \in G' : u \text{ does not dominate } w\}$, $w \in G^0$. Note that G^0 is symmetric because any two vertices of G^0 are dominated by a common vertex of G'. For F one-one the argument is as before. When F is not one-one the existence of three mutually exclusive symmetric definable relations on G^0, none of which contains I_G, is a contradiction. If v is dominated by just two vertices from each copy of $\overline{K}_4$ then the argument is easier, because we can see at once that there are too many 2-types in G^0.

As a second illustration consider the case $m = 3$, $n = 2$. Let $|G'| = \{v'_{i,j} : 1 \geqslant i \leqslant 3, 1 \leqslant j \leqslant 2\}$ where $v'_{i,j}$ and $v'_{k,\ell}$ are linked iff $i \neq k$.

Suppose v is dominated by $v'_{1,1}, v'_{2,1}$ and by no other vertex of G'. Let v fall in the derived diagraph of G^0. Define $F(w) = \{u \in G' : u \text{ dominates } w\}$, $w \in G^0$. Then F is a definable mapping of $|G^0|$ into $[G']^2$. Consider $v_1, v_2, v_3, v_4 \in G^0$ such that $F(v_1) = \{v'_{1,1}, v'_{2,2}\}$, $F(v_2) = \{v'_{1,1}, v'_{3,2}\}$,

$F(v_3) = \{v'_{1,2}, v'_{2,2}\}$, $F(v_4) = \{v'_{2,2}, v'_{3,2}\}$. Then $tp(v, v_i)$ $(1 \leq i \leq 4)$ are distinct 2-types none of which is the converse of any other, contradiction.

The only other possibility we need consider is that in which v is dominated by $v'_{1,1}$, $v'_{2,1}$, $v'_{3,1}$ and by no other vertex of G'. Then $|F(x) \cap F(y)| = i$ $(i \leq 3)$ are mutually incompatible symmetric relations definable on $|G^u|$. From 1.1

$$|F(x) \cap F(y)| = 3 \equiv x = y \ (x, y \in G^0) .$$

But this means there is a definable one-one correspondence between vertices of G^0 and copies of K_3 in G'. Hence G' is symmetric and so three mutually incompatible binary relations definable on $|G^0|$ are too many. This completes the case $m = 3$, $n = 2$.

The techniques used are sufficient for the general case.

(3) Suppose that $'G \approx G'$ and that G' contains both a pair of unlinked vertices and a pair of double-linked vertices. Then each $'v \in {'G}$ dominates a vertex in G^u and one in G^d. Thus every $v \in |G^u| \cup |G^d|$ is dominated by some vertex in $'G$. There is a vertex v_1 dominated by G' since $G' \approx {'G}$. Clearly, $v_1 \notin |G^u| \cup |G^d|$ and so $v_1 \in {'G}$. Also, v_1 cannot be dominated by any more vertices, whence $'G \in \mathcal{S}$. Finally, since G' dominates one vertex of $'G$, G' dominates $'G$, which yields $G \approx C[G']$.

(4) Let v'_0, $'v_0$ denote the least vertices of G', $'G$ respectively. The homogeneity of G yields an automorphism f taking v_0, v'_0 into $'v_0, v_0$ respectively. Since $f(|'(G')|) = |('G)'|$ we have the desired conclusion.

2. PLAN OF PROOF. The two theorems stated in the Introduction will be established by proving the special cases set out in the following table. Each line of the table consists of a condition on G in the left column, and a collection of finite homogeneous digraphs in the right column such that, if G is a finite homogeneous digraph satisfying the condition on the left, then G, or one of G and $\overline{G}$ in the case of lines 1, 2 and 8, is isomorphic to one of the digraphs on the right.

TABLE OF CASES

1.	$G' = \emptyset$	$\overline{K}_n$ $(n \geq 1)$ $K_m[\overline{K}_n]$ $(m, n \geq 2)$ P $K_3 \times K_3$
2.	$G' \approx \overline{K}_1$	$S[C]$ $(S \in \mathcal{S})$ $K_n[D]$ $(n \geq 1)$
3.	$G' \approx \overline{K}_2$	$K_n[C[\overline{K}_2]]$ $(n \geq 1)$ $\overline{D}[\overline{K}_2]$ H_1
4.	$G' \approx \overline{K}_n$ $(n > 2)$	$K_m[C[\overline{K}_n]]$ $(m \geq 1)$ $\overline{D}[\overline{K}_n]$
5.	$G' \approx K_m[\overline{K}_n]$ $(m, n \geq 2)$	$C[K_m[\overline{K}_n]]$
6.	$G' \approx P$	$C[P]$
7.	$G' \approx K_3 \times K_3$	$C[K_3 \times K_3]$
8.	$G'' \approx \overline{K}_1$	$K_n[H_0]$ $(n \geq 1)$ H_2
9.	$G'' \approx \overline{K}_2$	$\overline{H}_0[\overline{K}_n]$
10.	$G'' \approx \overline{K}_n$ $(n > 2)$	$\overline{H}_0[\overline{K}_n]$
11.	$G'' \approx K_m[\overline{K}_n]$ $(m, n \geq 2)$, P, or $K_3 \times K_3$	none
12.	$G''' \approx \overline{K}_1$	none
13.	$G''' \approx \overline{K}_n$ $(n \geq 2)$	none

Assume that the table is valid, we claim that for every finite homogeneous digraph G either G or $\overline{G}$ appears in the right-hand column. If $\mathrm{rank}(G) = 0$ the claim follows from line 1. If $\mathrm{rank}(G) = 1$ the claim follows from lines 2 through 7 because, by line 1, $\mathrm{rank}(G') = 0$ and $|G'| \neq \emptyset$ iff either G or $\overline{G}$

satisfies the condition on G in one of the lines 2 through 7. Similarly lines 8 through 11 establish the claim for $\mathrm{rank}(G) = 2$, because $\mathrm{rank}(G'') = 0$ and $|G''| \neq \emptyset$ iff either G or $\overline{G}$ satisfies the condition on G in one of the lines 8 through 11. Further, from lines 8 through 11 we see that, if $\mathrm{rank}(G) = 2$, then either G'' or $\overline{G}''$ is isomorphic to $\overline{K}_n$ for some $n \geq 1$. Thus lines 12 and 13 show that there are no finite homogeneous digraphs of rank 3.

It is a simple matter to show that the digraphs in the right-hand column are all homogeneous, thus in the following sections it will be enough to show that in each line there are no omissions in the right-hand column. The first line of the table is the theorem of Gardiner [2] mentioned in the Introduction. The other lines are verified in the remaining sections of the paper.

3. THE CASE $G' \approx \overline{K}_n$ $(n \geq 1)$. Suppose first that $G' \approx \overline{K}_1$ and let $f\colon |G| \to |G|$ be the permutation defined by letting $f(v)$ be the unique vertex dominated by v. Let $f(v_0)$, $f^2(v_0)$, ... be denoted v_1, v_2, ... respectively and m be the least positive integer such that $v_m = v_0$. If $0 < i < j < m$ then $tp(v_0, v_i) \neq tp(v_0, v_j)$ and $tp(v_0, v_1)$, $tp(v_0, v_{m-1})$ are the two possible asymmetric 2-types. If $m > 3$ then $tp(v_0, v_2) = tp(v_{m-2}, v_0)$ is symmetric when $2 = m - 2$, i.e. $m = 4$. Thus two cases must be considered: $m = 3$ and $m = 4$. Suppose $m = 3$ and that $v \in |G| - \{v_0, v_1, v_2\}$. For $i < 3$ $tp(v, v_i)$ is symmetric. Hence without loss of generality $tp(v, v_0) = tp(v, v_1)$, i.e. there is a v-automorphism taking v_0 onto v_1. This automorphism takes v_1 onto v_2 whence $tp(v, v_2) = tp(v, v_1)$. Thus $tp(v, v_0) = tp(v, v_1) = tp(v, v_2)$ and similarly $tp(v_0, v) = tp(v_1, v) = tp(v_2, v)$ which means that $y \in \{x, f(x), f^2(x)\}$ is a congruence relation on G. Therefore when $m = 3$ $G \approx S[C]$ for some $S \in \mathcal{S}$. If $m = 4$ then $G|\{v_i : i < 4\}$ is isomorphic to D or $\overline{D}$. Without loss suppose $G|\{v_i : i < 4\} \approx D$ then $y \in \{x, f(x), f^2(x), f^3(x)\}$ is a congruence relation on G. Therefore, $G \approx K_m[D]$ for some $m \geq 1$. This completes the case $G' \approx \overline{K}_1$.

Next suppose $G' \approx \overline{K}_2$. There are two cases according as $'G \approx K_2$ or $'G \approx \overline{K}_2$. Consider first that in which $'G \approx \overline{K}_2$. Let $G' = \{v_1', v_2'\}$ and $'G = \{'v_1, 'v_2\}$. If G' dominates $'v_1$ or $'v_2$ then G' dominates $'G$, and we see that there exists a unique vertex v_1 in G^u which dominates G' and is dominated by $'G$. Now "x dominates or is dominated by y" is a congruence relation on G. The restriction of G to any congruence class is isomorphic to $C[\overline{K}_2]$, whence $G \approx K_m[C[\overline{K}_2]]$ for some $m \geq 1$. It remains to consider the cases in which neither $'v_1$ nor $'v_2$ is dominated by G'. Let v_1 be a vertex which is dominated by G'. Since there exists $v \in G^u$ dominated by $'v_1$, every $v \in G^u$

is dominated by either $'v_1$ or $'v_2$. Hence $v_1 \in G^d$. Since there is an automorphism of G which interchanges v_0 and v_1, v_1 dominates $'G$. Thus every vertex in G^d dominates $'G$ and is dominated by G'. Further $\|G^d\| \leq 2$. If $\|G^d\| = 2$ it is easy to see that $G \approx \overline{D}[\overline{K}_2]$. If $\|G^d\| = 1$ it is tedious but not difficult to see that $G \approx H_1$. This completes the case $'G \approx \overline{K}_2 \approx G'$.

Now suppose $G' \approx \overline{K}_2$ and $'G \approx K_2$. Let $'v_1$, $'v_2$, v_1', v_2' be as before. If there exists $v_1 \in G^u$ dominated by $'G$, then $G^u = \{v_1\}$. Further, since $tp(v_0, v_1) = tp(v_1', v_2')$ there exists $v_2 \in G^d$ which dominates G' and $G^d = \{v_2\}$. Thus $\|G\| = 7$ and at the same time each vertex has a unique mate to which it is unlinked. This contradiction shows that v_0 is the unique vertex dominated by $'G$. Hence there are $v_1^u, v_2^u \in G^u$ dominated by $'v_1$, $'v_2$ respectively. Similarly, there are $v_1^d, v_2^d \in G^d$ dominating v_1', v_2' respectively. Further, it is clear that v_i', $'v_i$, v_i^u, v_i^d $(i = 1, 2)$ together with v_0 are all the vertices of G. Since $\|G^u\| = 2$, $tp(v_1^u, v_2^u)$ is symmetric. If v_1^u, v_2^u are unlinked there is a unique v dominating $\{v_1^u, v_2^u\}$. If v_1^u, v_2^u are double-linked there is a unique v dominated by $\{v_1^u, v_2^u\}$. In either case v depends uniquely on v_0 which contradicts our finding that each of the derived digraphs has cardinality 2. This completes the case $G' \approx \overline{K}_2$.

Now suppose that $G' \approx \overline{K}_n$ with $n > 2$. Let $G' = \{v_1', \ldots, v_n'\}$ with $'G = \{'v_1, \ldots, 'v_n\}$. Suppose there exists v dominated by v_1' and v_2', then by 1.2(1) v is dominated by at least $n - 1$ vertices of G'. Hence $'G$ has a substructure isomorphic to $\overline{K}_{n-1}$ which means that $'G \approx \overline{K}_n$. Thus there are just two possibilities $'G \approx \overline{K}_n$ and $'G \approx K_n$.

3.1. $G' \approx 'G \approx \overline{K}_n$ $(n > 2)$. Since $'G \approx G'$ there exists v_1 dominated by G' and if $v_1 \notin 'G$ then v_1 dominates $'G$. There are four subcases.

3.1.1. For some i, $1 \leq i \leq n$, $'v_i$ is dominated by G'. Then G' dominates $'G$ and $G \approx K_m[C[\overline{K}_n]]$ for some $m \geq 1$.

3.1.2. There is a unique $v \in G$ dominating $'G$ and dominated by G'. Since each $'v_i$ dominates $n - 1$ vertices of G^u, every vertex in G^u is dominated by at least one vertex of $'G$. Hence $v \in G^d$ which means that $G^d = \{v\}$. Thus every vertex has a unique mate to which it is double-linked. The mate of $'v_i$ must be in G' and so we may suppose without loss that v_i' is the mate of $'v_i$. Let $u \in G^u$; then u is not dominated by $'G$ since v is unique. Thus by 1.2(1) u is dominated by either one or $n - 1$ vertices of $'G$. In either case there is a 2-type t such that $tp(x, u) = t$ has a unique solution for $x \in 'G$. Since u has its mate $u^* \in G^u$ there is a definable mapping F of $|G^u|$ into $|'G|^2$ defined by $F(u) = \langle v, w\rangle$ where v, w are the unique vertices of $'G$ such that $tp(v, u) = tp(w, u^*) = t$. Since u, u^* are double-linked $v \neq w$, whence F is onto $|'G|^2 - I_G$. From this it is clear that there are more than four 2-types in G^u, contradiction.

3.1.3. There are exactly m vertices dominated by G' and dominating $'G$ where $1 < m < n$. By the same argument as before the vertices dominated by G' and dominating $'G$ are precisely those in G^d. Clearly, $G^d \approx \overline{K}_m$. By considering an automorphism of G taking v_0 into G^d we see that $G^u \approx \overline{K}_{m-1}$ and that each vertex of G^u dominates G' and is dominated by $'G$. No vertex in $'G$ dominates a vertex in G'. Otherwise G embeds T contradicting $G' \approx \overline{K}_n$. But now it is clear that each vertex in $'G$ dominates only m vertices, contradiction.

3.1.4. There are n vertices dominated by G' and dominating $'G$. In this case it is easy to see that $G \approx \overline{D}[\overline{K}_n]$.

3.2. $'G \approx \overline{G'} \approx K_n$ $(n > 2)$. Any vertex dominates at most one vertex in $'G$. Hence there exists v_1 in either G^u or G^d dominated by v_1'. Without loss suppose $v_1 \in G^u$. Since there is an automorphism switching v_0 and v_1, v_1 dominates some vertex $'f(v_1)$ in $'G$. This vertex is unique as is $f'(v_1) = v_1'$ the vertex in G' dominating v_1. Thus $'f(x) = 'f(y)$ and $f'(x) = f'(y)$ define equivalence relations on G^u. Each of these equivalence relations has n equivalence classes and so an equivalence class of one of the equivalence relations has the same power as an equivalence class of the other. Let m be the cardinality of the equivalence classes. Since $'v_1$ dominates $n - 1$ vertices in G^u, $\|G^u\| \leq n(n - 1)$ and so $m \leq n - 1$. By choosing G^d instead of G^u above if necessary we can suppose $m > 1$ except in case $n = 3$ and v_1' dominates just one vertex in each of $'G$, G^u, and G^d. We now examine this special case. Since G embeds $\overline{K}_3$, G^u embeds $\overline{K}_2$. Also $\|G^u\| = 3$ and so $G^u \approx \overline{K}_3$ which means that being unlinked is an equivalence relation with equivalence classes of power 4. However, $\|G^d\| = 3$ and so $\|G\| = 13$ which is a contradiction. Thus we may assume $m > 1$. Consider a typical vertex $v_1 \in G^u$ then $f'(v_1)$ is the unique vertex in G' dominating v_1. Also, the mapping

$$v_1 \to \{f'(v): v \in G^u \ \&\ 'f(v) = 'f(v_1)\} \subset |G'|$$

from G^u into $[G']^m$ is definable. It follows that more than four 2-types are realized in G^u. This contradiction completes 3.2.

4. THE CASE $G' \approx K_m[\overline{K}_n]$ $(m, n \geq 2)$. Notice that G does not embed T and so $'G$ is symmetric. We may suppose that $'G$ is not isomorphic to either K_{mn} or $\overline{K}_{mn}$ because this case may be subsumed under §3. By 1.2(3) if $'G \approx G'$ then

$G \approx C[K_m[\overline{K}_n]]$. Thus below we explore the case $'G \not\approx G'$.

4.1. $m = n = 2$. Then $'G \approx \overline{G'}$. Each $'v \in {'G}$ dominates exactly one vertex in G^u and two in G^d. Each $v' \in G'$ is dominated by exactly two vertices in G^u and one in G^d. Hence $\|G^u\|, \|G^d\| \leq 4$ and by homogeneity $\|G^u\|, \|G^d\| \in \{2, 4\}$. If $\|G^u\| = 2$ then $G^u \approx K_2$ dominates G' and we see that there exists $v \in G^d$ dominating G'. Therefore every $v \in G^d$ dominates G' and there are at least five vertices dominating G', contradiction. Thus $\|G^u\| = 4$ and there is a definable one-one correspondence between $|G^u|$ and $|'G|$, which means that G^u is isomorphic to one of G' and $\overline{G'}$. Suppose $G^u \approx G'$ then some v dominates G^u. From our initial remarks $v \notin |G^d| \cup |'G|$. If $v \in G'$ then G' dominates G^u contradicting $'G \not\approx G'$. In any case we have a contradiction and similarly if $G^u \approx \overline{G'}$.

4.2. $m = 2$, $n = 3$. There are three subcases:

4.2.1. $'G \approx \overline{G'}$. Let $|'G| = \{'v_{i,j}: 1 \leq i \leq 2,\ 1 \leq j \leq 3\}$ where $'v_{i,j}$, $'v_{k,\ell}$ are linked if and only if $i = k$ and $j \neq \ell$. There exists v dominating $'v_{1,i}$ and $'v_{2,j}$ and no other vertex of $'G$. Thus whichever of G', G^u, G^d v is a member of has power divisible by 9. Hence one of $\|G^u\|, \|G^d\|$ is divisible by 9. Each $'v_{i,j}$ dominates exactly two vertices of G^u. Thus there are 12 "arrows" from vertices in $'G$ to vertices in G^u ignoring any double linkages. Therefore $\|G^u\|$ divides 12. Similarly $\|G^d\|$ divides 12. Contradiction.

4.2.2. $'G \approx K_3[\overline{K}_2]$. Each $v' \in G'$ is dominated by a unique $v \in G^u$, whence $\|G^u\|$ divides $\|G'\| = 6$. It is easy to eliminate $\|G^u\| = 3$, because then each $v \in G^u$ dominates exactly two vertices of G' which yields $\|G^u\| \geq 6$ because any pair in G' has at least 6 conjugates. If $\|G^u\| = 1$, every $v \in G$ has a unique mate $u \in G$ to which it is unlinked. This is absurd since G embeds $\overline{K}_3$.

Suppose $\|G^u\| = 2$. Then $G^u \cup \{v_0\} \approx \overline{K}_3$ since G embeds $\overline{K}_3$. Therefore being unlinked is an equivalence relation E partitioning G into copies of $\overline{K}_3$. For each copy H of $\overline{K}_2$ in $'G$ there is a unique $v_H \in G$, necessarily from G^d since G' and G^u are clearly closed under E, such that $G \cup \{v_H\}$ is a copy of $\overline{K}_3$. The elements v_H constitute all of G^d, whence $\|G^d\| = 3$. Since $G' = K_2[\overline{K}_3]$, each $'v \in {'G}$ dominates a copy of $\overline{K}_3$ in G^d. Hence $G^d \approx \overline{K}_3$ is also closed under E, contradiction.

Finally suppose $\|G^u\| = 6$. Each $v' \in G'$ is dominated by a unique $v \in G^u$, whence there is a definable bijection between G' and G^u. Thus $G^u \approx K_2[\overline{K}_3]$ or $G^u \approx \overline{K}_2[K_3]$. Each $'v \in {'G}$ dominates two unlinked vertices of G^u. If $G^u \approx \overline{K}_2[K_3]$ then $'v \in G$ corresponds to a unique $v \in G^u$ which is unlinked to both the vertices in G^u dominated by $'v$. Thus there is a definable bijection between $'G \approx K_3[\overline{K}_2]$ and $G^u \approx K_2[\overline{K}_3]$, contradiction.

4.2.3. $'G \approx \overline{K}_3[K_2]$. Each $v' \in G'$ is dominated by a unique $v \in G^d$, whence $\|G^d\|$ divides $\|G'\| = 6$. Since each $'v \in {}'G$ dominates a copy of $\overline{K}_3$ in G^d, $\|G^d\| \geq 3$. If $\|G^d\| = 3$, then each $v \in G^d$ dominates two vertices in G', whence $\|G^d\| \geq 6$. Thus $\|G^d\| = 6$ and we have a definable bijection between G^d and G'. Since G^d embeds $\overline{K}_3$, we have $G_d \approx K_2[\overline{K}_3]$. Now there is a v_0-definable equivalence relation E on $'G$ defined by

$$xEy \equiv x, y \text{ dominate the same copy of } \overline{K}_3 \text{ in } G^d .$$

Since E has two classes we have a contradiction.

4.3. $m = 2$ and $n \geq 4$, or $m \geq 3$. As remarked above $'G$ is symmetric and we may suppose that it contains a linked pair of vertices. Thus there is a vertex v dominated by two linked vertices of G'. From 1.2(2) v is dominated by at least two of the m disjoint copies of $\overline{K}_n$ in G'. Hence $'G$ embeds $K_2[\overline{K}_n]$ and so has the form $K_p[\overline{K}_q]$ with $q \geq n$. (When $n = 2$ it is also possible for $K_3 \times K_3$ to embed $K_2[\overline{K}_n]$. But then $|{}'G| = |G'| = 2m$ is even, whence $'G \approx K_3 \times K_3$ is impossible.) Considering $\tilde{G}$ we see that either the case reduces to 4.1 or 4.2 or $n \geq q$. Thus we may suppose that $'G \approx G'$ which yields $G \approx C[K_m[\overline{K}_n]]$ as noted above

5. THE CASES $G' \approx P$, $K_3 \times K_3$. Consider first the case in which $G' \approx P$. Then $'G$ is symmetric of cardinality 5 and so is isomorphic to one of K_5, $\overline{K}_5$, and P where we recall that P is self-complementary. The possibilities K_5 and $\overline{K}_5$ are excluded by §3 above, whence $'G \approx G' \approx P$. From 1.2(3) $G \approx C[P]$.

When $G' \approx K_3 \times K_3$ we obtain $G \approx C[K_3 \times K_3]$ by a similar argument.

6. THE CASE $G'' \approx \overline{K}_1$. From above we know that G' or its complement has one of the forms $S[C]$, with $S \in \mathcal{S}$, or $K_n[D]$ $(n \geq 1)$.

6.1. $G' \approx C$. It is easy to see that $\|{}'G\| = 3$ and that one vertex in $'G$ dominates another because G embeds T. Thus $'G \approx C$ and there is a vertex v_1 dominated by G'. If $v_1 \in {}'G$ then its in-degree is > 3, contradiction. Thus $v_1 \in |G^u| \cup |G^d|$ and v_1 dominates $'G$. Since G embeds T each $'v \in {}'G$ dominates $v' \in G'$, and, since $'v$ dominates exactly three vertices, v' is unique. Thus v_0 is the unique vertex dominated by $'G$ whence v_1 is unique. If $v_1 \in G^u$, we easily see that $G \approx K_n[H_0]$ $(n \geq 1)$. If $v_1 \in G^d$, we obtain

$G \approx \overline{K}_n[\overline{H}_0]$ $(n \geq 1)$.

Before pursuing the other subcases some general remarks are in order. Let $v_1, v_2 \in G$ and v_1 dominate v_2, then there is a unique $v_3 \in G$, which we shall denote by $f(v_1, v_2)$, dominated by both v_1 and v_2. It follows that each ${}'v \in {}'G$ dominates a unique $v' \in G'$. This tells us that ${}'G \approx G'$ or ${}'G \approx \overline{G'}$. Further, if ${}'G$ embeds C then that copy of C dominates v_0 and so every copy of C dominates v_0 and so every copy of C in G dominates a unique vertex.

6.2. $G' \approx D$. For proof by contradiction suppose ${}'G \approx \overline{D}$. Let ${}'v \in {}'G$, then ${}'v$ dominates v_0, ${}'w \in G'$, $v' \in G'$, $v^u \in G^u$ and no others. Since ${}'w$ dominates v_0 which dominates v', we see that v' dominates v^u and v^u dominates ${}'w$. Thus v^u is dominated by a vertex in G'. Now $\|G^u\| > 1$, otherwise ${}'G$ dominates more than one vertex. If $\|G^u\| = 2$ then v^u is dominated by two unlinked vertices, contradicting ${}'G \approx \overline{D}$. Since ${}'v$ dominates unique $v^u \in G^u$ the only other possibility is $\|G^u\| = 4$. In this case ${}'v \rightarrow v^u$ is a definable one-one correspondence between ${}'G$ and G^u whence $G^u \approx D$ or $\overline{D}$. Similarly, $G^d \approx D$ or $\overline{D}$. Without loss $G^u \approx D$. There is a unique $v_1 \in G$ dominating G^u, which means that one of G', G^d, ${}'G$ is a singleton. This contradiction shows that ${}'G \approx D$. Let v_1 be the unique vertex dominated by G'. Since $G' \approx D$ each ${}'v \in {}'G$ dominates a unique vertex ${}'g({}'v) \in G^u$, whence each vertex of G^u is dominated by some vertex of ${}'G$. It follows that $|G^d| = \{v_1\}$ which means that every vertex has a unique mate to which it is double-linked. Every ${}'v \in {}'G$ has its mate in G'.

For a contradiction argument suppose ${}'g$ is one-one, then $G^u \approx D$ or $\overline{D}$. From the automorphism interchanging v_0 and v_1, each $v' \in G'$ dominates a unique vertex $g'(v') \in G^u$ and g' is one-one. Now each $v \in G^u$ is dominated by precisely three vertices: one in ${}'G$, one in G', and one in G^u. This contradiction shows that ${}'g$ is not one-one.

If ${}'v, {}'w \in {}'G$ and ${}'v$ dominates ${}'w$, then ${}'v$ and ${}'w$ dominate $f({}'v, {}'w) = v_0$ and this is the unique vertex dominated by both ${}'v$ and ${}'w$, whence ${}'g({}'v) \neq {}'g({}'w)$. It follows that $\|G^u\| = 2$ and so $G^u \approx K_2$. Further, each ${}'v \in {}'G$ dominates one vertex in G^u and is dominated by the other. Detailed consideration of the possibilities shows that $G \approx H_2$.

For the rest we want to show that all other subcases lead to contradiction. Suppose ${}'G \approx S[C]$ where S is symmetric and has both unlinked pairs and double-linked pairs, then any two copies of C which are either unlinked or double-linked dominate the same vertex. Since $G' \approx {}'G$ or $\overline{{}'G}$ consists of copies of C any two of which are either unlinked or double-linked, G' dominates some vertex, whence $G' \approx {}'G$. This contradicts 1.2(3). A similar argument disposes of all subcases except those in which $G' \approx K_n[D] \approx \overline{{}'G}$, $G' \approx \overline{K}_n[C] \approx {}'G$, or $G' \approx \overline{K}_n[C] \approx \overline{{}'G}$ with $n > 1$. We now consider these remaining cases in turn.

6.3. $G' \approx K_n[D] \approx \overline{{}'G}$ $(n > 1)$. Each ${}'v \in {}'G$ dominates a unique vertex $g({}'v) \in G^u$. If ${}'v$ dominates ${}'w \in {}'G$ then $g({}'v) \neq g({}'w)$, because otherwise $f({}'v, {}'w) = g({}'v) = g({}'w) = v_0$, contradiction. It follows that either g is one-one, or $g({}'v) = g({}'w)$ if and only if ${}'v$ and ${}'w$ are double-linked in ${}'G$. In the latter case, considering the v_0-automorphisms of ${}'G$, we see that $G^u \in \mathcal{S}$. But since ${}'G \approx \overline{K}_n[\overline{D}]$ each $v' \in G'$ is dominated by a copy of $\overline{D}$ in G^u. We conclude that g is one-one which tells us that $G^u \approx {}'G$ or G'. Similarly, $G^d \approx G'$ or ${}'G$. Without loss suppose $G^u \approx {}'G$ then there exists unique $v_1 \in G$ dominated by G^u. But then one of G', ${}'G$, G^d should be a singleton, contradiction.

6.4. ${}'G \approx G' \approx \overline{K}_n[C]$ $(n > 1)$. Since ${}'G \approx G'$ and each $v^u \in G^u$ is dominated by some ${}'v \in {}'G$, there exists unique $v_1 \in G^d$, and G' dominates v_1, and v_1 dominates ${}'G$. Thus each $v \in G$ has a unique mate to which it is double-linked, and each ${}'v \in {}'G$ has mate in G', and each $v^u \in G^u$ has mate in G^u. Now each ${}'v \in {}'G$ dominates a copy of $\overline{K}_{n-1}[C]$ and distinct ${}'v$ dominate disjoint copies of $\overline{K}_{n-1}[C]$. Hence $\|G^u\| \geq 9$. Since each $v^u \in G^u$ has its mate in G^u, $(G^u)' \approx {}'(G^u)$. Also, G^u being a subdigraph of G, we have $(G^u)' \approx \overline{K}_m$ $(m \leq n)$ or $(G^u)' \approx \overline{K}_m[C]$ $(m < n)$. However, previous cases show there is no homogeneous G satisfying all of: $G' \approx {}'G \approx \overline{K}_m$, $\|G\| \geq 9$, G embeds C, and there is a unique v_1 dominating ${}'G$ and dominated by G'. But if $(G^u)'$ embeds C then a copy of C in G^u is dominated by a vertex in G^u as well as by one in ${}'G$. This means that $\|{}''G\| > 1$ which contradicts ${}'G \approx \overline{K}_n[C]$.

6.5. $G' \approx \overline{K}_n[C] \approx \overline{{}'G}$ $(n > 1)$. Since $G' \approx \overline{K}_n[C]$ each ${}'v \in {}'G$ dominates a copy of $\overline{K}_{n-1}[C]$ in G^u. Hence $G^u \notin \mathcal{S}$ and similarly $G^d \notin \mathcal{S}$. However, each of the n copies of C in G' dominates a unique vertex of G and different copies dominate different vertices. Thus one of ${}'G$, G^u, and G^d is isomorphic to K_n or $\overline{K}_n$, contradiction.

7. THE CASE $G'' \approx \overline{K}_2$. From §3, G' is isomorphic to one of $K_n[C[\overline{K}_2]]$ $(n \geq 1)$, $\overline{D}[\overline{K}_2]$, and H_1.

7.1. $G' \approx K_n[C[\overline{K}_2]]$. Since $('G)' \approx '(G') \approx \overline{K}_2$ and $\|'G\| = \|G'\| = 6n$ we have $'G \approx G'$. From 1.2(3), $n = 1$. Following the line of 6.1 we easily see that $G \approx \overline{H}_0[\overline{K}_2]$.

7.2. $G' \approx \overline{D}[\overline{K}_2]$ or H_1. Since $('G)' \approx '(G') \approx \overline{K}_2$ and $\|'G\| = \|G'\| = 8$ we have $'G \approx \overline{D}[\overline{K}_2]$ or H_1. From 1.2(3) $G' \approx {}'G$ and so it is sufficient to consider the case in which $G' \approx H_1$ and $'G \approx \overline{D}[\overline{K}_2]$. For any vertices v, w such that v dominates w there are unique vertices x,y unlinked such that $\{v, w\}$ dominates $\{x, y\}$. Hence each $'v \in {}'G$ dominates a unique pair $\{x', y'\}$ of unlinked vertices in G'. Also $\{x', y'\}$ dominates a unique $v' \in G'$. Thus there is a definable one-one correspondence between $'G$ and G'. Under this correspondence $'G$ and G' have the same automorphisms which is impossible.

8. THE REMAINING CASES. If $G'' \approx \overline{K}_n$ $(n > 2)$, we can proceed as in §7 except that $G' \approx H_1$ is no longer possible. The only possibility for G is $\overline{H}_0[\overline{K}_n]$. This completes the verification of the first ten lines in the Table.

Now suppose $G'' \approx K_m[\overline{K}_n]$ $(m, n \geq 2)$ then $G' \approx C[K_m[\overline{K}_n]]$. Thus $('G)' \approx {}'(G') \approx K_m[\overline{K}_n]$, whence $'G \approx C[K_m[\overline{K}_n]]$. This contradicts 1.2(3). The same argument works when $G'' \approx P$ or $K_3 \times K_3$.

Let $G''' \approx \overline{K}_1$ then G' or $\overline{G'} \approx K_n[H_0]$ $(n \geq 1)$ or H_2. Consider first the case in which $G' \approx K_n[H_0]$. Since $('G)' \approx {}'(G') \approx C$ and $\|'G\| = \|G'\| = 8n$, $'G \approx K_n[H_0]$ or $\overline{K}_n[\overline{H}_0]$.

8.1. $G' \approx K_n[H_0]$ $(n > 1)$. Notice that for any vertices v, w such that v dominates w there is a unique copy of C dominated by both v and w. Therefore each $'v \in {}'G$ determines a unique $v' \in G'$ such that there is a copy of C in G' dominated by both $'v$ and v'. It follows that there are five 3-types of triples $\langle v, w, x\rangle$ in G such that v dominates w and w dominates x. Contradiction.

8.2. $G' \approx H_0 \approx \overline{'G}$. Let $v \in {}'G$, then as in 8.1 there is a unique copy of C, C_v say, in G' dominated by v. Also, there is a unique $w = f(v) \in G^u$ dominated by v and C_v. The subdigraph G' dominating w is homogeneous and cannot be all of G' and so is just C_v. Thus w determines C_v uniquely and hence determines v' the unique member of G' which dominates C_v. Since $'G$ contains no unlinked pair w dominates no vertex in G'. Also, there are at least four 3-types of triples $\langle v_0, w, x\rangle$ with $x \in {}'G$. This contradicts there being only three possible 2-types for pairs $\langle w, x\rangle$.

8.3. $'G \approx G' \approx H_0$. As in the previous subcase for each $v \in {}'G$ there is a unique $w = f(v) \in G^u$ such that v dominates w, and a unique copy C_v of C in G' dominated by v and dominating $f(v)$. As before the vertices of C_v are the only ones in G' dominating $f(v)$, whence $\|G^u\| \geq 8$. Therefore f is a definable one-one correspondence which means that $G^u \approx H_0$ or $\overline{H}_0$. Since $'G \approx G'$ there exists $v_1 \in G^d$ dominated by G' and dominating $'G$, whence no vertex in G^d dominates a vertex in G^u. But $f(v)$ is dominated by three vertices from G^u, C_v, and v, which makes seven vertices in all. This contradiction completes the case.

8.4. The case $G' \approx H_2$. Since $('G)' \approx {}'(G') \approx D$ and $\|'G\| = \|G'\| = 12$, $'G \approx H_2$. Now 1.2(3) yields a contradiction.

REFERENCES

[1] Fraissé, R., Sur l'extension aux relations de quelques propriétés des ordres, Ann. Sci. École Norm. Sup. 71 (1954), 361-388.

[2] Gardiner, A., Homogeneous graphs, J. Combinatorial Theory Ser. B 20 (1976), 94-102.

[3] Gardiner, A., Homogeneity conditions in graphs, J. Combinatorial Theory Ser. B 24 (1978), 301-310.

[4] Henson, C. W., A family of countable homogeneous graphs, Pacific J. Math. 38 (1971), 69-83.

[5] Henson, C. W., Countable homogeneous relational structures and categorical theories, J. Symbolic Logic 37 (1972), 494-500.

[6] Jonsson, B., Universal relational systems, Math. Scand. 4 (1956), 193-208.

[7] Jonsson, B., Homogeneous universal relational systems, Math. Scand. 8 (1960), 137-142.

[8] Lachlan, A. H., Countable ultrahomogeneous tournaments, Amer. Math. Soc. Abstracts 1 (1980), 80T-A17.

[9] Lachlan, A. H., and Woodrow, R. E., Countable ultrahomogeneous graphs, Trans. Amer. Math. Soc. 262 (1980), 51-94.

[10] Morley, M., and Vaught, R., Homogeneous universal models, Math. Scand. 11 (1962), 37-57.

[11] Rado, R., Universal graphs and universal functions, Acta Arith. 9 (1964), 331-340.

[12] Schmerl, J., Countable homogeneous partially ordered sets, Algebra Universalis 9 (1979), 317-321.

[13] Sheehan, J., Smoothly embeddable subgraphs, J. London Math. Soc. 9 (1974), 212-218.

[14] Woodrow, R. E., Theories with a Finite Number of Countable Models and a Small Language (Simon Fraser University, British Columbia, Canada, 1976).

[15] Woodrow, R. E., There are four countable ultrahomogeneous graphs without triangles, J. Combinatorial Theory Ser. B 27 (1979), 168-179.

PROCEEDINGS OF THE HERBRAND SYMPOSIUM
LOGIC COLLOQUIUM '81, J. Stern (editor)

BOREL SETS AND THE ANALYTICAL HIERARCHY

A. Louveau
Université Paris VI and UCLA

In this paper, we use Moschovakis' strategic basis theorem (see Moschovakis [1980]), to relate boldface and lightface pointclasses in the projective hierarchy. The main result is that, assuming projective determinacy, any Borel and Δ^1_{2n+1} set is Δ^1_1 in a Δ^1_{2n+1} real. This result, together with similar ones, for all levels of the projective hierarchy, are obtained in Section 1. In Section 2, we derive some applications to basis and uniformization problems.

Case $n = 1$ of the above result is due to Kechris [1978a], using strong set theoretical hypothesis, together with deep properties of Σ^1_3 sets. The present treatment owes much to discussions I had with him when he was visiting at the University Paris VII, during 1978-1979. The main result was obtained the very day my son Pierre was born, and I'd like to dedicate this paper both to him and to my wife, Lise.

§1. THE BASIC STRUCTURAL RESULT

Let Γ and Λ be two lightface classes of subsets of ω^ω, with $\Gamma \subsetneq \Lambda$. We let $\underset{\sim}{\Gamma} = \bigcup\{\Gamma(\alpha), \alpha \in \omega^\omega\}$ be the corresponding boldface class, and define $\Gamma(\Lambda)$ to be $\bigcup\{\Gamma(\alpha), \alpha \in \Lambda\}$, where $\alpha \in \Lambda$ means $\{\alpha\} \in \Lambda$ (for the Λ's we are interested in, it is equivalent to $\alpha \in \Lambda$, as a subset of ω^2). For closed enough Λ's, it is clear that any set in $\Gamma(\Lambda)$ is both in Λ and in $\underset{\sim}{\Gamma}$. What we want to do is to investigate the converse property, i.e. to study the pairs (Γ, Λ) for which $\Lambda \cap \underset{\sim}{\Gamma} = \Gamma(\Lambda)$. In the following, such a pair is referred to as an <u>effective pair</u>.

The first result concerning effective pairs is in Louveau [1980a] : For $\xi < \omega_1^{ck}$, the pairs $\langle \Sigma^0_\xi, \Delta^1_1 \rangle$ and $\langle \Pi^0_\xi, \Delta^1_1 \rangle$ are effective. This result has been extended to all "lightface" Wadge classes of Borel sets (Louveau [1981]), and by Kechris [1978a] who proves, as said above, that $\langle \Delta^1_1, \Delta^1_3 \rangle$ is an effective pair, using AD[L($\mathbb{R}$)].

We shall use here a different, and quite simpler, approach, adapted to Δ^1_{2n+1} sets, and prove, assuming $\underset{\sim}{\Delta}^1_{2n}$-determinacy, that the pairs $\langle \Delta^1_{2n-1}, \Delta^1_{2n+1} \rangle$, and $\langle \Sigma^1_k, \Delta^1_{2n+1} \rangle$, $\langle \Pi^1_k, \Delta^1_{2n+1} \rangle$, $\langle \Delta^1_k, \Delta^1_{2n+1} \rangle$, for $k < 2n-1$, are all effective pairs. The main ingredients, together whith Moschovakis' basis result, are a Wadge-type game and Addison's original idea for proving the separation property for Σ^1_3 sets. $\underset{\sim}{\Delta}^1_{2n}$ -determinacy is assumed henceforth.

THEOREM 1. Let A be a $\underset{\sim}{\Pi}^1_{2n-1}$ set in Δ^1_{2n+1}. Then A is Σ^1_{2n} in a Δ^1_{2n+1} real.

LEMMA 2. Suppose A is a Π^1_{2n} set, B is a Σ^1_{2n+1} set, and A can be separated from B by a $\underset{\sim}{\Pi}^1_{2n-1}$ set. Then there is some separating set in $\Pi^1_{2n-1}(\Delta^1_{2n+1})$.

Proof. Although it is not clear at first sight, the indexes in the lemma are carefully chosen. The result is strongly asymmetric, but will be enough in the applications. First, choose $B' \subset \omega^\omega \times \omega^\omega$, B' in Π^1_{2n}, such that $\alpha \in B \longleftrightarrow \exists\beta\ (\alpha,\beta) \in B'$. Let $P_{2n-1} \subseteq \omega^\omega$ be some complete Π^1_{2n-1} set, and consider the following Wadge-type game : I and II play integers, but II may pass, as long as he constructs a real. If I plays $\langle\alpha,\beta\rangle$ and II plays γ, II wins the game if $\alpha \in A \to \gamma \in P_{2n-1}$ and $(\alpha,\beta) \in B' \to \gamma \notin P_{2n-1}$. This is clearly a Σ^1_{2n} game for II. Moreover II wins this game : If C is a $\underset{\sim}{\Pi}^1_{2n-1}$ set separating A from B, and f a continuous function reducing C to P_{2n-1}, II simply follows f. By Moschovakis' strategic basis theorem, II has a Δ^1_{2n+1} winning strategy σ_0. Let then $C = \{\alpha : \forall\beta\ \sigma_0 * \langle\alpha,\beta\rangle \in P_{2n-1}\}$. C is $\Pi^1_{2n-1}(\sigma_0)$, and clearly C separates A from B.

Proof of Theorem 1. Let A be some set in $\underset{\sim}{\Pi}^1_{2n-1} \cap \Delta^1_{2n+1}$, and let $A' \in \Pi^1_{2n}$ be such that $\alpha \in A \longleftrightarrow \exists\beta\ (\alpha,\beta) \in A'$. Let B be defined by $(\alpha,\beta) \in B \longleftrightarrow \alpha \notin A$. A' is Π^1_{2n}, B is Σ^1_{2n+1} and A' can be separated from B by the $\underset{\sim}{\Pi}^1_{2n-1}$ set $A \times \omega^\omega$. Applying Lemma 2 to the pair (A',B) gives a $\Pi^1_{2n-1}(\Delta^1_{2n+1})$ set C separating A' from B. Finally $\alpha \in A \longleftrightarrow \exists\beta\ (\alpha,\beta) \in C$, so that $A \in \Sigma^1_{2n}(\Delta^1_{2n+1})$.

THEOREM 3. The pair $\langle\Delta^1_{2n-1},\Delta^1_{2n+1}\rangle$ is effective.

Proof. The result is obtained by repeated use of Theorem 1 and Lemma 2. Starting with a set A in $\Delta^1_{2n+1} \cap \underset{\sim}{\Delta}^1_{2n-1}$, we can first use Theorem 1 : A is $\Sigma^1_{2n}(\Delta^1_{2n+1})$. Applying it to the complement of A, we obtain that A is $\Delta^1_{2n}(\Delta^1_{2n+1})$. But then A is $\Pi^1_{2n}(\Delta^1_{2n+1})$ and is separable from its complement by a $\underset{\sim}{\Pi}^1_{2n-1}$ set. By Lemma 2 (relativized), A is $\Pi^1_{2n-1}(\Delta^1_{2n+1})$. Doing the same for the complement of A gives that $A \in \Delta^1_{2n-1}(\Delta^1_{2n+1})$.

Before going to the pairs $\langle\Sigma^1_k,\Delta^1_{2n+1}\rangle$, let us quote an immediate consequence of Theorem 1.

COROLLARY 4. Let A be a nonempty $\underset{\sim}{\Pi}^1_{2n-1}$ and Δ^1_{2n+1} set. Then A has a Δ^1_{2n+1} member.

Proof. By Theorem 1, A is $\Sigma^1_{2n}(\Delta^1_{2n+1})$, so by the ordinary basis theorem for Σ^1_{2n} sets, has a $\Delta^1_{2n}(\Delta^1_{2n+1})(=\Delta^1_{2n+1})$ member.

THEOREM 5. Let $k<2n-1$. Then the pairs $\langle\Sigma^1_k,\Delta^1_{2n+1}\rangle$, $\langle\Pi^1_k,\Delta^1_{2n+1}\rangle$ and $\langle\Delta^1_k,\Delta^1_{2n+1}\rangle$ are effective.

Proof. It is enough to prove it for Σ^1_k. Let $G\subseteq\omega^\omega\times\omega^\omega$ be a universal Σ^1_k set, and let $A\in\mathbf{\Sigma}^1_k\cap\Delta^1_{2n+1}$. Consider $C=\{\alpha : A=G_\alpha\}$. Because A and G are $\mathbf{\Sigma}^1_k$, C is $\mathbf{\Pi}^1_{k+1}$, so is $\mathbf{\Pi}^1_{2n-1}(k<2n-1)$. We claim C is Δ^1_{2n+1}. The reason is that A is $\mathbf{\Delta}^1_{2n-1}$ and Δ^1_{2n+1}, so is $\Delta^1_{2n-1}(\Delta^1_{2n+1})$ by Theorem 3. This clearly implies that C is Δ^1_{2n+1}. Then the basis result (Corollary 4) gives a Δ^1_{2n+1} member α in C, so that A is $\Sigma^1_k(\Delta^1_{2n+1})$.

As a first application of these results, let us quote the following corollary, which answers a conjecture of Kechris (1978b).

COROLLARY 6. Assume PD. For every $m\geqslant 3$, if $A\subseteq\omega_1$ is Δ^1_m in the codes, then A is $\Pi^1_1(\Delta^1_m)$ in the codes.

Proof. The even case is a result of Kechris (1978b). For the odd case, first remark that A is $\mathbf{\Pi}^1_1$ in the codes. So if $m>3$, we can apply Corollary 5 to get the result. Finally if $m=3$, we can remark that ω_1-A is also $\mathbf{\Pi}^1_1$ and Δ^1_3 in the codes, so is $\Sigma^1_2(\Delta^1_3)$ by Theorem 1. This gives that A is $\Pi^1_2(\Delta^1_3)$ and $\mathbf{\Pi}^1_1$ in the codes, so is $\Pi^1_1(\Delta^1_3)$ by Lemma 2.

REMARKS. The preceding results do not give a complete picture : We do not know if $\langle\Pi^1_{2n-1},\Delta^1_{2n+1}\rangle$ and $\langle\Pi^1_{2n},\Delta^1_{2n+1}\rangle$ are effective pairs. Theorem 1 is an intermediate result, which seems optimal using this Wadge-type technique. Analogously, we do not know if the basis result extends to $\mathbf{\Sigma}^1_{2n}\cap\Delta^1_{2n+1}$ sets. Even for $n=1$, we do not know whether any nonempty $\mathbf{\Sigma}^1_1\cap\Delta^1_3$ set contains a Δ^1_3 member.

§2. APPLICATIONS

The results presented in Section 1 can be applied to a wide class of basis or uniformization problems. It would be too long to discuss all of them (some general discussion may be found in the appendix of Louveau (1980b)), so we shall only look at some archetypical example : The generalization of the Arsenin-Kunugui and Saint-Raymond results on Borel sets with σ-compact sections. (See Moschovakis (1980) for a presentation of these results.) (The Arsenin-Kunugui result asserts that any Borel set with K_σ sections can be uniformized by a Borel function, and Saint-Raymond's result asserts that any Borel set with K_σ sections is the union of countably many Borel sets with compact sections.) We obtain the following generalization, again assuming $\underset{\sim}{\Delta}^1_{2n}$- determinacy, for $n \geq 1$.

THEOREM 7.

(a) Any $\underset{\sim}{\Delta}^1_{2n+1}$ set with $\underset{\sim}{\Pi}^1_{2n-1}$ sections can be uniformized by a $\underset{\sim}{\Delta}^1_{2n+1}$ function. (As $K_\sigma \subseteq \underset{\sim}{\Pi}^1_{2n-1}$, for $n \geq 1$, this extends in particular to all levels the Arsenin-Kunugui result.)

(b) If A is $\underset{\sim}{\Delta}^1_{2n+1}$ with K_σ sections, A is the countable union of $\underset{\sim}{\Delta}^1_{2n+1}$ sets with compact sections.

Proof. (a) is immediate from Corollary 4 and Δ^1_{2n+1}-selection theorem. (b) is obtained by arguing on each section. If the section is $\Delta^1_{2n+1}(x)$, then as it is $\underset{\sim}{\Delta}^1_1$ it is $\Delta^1_1(\Delta^1_{2n+1}(x))$ by Theorem 3, and then by applying the effective version of Saint-Raymond's result (Louveau [1977]), the section is the union of a countable sequence of compacts sets which is uniformly $\Delta^1_1(\Delta^1_{2n+1}(x))$. The Δ^1_{2n+1}-selection theorem then gives the result.

Remark that the preceding proof gives also effective versions of the result, and may be applied to $\underset{\sim}{\Delta}^1_{2n+1}$ sets with sections having different properties as well.

For more general uniformization problems, the results in Section 1 may not be sufficient, but the technique may be still used. It is in particular the case when one wants to extend Moschovakis' strategic basis theorem to extended notions of "winning strategies".

Let $A \subseteq \omega^\omega \times \omega^\omega$, and Γ be some class. Let us say that a function $f : \omega^\omega \to \omega^\omega$ is a Γ-winning strategy for I (resp. II) in the game with payoff A if Graph $f \in \Gamma$, and $\{(f(\alpha),\alpha),\ \alpha \in \omega^\omega\} \subseteq A$ (resp. $\{(\alpha,f(\alpha)),\ \alpha \in \omega^\omega\} \cap A = \phi$). The interest in such extended notions of strategies comes from a result of Kunen that AD is equivalent to the analogous statement $AD(\underset{\sim}{\Delta}^1_1)$, where the usual notion of strategy is replaced by that of $\underset{\sim}{\Delta}^1_1$-strategy.

It is natural to ask whether the strategic basis result is still true with this extended notion of strategy. The answer is positive,and in some sense generalizes Theorem 7 (a).

THEOREM 8. Assume $\underset{\sim}{\Delta}^1_{2n}$-determinacy. Let Γ be any of the pointclasses Δ^1_{2k-1}, Π^1_{2k-1} or Σ^1_{2k}, for $k \leq n$. Let A be a Σ^1_{2n} set. If player I has a $\underset{\sim}{\Gamma}$-winning strategy in the game with payoff A, he has a $\Gamma(\Delta^1_{2n+1})$-winning strategy.

Proof of the case $\Gamma = \Pi^1_{2k-1}$ or Σ^1_{2k}. For simplicity, take $n = 1$ and $\Gamma = \Pi^1_1$, the general case being the same. So let A be a Σ^1_2 set, and f a $\underset{\sim}{\Pi}^1_1$-winning strategy for Player I in the corresponding game. By the uniform parametrization property of Π^1_1, there is a Π^1_1 set $G \subseteq \omega^\omega \times \omega^\omega$ such that for every Π^1_1 set $H \subseteq \omega^\omega \times \omega^\omega \times \omega^\omega$, there is a total recursive function $h : \omega^\omega \times \omega^\omega \to \omega^\omega$ such that

$$(\alpha,\beta,\gamma) \in H \longleftrightarrow (h(\alpha,\beta),\gamma) \in G.$$

Let $G^* \in \Pi^1_1$ uniformize G, and define $(\alpha,\beta) \in A^* \longleftrightarrow \exists\delta[(\alpha,\delta) \in G^* \vee (\delta,\beta) \in A]$. A^* is a Σ^1_2 set. We claim that Player I has a $\underset{\sim}{\Sigma}^0_1$-winning strategy in the game with payoff A^*: Let $C \in \Pi^1_1$ and α_0 be such that $(\alpha_0,\beta,\alpha) \in C \longleftrightarrow \alpha = f(\beta)$. Then for some recursive function h $\alpha = f(\beta) \longleftrightarrow (h(\alpha_0,\beta),\alpha) \in G$, so

$$\alpha = f(\beta) \longleftrightarrow (h(\alpha_0,\beta),\alpha) \in G^*.$$

Let $g(\beta) = h(\alpha_0,\beta)$. The function g is continuous, and moreover for each β there is a δ, namely $f(\beta)$, such that $(g(\beta),\delta) \in G^*$ and $(\delta,\beta) \in A$. So $(g(\beta),\beta) \in A^*$ and g is a $\underset{\sim}{\Sigma}^0_1$ winning strategy for I in the game A^*.

<u>Claim</u>. If A is Σ^1_{2n} and I has a $\underset{\sim}{\Sigma}^0_1$ winning strategy in this game, I has a $\Sigma^0_1(\Delta^1_{2n+1})$-winning strategy.

Assuming this claim, there is some $\varepsilon \in \Delta^1_3$ such that $\{\varepsilon\}$ is a $\underset{\sim}{\Sigma}^0_1$-winning strategy for I in A^*. Let then $(\beta,\alpha) \in D \longleftrightarrow G^*(\{\varepsilon\}(\beta),\alpha)$. D is a $\Pi^1_1(\varepsilon)$ graph, say of a function k. We claim k is a winning $\Pi^1_1(\Delta^1_3)$-strategy for I in A : First for each β $(\{\varepsilon\}(\beta),\beta) \in A^*$, so there is an α such that $(\{\varepsilon\}(\beta),\alpha) \in G^*$, i.e. k is total ; moreover this α satisfies $(\alpha,\beta) \in A$ by the definition of A^*.

So it just remains to prove the claim. But for this it is enough to consider the following game A' : I passes or plays integers, II plays integers, I eventually produces α and II produces β. I wins if $(\alpha,\beta) \in A$. It is clearly a Σ^1_{2n} game, and a continuous winning strategy for I in A gives a usual strategy for I in A'. By the strategic basis theorem, I has a Δ^1_{2n+1} winning strategy in A', which in turn gives a $\Sigma^0_1(\Delta^1_{2n+1})$-winning strategy for I in A.

<u>Proof for</u> $\Gamma = \Delta^1_{2k-1}$. Again let $n = 1$, and $\Gamma = \Delta^1_1$. Now A is Σ^1_2 and f is a Borel function such that for all β $(f(\beta),\beta) \in A$. Let G, G^* be as before, and consider now the following game A^* :

$$(\alpha,\beta) \in A^* \leftrightarrow \forall x(G^*((\alpha)_0,x) \leftrightarrow \neg G((\alpha)_1,x) \wedge \exists\delta[((\alpha)_0,\delta) \in G^* \wedge (\delta,\beta) \in A]$$

A^* is Σ^1_2, because

$$(\alpha,\beta) \in A^* \leftrightarrow \forall x(G^*((\alpha)_0,x) \vee G((\alpha)_1,x)) \wedge$$
$$\wedge \exists\delta[((\alpha)_0,\delta) \in G^* \wedge ((\alpha)_1,\delta) \notin G \wedge (\delta,\beta) \in A].$$

Now let F,F' in Π^1_1, and $\alpha_0 \in \omega^\omega$ be such that α_0 is a Borel code for f and

$$(\alpha_0,\beta,\delta) \in F \leftrightarrow f_{\alpha_0}(\beta) = \delta$$
$$(\alpha_0,\beta,\delta) \in F' \leftrightarrow f_{\alpha_0}(\beta) \neq \delta.$$

Let h_0 and h_1 be two total recursive functions such that

$$(\alpha_0,\beta,\delta) \in F' \leftrightarrow (h_0(\alpha_0,\beta),\gamma) \in G \leftrightarrow (h_0(\alpha_0,\beta),\gamma) \in G^*$$
$$(\alpha_0,\beta,\gamma) \in F' \leftrightarrow (h_1(\alpha_0,\beta),\gamma) \in G.$$

It is immediate to check that the continuous function $g(\beta) = (h_0(\alpha_0,\beta),h_1(\alpha_0,\beta))$ is a continuous winning strategy for I in A^*. So there is some $\varepsilon \in \Delta^1_3$ such that $\{\varepsilon\}$ is a continuous winning strategy for I in A^*. Let then $k(\beta) = \delta \longleftrightarrow ((\{\varepsilon\}(\beta))_0,\delta) \in G^*$.

It is easily checked, as before, that k is total and for every β $(k(\beta),\beta)$ $\beta \in A$. Moreover $k(\beta) = \delta \longleftrightarrow ((\{\varepsilon\}(\beta))_0, \delta) \in G^*$ and $k(\beta) \neq \delta \longleftrightarrow ((\{\varepsilon\}(\beta))_1, \delta) \in G$ so that k is a $\Delta^1_1(\Delta^1_3)$ winning strategy.

The uniformization result which can be extracted from this result has a different flavor than Theorem 7 (a) :

COROLLARY 9. Let A be a Σ^1_{2n} set in $\omega^\omega \times \omega^\omega$. If A can be uniformized by a $\underset{\sim}{\Delta}^1_{2k-1}$ set, for $k \leq n$, then A can be uniformized by a $\Delta^1_{2k-1}(\Delta^1_{2n+1})$ set.

<u>Proof</u>. From the hypothesis, the projection B of A is Σ^1_{2n} and $\underset{\sim}{\Delta}^1_{2k-1}$. By Theorems 3 and 5, B is $\Sigma^1_{2n}(\Delta^1_{2n+1})$. But then A^* defined by $(\alpha,\beta) \in A^* \longleftrightarrow (\alpha,\beta) \in A \vee \alpha \notin B$ is $\Sigma^1_{2n}(\Delta^1_{2n+1})$, and moreover I has a $\underset{\sim}{\Delta}^1_{2k-1}$ winning strategy in A^*. By the preceding theorem, I has a $\Delta^1_{2k-1}(\Delta^1_{2n+1})$ winning strategy say h, and then C defined by $(\alpha,\beta) \in C \longleftrightarrow \alpha \in B \wedge h(\alpha) = \beta$ is a $\Delta^1_{2k-1}(\Delta^1_{2n+1})$ set which uniformizes A.

REFERENCES :

A.S. Kechris

[1978a] "A basis result for Δ^1_3 Borel sets", Sept. 1978 (mimeographed notes).

[1978b] "Countable ordinals and the analytical hierarchy III", Oct. 1978 (mimeographed notes).

A. Louveau

[1977] "Recursivity and compactness", in "Higher Set Theory", Proceedings of the Oberwolfach Conference, Lecture Notes in Math. 669, Springer Verlag, Heidelberg, p. 303-337.

[1980a] "A separation theorem for Σ_1^1 sets" Trans. Amer. Math. Soc., 260, p. 363-378.

[1980b] "Ensembles analytiques et boréliens dans les espaces produits", Astérisques, vol. 78.

[1981] "Some properties of the Wadge hierarchy of Borel sets", to appear in Cabal Seminar, 1980-1981 (A.S. Kechris, D.A. Martin, Y.N. Moschovakis, eds.).

Y.N. Moschovakis

[1980] "Descriptive Set Theory", North-Holland, New York.

PROCEEDINGS OF THE HERBRAND SYMPOSIUM
LOGIC COLLOQUIUM '81, J. Stern (editor)

STONE DUALITY FOR FIRST ORDER LOGIC*

by M. Makkai**
McGill University, Montreal

§1. INTRODUCTION

It is my purpose to present a very general theory concerning the relationship of syntax and semantics of first order logic. The theory has a close formal relationship to the Stone duality theory for Boolean algebras, as will be pointed out in detail below. It subsumes the Gödel completeness theorem, which occupies a place in it that is analogous to that of the Stone representation inside Stone duality. The theory makes an essential use of ultraproducts.

The theory is formulated in the language of category theory. In fact, categories appear in it in three ways : (1) (first order) theories themselves are made into categories (pretoposes), (2) the collection of models of a fixed theory is made into a category and endowed with an additional structure derived from ultraproducts, resulting in something called the ultracategory of models, and (3) pretoposes on the one hand, and ultracategories on the other, are organized into categories (actually : 2-categories), and the main result is stated, in its final form, in terms of a comparison between these 2-categories.

The frequent appearance of categories and related concepts expressing basic formal relationships between mathematical objects on several 'levels' is, in my opinion, the chief indication of the importance of category theory. I think, the present theory supports this idea.

*This paper is an expanded version of a talk given at the Herbrand Symposium. A full writeup of the results reported on here is submitted for publication elsewhere [5].

**During the work on this paper, the author enjoyed the hospitality of the Institute for Advanced Studies, The Hebrew University of Jerusalem. He is supported by a grant of the Natural Sciences and Engineering Research Council of Canada.

In this summary, I will not try to arrive at precise formulations as quickly as possible, or ever at all in some cases. Rather, I will try to make the formal aspects and the motivation clear.

§2. RECOVERING A THEORY FROM THE ULTRACATEGORY OF ITS MODELS

To indicate the core of the theory, I sketch, with a minimum of categorical terminology, how to endow the collection of models of a theory with a natural abstract structure so that from the resulting entity one can fully recover the theory as a syntactical structure.

Consider a first order theory T, in the ordinary sense. The models of T form a category Mod T with morphisms the elementary embeddings. One observes that every formula ϕ of the theory, say with one free variable, gives rise to a functor $[\phi]$ from Mod T to Set, the category of all sets and functions : $[\phi]$ associates with any model M the extension $\phi(M)$ of ϕ in M ; with any morphism $h : M \to N$ the map $a \mapsto h(a)$. Since h is elementary, the latter is a function $\phi(M) \longrightarrow \phi(N)$. Let us call functors from Mod T to SET of the form $[\phi]$ <u>standard</u>. We would like to find properties of functors from Mod T to SET that are distinctive of standard functors.

By Los' theorem, ultraproducts of models of T are again models of T. Moreover, the operation of taking ultraproducts, with a given ultrafilter U on a set I, can be construed as a functor $(\text{Mod T})^I \longrightarrow \text{Mod T}$, by defining the ultraproduct of elementary maps in a natural way. Also, and actually in the first place, one has ultraproduct functors

$$[U] : \text{SET}^I \longrightarrow \text{SET}.$$

By Los' theorem again, the standard functors are seen to <u>preserve ultraproducts</u>, at least up to isomorphism ; the precise statement of this asserts the existence of a natural isomorphism of two composite functors.

There are canonically defined maps between various ultraproducts of sets ; the simplest one is the well-known 'diagonal' ('constant') map of a set into an ultrapower of it. A general notion of such canonical maps, called <u>ultramorphisms</u>, is the main new concept of the present work. We can lift ultramorphisms from SET to Mod T (think of the canonical elementary embedding of a model into an ultrapower of it) and talk about functors from Mod T to SET <u>preserving</u> ultramorphisms. Standard functors are readily seen to preserve ultramorphisms. The main part of the content of the main result is that the structure preserving functors from Mod T to SET, with respect to all the aforementioned structure put on Mod T and SET, will be essentially just the standard ones.

Here we arrive at a point convenient for motivating why one should consider the theory itself, in its syntactical quality, a category. The structure preserving functors Mod T ⟶ SET mentioned above form, in a natural way, a category (call it C) ; the morphisms of C will be natural transformations. A more precise statement of the result says that the correspondence $\phi \longrightarrow [\phi]$ establishes an equivalence ('isomorphism') of T (as a category) and C. For this statement to make sense, we have to construe T as a category in the first place.

There does not seem to be any way of organizing the models of a theory into an abstract structure other than introducing a category of models, possibly with additional structure. Once we have introduced categories on this level, we are stuck with them and theories necessarily have to be identified with categories. In short, categorical logic seems invevitable.

The appropriate notion is that of <u>pretopos</u> (due to A. Grothendieck [1], see also [3]).

§3. PRETOPOSES

My whole story is based, in more than one way, on the basic notions of <u>limit</u> and <u>colimit</u> in a category.

Given a pair of objects A,B in a category, the specification of a <u>product</u> of A and B (if exists) consists of a further object, denoted $A \times B$ (although it is not uniquely given), and two morphisms as shown :

$$\begin{array}{c} A \times B \\ \pi_A \swarrow \searrow \pi_B \\ A \qquad\qquad B \end{array}$$
(*)

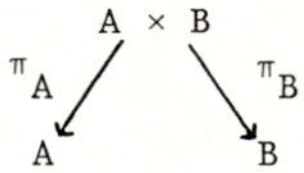

with the following <u>universal property</u> : whenever

$$\begin{array}{c} A \qquad\qquad B \\ \nwarrow p_A \quad p_B \nearrow \\ C \end{array}$$

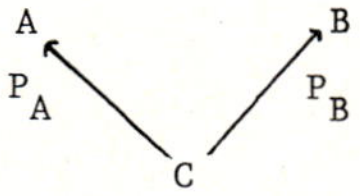

is any pair of morphisms as show, there is a <u>unique morphism</u> $C \xrightarrow{f} A \times B$ such that the two triangles in the diagram

$$\begin{array}{c} A \times B \\ \pi_A \nearrow \uparrow \nwarrow \pi_B \\ A \quad\; f \;\quad B \\ \nwarrow p_A \;\; p_B \nearrow \\ C \end{array}$$

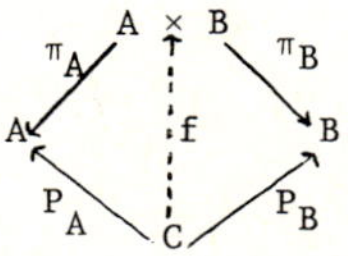

commute. If this holds, the diagram (*) is called a <u>product diagram</u>. Products are unique "up to isomorphism" ; this fact makes it possible to consider 'product' a single-valued operation for all practical purposes. Products are a special case of the notion of <u>limit</u> (or left limit, or inverse limit) (see [2]). The general notion starts with an arbitrary collection of objects and morphisms (a <u>diagram</u>) instead of just the pair A, B of objects (and no morphisms). The notion of <u>colimit</u> is the exact dual : reverse all arrows.

Ultraproducts are combinations of (infinite) products and (directed) colimits in SET ; in fact, we have the equality

$$\prod_{i\in I} A_i/U = \underset{P\in U^{op}}{\underrightarrow{\mathrm{colim}}} \prod_{i\in P} A_i .$$

Limits and colimits exist in SET, and can be computed in standard ways, e.g. product is Cartesian product.

Now, we concentrate on limits and colimits of <u>finite</u> diagrams in SET. Some of these "logical" operations are "first order". Ultraproducts give a convenient criterion to decide which are and which are not.

Ultraproducts <u>preserve</u> finite products : if, for each $i\in I$,

$$A_i \xleftarrow{\;f_i\;} C_i \xrightarrow{\;g_i\;} B_i$$

is a product diagram in SET, then so is

$$\prod A_i/U \xleftarrow{\;\prod f_i/U\;} \prod C_i/U \xrightarrow{\;\prod g_i/U\;} \prod B_i/U \ ;$$

this is readily verified. It turns out that all finite limits are preserved by ultraproducts. On the colimit side, the situation is not as nice. Coproducts (the dual of products ; disjoint sums in SET) are preserved. Also, a special kind of coequalizer (the colimit of a diagram of the form $A \underset{g}{\overset{f}{\rightrightarrows}} B$) is : if the pair of morphisms forms an <u>equivalence relation</u> (in the sense that the map $a \longmapsto \langle f(a), g(a)\rangle$ is a bijection of A onto an equivalence relation on B), then the coequalizer is preserved by ultraproducts. I note that the notion of "equivalence relation" can be appropriately defined in any category having finite limits.

Abstracting, we define a <u>pretopos</u> as a category having finite limits, finite coproducts (including the 'empty coproduct') and coequalizers of equivalence relations with these operations satisfying some simple algebraic conditions ("identities"). I do not list the "identities", but point out a theorem that "explains" them, just as the completeness theorem "explains" the syntactical rules of inference in logic.

SET is a pretopos, the standard pretopos ; so is any Cartesian power SET^I (I any set) as well as all subcategories of these that are closed under the pretopos operations. Call any such category (as well as isomorphic categories) a representable pretopos. Of course, the "identities" are chosen so that they are true in SET, and as a consequence, in all representable pretoposes.

Gödel-Deligne-Joyal Representation Theorem. Every small pretopos is representable.

This theorem is to be compared to the Stone representation theorem ; and also, to Gödel's completeness theorem (see below).

Note that the theorem gives an alternative definition of 'pretopos'.

The proper notion of morphism between pretoposes is that of an elementary functor (called logical functor in [3]), one that preserves the pretopos operations.

A word about the connections with ordinary formulations in logic. There is an essentially one-to-one correspondence between first order theories on the one hand and pretoposes on the other. Models of a theory correspond to elementary functors from the corresponding pretopos to SET, and in fact, elementary functors in general correspond to interpretations suitably defined. The pretopos corresponding to a theory (its pretopos completion) is obtainded in a simple syntactical way, much the same as the Lindenbaum-Tarski algebra of the theory. Once one establishes these translations, the representation theorem becomes equivalent to Gödel's completeness theorem ; see [3] and also the last section of [4].

A final point in this section. We have found that SET carries a pretopos structure as well as the "ultrastructure" consisting of the ultraproduct functors ("infinitary operations"). We have that these two structures commute with each other ; this fact can be succintly expressed as

Los' Theorem. Every ultraproduct functor

$$[U] : SET^I \longrightarrow SET$$

is elementary.

§4. THE STONE ADJUNCTION

Two functors being adjoint is an approximation of their being inverses of each other. Given functors :

$$\mathcal{C} \underset{F}{\overset{G}{\rightleftarrows}} \mathcal{D}\,,$$

we say F is a <u>left adjoint</u> of G (notation $F \dashv G$), if there are natural transformations

$$\eta : 1_{\mathcal{D}} \longrightarrow GF$$

$$\varepsilon : FG \longrightarrow 1_{\mathcal{C}}$$

such that the composites

$$G \xrightarrow{\eta \circ 1_G} GFG \xrightarrow{1_G \circ \varepsilon} G$$

$$F \xrightarrow[1_F \circ \eta]{} FGF \xrightarrow[\varepsilon \circ 1_F]{} F$$

are the respective identities 1_G, 1_F ; the pair (η,ε) is an <u>adjunction</u> of F and G. One indication of the integrity of this notion is that for any G (say) there is essentially only one, if any, adjoint F, "the best approximation of the inverse of G". If η and ε are isomorphisms (have inverses), we get the notion of (F,G) (or F, or G) being an <u>equivalence</u>, the proper notion of "isomorphism" for categories.

Another notion : given a category C, and any object A_o in C, we have the so-called (contravariant) hom-functor

$$\underbrace{\mathcal{C}^{op}}_{\text{opposite of } C} \xrightarrow{\mathrm{Hom}(-,A_o)} \mathrm{SET}$$

$$B \longmapsto \mathrm{Hom}(B,A_o) = \text{set of morphisms } B \to A_o \text{ in } \mathcal{C}$$

$$\begin{array}{c} B \\ f \downarrow \\ C \end{array} \longmapsto \begin{array}{c} \mathrm{Hom}(B,A_o) \\ \uparrow \ \mathrm{Hom}(f,A_o) \\ \mathrm{Hom}(C,A_o) \end{array} \quad \left(\begin{array}{ccc} B \xrightarrow{f} C & \to & A_o \\ & \updownarrow & \\ C & \longrightarrow & A_o \end{array} \right)$$

($\mathrm{Hom}(f,A_o)$: composition with f)

We have the following "self-adjointness"

$$\left.\mathrm{SET}^{op} \underset{\substack{F = \mathrm{Hom}(-,A_o) \\ F \dashv G}}{\overset{G = \mathrm{Hom}(-,A_o)}{\rightleftarrows}} \mathrm{SET} \quad \right\} \qquad (**)$$

with any set A_o.

I will now take a look at the Stone duality for Boolean algebras and Stone spaces. This view, and in particular, what I call "Stone adjunction" below, are described in precise general terms with illuminating theorems in [6] ; see especially Proposition 2.7 there.

Let *Boole* be the category of Boolean algebras (B.a.'s), $\underline{\text{Stone}}$ the category of Stone spaces (compact totally disconnected spaces). One has a particular B.a., the 2-element one

$$\mathit{2} \in |\mathit{Boole}|$$

and also, the 2-element discrete space

$$\underline{2} \in |\underline{\text{Stone}}| \;;$$

the underlying sets of these are identical,

$$|\mathit{2}| = |\underline{2}| = 2.$$

Moreover, the B.a. structure on 2 commutes with the space structure in the sense that all Boolean operations are continuous.

Let $\underline{E}$ be any Stone space (or, in fact, for now, any topological space). Consider the set of continuous maps from $\underline{E}$ to $\underline{2}$:

$$\text{Hom}\ (\underline{E},\ \underline{2}). \tag{***}$$

This is a subset of the set

$$\text{Hom}\ (E,2) \qquad (E = |\underline{E}| = \text{underlying set of } \underline{E}).$$

Since 2 carries the B.A. structure $\mathit{2}$, the last set carries the structure of the product (power) algebra ; we obtain the B.a.

$$\mathit{Hom}\ (E,2) \in |\mathit{Boole}|.$$

Now, the subset (***) turns out to be closed under the operations of the last B.a., directly because the Boolean operations on $\mathit{2}$ are continuous. Hence the subset determines a Boolean (sub-) algebra

$$\mathit{Hom}\ (\underline{E},\underline{2}) \in |\mathit{Boole}|.$$

What we have done is indicate the object function of the functor F in the pair

$$\mathit{Boole}^{op} \underset{F = \mathit{Hom}(-,\underline{2})}{\overset{G = \underline{\text{Hom}}(-,\mathit{2})}{\rightleftarrows}} \underline{\text{Stone}}$$

In an entirely symmetric fashion, we obtain G. It turns out that F is a left adjoint to G ; the adjunction is 'induced' by the one in (**). I emphasize the general nature of this construction ; nothing was used except some very general closure properties of *Boole* and Stone, as well as the fact that "2 commutes with 2". What is not purely formal is the

Stone Duality Theorem. The 'Stone adjunction' is an equivalence of categories.

All information usually associated with Stone duality is contained in this formulation.

Of course, one has the usual logical associations (which are helpful for our next step) : a B.a. is the same as a theory in propositional logic, a homomorphism $B \to 2$ (or ultrafilter) is a "model" of B, Hom $(B,2)$ is the "space of models" of B.

It is fair to say that the specific Stone adjunction considered is completely given by the pair

$$(2,\ \underline{2})$$

of 'commuting' structures.

Next, we replace 2 by SET, and we consider an analog of the Stone adjunction in the new situation. Strictly speaking, this new 'Stone adjunction' does not fall in the general framework of [6], basically because the theory in [6] refers to structures (algebras) with underlying sets, and now we have, rather, underlying categories. It is plausible that a useful general theory of Stone adjunctions, actually with underlying objects in an arbitrary Cartesian closed category, could be developped. Until such a theory has been given, the 'Stone adjunction' next considered remains an analog, rather than a special case, of situations considered in the literature.

The 'Stone adjunction' for predicate logic is based on the pair

$$(\mathit{Set},\ \underline{\mathrm{SET}})$$

of 'commuting' structures. Here *Set* is the standard pretopos SET, SET is the standard ultracategory ; this is the category SET, together with the infinitary operations : the ultraproduct functors, plus some other things called ultramorphisms explained in the next section. The phrase that these 'commute' is taken to mean the fact expressed by Los' theorem formulated above.

Let me elaborate. The category *Boole* is now replaced by the 2-category *PT* of all pretoposes (without going into details, a 2-category *C* is a category with additional structure that makes the set of morphisms $A \longrightarrow B$, for any pair of objects A, B, into the set of objects of a category ; the morphisms of the latter are called 2-cells of $\mathcal{C}$; composition of morphisms of $\mathcal{C}$ then has to be appropriately functorial, etc. ; the category of all categories is a 2-category ; see [2]) : objects of $\mathcal{PT}$ are the pretoposes, morphisms are elementary functors, 2-cells are all (appropriate) natural transformations.

The category Stone is replaced by UC, the 2-category of ultracategories. A pre-ultracategory (p-u.c.) $\underline{K}$ is a category K together with (arbitrary) 'ultraproduct' functors

$$[U] : K^I \longrightarrow K$$

associated with all ultrafilters (I,U). A pre-ultrafunctor (p-u-f.) is a functor preserving ultraproducts 'up to specified isomorphisms" : a p-u.f. $X : \underline{K} \longrightarrow \underline{S}$ between p-u.c.'s is a functor $X : K \longrightarrow S$ together with a natural isomorphism $[X,U]$ associated with every ultrafilter U as shown :

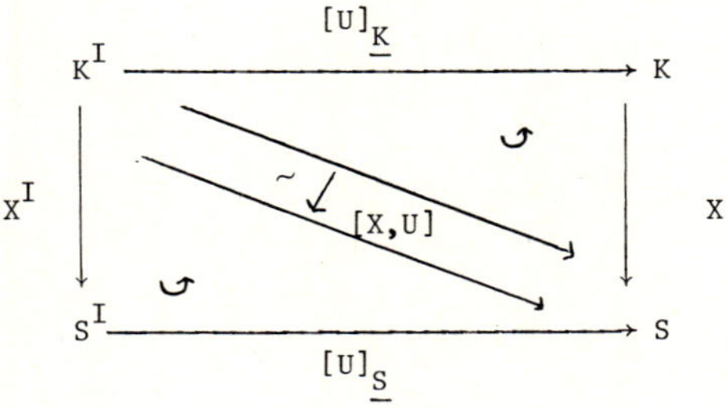

Ultracategories carry some additional structure explained in the next section ; ultrafunctors are p-u.f.'s respecting (in a natural sense) this additional structure as well. These last two kinds of things form the objects and the morphisms of UC. The 2-cells of UC are ultratransformations ; these are natural transformations between the functor-parts of the ultrafunctors such that they respect ultraproducts in a natural sense (the additional structure does not enter here).

The Stone-adjunction takes place between 2-functors

$$\mathcal{PT}^{op} \underset{F = \mathcal{H}om(-,\underline{SET})}{\overset{G = \underline{Hom}(-,\mathcal{S}et) = \underline{Mod}(-)}{\rightleftarrows}} \underline{UC}$$

Here $\underline{Hom}(\mathcal{T}, \mathcal{S}et) = \underline{Mod}(\mathcal{T})$, for any pretopos $\mathcal{T}$, is the ultracategory of all elementary functors $\mathcal{T} \longrightarrow \mathcal{S}et$; if $\mathcal{T}$ corresponds to the theory T, this is the ultracategory of all models of T, mentioned in Section 2. $\underline{Mod}(\mathcal{T})$ is defined in the same canonical way as Hom $(B,2)$ was in the case of the original Stone adjunction. *Hom* (K, SET), for any ultracategory K, is the category (a pretopos) of all ultrafunctors and ultratransformations from K to SET ; this is similarly canonically defined as *Hom* (E,2) above. The "counit" ε of the adjunction (η,ε) of F and G,

$$\varepsilon : FG \longrightarrow 1_{PT}$$

(a 2-transformations between 2-functors)

gives, for every pretopos $T \in |PT|$, an elementary functor

$$\varepsilon_T : T \longrightarrow \underbrace{FG(T)}_{Hom(\text{Mod } T,\ \text{SET})} ;$$

this is essentially the "functor"

$$\phi \longrightarrow [\phi]$$

mentioned in Section 2.

Theorem. For any small pretopos T, ε_T is an equivalence of categories.

The 'faithfulness' (conservativeness) of ε_T is equivalent to the Representation Theorem, hence to Gödel's Completeness Theorem ; it is in this sense that the theorem subsumes completeness.

An (ordinary 1-categorical) adjunction (η,ε) of

$$C \underset{F}{\overset{G}{\rightleftarrows}} D$$

where ε is an isomorphism is called a reflection ; this implies that G is full and faithful, hence C appears as a full subcategory of D. There is a similar conclusion in our (2-categorical) case ; one consequence is the

Corollary. Two small pretoposes T_1 and T_2 are equivalent categories if and only if there is an ultrafunctor Mod $T_1 \longrightarrow$ Mod T_2 whose functor part is an equivalence.

Another consequence is the following result that was already shown in [MR].

Corollary. If the elementary functor $I : T_1 \longrightarrow T_2$ (between small pretoposes) induces an equivalence Mod $T_2 \longrightarrow$ Mod T_1, then I is an equivalence as well.

Our result fails to have the nice symmetry of the Stone theorem ; we do not get that η is also an isomorphism (equivalence). The situation is like starting with say Comp, the category of all compact spaces, in place of Stone. Setting up the Stone adjunction with Comp gives $\eta_{\underline{E}}$ which is a homeomorphism just in case $\underline{E}$ is totally disconnected, as it is easily seen. Having realized this, the "perfect duality" of Stone's theorem is obtained by cutting down Comp to Stone. We can do the same in our case except that we have no simple way of expressing when $\eta_{\underline{K}}$ is an equivalence of categories.

§5. ULTRAMORPHISMS

There are 'canonical maps' one can define once one has ultraproducts around. The most well-known one is the constant (or diagonal) map of a set into an ultrapower of it :

$$A \xrightarrow[\delta_A]{} A^U = \Pi A/U$$

$$a \longmapsto \langle a\rangle/U.$$

One way of expressing canonicity is to say that this is 'natural' in the sense that if $A \xrightarrow{f} B$ is any map, we have that the diagram

$$\begin{array}{ccc} A & \xrightarrow{\delta_A} & A^U \\ {\scriptstyle f}\downarrow & & \downarrow{\scriptstyle f^U} \\ B & \xrightarrow[\delta_B]{} & B^U \end{array}$$

commutes. A better (because stronger) way is to say that δ_A is defined solely on the basis of the universal properties of the (infinite) products and colimits underlying the definition of ultraproduct. It is well-known that δ lifts to models of a theory, and retains the same naturality. It is an important point however that for models, ultraproducts do not have the same product-colimit definition as for sets simply because Mod T may not have products.

When we think of additional structure for Mod T, $\delta = \delta^{(U)}$ is a prime candidate. The effect of adding δ will be felt when we look at structure-preserving maps : the ultrafunctors will have to preserve δ ; this phrase has a uniquely recoverable natural meaning.

Unfortunately, however, the $\delta^{(U)}$'s are not enough. I will give the 'abstract' definition of the required generalization. The better (more restrictive) definition gives ultramorphisms as maps definable on the basis of the universal properties of limits and colimits in SET. It is interesting that (infinite) limits more complicated than products appear here. For these matters see [5].

An ultragraph Γ is given by

(i) a small (directed) graph whose set of nodes is partitioned into two disjoint sets Γ^f, Γ^b, the sets of free and bound nodes, respectively ; and

(ii) an assignment of a 'formal ultraproduct', i.e. a triple $\langle I_\beta, U_\beta, g_\beta \rangle$, to all bound nodes β such that U_β is an ultra-filter on I_β, and g_β is a function from I_β into Γ^f. An ultradiagram of type Γ in a p-u.c. $\underline{S}$ is a diagram $A : \Gamma \longrightarrow S$ satisfying

$$A(\beta) = \prod_{i \in I_\beta} A(g_\beta(i))/U_\beta$$

("the formal ultraproduct becomes a real one") for all $\beta \in \Gamma^b$.

The notation $A : \Gamma \longrightarrow \underline{S}$ indicates that A is an ultradiagram of type Γ in $\underline{S}$. A morphism of ultradiagrams

$$A, B : \Gamma \rightrightarrows \underline{S}$$

is a natural transformation $\Phi : A \longrightarrow B$ between A and B as diagrams satisfying the additional condition

$$\Phi_\beta = \prod_{i \in I_\beta} \Phi_{g_\beta(i)}/U_\beta$$

for all $\beta \in \Gamma^b$. The above specifications define the category $\mathrm{Hom}(\Gamma, \underline{S})$ of ultradiagrams of type Γ in $\underline{S}$.

Let Γ be an ultragraph, k and ℓ two distinguished nodes of Γ, $\Gamma^* = (\Gamma, k, \ell)$. Let S be a pre-ultracategory. k defines the functor

$$(k) : \mathrm{Hom}(\Gamma, \underline{S}) \longrightarrow S$$

'evaluation at k' :

$$A \longmapsto A(k)$$

$$\Phi \Big\downarrow \begin{matrix} A \\ \\ B \end{matrix} \longmapsto \Phi_k$$

similarly for (ℓ). An ultramorphism of type Γ^* in $\underline{S}$ is a natural transformation

$$\delta : (k) \longrightarrow (\ell).$$

The ultramorphisms in $\underline{\mathrm{SET}}$ play a special role ; in fact, the class of all of them, Δ, is part of the similarity type of ultracategories. An ultracategory $\underline{K}$ is a p-u.c. $\underline{K}$ together with an (arbitrary) ultramorphism $\delta^{\underline{K}}$ in $\underline{K}$ associated with every $\delta \in \Delta$ such that $\delta^{\underline{K}}$ is of the same type Γ^* as δ is. <u>SET</u> is the <u>standard ultracategory</u>, with $\delta^{\underline{SET}} = \delta$.

Let $X : K \longrightarrow \underline{S}$ be a pre-ultrafunctor ; for simplicity, assume that all $[X,U]$ are identities. X is an <u>ultrafunctor</u> if for all $\delta \in \Delta$, we have the identity

$$(\delta^{\underline{K}})_{A} = (\delta^{\underline{S}})_{X \circ A}$$

for all ultradiagrams A of the appropriate type. For a general pre-ultrafunctor, a straightforward modification is needed since then $X \circ A$ is not necessarily literally an ultradiagram.

Returning to the 'diagonal' map : now Γ is $\{k, \ell\}$, k free, ℓ bound ; one associates with ℓ the triple $\langle I_\beta, U_\beta, g_\beta \rangle = \langle I, U, g \rangle$, with (of course) g the constant function $I \longrightarrow \{k\}$; an ultra-diagram of type Γ is given by its value $A = A(k)$ (since $A(\ell)$ is necessarily the ultrapower A^U) ; etc.

§6. ON THE PROOF.

<u>Fact</u>. Let $\varepsilon : T \longrightarrow T'$ be an elementary functor between pretoposes. Then ε is an equivalence iff

(i) ε is conservative,

(ii) ε is subobject-full, and

(iii) every object of T' has a cover via ε.

<u>Explanation</u>. ε conservative means that ε induces a 1-1 map $\mathrm{Sub}_T(A) \rightarrow \mathrm{Sub}_{T'}(\varepsilon A)$, with $\mathrm{Sub}_T(A)$ the lattice of subobjects of A in T. ε subobject-full means that the same map is onto.

To explain (iii), I first say this.

A <u>partial</u> (A-)<u>cover of</u> X (<u>via</u> ε) is a pair Φ, Ψ :

$$\begin{array}{ccc} Y & \xrightarrow{\Psi} & \varepsilon A \\ {\scriptstyle \Phi}\downarrow & & \\ X & & \end{array}$$

with $A \in |T|$ and such that Ψ is a monomorphism. A <u>cover of</u> X (<u>via</u> ε) is a finite family of partial covers (Φ_i, Ψ_i) of X such that the only monomorphisms

$$Z \xrightarrow{f} X$$

such that each Φ_i factors through f are isomorphisms.

Recall that the Theorem says that

$$\varepsilon = \varepsilon_T : T \longrightarrow T' = Hom\ (\underline{\text{Mod}}\ T,\ \underline{\text{SET}})$$

is an equivalence for any small pretopos T. We show that ε satisfies (i), (ii), (iii). (i) is essentially the Representation Theorem (and no ultraproducts enter into its proof). (ii) uses the ultraproduct-structure on Mod T, but not the ultramorphisms, and I won't say more about this.
I'd like to sketch how ultramorphisms enter into the proof of (iii).

Let X be any object of T', i.e. an ultrafunctor $X : \underline{\text{Mod}}\ T \longrightarrow \underline{\text{SET}}$. For simplicity, assume that all $[X,U]$ are identities. Unraveling the definition of 'cover', we obtain the following.
A partial A-cover of X is a family $\Sigma = \langle \Sigma_M : M \in |\text{Mod } T| \rangle$ of subsets

$$\Sigma_M \subset M(A) \times X(M)$$

such that the following are satisfied :

(i) whenever $\langle a,x\rangle \in \Sigma_M$, and $h : M \longrightarrow N$ is a morphism in Mod T, then $\langle h_A(a), X(h)(x)\rangle \in \Sigma_N$;

(ii) whenever $\langle a_i, x_i\rangle \in \Sigma_{M_i}$ for all $i \in P$, $P \subset I$, $P \in U$, U an ultrafilter on I, then $\langle\langle a_i\rangle/U, \langle x_i\rangle/U\rangle \in \Sigma_{\Pi M_i/U}$;

(iii) whenever (I,V) is an ultrafilter, and $\langle\langle a_i\rangle/V, \langle x_i\rangle/V\rangle \in \Sigma_{\Pi M_i/V}$, then there is $P \in V$ such that $\langle a_i, x_i\rangle \in \Sigma_{M_i}$ for all $i \in P$;

(iv) $\langle a,x\rangle \in \Sigma_M$ and $\langle a,x'\rangle \in \Sigma_M$ imply $x = x'$.

A <u>cover</u> of X is a finite set $\{\Sigma^i : i < n\}$ of partial covers such that

(v) for every $M \in |\text{Mod } T|$, and every $x \in X(M)$ there is $i < n$ and $a \in M(A_i)$ such that $\langle a,x\rangle \in \Sigma^i_M$.

The proof of the existence of a cover proceeds as follows. <u>First</u>, we show that for every $M \in |\text{Mod } T|$ and $x \in X(M)$ there are $A \in |T|$ and $a \in M(A)$ such that (A,a) is a <u>support</u> of x in the sense that

$$(1)\quad \begin{cases} \text{for any } N \in |\text{Mod } T|, \text{ and } h_1,h_2 : M \rightrightarrows N, \\ \text{if } (h_1)_A(a) = (h_2)_A(a), \text{ then } (Xh_1)(x) = (Xh_2)(x). \end{cases}$$

The proof of this uses that X preserves the diagonal (the simplest) ultramorphism.

<u>Second</u>, we show that

(2) $\begin{cases} \text{whenever } (A,a_o) \text{ is a support of } x_o \ (x_o \in X(M_o)), \text{ then there is} \\ \text{a partial cover } \Sigma \text{ of } X \text{ such that } \langle a_o, x_o\rangle \in \Sigma_{M_o}. \end{cases}$

Note that having done so we have a "cover" but with possibly infinitely (proper-class-many) partial covers instead of finitely many.

This is the part of the proof that uses complicated ultramorphisms. We are to construct a family $\Sigma = \langle \Sigma_M : |\text{Mod } T| \rangle$ satisfying (i) - (iv) and

(vi) $\langle a_o, x_o\rangle \in \Sigma_{M_o}$.

First, we do something slightly less : we select a <u>set</u> K^* of models M, morphisms h and ultrafilters U and V for which we ensure the conditions ; having done the construction for an arbitrary set K^* instead of <u>all</u> models, etc., we'll argue that the original task also can be done. Having now K^*, we want to build $\Sigma^* = \langle \Sigma_M^* : M \in K^* \rangle$. For each $M \in K^*$, we throw in more and more pairs $\langle a, K\rangle$ $(a \in M(A), x \in X(M))$ in an effort to satisfy (i) - (iii) and (vi). Of course, we start with throwing $\langle a_o, x_o\rangle$ into $\Sigma^*_{M_o}$. At any stage, conditions (i) and (ii) are honored by simply throwing in necessary points. Satisfying (iii), however, requires a choice of a set $P \in V$. Repeating those steps transfinitely often, including choices "$P \in V$", we end up with a family Σ^* satisfying (i) - (iii) and (vi). We might have failed to satisfy (iv) though. We now make the assumption that, indeed, at all possible series of choices '$P \in V$' we fail. Using this assumption we are able to build two ultramorphisms, and show that the fact that X preserves these leads to a contradiction to the assumption that (A,a_o) is a support of x_o. The ultragraphs underlying these ultramorphisms are about the same size as K^* itself, and they also use an ultraproduct by an ultrafilter further and above those considered in K^*. The ultragraphs used code, in a sense, the procedure of searching for the right choices of '$P \in V$' mentioned above.

The consideration showing that doing the task (2) for arbitrary bounded K^* suffices uses the Keisler-Shelah Isomorphism Theorem : elementary equivalent models have isomorphic ultrapowers. We show, using this theorem, that, for a fixed cardinal κ, if Σ_1^* and Σ_2^* both satisfy the conditions with some suitable K^*, then the fact that $(\Sigma_1^*)_M = (\Sigma_2)^*_M$ for all M of cardinality less than $< \kappa$ implies that $\Sigma_1^* = \Sigma_2^*$. It is easy to see that now we can put together a Σ as required in (2).

The <u>third</u>, and final, step is to reduce the infinite cover to a finite cover ; this is done by something resembling a compactness argument, and also, by the 'boundedness' just mentioned.

REFERENCES

[1] M. Artin, A. Grothendieck and J.L. Verdier, Theorie des Topos et Cohomologie Etale des Schemas, Springer Lecture Notes in Math., vol.'s 269 and 270, 1972.

[2] S. Mac Lane, Categories for the Working Mathematician, Springer-Verlag, 1971.

[3] M. Makkai and G.E. Reyes, First order categorical logic, Springer Lecture Notes in Math., vol. 611, 1977.

[4] M. Makkai, Full continuous embeddings of toposes, Transactions A.M.S.

[5] M. Makkai, Stone duality for first order logic, submitted to Advances in Math.

[6] J. Lambek and B.A. Rattray, A general Stone-Gelfand duality, Transactions A.M.S., 248 (1979), 1-35.

PROCEEDINGS OF THE HERBRAND SYMPOSIUM
LOGIC COLLOQUIUM '81, J. Stern (editor)

DEGREES OF MODELS OF TRUE ARITHMETIC

by David Marker*

Yale University

§1. INTRODUCTION

If M is a countable model of Peano Arithmetic, we will consider M to be of the form $(\omega, \oplus, \odot)$; where $\oplus$ and $\odot$ are binary functions and M is simply equality for ω. By deg (M) we mean the supremum of the degrees of $\{<n_0,n_1,n_2>: M \models n_0 \oplus n_1 = n_2\}$ and $\{<n_0,n_1,n_2>: M \models n_0 \odot n_1 = n_2\}$, i.e., deg (M) is the degree of the basic diagram of M. Note that if $M = (\omega, \oplus, \odot, \circledeq)$ and deg $(M) \leq \underset{\sim}{d}$, then the usual Henkin process allows us to pass to $N = (\omega, \oplus', \odot')$, where equality in N is equality for ω, $M \equiv N$, and $\deg(N) \leq \underset{\sim}{d}$. Further if M is nonstandard so is N.

There is a great deal known about $D_0 = \{\underset{\sim}{d}$: there is $M \models PA$ nonstandard, with $\deg(M) = \underset{\sim}{d}\}$. The following are among the more striking results.

1) The usual Henkin argument shows there is $\underset{\sim}{d} \in D_0$ with $\underset{\sim}{d} \leq 0'$.

2) Tennenbaum [T] showed that $0 \notin D_0$. (i.e., There is no recursive non standard $M \models PA$.)

3) Shoenfield [S] used Kreisel's basis theorem to find $\underset{\sim}{d} \in D_0$ such that $\underset{\sim}{d} < 0'$.

4) Jockusch and Soare [JS] improved Shoenfield's result by exhibiting $\underset{\sim}{d} \in D_0$ with $\underset{\sim}{d}' = 0'$. Their proof used the following classification theorem (see Simpson [Si]).

5) $\underset{\sim}{d} \in D_0$ iff there is a complete $T \supseteq PA$ with $T \equiv_T d$ iff $\underset{\sim}{d}$ separates a pair of effectively inseperable r.e. sets.

On the other hand $D_1 = \{\underset{\sim}{d}$: There is a nonstandard $M \models Th(N)$ with $\deg(M) = \underset{\sim}{d}\}$ is less understood. There are however several analogies with the above results on D_0.

1') As $Th(N) \equiv_T 0^{(\omega)}$, an effective Henkin argument yields $\underset{\sim}{d} \in D_1$, $\underset{\sim}{d} \leq 0^{(\omega)}$.

* This paper represents work to be submitted in partial fulfilment of the requirements for the degree of Doctor of Philosophy from Yale University.

2') Feferman [F] showed that if $\underset{\sim}{d} \in D_1$, then for any $n \in \omega$, $0^{(n)} \leq \underset{\sim}{d}$, (i.e., every arithmetic set is recursive in $\underset{\sim}{d}$). For given $X \subseteq \omega$, there is a formula $\varphi(v)$ in the language of PA such that if $M \models Th(N)$ and $n \in \omega$, then $n \in X$ iff $M \models \varphi(\underline{n})$†. Let $\Gamma(v) = \{p_{\underline{n}} | v \leftrightarrow \varphi(\underline{n}) : n \in \omega\}$ where $p_{\underline{n}}$ denotes the nth prime number. As $\Gamma(v)$ is recursive, consistent and of bounded complexity, there is $a \in M$ realizing $\Gamma(v)$ (see for example Macintyre [M]). Then $n \in X$ iff $M \models p_{\underline{n}} | a$, so $X \leq deg(M)$.

3') By a clever diagonalization argument Knight [K] constructed $\underset{\sim}{d} \in D_1$ s.t. $\underset{\sim}{d} < 0^{(\omega)}$. Knight conjectured that Feferman's restriction is the only restriction on D_1 and in fact $D_1 = \{\underset{\sim}{d}$: For all $n \in \omega \, \underset{\sim}{d} \geq 0^{(n)}$.

Herein we make a modest contribution toward Knight's conjecture by showing that $D_1 \supseteq \{d'$: For all $n \in \omega$ $\underset{\sim}{d} \geq 0^{(n)}\}$. From this we are able to deduce Knight's result and a $Th(N)$ analogue of the Jockusch-Soare theorem for PA. Our proof uses ideas from Harrington's [H] construction of a nonstandard $M \models PA$ with $deg(M) \leq 0'$ and $Th(N) \equiv 0^{(\omega)}$.

We would like to thank Angus Macintyre and Steve Brackin for numerous conversations on Harrington's theorem, and Carl Jockusch for bringing Knight's conjecture to our attention.

§2. THE MAIN THEOREM

We heavily use the following fact from degree theory.

Fact I : If for every $n \in \omega$ $\underset{\sim}{d} > 0^{(n)}$, then $0^{(\omega)} \leq \underset{\sim}{d}''$.

Proof : See Epstein [E].

We may now prove the main theorem.

Theorem 2 : If for every $n \in \omega$ $\underset{\sim}{d} \geq 0^{(n)}$, then there is a nonstandard $M \models Th(N)$ s.t. $deg(M) \leq \underset{\sim}{d}'$.

Proof : We will describe a three worker construction which produces a complete, consistent, Henkinized theory T s.t. M is the canonical model of T. Worker 1 will use oracle d' and produce the Σ_1^0-diagram of M. Worker 2 uses oracle d'' and builds the Σ_2^0-diagram of M. Worker 3 uses oracle d''' and constructs the full diagram of M. By fact 1 $0^{(\omega)} \leq_T d''$, so both workers 2 and 3 have access to $Th(N)$.

Let L be the language of Peano Arithmetic. Let $C = \{c_i : i \in \omega\}$ be a set of Henkin constants. For $i = 1,2$ let $\{\varphi_j^i : j \in \omega\}$ be a recursive listing of all.

† Here $\underline{n}$ is the term $1+1+\ldots+1$ (n times).

Σ_i^0-sentences in $L(C)$. By $b\,\Sigma_n^0$ we denote all Boolean combinations of Σ_n^0-formulas. $B_n(S)$ will denote all $b\,\Sigma_n^0$ consequences of S allowing no constants from C not already mentioned in S. For $i = 1,2$ let $\{\Gamma_j^i(v) : j \in \omega\}$ list all sets Γ of $b\,\Sigma_i^0$ formulas allowing one free variable v and only finitely many constants from C, such that Γ is r.e. in d.

We will assume, via the recursion theorem, that each worker knows the others'strategies. That is, we actually describe a recursive function g s.t. if x is an index for the strategy used by worker n+1, $g(x)$ will be an index for the strategy used by worker n[x and g(x) are indicies for Turing machines with oracle calls]. By the recursion theorem there is a natural number e s.t. for any oracle A $W_e^A = W_{g(e)}^A$. We use a strategy given by index e.

By any stage s worker j will have committed himself to a finite conjunction T_s^j chosen from $\{\varphi_i^j, \neg\varphi_i^j : i \leq s\}$. T^j will denote $\{T_s^j : s \in \omega\}$. We will arrange things to that $T^1 \subseteq T^2 \subseteq T^3$, $Th(N) \subseteq T^3$, and T^3 is complete, consistent, and Henkinized.

As $d^{(n+1)}$ is r.e. in $d^{(n)}$, worker n may enumerate worker n+1's oracle and approximate the actions of worker n+1. For $n = 1,2$ let $K_{i,j}^{n+1}$ $i \leq j$ denote player n's approximation to T_i^{n+1} based on the jth approximation to $d^{(n+1)}$. To insure our computation of $K_{i,j}^{n+1}$ converges, we continue enumerating $d^{(n+1)}$ and dovetail computations using better approximation to $d^{(n+1)}$. As worker n+1's computations of T_i^{n+1} $i \leq j$ converge, we will eventually find a convergent computation of $K_{i,j}^{n+1}$ $i \leq j$. In time we will have enumerated enough of $d^{(n+1)}$ to correctly answer all of worker n+1's stage i oracle queries. Thus for each i there is an s such that, for any $s' \geq s$ $K_{i,s}^{n+1} = T_i^{n+1}$.

For any s and $i<s$ worker 2 will form a set $U_i^2(s)$ which is r.e. in d. (To be perfectly correct worker 2 will find $u_i^2(s) \in \omega$ which is an index for $U_i^2(s)$ in d. We will identify $U_i^2(s)$ with $u_i^2(s)$ whenever no significant confusion arises.) $U_i^2(s)$ will contain $B_2(Th(N) \wedge K_{i,s}^3)$. Note that if $K_{i,s}^3 \subseteq \Sigma_m^0$, then $B_2(Th(N) \wedge K_{i,s}^3) = \{\varphi \in b\Sigma_2^0(\bar{c}) : Th_{\Pi_{m+3}^0}(N) \vdash \forall\bar{c}(K_{i,s}^3 \rightarrow \varphi)\}$, where $\bar{c}$ are the C constants in $K_{i,s}^3$. As $Th_{\Pi_{m+3}^0}(N) <_T d$, $B_2(Th(N) \wedge K_{i,s}^3)$ is r.e. in d.

If $\Gamma_i^2(v)$ ends up consistent with the full diagram of M, there will be a large s' and $c \in C$ such that for any $s \geq s'$ $U_i^2(s) \supseteq \Gamma_i^2(c)$. We will arrange things so that for each i, there is a t so that if $s \geq t$ $U_i^2(s) = U_i^2(t)$ (in fact $u_i^2(s) = u_i^2(t)$). U_i^2 will denote $\lim_s U_i^2(s)$.

Worker 1 will calculate $K_{i,s}^2$ approximating T_i^2. He will also calculate $\delta_j(s,t)$, an approximation to $u_j^2(s)$ based on his t^{th} approximation to d". $\Delta_j(s,t)$ will be the set r.e. in d with index $\delta_j(s,t)$. At some stage t worker 1 will have enumerated a sufficient portion of d" to guarantee that for any $t' \geq t$ $\delta_j(s,t') = u_j^2(s)$. As above, whenever possible we suppress $\delta_j(s,t)$ and concentrate on $\Delta_j(s,t)$.

For $i<s$ worker 1 will form a set $U_i^1(s)$ r.e. in d (again, we really find an index $u_i^1(s)$ for $U_i^1(s)$). $U_i^1(s)$ will contain $B_1(Th(N) \cup K_{i,s}^2 \cup \Delta_1(i,s) \cup \ldots \cup \Delta_{i-1}(i,s))$. Note that, as above, if X is r.e. in d, so is $B_1(X)$, and in fact the function which takes an index for x to an index for $B_1(X)$ can be computed in d'. If $\Gamma_i^1(v)$ is eventually consistent with the full diagram of M, there will be a constant $c \in C$ and a stage a such that if $s' \geq s$, then $U_i^1(s') \supseteq T_i^1(c)$. Again we will arrange things so that $\lim_s u_i^1(s)$ exist. U_i^1 will denote $\lim_s U_i^1(s)$.

The basic ideas :

Before detailing the construction, we should outline the ideas behind it.

There would be no difficulties involved if we only had to maintain the consistency of T^1, T^2 and T^3. Worker 1 would make sure $Th_{\Sigma_4^0}(N) \cup T_s^1$ is consistent for each s. As $T' \leq_T d'$ and worker 2 has oracle d", worker 2 could maintain consistency of $T^1 \cup Th(N) \cup T_s^2$. (Note that as $T \subseteq \Sigma_1^0$ and $Th_{\Sigma_4^0}(N) \cup T_s^1$ is consistent for each s $Th(N) \cup T^1$ is consistent.) Similarly worker 3 could ensure $Th(N) \cup T^1 \cup T^2 \cup T_s^3$ is consistent, while completing T^3.

The difficulty arises in ensuring that T^3 is Henkinized. For example, suppose worker 2 writes down $\exists x \forall y\, \varphi(x,y,\bar{c})$, where φ is open. In order to provide a witness for $\forall y\, \varphi(x,y,\bar{c})$ we must ensure that worker 1 has set aside a constant c_0 such that $\exists y\, \varphi(c_0,y,\bar{c}) \notin T^1$. Similarly, if worker 3 writes down $\exists x \forall y \exists z\, \varphi(x,y,z,\bar{c})$, worker 2 must set aside a c_0 s.t. $\exists y \forall z\, \varphi\, (c_0,y,z,\bar{c}) \notin T^2$ and for any c_1 worker 1 must ensure $\exists z\, \varphi(c_0,c_1,z,\bar{c}) \in T^3$.

This difficulty is overcome by our approximation procedures. If worker 2 writes down $\exists x \forall y\, \varphi(x,y,\bar{c})$, there will be a later stage where worker 1 believes that worker 2 wants a witness to $\forall y\, \varphi(x,y,\bar{c})$. At this point worker 1 will find a c_0 such that none of the workers could have considered c_0 by this point and set c_0 aside as a witness (i.e., worker 1 will not write down $\exists y\, \varphi(c_0,y,\bar{c})$). As worker 2 will realize that worker 1 has done this, worker 2 may at some point write down $\forall y\, \varphi(c_0,y,\bar{c})$. From this point on everyone is committed to this choice.

Providing a witness for a Σ_n^0-sentence $\exists x\, \psi(x)$ where $n>2$, is a bit more complex. First, Worker 2 must provide a witness for the Σ_2^0-consequences of $\psi(x)$.

Secondly this witness must also have been provided by worker 1 as a witness to the Σ_1^0-consequences of $\psi(x)$. To ensure this occurs workers 1 and 2 attempt to partially saturate the model. Namely ; if they believe it is consistent with the actions of higher level workers, they will set aside a witness for the type $T_i^j(v)$.

Witnessing Γ_i^j is given priority over witnessing T_k^j for $i<k$, to make sure all guesses settle down. If it's consistent for worker 2 to realize $\Gamma_i^2(v)$, then it must be consistent for worker 1 to realize $B_1(\Gamma_i^2(v))$. So worker 1 will have done so. Thus worker 2 will be able to choose a witness. Similarly if it's consistent for worker 3 to write down $\exists\psi(x)$, then it must have been consistent for worker 2 to realize $B_2(\psi(x))$. Hence worker 2 must have done so.

One final point should be made. As worker 1 does not have access to all of $Th(N)$, it can not compute the consequences of formulas of arbitrary complexity. For this reason we have the intermediate worker 2, which, while it could maintain consistency with $Th(N)$, restricts itself to producing the Σ_2^0-theory to keep worker 1 happy. Worker 3 may, of course, produce sentences of arbitrary complexity, since worker 2 has the resources to compute their consequences.

<u>The construction</u>

<u>Worker 1</u> : Stage s.

Worker 1 first enumerates a bit more of d". As worker 1 knows worker 2 will by playing to maintain consistency, worker 1 will enumerate enough of d" to ensure that the computations of $K_{i,s}$ and $\Delta_1(i,s) \ldots \Delta_{i-1}(i,s)$ converge for $i \leq s$, and that $K_{i,s}^2 \cup \Delta_1(i,s) \cup \ldots \cup \Delta_{i-1}(i,s)$ is consistent for $i \leq s$. Here by consistent we mean consistent with $Th(N)$. At first it may seem that this is beyond worker 1's abilities as he only had d' as an oracle. But if $X \subseteq b\Sigma_2^0$ then $X \cup Th(N)$ is inconsistent iff $X \cup Th_{\Sigma_4^0}(N)$ is inconsistent. As $Th_{\Sigma_4^0}(N) < d$, this can be checked effectively in d'.

If φ_s^1 is atomic, worker 1 sets $T_s^1 = T_{s-1}^1 \wedge \varphi_s^1$ or $T_s^1 = T_{s-1}^1 \wedge \neg\varphi_s^1$ to maintain the consistency of $Th_{\Sigma_4^0}(N) \cup K_{i,s}^2 \cup \Delta_1(i,s) \cup \ldots \cup \Delta_{i-1}(i,s) \cup U_1^1(s-1) \cup \ldots \cup U_{i-1}^1(s-1) \cup T_s^1, i \leq j$ for as large an j as possible. If φ_s^1 is not atomic, worker 1 sets $T_s^1 = T_{s-1}^1 \wedge \varphi_s^1$ or $T_s^1 = T_{s-1}^1$ to maintain consistency for the maximal possible j.

We next define $U_i^1(s)$, $i<s$. Let χ_i^s denote

$K_{i,s}^2 \cup \Delta_1(i,s) \cup \ldots \cup \Delta_{i-1}(i,s) \cup Th_{\Sigma_4^0}(N) \cup U_1^1(s) \cup \ldots \cup U_{i-1}^1(s) \cup T_i^1$.

<u>Case 1</u> : $T_s^1 \cup \chi_i^s \cup \Gamma_i^1(v) \cup U_i^1(s-1)$ is consistent.

a) If $U_i^1(s-1)$ contains a realization of $\Gamma_i^1(v)$, then set $U_i^1(s) = B_1(\chi_i^s \cup U_i^1(s-1))$.

b) If $U_i^1(s-1)$ does not realize $\Gamma_i^1(v)$, find a constant c which worker 1 could not have used by this stage. Let $U_i^1(s) = B_1(\chi_i^s \cup U_i^1(s-1) \cup \Gamma_i^1(c))$.

Case 2 : $T_s^1 \cup \chi_i^s \cup \Gamma_i^1(v)$ is consistent, but $\chi_i^s \cup \Gamma_i^1(v) \cup U_i^1(s-1)$ is inconsistent. Let $U_i^1 = B_1(\chi_i^s \cup \Gamma_i^1(c))$, where c is a constant which worker 1 could not have used by this stage.

Case 3 : $T_s^1 \cup \chi_i^s \cup \Gamma_i^1(v)$ is inconsistent.

a) $T_s^1 \cup \chi_i^s \cup U_i^1(s-1)$ is consistent. Let $U_i^1(s) = B_1(\chi_i^s \cup U_i^1(s-1))$.

b) Otherwise let $U_i^1(s) = B_1(\chi_i^s)$.

This concludes worker 1's construction.

There are several things to notice.

Our program is to maintain consistency, while if at all possible, realizing the types $\Gamma^1(v)$. We give priority to realizing types of lower index. We also give priority to our earlier witnesses.

By induction it is easily seen that each $U_i^1(s)$ is r.e. in d.

Suppose at stage s_0, $U_1^1, \ldots, U_{i-1}^1$ have settled down and for $s \geq s_0$ $K_{j,s}^2 = T_j^2$ and $U_k^2(j) = \Delta_k(j,s)$ for all $k < j \leq s$. Then from this stage onward χ_i^s is fixed. Since for $s > s_0$ T_s^1 will be chosen to maintain consistency with $U_i^1(s_0)$, once we make the decision to omit $\Gamma_i^1(v)$ or realize it with c, we will never be forced to back down from this choice. Thus for $s \geq s_0+1$ $U_i^1(s) = U_i^1(s_0+1)$. Thus $\lim_s U_i^1(s)$ exists.

Worker 2 : At stage s will be in either the "active" mode or the "waiting" mode.

Worker 2 knows that for each i there is an s such that if $s' \geq s$ $u_i^1(s') = u_i^1(s)$. The condition $\forall s' \geq u_i^1(s') = u_i^1(s)$ is recursive in d", so worker 2 may calculate indicies for $U_1^1, \ldots, U_{s-1}^1$.

Worker 2 enumerates more of d"' and calculates $K_{i,s}^3$. As worker 3 is maintaining consistency, worker 2 may enumerate enough to d"' to ensure the consistency of $T^1 \cup U_1^1 \cup \ldots \cup U_{s-1}^1 \cup T_{s-1}^2 \cup K_{i,s}^3 \cup Th(N)$ for $i \leq s$.

Case I : Worker 2 is in the active mode.

Worker 2 consider the next φ_i^2 which has not been attended to, and sets $T_s^2 = T_{s-1}^2 \wedge \varphi_i^2$ or $T_s^2 = T_{s-1}^2$ to keep $Th(N) \cup T^1 \cup T_s^2 \cup U_1^1 \cup \ldots \cup U_{s-1}^1 \cup K_{i,s}^3 \cup U_1^2(s-1) \cup \ldots \cup U_{i-1}^2(s-1)$, $i<j$ consistent for as large as j as possible.

We now must define $U_j^2(s)$ $j<s$. Let χ_i^s denote $K_{i,s}^3 \cup U_1^1(s) \cup \ldots \cup U_{i-1}^1(s)$. Let ψ_s denote $T^1 \cup U_1^1 \cup \ldots \cup U_{s-1}^1 \cup T_s^2$.

<u>Case 1</u> : $\psi_s \cup \chi_i^s \cup \Gamma_i^2(v) \cup U_i^2(s-1)$ is consistent.

a) If $U_i^2(s-1)$ contains a witness for $\Gamma_i^2(v)$, then we set $U_i^2(s) = B_2(\chi_i^s \cup U_i^2(s-1))$.

b) If $U_i^2(s-1)$ contains no realization of $\Gamma_i^2(v)$, we calculate $B_1(T_{s-1}^2 \cup \Gamma_i^2(v) \cup U_1^2(s-1) \cup \ldots \cup U_{i-1}^2(s-1))$. This is r.e. in d and thus is $\Gamma_\ell(v)$ for some $\ell \in \omega$. In d'' we may calculate ℓ. Worker 2 sets $U_j^2(s) = B_2(\chi_j^s)$ for $i \leq j < s$ and goes into the waiting mode to find a witness for Γ_ℓ^1.

<u>Case 2</u> : $\psi_s \cup \chi_i^s \cup \Gamma_i^2(v)$ is consistent, but $\chi_i^s \cup \Gamma_i^2(v) \cup U_i^2(s-1)$ is inconsistent. In the case we act just as in 1b).

<u>Case 3</u> : $\psi_s \cup \chi_i^s \cup T_i^2(v)$ is inconsistent.

a) $\psi_s \cup \chi_i^s \cup U_i^2(s-1)$ is consistent. Let $U_i^2(s) = B_2(\chi_i^s \cup U_i^2(s-1))$.

b) Otherwise set $U_i^2(s) = B_2(\chi_i^s)$.

<u>Case II</u> : Player 2, is waiting for a witness to $\Gamma_\ell^1(v)$ which was demanded at stage s_0.

If $s \leq \ell$, then set $T_s^2 = T_{s_0}^2$, $U_j^2(s) = U_j^2(s_0)$, $U_i^2(s) = \phi$ for $s_0 \leq i < s$. If $s > \ell$, we look at U_ℓ^1 to see if it contains a witness c for $\Gamma_\ell^1(v)$. If it does we set $U_j^2(s) = B_1(\chi_i^{s_0} \cup \Gamma_\ell^2(c))$ for $i \leq j < s$, and return to the active mode.

This concludes worker 2's construction.

Several observations : Again we maintain consistency while attempting to realize any consistent type.

If we shift into the waiting mode we will enentually shift our of it.

As before once $U_1^2, \ldots, U_{i-1}^2$ and $K_{i,s}^3$ settle down U_i^2 will settle down.

At stage s worker 2 need only know about $U_1^1, \ldots, U_{s-1}^1$. As these are computed without reference to $U_i^1(s)$ $i < s$, we avoid circularity in our use of the recursion theorem.

Workers 1 and 2 have guaranteed that for $i = 1,2$ if $\Gamma_j^i(v)$ is consistent with the full diagram of M, there is a realization of it.

Worker 3 : Worker 3 will be in either the active mode, the waiting mode, or the passive mode.

If worker 3 is active or waiting and worker 2 shifts from the active to the waiting mode, then worker 3 shifts into the passive mode. When worker 2 returns to the active mode, worker 3 returns to the mode he was previously in.

Case I : Worker 3 is passive.
Worker 3 sets $T^3_s = T^3_{s-1}$.

Case II : Worker 3 is active.
Worker 3 considers the first φ^3_i which has not yet been attended to, and sets $T^2_s = T^2_{s-1} \wedge \pm \varphi^3_i$ to maintain the consistency of $Th(N) \cup T^2 \cup T^3_s \cup \ldots \cup U^2_{s-1}$. If we have added $\exists x\psi(x)$ we switch to the waiting mode to find a witness for $\psi(x)$.

Case III : Worker 3 is waiting for a witness to $\psi(x)$, where $\exists x\psi(x)$ was added at stage s_0.
Worker 3 calculates ℓ s.t. $\Gamma^2_\ell(v) = B_2(T^3_{s_C} \wedge \psi(v))$. If $s \leqslant \ell$, we let $T^3_s = T^3_{s_0}$.
If $s > \ell$, by our construction there is a witness c to $\Gamma^2_\ell(v)$ in U^2_ℓ. Let $T^3_s = T^3_{s-1} \wedge \psi(c)$.

This concludes worker 3's construction.

We observe that : The passive mode ensures that worker 3 remains inactive during thd period when worker 3 is not paying attention to it's actions.

As worker 2 always leaves the waiting mode, worker 3 will eventually leave the passive mode.

Worker 3 will always return form the waiting mode to the active mode.

Player 3 ensures that T^3 is complete, consistent, $T^3 \supseteq Th(N)$ and Henkinized.

From T^1 we pass to M, the canonical model of T^3, effectively in d'. Thus $\deg(M) \leqslant \underset{\sim}{d}'$. The only thing we have not yet shown is that M is nonstandard. But this is easy since $\Gamma(v) = \{v \neq \underline{n} : n \in \omega\}$ is consistent and thus realized by some $a \in M$. //

We have not yet hown the result we claimed in the introduction as we are only sure of $\underset{\sim}{e} \in D_1$ with $\underset{\sim}{e} \leqslant \underset{\sim}{d}'$. That we can guarantee $\underset{\sim}{d}' \in D_1$ follows from the next lemma.

Lemma 3 : If $M = (\omega, \quad , \oplus) = PA$, where equality for M is equality on ω, and $\deg(M) \leqslant \underset{\sim}{d}$, then there is $N \quad M$ s.t. $\deg(N) = \underset{\sim}{d}$.

Proof : Let $f : \omega \longrightarrow \omega$ be a bijection s.t. f is computable in $\underset{\sim}{d}$ and $m \in d$ iff $M = f(m)$ is even. Define $\boxed{+}$ and $\boxed{\cdot}$ on ω by

$$i \boxed{+} j = k \text{ iff } M = f(i) \quad f(j) = f(k)$$

$$i \boxed{\cdot} j = k \text{ iff } M = f(i) \odot f(j) = f(k).$$

Let $N = (\omega, \boxed{+}, \boxed{\cdot})$. Then $f \to N \to M$ is an isomorphism and $m \in d$ iff $N \models m$ is even. So $\underset{\sim}{d} \leqslant \deg(N)$. But $\deg(N) \leqslant \underset{\sim}{f} \vee \deg(M) \leqslant \underset{\sim}{d}$. So $\deg(N) = \underset{\sim}{d}$. //

Corollary 4 : $D_1 \supseteq \{\underset{\sim}{d}' : \text{For every } n \in \omega \ \underset{\sim}{d} \geqslant 0^{(n)}\}$.

Proof : Clear from theorem 2 and lemma 3. //

To derive Knight's theorem and the Jockusch-Soare analogue we need the following fact from degree theory.

Fact 5 : There is $\underset{\sim}{e}$ s.t. for all $n \in \omega \ \underset{\sim}{e} \geqslant 0^{(n)}$ and $\underset{\sim}{e}'' = 0^{(\omega)}$.

Proof : See Epstein [E]. //

Corollary 6 : 1) (Knight) There is $\underset{\sim}{d} \in D_1$ with $\underset{\sim}{d} < 0^{(\omega)}$.

2) There is $\underset{\sim}{d} \in D_1$ with $\underset{\sim}{d}' = 0^{(\omega)}$.

Proof : Let $\underset{\sim}{e}$ be as in Fact 5. Apply corollary 4, and let $\underset{\sim}{d} = \underset{\sim}{e}'$. //

§3. CONCLUSION

Let us conclude with two questions and a conjecture.

Question 1 : If for all $n \in \omega \ \underset{\sim}{d} \geqslant 0^{(n)}$ and $\underset{\sim}{d}' \geqslant 0^{(\omega)}$, is there an $\underset{\sim}{e}$ such that for all $n \in \omega \ \underset{\sim}{e} \geqslant 0^{(n)}$ and $\underset{\sim}{e}' \leqslant \underset{\sim}{d}$?

Conjecture : If for all $n \in \omega \ \underset{\sim}{d} \geqslant 0^{(n)}$ and $\underset{\sim}{d}' \geqslant 0^{(\omega)}$, then $\underset{\sim}{d} \in D_1$. A positive solution to question 1 would imply this conjecture.

Question 2 : Is there $\underset{\sim}{d} \in D_1$ s.t. $\underset{\sim}{d}' \ngeqslant 0^{(\omega)}$?

Knight's conjecture would imply the above conjecture and a positive solution to question 2.

REFERENCES :

[E] R. Epstein. Degrees of Unsolvability : Structure and Theory. Lecture Notes in Mathematics no. 759, Springer-Verlag, Berlin, 1979.

[F] S. Feferman, Arithmetically definable models of formalizable arithmetic, Notices AMS 5 (1958).

[H] L. Harrington, Building arithmetical models of PA, handwritten notes.

[JS] C. Jockusch-R. Soare, Π_1^0 - classes and degrees of theories, Transactions AMS 173 (1972).

[K] J. Knight, A nonstandard model of arithmetic of degree less than that of true arithmetic, handwritten notes.

[M] A. Macintyre, The complexity of types in field theory, in Logic Year 1979-1980 (ed. M. Lerman, J. Schmerl and R. Soare), Lecture Notes in Mathematics no. 859, Springer-Verlag, Berlin, 1981.

[S] J. Shoenfield, Degress of Models, JSL 25 (1960).

[Si] S. Simpson, Degrees of unsolvability : a survey of results, in Handbook of Mathematical Logic (ed, J. Barwise), North Holland, Amsterdam, 1978.

[T] S. Tennenbaum, Non-archimedean models for arithmetic, Notices AMS 6 (1959).

PROCEEDINGS OF THE HERBRAND SYMPOSIUM
LOGIC COLLOQUIUM '81, J. Stern (editor)

FIFTY YEARS OF DEDUCTION THEOREMS

by Jean Porte

Université des Sciences et de la Technologie d'Alger

1. Jacques Herbrand gave in his thesis ([21], see [22] pp. 90-91) the first known statement, with proof, of the classical deduction theorem for an axiomatization of the first-order predicate calculus. By keeping only the propositional part of the theorem and of its proof, we obtain a version of them which is pratically identical with what we can find in modern textbooks.

It must not be forgotten that Tarski already knew the result for the classical propositional calculus (PC), since he used it as a primitive notion in [46] and in [47] (see also the discussion in the notes of [48] and [49]). Jaśkowski as well used the deduction theorem as a kind of "rule" in [24] (written several years before its publication).

The theorem can, in modern terms, be stated as follows :

$$x_1,\dots x_n, y \vdash_{PC} z \Rightarrow x_1,\dots x_n \vdash_{PC} y\to z \qquad (1)$$

where $n>0$, $x_1,\dots x_n$, y, z are formulas, $\to$ is the implication of the formal system under consideration, PC is this formal system, $\vdash$ is the deducibility relation of PC, and $\Rightarrow$ is a metamathematical implication.

It is to be remarked that the converse of (1) :

$$x_1,\dots x_n, y \vdash_{PC} z \Leftarrow x_1,\dots x_n \vdash_{PC} y\to z \qquad (2)$$

is equivalent to the detachment property of $\to$, namely

$$x,\ x\to y \vdash y \qquad (3)$$

The theorem is not true, in form (1), for the most usual axiomatizations of the predicate calculus (with the rule of generalization), when hypotheses may contain free variables. See for instances [23] and [28].

But in this short survey I intend only to stress the chief lines of research which have led to generalizations of the classical deduction theorem in various propositional calculi.

2. It is immediate that we can replace in (1) PC by the intuitionistic propositional calculus (IC), or even (with a small modification of Herbrand's proof) by the implicational part of IC.

In 1968, Witold A. Pogorzelski found the minimal subsystem of implicational IC in which the classical deduction theorem holds ; see [36]. It must be remarked that the minimal system in which both (1) and (2) hold is the implicational IC itself.

For other propositional calculi, the ways of generalizing (1) are less simple ; there are chiefly two :

(I). To replace the deducibility relation $\vdash$ (identical with the "consequence" operation of Tarski [46] by a weaker relation.

(II). To replace $y \to z$ by a more complex function, $d(y,z)$, of the formulas y and z.

3. Generalizations of type (I) have been found by More space -kwei [29], and independently by Church [9] and [11] (see also [15]). Their importance is in the fact that they are one of the sources of the relevant logics studied in [1].

In (1), what stands at the left of the "turnstile", $\vdash$, is a <u>sequence</u> of formulas ; but in the ordinary treatment of deducibility (for PC or IC, as in [18]) only the <u>set</u> of those formulas plays any role, and moreover that set may be increased with formulas not actually used in the formal deduction. Those features are changed in order to achieve the notion of "relevant" deducibility. In the system $R_{\to}$ of [1] (identical with the system of Church [9] and [11]), the turnstile means that every formula written in the <u>sequence</u> placed on the left is actually used in order to construct a deduction of the formula placed on the right. Thus, in $R_{\to}$ we have, p and q being different propositional variables,

$$p, q \not\vdash p \qquad (4)$$

and even

$$p, p \not\vdash p \qquad (5)$$

Such restrictions of deducibility occur in all the relevant logic, with implication weaker than in $R_{\to}$. Even the semi-relevant logic RM, which accepts the deducibility statements of the form

$$x, x \vdash x \qquad (6)$$

(this is often called "the Mingle property"), rejects $p,q \vdash p$.

Various form of the deduction theorem for several relevant logics are studied in

[1]. They are complex, formulas $(x \wedge y) \to z$ and $x \to (y \to z)$ being non-equivalent in these logics.

4. Generalizations of type (II) have been introduced in [37]. It is easily proved that a sufficient condition for

$$x_1, \dots x_n, y \vdash z \quad \Rightarrow \quad x_1, \dots x_n \vdash d(y,z) \tag{7}$$

is the conjunction of the following statements :

$$\vdash d(x,x) \tag{8}$$

$$y \vdash d(x,y) \tag{9}$$

$$d(x,A_1), \dots d(x,A_k) \vdash d(x,B) \tag{10}$$

For every postulated rule of the form

$$A_1, \dots A_k \ / \ B .$$

It follows that, for the modal systems S4 and S5, axiomatized with the rules of material detachment and of generalization, we can take d as d_1 defined, L being necessity, by

$$d_1(y,z) = Ly \to z \tag{11}$$

while for Łukasiewicz's three-valued logic we can take d as d_3 defined by

$$d_3(y,z) = y \to (y \to z) \tag{12}$$

Those results were independently rediscovered and generalized by polish researchers: W.A. Pogorzelski [33] rediscovered (12) and generalized it to Łukasiewicz's n-valued logics, while Żarnecka-Biały [52], [53] rediscovered the results concerning S4 and S5 and found several variants of them.

It may be remarked that for S4 and S5 we can use as well d_2 defined by

$$d_2(y,z) = Ly \to Lz \tag{13}$$

Moreover the results with d_1 and d_2 can be extended partially to certain less known modal systems : d_1 and d_2 for E4 and E5, d_2 for K4 and K5. But only in S4 and S5 have we the converse of (7), i.e. the detachment property for d_1 and d_2. This fact allows to transform every statement of deducibility into an equivalent "thesishood" statement, the result being that the truth of a statement of deducibility in S4 or in S5 is a decidable problem ; see [38].

In [51], Tokarz has found a deduction theorem for the "mingle" system, RM, of [1], with a d_4 defined by

$$d_4(y,z) = (\neg(y \rightarrow \neg z) \vee (y \rightarrow z)) \wedge (\neg y \vee z) \quad (14)$$

But it is to be remarked that Tokarz's result is stated for a deducibility in the sense of Tarski's consequence, which is different from the "relevant deducibility", even in the case of RM, since Tokarz's (Tarski's) deducibility accepts $p,q \vdash p$, which is rejected by the "semi-relevant deducibility" of RM.

5. The foregoing brief sketch does not, obviously, exhaust the subject. Other lines of reserarch have been foollowed in several papers listed in the bibliography below.

Among the chief unsolved problems, I may state the following one : How to prove that no solution to (8)-(10) exists ? (barring of course uniteresting trivial solution like $d(x,y) = (x \rightarrow y) \rightarrow (x \rightarrow y)$, and particularly focusing on solution for which d has the detachment property). It may be conjectured that such a case, without non-trivial solution is the modal system T axiomatized with rules of material detachment and necessitation.

REFERENCES

(The deduction Theorem is mentioned in most textbooks of logic. The books listed here are only those which contain some novel views about the Theorem).

[1] Anderson, A.R., and N.D. Belnap, Entailment, Princeton University Press, Princeton, 1975.

[2] Barcan, R.C., "The deduction theorem in a functional calculus of first order based on strict implication", The Journal of Symbolic logic, 11 (1946), 115-118.

[3] Barcan Marcus, R.C., "Strict implication, deducibility and the deduction theorem", The Journal of Symbolic Logic, 18 (1953), 234-236.

[4] Bunder, M.W., Set Theory based on combinatory Logic, Dissertation, Amsterdam, 1969 - see [40].

[5] Bunder, M.W., "Alternative forms of propositional calculus for a given deduction theorem", Notre Dame Journal of Formal Logic, 20 (1979), 613-619.

[6] Bunder, M.W., "Deduction theorem for significance logics", Notre Dame Journal of Formal Logic, 20 (1979), 695-700.

[7] Burgess, J.P., "Quick completeness proof for some logic of conditionals", Notre Dame Journal of Formal Logic, 22 (1981), 76-84.

[8] Church, A., Review of Quine, "A short course in logic", The Journal of Symbolic Logic, 12 (1947), 60-62.

[9] Church, A., "The weak positive implicational propositional calculus" (abstract), The Journal of Symbolic Logic, 16 (1951), 238.

[10] Church, A., "Minimal logic" (abstract), The Journal of Symbolic Logic, 16 (1951), 239.

[11] Church, A., "The weak theory of implication", Kontrolliertes Denken. Untersuchungen zum logikkalkül und zur logik der Einzelwissenschaften, edited by A. Menne, A. Wihelmy, H. Angstl (Festgabe zum 60. Geburstag von Prof. W. Britzelmayr), rotaprint, Kommissions-Verlag Karl Alber, Munich, 1951, pp. 22-37. - see [15].

[12] Church, A., Introduction to Mathematical Logic, Princeton University Press, Princeton, 1956.

[13] Curry, H.B., "Generalizations of the deduction theorem", Proceedings International Congress of Mathematicians, Amsterdam 1954, vol. 2, pp. 399-400.

[14] Curry, H.B., "The interpretation of formalized implication", Theoria, 25 (1959), 1-26.

[14 bis] Curry, H.B., "The deduction theorem in the combinatory theory of restricted generality", Logique et Analyse, 3 (1960), 15-39.

[14 ter] Curry, H.B., "Basic verifiability in the combinatory logic of restricted generality", Essays on the Foundations of Mathematics (dedicated to Prof. A.H. Fraenkel on his 70th birthday), Magnes Press, The Hebrew University, Jerusalem, 1961 - pp. 165-189.

[15] Curry, H.B., and W. Craig, Review of [11], The Journal of Symbolic Logic, 18 (1953), 177-178.

[16] Curry, H.B., and R. Feys, Combinatory Logic, vol. 1, North-Holland, Amsterdam, 1958.

[17] Goddard, L., and R. Routley, The Logic of Significance and Context, Scottish Academic Press, Edinburgh, 1973.

[18] Gentzen, G., "Untersuchungen über das logische Schliessen", Mathematische Zeitschrift, 39 (1934), 176-210 and 405-431.

[19] Gentzen, G., Recherches sur la Déduction Logique, translation of [18] with supplementary notes by R. Feys and J. Ladrière, P.U.F., Paris, 1955.

[20] Herbrand, J., "Sur la théorie de la démonstration", Comptes-Rendus hebdomadaires des séances de l'Académie des Sciences de Paris, 186 (1928), 1274-1276. Also [22], pp. 21-23.

[21] Herbrand, J., "Recherches sur la théorie de la démonstration" (thèse), Prace Towarzystwa Naukowego Warszawskiego, Wydział 3, 33 (1931). - Also 22, pp. 35-153.

[22] Herbrand, J., Ecrits Logiques, ed. by J. van Heijenoort, Presses Universitaires de France, Paris, 1968.

[23] Hilbert, D., und P. Bernays, Grundlagen der Mathematik, vol. 1, Springer, Berlin, 1934.

[24] Jaśkowski, S., "On the rules of supposition in formal logic", Studia Logica, 1 (1934), 5-32. - Also in [26], pp. 232-258.

[25] Krolikowski, S.J., "A deduction theorem for rejection theses in Łukasiewicz's system of modal logic", Notre Dame Journal of Formal Logic, 20 (1979), 461-464.

[26] McGall, S., (editor) Polish Logic 1920-1939, The Clarendon Press, Oxford, 1967.

[27] Meredith, D., "Positive logic and λ-constants", Studia Logica, 37 (1978), 269-285.

[28] Montague, R., and L. Henkin, "On the definition of 'formal deduction'" The Journal of Symbolic Logic, 21 (1956), 129-136.

[29] Moh Shaw-Kwei, "The deduction theorem and two new logical systems", Methodos, 2 (1950), 56-75.

[30] Parry, W.T., "The Logic of C.I. Lewis", in [39] pp. 115-154.

[31] Perzanowski, J., "The deduction theorem for the modal propositional calculi formalized after the manner of Lemmon. Part I", Reports on Mathematical Logic, 1 (1973), 1-12.

[32] Perzanowski, J., and A. Wroński, "The deduction theorem for the system T of Feys-Wright", Zeszyty Naukowe Universytetu Jagiellońskiego, 265, Prace z Logiki, 6, (1971), 11-14.

[33] Pogorzelski, W.A., "The deduction theorem for Łukasiewicz many-valued propositional calculi", Studia Logica, 15 (1964), 7-19.

[34] Pogorzelski, W.A., "Przegląd twierdzeń o dedukcji dla rachunkow zdań", Studia Logica, 15 (1964), 163-178.

[35] Pogorzelski, W.A., "Schemat twierdzeń o dedukcji dla rachunku zdań", Studia Logica, 15 (1964), 181-187.

[36] Pogorzelski, W.A., "On the scope of the classical deduction theorem", The Journal of Symbolic Logic, 33 (1968), 77-81.

[37] Porte, J., "Quelques extensions du théorème de déduction ", Revista de la Unión Matemática Argentina, 20 (1960), 259-266.

[38] Porte, J., "The deducibilities of S5", Journal of Philosophical Logic, forthcoming.

[39] Schilpp, P.A., editor, The Philosophy of C.I. Lewis, Cambridge University Press, Cambridge, England, 1968.

[40] Seldin, J.P., Review of [4], The Journal of Symbolic Logic, 35 (1970), 147-148.

[41] Smiley, T.J., "Entailment and deducibility", Proceedings of the Aristotelian Society, 59 (1958-1959), 223-254.

[42] Surma, S.J., "Twierdzenia o dedukcji niewprost", Studia Logica, 20 (1967), 151-166.

[43] Surma, S.J., "Twierdzenia o dedukcji dla implikacji zstępujących", Studia Logica, 22 (1968), 61-77.

[44] Surma, S.J., "Deduction theorems in modal systems constructed by Gödel's method", Zeszyty Naukowe Universytetu Jagiellońskiego, 265, Prace z Logiki, 6, (1971), 69-83.

[45] Surma, S.J., "The deduction theorems valid in certain fragments of the Lewis system S2 and the system T of Feys-von Wright", Studia Logica, 31 (1972), 127-136.

[46] Tarski, A., "Über einige fundamentale Begriffe der Metemanthematik", Sprawozdania z posiedzen Towarzystwa Naukowego Warszawskiego, Wydział 3, 23 (1930), 22-29. French translaltion in [49], pp. 35-43. English translation in [48], pp. 30-37.

[47] Tarski, A., "Grundzüge des Systemenkalkül", Fundamenta Mathematica, 25 (1935) 503-526, and 26 (1936), 283-301. French translation in [50], pp. 69-108. English translation in [48], pp. 342-383.

[48] Tarski, A., Logic, Semantics, Metamathematics ; paper from 1923 to 1938, The Clarendon Press, Oxford, 1956. (edited by J.H. Woodger).

[49] Tarski, A., Logique Sémantique, Métamathématique, 1923-1944, (edited by G. Granger), vol. 1, A. Colin, Paris, 1972.

[50] - id - vol. 2, A. Colin, Paris, 1974.

[51] Tokarz, M., "Deduction theorem for RM and its extensions", Studia Logica, 38 (1979), 105-111.

[52] Żarnecka-Biały, E., "A note on deduction theorem for Gödel's propositional calculus G4", Studia Logica, 23 (1968), 35-40.

[53] Żarnecka-Biały, E., "The deduction theorem for Gödel's propositional calculus G5", Zeszyty Naukowe Universytetu Jagiellońskiego, 233, Prace z Logiki, 5, (1970), 77-78.

[54] Zeman, J.J., "The deduction theorem in S4, S4.2, and S5", Notre Dame Journal of Formal Logic, 8 (1967), 56-60.

PROCEEDINGS OF THE HERBRAND SYMPOSIUM
LOGIC COLLOQUIUM '81, J. Stern (editor)

BOUNDING GENERALIZED RECURSIVE FUNCTIONS OF ORDINALS BY EFFECTIVE FUNCTORS ; A COMPLEMENT TO THE GIRARD THEOREM

J.P. Ressayre *

C.N.R.S., U.E.R. de Maths, Tour 45-55, 5e étage,
Université de Paris 7, 2 place Jussieu 75231 Paris Cedex 05

We give a new proof, and some extensions, of Girard's theorem on the growth of primitive recursive functors from ordinals to ordinals.

INTRODUCTION

Functors from ordinals to ordinals are a basic tool of Girard's Π^1_2 - Logic. We first recall the most basic facts concerning these functors. A detailed exposition is made in (G1, Ch. I).

Denote by OL be the category of linear orderings and order preserving maps ; let ON be the restriction of OL to ordinals, and $ON\upharpoonright\omega$ the restriction to integers. The fact that OL is closed under directed limits, and that every linear ordering is a directed limit of finite orderings, allows to extend every functor $F : ON\upharpoonright\omega \to OL$ to a functor $\bar{F} : OL \to OL$, commuting to directed limits :

I.1. Definition of $\bar{F}$ - a - In order to define $\bar{F}(x)$, where $x \in OL$, one chooses any directed system $(x_i, f_{ij})_{i<j\in I}$ of finite $x_i \in OL$, whose limit is x (such a system exists, because x is the union of its finite suborderings !). Since the x_i's are finite, one can assume they are integers ; then since F is a functor over $ON\upharpoonright\omega$, $(F(x_i), F(f_{ij}))_{i<j\in I}$ is defined and is a directed system in OL. One then puts

$$\bar{F}(x) =_{df} \varinjlim (F(x_i), F(f_{ij}))_{i<j\in I} .$$

If f is a morphism of OL, that is an order preserving map : $x \to x'$ with $x, x' \in OL$, then one defines $\bar{F}(f) : \bar{F}(x) \to \bar{F}(x')$ in a similar way, by choosing a directed system S_f in $ON\upharpoonright\omega$, such that $\varinjlim S_f = f$, and putting

$$\bar{F}(f) =_{df} \varinjlim F(S_f).$$

- b - It is easy to see that the functor $\bar{F}$ commutes to $\varinjlim$, that is $\bar{F}(\varinjlim S) = \varinjlim \bar{F}(S)$, for every directed family S. This property

---

* An essential step of the work exposed here was done not by the expositor, but by L. Harrington -see the last remark of § II. The author is also indebted in J. Van de Wiele, who pointed out an error, and provided the result which allows to correct it -see th. III-4. And the overall debt of this work to Girard's Π^1_2 - Logic will be obvious, but in addition the author wants to thank Girard for his patience and for stimulating conversations.

characterizes the functor $\bar{F}$ up to canonical isomorphism : if G is any other functor : $OL \to OL$ such that G commutes to $\underrightarrow{\lim}$ and $G \upharpoonright (ON \upharpoonright \omega) = F$, then there is a unique and canonical isomorphism between $\bar{F}$ anf G.

This starts to explain why such functors are a powerful tool : although they operate on the whole class OL, they are determined by their restriction to $ON \upharpoonright \omega$ - hence by a finite amount of information, if this restriction can be coded in a recursive way.

- c - Given $G = OL \to OL$, here is the practical criterion to decide whether G commutes to $\underrightarrow{\lim}$. If $x \in OL$ and $z \in G(x)$, we say that z has a finite support if there exists $p \in \omega$ and an order morphism $f : p \to x$ such that $z \in$ range of $G(f)$; and we say that G has the finite support property if $\forall x \in OL$ $\forall z \in G(x)$, z has a finite support (this property is the categorical analog of the property $G(x) = \bigcup_{X \text{ finite} \subset x} G(x)$, when G is any function from sets to sets). Then

G commutes to $\underrightarrow{\lim}$ iff G has the finite support property.

Let us now consider $F : ON \upharpoonright \omega \to ON$, and try to extend F to a functor from ON to ON (instead of OL) :

I.2. Remarks - a - If $\bar{F}(x)$ is well ordered for every well ordering x, then the restriction of $\bar{F}$ to well-orderings is isomorphic to a functor from ON to ON, which extends F and which we still denote by F : for $x \in On$, F(x) is defined by $F(x) = |\bar{F}(x)|$ (where $|R|$ denotes the type of the ordering R, hence $|R| \in On$ for each well ordering R). And F(f) is defined in a similar way. Since F is essentially isomorphic to a restriction of $\bar{F}$, F also commutes to $\underrightarrow{\lim}$.

- b - If $F = ON \upharpoonright \omega \to ON$ has some extension to a functor $G : ON \to ON$ (not necessarily commuting to $\underrightarrow{\lim}$) then it is easy to see that $|\bar{F}(x)| \leqslant G(x)$ for every $x \in On$; hence $\bar{F}(x)$ is well ordered for every well ordered x. By (a), it implies that F has an extension to a functor $F : ON \to ON$ commuting to $\underrightarrow{\lim}$, and $|\bar{F}(x)| \leqslant G(x)$ implies $F(x) \leqslant G(x)$ for all $x \in On$.

- c - But there are trivial examples of $F : ON \upharpoonright \omega \to ON$, such that $\bar{F}(\omega)$ already is not well ordered; in such cases F has no extension to a $G : ON \to ON$.

In fact :

I.3. Theorem (Girard). The set of (integers coding) all primitive recursive functors : $ON \upharpoonright \omega \to ON \upharpoonright \omega$, which do have an extension from ON to ON, is Π^1_2 - complete.

I.4. Remarks and notations - a - Here we assume all hereditarily finite sets, in particular all elements of $ON \upharpoonright \omega$, to be coded by integers in the usual way ; and we say that $F : ON \upharpoonright \omega \to ON \upharpoonright \omega$ is p.r. if the function : $\omega \to \omega$, which represents F via the coding, is p.r..

- b - Hence forth we shall be interested only in functors from ON to ON which commute to $\underrightarrow{\lim}$, hence by I.1.-b- are completely determined by their restriction to $ON \upharpoonright \omega$. This induces the following abusive notations, for a functor F from OL to OL or ON to ON : we say that F is a primitive recursive functor iff F(n) is finite for all $n \in \omega$, $F \upharpoonright (ON \upharpoonright \omega)$ is p.r. (in the sense indicated in (a)) and F commutes to $\varinjlim$. Similarly, if α is an admissible ordinal, we shall say that F is α-finite iff $F \upharpoonright (ON \upharpoonright \omega)$ is α-finite (in other words, is an element of L_α) and F commutes to $\varinjlim$.

\- c - The above notation should not induce a confusion between these p.r. functors from ON to ON, and set - p.r. functions from On to On, in the sense of Jensen-Karp : in the case of functors, the primitive recursion is over ω, whereas in set - p.r., it is over On - a big difference !

\- d - Th. I.3 is a Π^1_2 analogue of the Π^1_1- completeness of O, tied to the "β-completeness theorem" of (G2). It is one of a host of results of Girard and collaborators which show the relevance for Logic of the study of these functors : see the bibliography in (G1) , (G3).

It is easy to see that all functors $ON \to ON$ are non decreasing functions; Girard started to study the growth of these functions. The present work is a sequel of his study. The question is tied to the study of admissible ordinals and ordinal recursive functions ; so we shall assume familiarity with their basic theory : for all unexplained notations and properties (especially in § III), see (B). We only recall that an α-recursive function is a function : $\alpha \to \alpha$ which is $\underset{\sim}{\Sigma}$ - definable inside L_α ; that KP denotes the Kripke-Platek axioms of set theory, that the transitive sets satisfying KP are called the admissible sets, and the ordinals α such that $L\alpha$ is admissible are called the admissible ordinals. We recall Girard's results.

I.5. Theorem (Girard) If α is admissible and F is an α-finite functor : $ON \to ON$, then α is closed under the function : $\gamma \mapsto F(\gamma)$, and the restriction to α of this function is α -recursive.

Proof - Because of its canonical character, the application $x \mapsto \bar{F}(x)$ is easily shown to be $\underset{\sim}{\Sigma}$ inside L_α. Then the result follows from $F(x) = |\bar{F}(x)|$ (R.I.2.a) and the fact that the application which to every well ordering $R \in L_\alpha$ associates its type $|R| \in \alpha$ is Σ, inside every admissible L_α.

Conversely Girard, (G.4), showed the surprising fact that for quite a large class of ordinals of the form α^+ (= the next admissible ordinal to α), every α^+-recursive function is eventually bounded by a primitive recursive functor. In the particular case $\alpha = \omega$, another proof of his results has been given by Masseron, (M). And similar results have been obtained by Van de Wiele, (VdW) when the class of α^+-recursive function is replaced by other classes of generalized recursive functions ; with a remarkable corollary concerning $(\infty,0)$-recursive functions. We shall give a new proof of Girard's results and some extensions of them. In order to state the results precisely, let us say that an ordinal α is β-definable iff there is a Σ^1_1 sentence $\exists R\theta(R)$ of set theory, such that α is the only ordinal which satisfies $\exists R\theta(R)$ (that is, which can be enriched to a model $(\alpha, \in\upharpoonright\alpha, X)$ of $\theta(R)$; this terminology follows (G.4)).

I.6. Theorem (Girard ; see also (M), in the case $\alpha = \omega$). If every ordinal $\leq \alpha$ is β-definable, then (1) $\alpha^+ = \sup F(\alpha)$, where F ranges over all p.r. functors : $ON \to ON$. (2) For every α^+-recursive function $f : \alpha^+ \to \alpha^+$, there is a p.r. functor $F : ON \to ON$ such that $f(\gamma) \leq F(\gamma)$ for all $\gamma \in [\alpha, \alpha^+[$.

I.7. Remarks - a - Part (1) of the theorem is an immediate consequence of Part (2).

- b - Let s_o denote the first non β-definable ordinal ; Girard showed that the conclusion of I.6. is false when $\alpha = s_o$. So s_o is the first ordinal α for which his theorem does not hold. We shall see in III.10. that s_o is the first $\underset{\sim}{\Sigma}^1_1$ -reflecting ordinal (see (A-R) for this notion) .

- c - The conclusion of (2) cannot be extended to $\gamma < \alpha$, as is obvious in the case $\alpha = \omega$: firstly, a function f which is $\underset{\sim}{\Sigma}$ in L_{ω^+} can take infinite values even on integers, hence cannot be bounded by $F : \omega \to \omega$. Secundly even if we assume that f maps ω into ω, $f \restriction \omega$ can be any Π^1_1 function, hence can't be bounded by a p.r. F.

- d - Girard's result is surprising because, as we remarked in I.4.c., even if f is only set - p.r., then f can use induction over all ordinals $< \alpha^+$, whereas a p.r. functor F is determined by its restriction to $ON \restriction \omega$, whose definition uses only induction over ω ! One can consider I.6. as a Π^1_2 analogue of Spector's boundedness theorem, saying that every Σ^1_1 ordinal is actually p.r..

In Girard's work, Th. I.6. has a rather indirect proof : it appears as a corollary to a proof theoretic analysis of monotone first order inductive definitions, based on a "functorial" cut elimination theorem which uses Girard's functor Λ . In § II, we expose a new proof of I.6., in which it appears as a corollary to the β-completeness theorem. Since this basic result has a simple proof, we thus give a more direct argument. We shall also give some complements to Th. I.6., which we now summarize.

I.8. Theorem For every ordinal α, the following properties are equivalent :

(o) α is β-definable, moreover every ordinal less than α is Σ-definable from α inside L_{α^+} (that is, $\underset{\sim}{\Sigma}$-definable in L_{α^+}, using α as only parameter).

(1) $\alpha^+ = \sup F(\alpha)$, where F ranges over all p.r. functors : $ON \to ON$.

(2) For every α^+-recursive function f, there is a p.r. functor $F : ON \to ON$ such that $f(\gamma) \leq F(\gamma)$ for all $\gamma \in [\alpha, \alpha^+[$.

I.9. Corollary The class of ordinals α for which the conclusion of Girard's boundedness theorem I.6. holds, is cofinal in the first stable ordinal, σ_o.

In a sense, the first $\underset{\sim}{\Sigma}^1_1$ -reflecting ordinal s_o (see R.I.7.b), and the first stable ordinal σ_o, are quite large countable ordinals. So the conclusion of Girard's boundedness theorem holds every where on a large initial segment s_o of $\aleph_1$, and then holds cofinally in a still much larger initial segment σ_o.

I.10. Notations - a - Let us call total functors all functors $F : ON \to ON$, and partial functors all functors $F : ON \to OL$; for $\alpha \in On$, we then say that $F(\alpha)$ is undefined if $F(\alpha)$ is not well ordered ; and $F(\alpha)$ is defined if $|F(\alpha)| \in On$, in which case we identify $F(\alpha)$ and $|F(\alpha)|$. (These definitions are inspired by analogy with total and partial recursive functions).

- b - For α in On, let $PAD(\alpha)$ denote $\sup\{F(\alpha) : F$ is a total p.r. functor$\}$ (this notation was introduced by Girard, (G.4.)., using the initials of the words "prochain admissible"). And let $PAD'(\alpha)$ denote $\sup\{F(\alpha) : F$ is a partial p.r. functor, such that $F(\alpha)$ is defined$\}$.

It is natural to investigate the relations between properties involving total functors, and corresponding properties and constructions involving partial functors. To begin with, Girard asked the following questions : -1- If F is any partial p.r. functor, is F always bounded by a total p.r. functor, or by the sup of these functors ? -2- What are the possible relations between $PAD(\alpha)$, $PAD'(\alpha)$ and α^+ ?

We shall answer negatively the first question, see IV.7. And we shall give a complete answer to the secund question (see IV.6 and IV.8).

These results are exposed in § IV ; in § III, we gather the facts and results on definability of ordinals that we need. And in § V, we examine the results which one obtains when one does not restrict oneself to functors F such that $F\restriction(ON\restriction\omega)$ is p.r..

§ II - A PROOF OF GIRARD'S BOUNDENESS THEOREM

We let $\mathcal{L}$ be a recursive language containing $\in$ and the distinguished constant C ; for $\gamma \in \mathrm{Cn}$, we call γ-model of $\mathcal{L}$ any model M of $\mathcal{L}$ such that $(C_M, \in^{(M)} \restriction C_M) \simeq (\gamma, \in \restriction \gamma)$, where C_M denotes $\{a \in M : M \models a \in C\}$. If T is a theory and ψ a sentence in $\mathcal{L}$, we write $T \vdash_\gamma \psi$ if ψ is true in all γ-models of T, and $T \vdash_{\leqslant\gamma} \psi$ if $T \vdash_{\gamma'} \psi$ holds for all $\gamma' \leqslant \gamma$.

II.1. <u>Theorem</u> (Girard) - <u>Let</u> T <u>be any recursive theory in</u> $\mathcal{L}$; <u>for each sentence</u> ψ <u>there is a functor</u> F_ψ : ON $\to$ OL <u>such that</u>

-i- <u>the application</u> : $\psi \mapsto F_\psi \restriction (\mathrm{ON} \restriction \omega)$ <u>can be coded in a p.r. way</u>
-ii- <u>for each</u> ψ, F_ψ <u>commutes to</u> $\varinjlim$
-iii- <u>for all</u> γ, $T \vdash_{\leqslant\gamma} \psi$ <u>iff</u> $|F_\psi(\gamma)| \in \mathrm{On}$.
-iv- <u>for all</u> $n \in \omega$, $T \vdash_{\leqslant n} \psi$ <u>iff</u> $F_\psi(n)$ <u>is finite</u>.

The Π^1_2-completeness theorem I.3 is an easy consequence of II.1 and of the well known theorem that $\{\psi \in \mathcal{L} : \vdash_\gamma \psi$ holds for all $\gamma \in \mathrm{On}\}$ is a Π^1_2 complete set. For the sake of completeness and since it is essential for the sequel, we shall give a proof of II.1.

For any letter ℓ, the notation $\bar{\ell}_n$ stands for the finite sequence $\ell_o \ldots \ell_{n-1}$; and $\bar{\ell}$ stands for the infinite sequence $\ell_o \ldots \ell_n \ldots$. If f is any function, $f\,\bar{\ell}_n$ stands for $f(\ell_o) \ldots f(\ell_{n-1})$, and similarly for $f\,\bar{\ell}$. Thus e.g., if $\theta(v_o \ldots v_n)$ is a formula, it can also be written $\theta(\bar{v}_{n+1})$ or $\theta(\bar{v}_n, v_n)$, etc...

We let $h_n(v_o \ldots v_n) = h_n(\bar{v}_n, v_n)$, $n \in \omega$, enumerate all Henkin axioms for $\mathcal{L}$ of the form $\exists u \in C\theta(\bar{v}_n u) \to \theta(\bar{v}_n v_n)$, where $\theta \in \mathcal{L}$. Hence $\bar{h}_n$ denotes the n first Henkin axioms, and $\bar{h}$ all these axioms.

For any sequence $\bar{\alpha} \in \gamma^\omega$ or $\bar{\alpha}_n \in \gamma^{<\omega}$, we call <u>diagram of the sequence</u> its diagram for the relation $\in \restriction \gamma$.

We fix a theory T and a sentence ψ in $\mathcal{L}$. Suppose that M is a $\leqslant \gamma$-model of $T + \neg\psi$. Any model can be enriched to a model of the Henkin axioms $\bar{h}$. So there is a sequence $\bar{\alpha} \in \gamma^\omega$ such that $M \models \bar{h}(\alpha)$, hence the theory $T + \neg\psi + \bar{h}(\bar{\alpha})$ is consistent with the diagram of $\bar{\alpha}$.

Conversely if $\bar{\alpha} \in \gamma^\omega$ is given with the above property, by the Henkin lemma there exists a model M' of $T + \neg\psi$ + diagram of $\bar{\alpha}$, such that $C_{M'} = \{\alpha_n ; n \in \omega\}$. Then if γ' = order type of $\gamma \restriction \{\alpha_n ; n \in \omega\}$, $\gamma' \leqslant \gamma$

and M' is isomorphic to a γ'-model ; so $T + \neg\psi$ has a $\leqslant \gamma$-model.

Thus we have shown that to find a $\leqslant \gamma$-model of $T + \neg\psi$ is the same as to find a sequence $\bar{\alpha} \in \gamma^{\omega}$ with the above property ; and this reduces to find an infinite branch in a certain tree $\mathcal{T}_{\psi}(\gamma)$:

II.2 <u>Notations</u> - a - Write $\bar{\alpha}_n \prec \beta_p$ if $\bar{\alpha}_n$ is a restriction of $\bar{\beta}_p$, that is $n \leqslant p$ and $\bar{\alpha}_n = \bar{\beta}_n$; call a set $\mathcal{T} \subset \gamma^{<\omega}$ a <u>tree</u> if it is a tree for the $\prec$ relation, that is : $\bar{\alpha}_n \in \mathcal{T} \to \bar{\alpha}_m \in \mathcal{T}$ for all $m < n$. Call <u>infinite branches of</u> $\mathcal{T}$ all sequences $\bar{\alpha} \in \gamma^{\omega}$ such that $\bar{\alpha}_n \in \mathcal{T}$ for all n ; and say that $\mathcal{T}$ <u>is well founded</u> if it has no infinite branch, that is if it is well founded for the opposite of the $\prec$ relation.

- b - Let $\mathcal{T}_{\psi}(\gamma) =_{df} \{\bar{\alpha}_n \in \gamma^{<\omega} : \bar{\alpha}_n$ satisfies $(*)\}$, where $(*)$ is the condition :

$(*)$ $T + \neg\psi + \bar{h}_n(\bar{\alpha}_n)$ <u>is consistent with the diagram of $\bar{\alpha}_n$</u> .

The definition of $\mathcal{T}_{\psi}(\gamma)$ implies that it is a tree whose infinite branches are the sequences $\bar{\alpha} \in \gamma^{\omega}$ such that $T + \neg\psi + \bar{h}(\bar{\alpha})$ + diagram of $\bar{\alpha}$ is <u>finitely</u> consistent - hence consistent, by compactness. Then by the above discussion, we have :

(II.3) $T \vdash_{\leqslant\gamma} \psi$ <u>iff</u> $\mathcal{T}_{\psi}(\gamma)$ <u>is well founded</u>

Besides II.3, the crucial property of $\mathcal{T}_{\psi}(\gamma)$ is the following trivial observation - which implies all "functorial" properties of $\mathcal{T}_{\psi}(\gamma)$ used later :

(II.4) <u>if</u> f <u>is any order preserving map</u> : $\gamma \to \gamma'$, <u>then</u> $\bar{\alpha}_n \in \mathcal{T}_{\psi}(\gamma) \leftrightarrow f\,\bar{\alpha}_n \in \mathcal{T}_{\psi}(\gamma')$, <u>for all</u> $\bar{\alpha}_n \in \gamma^{<\omega}$.

For any order preserving map $f : \gamma \to \gamma'$, let $\mathcal{T}_{\psi}(f)$ denote the restriction to $\mathcal{T}_{\psi}(\gamma)$ of the application : $\bar{\alpha}_n \mapsto f\bar{\alpha}_n$ from $\gamma^{<\omega}$ to $\gamma'^{<\omega}$. From II.4 and the definition of $\mathcal{T}_{\psi}(f)$ follows :

(II.5.1) $\mathcal{T}_{\psi}(f)$ sends $\mathcal{T}_{\psi}(\gamma)$ into $\mathcal{T}_{\psi}(\gamma')$, and
$\mathcal{T}_{\psi}(g \circ f) = \mathcal{T}_{\psi}(g) \circ \mathcal{T}_{\psi}(f)$,

(II.5.2) $\mathcal{T}_{\psi}(f)$ preserves α , and its range is an <u>initial part</u> of $\mathcal{T}_{\psi}(\gamma')$ (that is, closed under restriction of sequences) ,

(II.5.3) for every $\bar{\alpha}_n \in \mathcal{T}_{\psi}(\gamma)$ there is $p \in \omega$ and $f : p \to \gamma$ such that $\bar{\alpha}_n \in$ range of $\mathcal{T}_{\psi}(f)$.

If the reader defines the suitable notion of morphism from a tree $\mathcal{T} \subset \gamma^{<\omega}$ to a tree $\mathcal{T}' \subset \gamma'^{<\omega}$, he will obtain a category of trees, such that II.5.1 and II.5.2 say that $\mathcal{T}_\psi(.)$ is a functor from ON into this category of trees ; and II.5.3 says that $\mathcal{T}_\psi(.)$ has the finite support property, hence commutes to $\underrightarrow{\lim}$. We want to convert this functor from ordinals to trees, into a functor $F_\psi : ON \to OL$; for this we use the Brouwer - Kleene procedure to convert trees into linear orderings :

II.6 Notation - For any tree $\mathcal{T} \subset \gamma^{<\omega}$, we denote by $BK(\mathcal{T})$ the ordering

$$\bar{\beta}_q < \bar{\alpha}_n \quad \text{iff} \quad \bar{\alpha}_n, \bar{\beta}_q \in \mathcal{T} \quad \text{and } \underline{\text{either}} \quad \bar{\alpha}_n \prec \bar{\beta}_q$$

$\underline{\text{or}}$ $\quad \exists p < \min(n,q)$ such that $\bar{\alpha}_p = \bar{\beta}_p$ and $\beta_p \in \alpha_p$

(this last condition says approximately that the sequence $\bar{\beta}_q$ stands more on the left of $\mathcal{T}$ than the sequence $\bar{\alpha}_p$ - see picture).

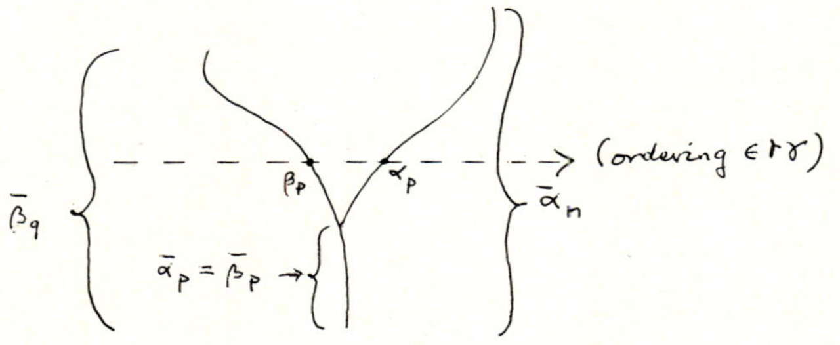

It is well known and easy to check that

(II.7.a) $BK(\mathcal{T})$ is a linear ordering having the same domain as $\mathcal{T}$,

(II.7.b) $\mathcal{T}$ is well founded iff $BK(\mathcal{T})$ is well ordered.

Define F_ψ by $F_\psi(\gamma) = BK(\mathcal{T}_\psi(\gamma))$, and $F_\psi(f) = \mathcal{T}_\psi(f)$ for every order preserving map $f = \gamma \to \gamma'$. From II.5.2 follows that the map $F_\psi(f) : \bar{\alpha}_n \mapsto f\,\alpha_n$ preserves the orderings $BK(\mathcal{T}_\psi(\gamma))$ and $BK(\mathcal{T}_\psi(\gamma'))$. Then from II.5.1 follows that F_ψ is a functor from ON into OL. And the finite support property II.5.3 of $\mathcal{T}_\psi$ immediately implies the same property for F_ψ, since $\mathcal{T}_\psi(\gamma)$ and $F_\psi(\gamma) = BK(\mathcal{T}_\psi(\gamma))$

have the same domain. So F_ψ commutes to $\underrightarrow{\lim}$; and II.3 + II.7.b show that

(II.1.iii) $T \vdash_{\leqslant\gamma} \psi$ iff $|F_\psi(\gamma)| \in On$.

However, by Koenig's lemma, for each $n \in \omega$ $\mathcal{T}_\psi(n)$ is well founded iff it is finite ; this and II.1.iii implies II.1.iv. Thus F_ψ has all the properties required by Th. II.1, except for the primitive recursiveness required by II.1.i. But let (t_n) be an increasing chain of finite parts of T, which is p.r. relative to T, and whose union is T. Let (p_n) be a p.r. enumeration of all proofs in the finitary first order logic of $\mathcal{L}$. And in the definition II.2.b of $\mathcal{T}_\psi(\gamma)$, replace the condition (*) (which decides whether $\bar{\alpha}_n \in \mathcal{T}_\psi(\gamma)$ or not) by

(*)' there is no proof $p(\bar{v}) \in \{ pi; i < n \}$ such that $p(\bar{\alpha}_n)$ derives a contradiction from the axioms ($t_n + \neg\psi + \bar{h}_n(\bar{\alpha}_n)$ + diagram of $\bar{\alpha}_n$).

Then all properties of $\mathcal{T}_\psi$ and F_ψ proved so far remain true. Moreover the condition $\bar{\alpha}_n \in \mathcal{T}_\psi(\gamma)$ is clearly p.r. relative to T and the diagram of γ. And it is clear that for every tree $\mathcal{T} \subset \gamma^{<\omega}$ the ordering $BK(\mathcal{T})$ is p.r. relative to $\mathcal{T}$. These facts easily imply II.1.c, which ends the proof of Th. II.1.

The p.r. character of the definition of F_ψ also yields strong absoluteness properties of this definition :

II.8 <u>Lemma</u> - a - <u>The application</u> : $X \mapsto F_\psi(X)$ <u>is</u> Σ : <u>it is given by a</u> Σ <u>formula</u> $\theta_\psi(v,w)$ <u>such that</u>, <u>for every</u> $X \in OL$, $F_\psi(X)$ <u>is the unique set</u> w <u>which satisfies</u> $\theta_\psi(X,w)$.
- b - <u>Moreover</u> $KP \vdash \forall X \in OL\ \exists!\, w\ \theta_\psi(X,w)$; <u>for every</u> ω-<u>model</u> M <u>of</u> KP <u>and every</u> $X \in M$ <u>such that</u> $M \models X \in OL$, <u>denote by</u> $F_\psi(X)^M$ <u>the only element of</u> M <u>which inside</u> M <u>satisfies</u> $\theta_\psi(X,w)$. <u>Then under canonical identifications of elements of</u> M <u>with sets in</u> V, $F_\psi(X)^{(M)} = F_\psi(X)$.

The above lemma is uniform in ψ ; to make this explicit, we can assume that $\mathcal{L}$ is a recursively presented language included in L_ω. Then if in addition for every model M of KP we identify the well founded part of M with the transitive set isomorphic to it, we have $L_\omega^{(M)} = L_\omega$ for every ω-model M of KP, so formulas of $\mathcal{L}$ are elements of M. Under these conventions we have a uniform version of L.II.8.

II.9 Lemma - a - <u>The application</u> : ψ, X $\mapsto F_\psi(X)$ ($\psi \in \mathcal{L}$, X$\in$ OL) <u>is</u> <u>Σ : it is given by a Σ formula</u> $\theta(u,v,w)$ <u>such that</u> $F_\psi(X)$ <u>is the unique set which satisfies</u> $\theta(\psi,X,w)$.

- b - <u>Moreover this definition is absolute with respect to ω-models of</u> KP : KP$\vdash \forall\psi\in\mathcal{L}\ \forall X\in OL\ \exists!\ w\ \theta(\psi,X,w)$ <u>and for any ω-model</u> M <u>of</u> KP, <u>any</u> $\psi\in\mathcal{L}$ <u>and any</u> X$\in$ M <u>such that</u> M$\models$X$\in$ OL, $F_\psi(X)^{(M)}$ <u>is the unique element which inside</u> M <u>satisfies</u> $\theta(\psi,X,w)$.

- c - <u>Let us denote by</u> F(u,v) <u>the</u> Σ <u>function defined in</u> KP <u>by</u> $\theta(u,v,w)$; <u>and for any</u> $\psi\in\mathcal{L}$ <u>denote by</u> $F_\psi(u)$ <u>the</u> Σ <u>function defined in</u> KP <u>by the formula</u> $\theta_\psi(v,w)$ <u>of</u> L.II.13. <u>Then for any fixed</u> ψ <u>we have</u>

$$KP\vdash \forall X\in OL\quad F(\psi,X) = F_\psi(X).$$

We now turn to the proof of Girard's boundedness theorem I.6 : we consider an infinite ordinal α such that every ordinal $\alpha' < \alpha$ is β-definable ; and an α^+-recursive function f. We shall exhibit a sentence $\varphi = \varphi(C)$ of $\mathcal{L}$, and a recursive theory T such that $|F(\gamma)| \in On$ for all $\gamma\in On$ and $f(\gamma) \leqslant F_\varphi(\gamma)$ for all $\gamma\in[\alpha,\alpha^+[$; where F_φ is the functor : ON$\to$OL given by Th. II.1. This shows I.6.(2), with F = F_φ . We first choose this theory T. By assumption f is $\underset{\sim}{\Sigma}$ definable inside L_{α^+} ; by a theorem of Kripke and Platek, one can assume that $\alpha_o \leqslant \alpha$, where α_o is the parameter of the $\underset{\sim}{\Sigma}$ definition of f (see Th. III.2 below). Then α_o is β-definable, by assumption on α ; let $\exists R_o\,\theta_o(R_o)$ and $\exists R\theta(R)$ be Σ^1_1 sentences which define α_o and α among ordinals. We incorporate individual constants a_o, a and relational constants R_o, R to our language $\mathcal{L}$, and we set

$$T = KP + a_o \leqslant a \leqslant C \wedge \theta_o(R_o)^{(a_o)} \wedge \theta(R)^{(a)} + \neg\exists y\in]a,C]\ (y \text{ is an admissible ordinal}).$$

II.10 <u>Claim</u> - <u>If</u> M <u>is a</u> γ-<u>model of</u> T, <u>then</u> $\gamma\in[\alpha,\alpha^+[$ <u>and</u> $a_o^{(M)} = \alpha_o$, $a^{(M)} = \alpha$.

<u>Proof</u> - Let M be a γ-model of T ; $C_M = \gamma$ and M $a_o < a < C$, so $a_o^{(M)}$, $a^{(M)}$ are standard ordinals. Since α_o, α are the only ordinals satisfying $\exists R_o\theta_o(R_o)$, $\exists R\theta(R)$, the fact $M\models \theta_o(R_o)^{(a_o)} \wedge \ldots \theta(R)^{(a)}$ implies that $a_o^{(M)} = \alpha_o$, $a^{(M)} = \alpha$ - hence $\alpha < \gamma$. Moreover,

$$M\models \neg\exists y\in]a,C]\ (y \text{ is an admissible ordinal})$$

implies $\alpha^+ \leqslant \gamma$. So $\gamma\in[\alpha,\alpha^+[$.

Our sentence φ such that $f(\gamma) \leqslant F_\varphi(\gamma)$ for all $\gamma \in [\alpha,\alpha^+[$ shall be a self referring sentence of the kind constructed by Gödel. To make self reference easier, we remark that every model of T contains a transitive part identified with L_ω, hence contains $\mathcal{L}$ since we assumed $\mathcal{L} \subset L_\omega$; and we go to a definitional extension of $\mathcal{L}$ in the theory T, such that for every $\psi \in \mathcal{L}$ there is a term $\ulcorner\psi\urcorner \in \mathcal{L}$ such that $\ulcorner\psi\urcorner^{(M)} = \psi$, for all $M \models T$.

We denote by A(u) the formula of $\mathcal{L}$:

$$A(u) = \neg\, |F(u,C)| < f(C) ;$$

where F(u,v) is the Σ function considered in L.II.9 (so by this lemma, $F(u,v)^{(M)}$ is the application : ψ, $X \mapsto F_\psi(X)$, in each ω-model M of T) ; and where the parameter α_0 of the $\underset{\sim}{\Sigma}$ definition of f is replaced by the constant a_0.

Using Gödel's trick, we obtain a sentence φ of $\mathcal{L}$, such that $T \vdash \varphi \leftrightarrow A(\ulcorner\varphi\urcorner)$. So

$$T \vdash \varphi \leftrightarrow \neg\, |F(\ulcorner\varphi\urcorner, C)| < f(C),$$

and using L.II.9 and the fact that $\ulcorner\varphi\urcorner^{(M)} = \varphi$ in all models M of T, we conclude :

II.11 - $T \vdash \varphi \leftrightarrow \neg\, |F_\varphi(C)| < f(C)$.

II.12 <u>Claim</u> - $T \vdash_\gamma \varphi$ <u>for all</u> γ

<u>Proof</u> - Suppose not : there is a γ-model M of $T + \neg\varphi(\gamma)$; then by II.11, $M \models |F_\varphi(\gamma)| < f(\gamma)$. And by Cl.II.10, $a^{(M)} = \alpha$, $a_0^{(M)} = \alpha_0$, and $\gamma \in [\alpha,\alpha^+[$; then since $\gamma \in M$, the truncation lemma of F. Ville-Barwise (see (B)) implies that $L_{\gamma^+} = L_{\alpha^+}$ is a transitive part of M, and then $f(\gamma)^{(M)}$ is $f(\gamma)$. Also, since $\gamma \geqslant \alpha$ and α is infinite, M is an ω-model of KP, so by L.II.8.b, $F_\varphi(\gamma)^{(M)} = F_\varphi(\gamma)$.

We thus have $f(\gamma)^{(M)} = f(\gamma)$, $F_\varphi(\gamma)^{(M)} = F_\varphi(\gamma)$ and $M \models |F_\varphi(\gamma)| < f(\gamma)$, which easily implies that $|F_\varphi(\gamma)| < f(\gamma)$. Thus $F_\varphi(\gamma)$ is a (true) well ordering, hence by Th. II.1, $T \vdash_{\leqslant\gamma} \varphi$. Assuming that $T \vdash_\gamma \varphi$ was false, we ended showing it to be true, and this contradiction proves the claim.

From Th. II.1 and the claim follows that $|F_\varphi(\gamma)| \in On$ for all γ : thus F_φ satisfies one of the two properties required by Th. I.6 for the functor F. Assume that F_φ does not satisfy the other : there is $\gamma \in [\alpha, \alpha^+[$ such that $f(\gamma) \leqslant |F_\varphi(\gamma)|$ is false. Then since we already proved $|F_\varphi(\gamma)| \in On$, it implies that $|F_\varphi(\gamma)| < f(\gamma)$. It is clear that the standard universe V can be expanded to a γ-model of T (taking $a_o^{(V)} = \alpha_o$, $a^{(V)} = \alpha$, and choosing $R_o^{(V)}$, $R^{(V)}$ suitably). Then by II.12, $V \models \varphi(\gamma)$, hence by II.11 $|F_\varphi(\gamma)| < f(\gamma)$. Assuming that the conclusion of Th. I.6 was not satisfied when $F = F_\varphi$, we proved that $|F_\varphi(\gamma)| < f(\gamma)$ is both true and false ; this ends our proof of Th. I.6.

Remark - The above proof should mostly be attributed to L. Harrington, since the crucial idea of using a self-referring sentence is due to him. For a detailed account of Harrington's contribution as well as for some complements to the above proof, see (R.1).

§ III - ON THE DEFINABILITY OF ORDINALS

We collect here the remarks and results on definability which we need. We first recall some basic facts, from III.1 to III.3.

III.1. Remarks - a - Assume that α is a limit ordinal, and γ is Σ_1 definable in L_α : there is $\theta(uv) \in \Delta_0$ such that γ is the only element which satisfies $\exists u\theta$ in L_α. Then in any end-extension M of L_α, γ still satisfies $\exists u\,\theta$, but there may be other elements which do so. However, γ remains Σ_1 definable in M, by the formula $\theta'(\gamma)$:

$$" \exists \delta [(L_\delta \models \exists u\ \theta(u\gamma)) \wedge \forall \delta' < \delta\ \ L_{\delta'} \models \neg \exists v \exists u\ \ \theta(uv)\ ]" .$$

- b - Uniformizing (a) with respect to a parameter x, one obtains : if $\psi(xv)$ is any Σ_1 formula whose interpretation in L_α is a partial function f, then there exists a Σ_1 formula $\psi'(xv)$ such that for any end extension M of L_α, $M \models \forall x\ \exists^{\leq 1} v\ \ \psi'(xv)$, and $\psi'^{(M)} \restriction L_\alpha = f$.

- c - It also follows from (a) that Σ_1 definability from a parameter c is Σ_1 definable in L_α : for γ is Σ_1 definable from c inside L_α iff L_α satisfies

$$\exists \delta \in O_n \exists \theta \in \Delta_0\ [(L_\delta \models \exists u\ \theta(u,\gamma,c)) \wedge \forall \delta' < \delta\ \ L_{\delta'} \models \neg \exists v \exists u\ \theta(u,v,c)]$$

III.2. Theorem (Kripke Platek) - Let γ be any ordinal ; for all $\alpha < \gamma$, the collection of all elements which are Σ_1 definable from parameters $< \alpha$ inside L_γ, is a set of the form L_δ, $\delta \leq \gamma$, and L_δ is a Σ - elementary submodel of L_γ.

III.3. Corollary (Kripke Platek) - a - Let α be any ordinal ; every ordinal $< \alpha^+$ is Σ -definable from parameters $\leq \alpha$ inside L_{α^+} .

- b - Let σ_0 be the first stable ordinal ($\longleftrightarrow_{df}$ L_{σ_0} is the smal-smallest Σ-elementary submodel of L). Then σ_0 is equal to the set of all ordinals which are Σ -definable in L.

Proof - See (B).

III.4. Theorem (van de Wiele) - Let α be any ordinal, and A be the set of ordinals which are Σ definable from α inside L_{α^+} ; if A is cofinal in α^+, then $A = \alpha^+$.

Proof - Otherwise let γ = first ordinal not in A ; $\gamma < \alpha^+$ and $L_{\alpha^+} \models (\forall\beta < \gamma)\ \beta$ is Σ_1 definable from α. By III.1.c., this is a Σ statement, hence by Σ reflection, there is $\delta < \alpha^+$ such that $L_\delta \models (\forall\beta<\gamma)\ \beta$ is Σ_1 definable from α. Since A is cofinal in α^+, there is $\delta' \in A$ such that $\delta \leqslant \delta'$. Then $L_{\delta'} \models (\forall\beta < \gamma)\ \beta$ is Σ_1 definable from α. On the other hand, γ is not Σ_1 definable from α inside $L_{\delta'}$ for, by III.1.a., it would remain so in L_{α^+}, contradicting the choice of γ. But thus γ is definable inside L_{α^+} as the first ordinal which is not Σ_1 definable from α inside $L_{\delta'}$. Since $\delta' \in A$, this implies $\gamma \in A$, a contradiction.

The next result is useful in order to correlate various ways of defining ordinals.

III.5. Lemma For any class $\mathcal{C}$ of infinite ordinals, the following conditions are equivalent :

(o) there is a Σ_1^1 sentence $\exists R\theta(R)$ such that $\mathcal{C} = \{\alpha \in On : (\alpha, \in\restriction\alpha) \models \exists R\theta(R)\}$;

(1) there is a recursive theory T, in an extension of the language $\mathcal{L}$, such that $\mathcal{C} = \{\alpha \in On :\ T \text{ has an } \alpha\text{-model}\}$

(2) there is a Π_1 formula $\varphi(v)$ such that $\mathcal{C} = \{\alpha \in On : L_{\alpha^+} \models \varphi(\alpha)\}$.

Proof - (o) → (1) is obvious, and (1) → (o) is a well known coding trick.

(1) → (2) - Given T as in (1), and $\alpha \in On$, for every $\gamma < \alpha$ we add a constant $\underline{\gamma}$ to the language of set theory, and let Δ_α be the conjunction of the diagram of $(\alpha, \in\restriction\alpha)$ and of the sentence $\forall x\,[x \in C \leftrightarrow \bigvee_{\gamma<\alpha} x = \underline{\gamma}]$; Δ_α is an $\mathcal{L}_{\infty\omega}$ sentence of L_{α^+} which has only α-models. Hence T has no α-model iff $\vdash \neg(T \wedge \Delta_\alpha)$. By the Barwise completeness theorem, the relation $\vdash$ is Σ expressible inside L_{α^+} ; so there is clearly a Σ formula $\psi(v)$ such that $\forall\alpha \in On$

$$L_{\alpha^+} \models \psi(\alpha) \quad \text{iff} \quad \vdash \neg (T \wedge \Delta_\alpha).$$

Hence T has an α-model iff $L_{\alpha^+} \models \neg\psi(\alpha)$.

(2) → (1) - Given $\varphi(v) \in \Pi_1$, we let T be KP + $\varphi(C)$. If M is an α-model of T, then by the truncation lemma L_{α^+} is a transitive part of M ; hence $M \models \varphi(\alpha)$ implies $L_{\alpha^+} \models \varphi(\alpha)$. Conversely, if $L_{\alpha^+} \models \varphi(\alpha)$, then (L_{α^+}, α) is a C-model of T. So $\alpha \in \mathcal{C} \leftrightarrow$ T has an α-model.

Our next goal is to compare β-definability with Σ_1 definability inside L :

III.6. Remarks - a - If α is β-definable then α is Σ_1-definable in L ; for let $\exists R\theta(R)$ be a Σ^1_1 formula defining α among all ordinals. By III.5., there is a Π_1 formula $\varphi(v)$ such that α is the only ordinal for which $L_{\alpha^+} \models \varphi(\alpha)$; then

$$\exists u\ [\ u = v^+ \wedge L_u \models \varphi(v)\ ] \quad \text{gives a } \Sigma_1 \text{ definition of } \alpha \text{ in L.}$$

- b - The converse of (a) is false. E.g., let α be the first ordinal such that $L_{\alpha^{++}} \models \alpha$ is uncountable ; this easily gives a Σ_1 definition of α in L. But in order for α to be β-definable, by III.5. we should be able to express this by a Π_1 formula $\varphi(\alpha)$ satisfied in L_{α^+}, which is not possible.

- c - By (a) and III.3.b, the collection of β-definable ordinals is strictly contained in σ_0 ; it is in addition cofinal to σ_0. To see this, let α_0 be any ordinal $< \sigma_0$. By III.3.b, α_0 is Σ_1 definable in L, by some formula $\exists u\theta(uv)$, $\theta(uv) \in \Delta_0$. Let α be the first limit ordinal such that $L_\alpha \models \exists v \exists u\ \theta$; then $\alpha > \alpha_0$, otherwise $L_\alpha \models \exists u\ \theta(u\gamma)$ for some $\gamma < \alpha_0$, hence $L \models \exists u\theta(u\gamma)$ and $\exists u\theta$ would not define α_0 in L. In addition, α is β-definable : for the condition "α = first ordinal such that $L_\alpha \models \exists v \exists u\ \theta$" is easily expressed by a Π_1 sentence $\varphi(\alpha)$ inside L_{α^+}, and we apply III.5..

- d - In addition to being β-definable, the ordinal $\alpha > \alpha_0$ constructed in (c) is such that every element of L_α is Σ_1 definable from α inside L_{α^+} : otherwise, by III.2. these elements form a set L_δ, $\delta < \alpha$, which is a Σ_1 submodel of L_α. Hence $L_\delta \models \exists v \exists u\ \theta$, and L_α would not be the first to satisfy $\exists v \exists u\ \theta$.

Thus α satisfies condition (o) of Th. I.8 ; we have thus shown that the ordinals α for which Girard's bounding theorem I.6 holds are cofinal in σ_0.

Next, we split the notion of β-definability into two parts, both of which will be used when studying partial p.r. functors :

Definition - α is β^- -definable if there is a Σ^1_1 sentence $\exists R\theta$, such that α is the first ordinal satisfying $\exists R\theta$. And α is β^*-definable if it is the last ordinal satisfying some Σ^1_1 sentence $\exists R\theta$.

It is immediate that β-definability $\leftrightarrow$ $(\beta^- + \beta^*)$-definability.

III.7. Proposition (Girard) α is β^--definable iff there is a p.r. functor $F : ON \to OL$ such that α is the first ordinal such that $|F(\alpha)| \notin On$.

Proof - If α is β^--definable, by III.5 there is a recursive theory T such that $T \vdash_{<\gamma} \perp$ iff $\gamma < \alpha$; then by Th. II.1, α is the first ordinal such that $|F_\perp(\alpha)| \notin On$.

III.8. Theorem (Girard) Let α be closed under partial p.r. functors : if F is a partial p.r. functor : $ON \to ON$, then $\gamma < \alpha \to F(\gamma) < \alpha$ whenever $F(\gamma)$ is defined. Then α is β^--definable iff α is β-definable. In particular β-definability and β^--definability are equivalent for all ordinals which are admissible or limit of admissible ordinals.

Proof - Assume that α is β^--definable : by III.7, there is a p.r. functor F such that α is the first ordinal for which $|F(\alpha)| \notin On$. By the closure hypothesis on α, we have $F''\alpha \subset \alpha$. This allows to give a β-definition of α : let T be $KP + (\forall u < C) F(u) < C + F(C) \notin On$; then if M is a γ-model of T, $F(C)^{(M)} = F(\gamma)$, so $F(\gamma)$ is not well ordered hence $\gamma > \alpha$. But for $\delta < \delta$, $M \models F(\delta) < C$ implies $F(\delta) < \delta$ hence $F(\delta)$ well ordered ; this implies $\gamma < \alpha$, hence $\gamma = \alpha$. So α is the only ordinal such that T has an α-model, and by III.5, α is β-definable.

III.9. Remark - Assume as above that α is the first ordinal such that $F(\alpha)$ is not defined (that is not a well ordering). But assume that α is not closed under the functor F ; then the above theory $T = KP + (\forall u < C) F(u) < C$ is not a β-definition of α, because

$\forall u < \alpha\, F(u) < \alpha$ is false. But one is tempted to believe that $T' = KP + \forall u < \alpha\, |F(u)| \in On$ is a β-definition of α. If this was true, β^--definability would always be equivalent to β-definability, which is not the case (see IV.7). The reason why T' is not always a β-definition of α is that even if $F(\alpha)$ is not well ordered, there sometimes exists <u>non standard</u> models of KP containing α, which "believe" that $F(\alpha)$ is well ordered ; such models allow T' to have γ-models, with $\gamma > \alpha$.

Let s_o denote the first non β-definable ordinal : it is the first α for which Girard's boundeness theorem I.6 does not hold ; to see this, we are going to study s_o.

<u>Fact 1</u> - If $\alpha < s_o$, then every ordinal $\gamma < \alpha$ is Σ-definable from α inside L_{α^+} .

<u>Proof</u> - By I.6., we know that $\alpha = \sup\{F(\alpha) : F$ total p.r. functor : $ON \to ON\}$. By Th. I.5, this implies that $\alpha = \sup\{\delta : \delta$ is Σ-definable from α in $L_{\alpha^+}\}$; which by III.4 implies Fact 1.

<u>Fact 2</u> - s_o is limit of admissible ordinals.

<u>Proof</u> - If not, let α be the last ordinal $< s_o$ which is admissible or limit of admissible ; so $\alpha < s_o < \alpha^+$, and α is β-definable. It is easy to see that if α is β-definable, then α^+ also is β-definable ; so $s_o \neq \alpha^+$, and $\alpha < s_o < \alpha^+$. Then combining a Σ_1 definition of s_o from α (which exists in L_{α^+}, by Fact 1), and a β-definition of α, one obtains a β-definition of s_o. This being absurd, Fact 2 is proved.

III.10 <u>Theorem</u> (Girard except (d)) - <u>Let</u> s_o <u>be the first non β-definable ordinal</u> - a - s_o <u>is the first non β^--definable ordinal</u>. - b - s_o <u>is the first ordinal</u> α <u>such that for every partial p.r. functor</u> $F : ON \to ON$, $\alpha \subset$ dom. $F \to \alpha \cup \{\alpha\} \subset$ dom. F. - c - s_o <u>is the first counter example to Girard's boundedness property I.6</u>. - d - s_o <u>is the first</u> Σ^1_1 <u>-reflecting ordinal</u>.

<u>Proof</u> - By Fact 2 and III.8, s_o is not β^--definable, for it would then be β-definable ; this implies (a) ; and (b) is equivalent, by III.7. The condition, for any p.r. functor F, that $F''s_o \subset s_o$,

is $\underset{\sim}{\Delta}_o$ inside $L_{s_o^+}$; then the admissibility of $L_{s_o^+}$ implies

$$s_o^+ > \sup\{F(s_o) : F \text{ p.r. functor such that } F'' s_o \subset s_o\}$$

and since s_o is limit of admissibles, $F'' s_o \subset s_o$ is true for every total p.r. F, hence

$$s_o^+ > \sup\{F(s_o) : F \text{ total p.r. functor} : ON \to ON\}$$

showing (c). To see (d), note that s_o is non β^--definable iff (for every Σ_1^1 sentence $\exists R\theta(R)$, $s_o \models \exists R\theta(R) \to \exists\gamma < s_o$, $\gamma \models \exists R\theta(R)$), in other words iff s_o is Σ_1^1 reflecting. We have to show that s_o is $\underset{\sim}{\Sigma}_1^1$ -reflecting, which is the same property asserted for formulas $\exists R\theta(R)$ which may contain parameters $< s_o$. But here the parameters, being β-definable, are easy to eliminate ; hence s_o is $\underset{\sim}{\Sigma}_1^1$ -reflecting. And if $\alpha < s_o$, α is β-definable, a fortion β^--definable, hence non Σ_1^1-reflecting, a fortion non $\underset{\sim}{\Sigma}_1^1$-reflecting.

III.11 <u>Remark</u> - s_o is β^*-definable : it is the last γ such that the theory KP + C is admissible + $\forall\alpha < C$ (α is β^--definable) has a γ-model.

III.12 <u>Theorem</u> - <u>If there is a well ordering</u> $\rho \subset \omega^2$ <u>of type</u> α, <u>such that</u> $\{\rho\}$ <u>is a</u> Π_1^1 <u>singleton, then</u> α <u>is</u> β^*-<u>definable</u>.

<u>Proof</u> - Suppose that $\rho \subset \omega^2$, $|\rho| = \alpha$ and ρ is the only relation on ω such that $\langle\omega, \rho, \mathcal{P}(\omega)\rangle$ satisfies the Π_1^1 sentence $\forall S\varphi(RS)$. Then it is known that $\rho \in L_{\alpha^+}$ (see (R2)) ; hence ρ is Σ definable in L_{α^+} as the only ρ such that

$(\Delta_\omega \wedge \text{diagramm of } \rho) \vdash \varphi(RS)$, where Δ_ω is as defined in the proof of III.5. Then α is the last ordinal such that $L_{\alpha^+} \models \varphi(\alpha)$, where $\varphi(\alpha)$ is the Π_1 sentence :

$$\forall\ \gamma < \alpha \neg [\exists\ \rho \subset \omega^2\ |\rho| = \gamma \wedge (\Delta_\omega \wedge \text{ diagram of } \rho) \vdash \varphi(RS)\ ].$$

Hence by III.5., α is β^*-definable.

III.13 <u>Remark</u> - Every ordinal $\alpha < s_o$ is Σ-definable from s_o inside $L_{s_o^+}$:

by Fact 2, $\alpha < s_o \to \alpha^+ < s_o$; and since α is β-definable, by III.5, there is a Π_1 formula $\varphi(v)$ such that α is the only ordinal for which $L_{\alpha^+} \models \varphi(\alpha)$. Then α is defined, in L_{s_o} hence in $L_{s_o^+}$, by $\theta(u)=$

$$\exists v\ \ [v = u^+ \wedge L_v \models \varphi(u)\ ].$$

§ IV - THE CONVERSE TO GIRARD'S BOUNDEDNESS THEOREM, AND EXTENSIONS TO PARTIAL FUNCTORS

We prove Th. I.8, which extends Girard's theorem I.6 to a necessary and sufficient condition; and we investigate the relations between total and partial p.r. functors.

IV.0 Notations - For every ordinal α, $\hat{\alpha}$ will denote the sup of all ordinals which are Σ definable from α inside L_{α^+}. And recall from § I that $PAD(\alpha) =_{df} \sup\{ F(\alpha) \mid F \text{ total p.r. functor} : ON \to ON \}$

$PAD'(\alpha) =_{df} \sup\{ F(\alpha) \mid F \text{ partial p.r.}: ON \to ON, \text{ such that } \alpha \in \text{dom } F \}$.

Remark - The application $\alpha \mapsto PAD(\alpha)$ can be extended to a functor $ON \to ON$, as was shown by Girard, (G.4). Since for every p.r. functor $F : ON \to ON$ we have $F(\alpha) < \alpha^+$, and since $PAD(\alpha) = \alpha^+$ for many α's, the functor PAD is not p.r. - in fact $PAD \restriction (ON \restriction \omega)$ is Π^1_2-complete.

Van de Wiele's theorem III.4 can be reformulated as follows.

IV.1 Theorem - The following three conditions are equivalent : (i) $\hat{\alpha} = \alpha^+$ (ii) every ordinal $\gamma < \alpha$ is Σ definable from α inside L_{α^+} (ii)' every ordinal $\gamma < \alpha^+$ is Σ definable from α inside L_{α^+}.

Proof - (i) $\to$ (ii)' is Th. III.4, (ii)' $\to$ (ii) is trivial, and (ii) $\to$ (i) follows from Th. III.2.

IV.2 Theorem - If α is β^--definable then $\hat{\alpha} = PAD'(\alpha)$.

Proof - From Th. I.5 follows that $\hat{\alpha} \geq PAD'(\alpha)$. In order to show $\hat{\alpha} \leq PAD'(\alpha)$, we consider any ordinal ν which is Σ-definable from α inside L_{α^+}, and we let $\sigma(\alpha v)$ be a Σ definition of ν from α. Moreover we let $\exists R\, \theta(R)$ be a Σ^1_1 sentence such that α is the first ordinal to satisfy it.

We consider the theory $T = KP + \theta(R)^{(C)}$, and choose a sentence φ such that

$$T \vdash \varphi \leftrightarrow \neg \exists v\, [\sigma(C,v) \wedge |F_\varphi(C)| < v\,].$$

Claim - $T \vdash_{\leqslant\alpha} \varphi$.

Proof - Assume that $\gamma \leqslant \alpha$ and M is a γ-model of T ; then $\gamma \geqslant \alpha$, since $\gamma \models \exists R\, \theta(R)$. So $\gamma = \alpha$, and by the truncation lemma, L_{α^+} is a transitive part of M. Then $\nu \in M$, and if we had $M \models \neg\varphi$, it would imply $|F_\varphi(\alpha)| < \nu$, hence by Th. II.1 $T \vdash_{\leqslant\alpha} \varphi$, while M was supposed an α-model of $T + \neg\varphi$. So $M \models \varphi$, and the claim is proved.

It implies that $|F_\varphi(\alpha)| \in On$ (by II.1) ; and that $\neg |F_\varphi(\alpha)| < \nu$, hence $|F_\varphi(\alpha)| \geqslant \nu$ (because V can be expanded to an α-model of T, hence of φ by the Claim, hence of $\neg\exists v\ [(\sigma,v) \wedge |F_\varphi(\alpha)| < v]$ by choice of φ).

IV.3 Theorem - If α is β-definable, then $\hat{\alpha} = PAD(\alpha) = PAD'(\alpha)$.

Proof - Similar to that of IV.2 and of I.6.

IV.4 Theorem - The following conditions are equivalent for every α :

(o) α is β-definable and $\hat{\alpha} = \alpha^+$

(o)' α is β^--definable and $\hat{\alpha} = \alpha^+$

(1) $PAD(\alpha) = \alpha^+$ (1)' $PAD'(\alpha) = \alpha^+$

(2) For every total α^+-recursive function f, there is a total p.r. function F such that $f(\gamma) < F(\gamma)$ for all $\gamma \in [\alpha,\alpha^+[$.

Note that if we drop conditions (o)' dans (1)', then by IV.1 this result is equivalent to Th. I.8.

Proof - (o) $\rightarrow$ (2) - Let f be a total α^+-recursive function ; By (o) and IV.1, every ordinal $<\alpha$ is Σ -definable from α inside L_{α^+}. Then by Th. III.2, we can assume that α is the only paramater in the $\underset{\sim}{\Sigma}$ definition of f. Let $\exists R\, \theta(R)$ be a Σ^1_1 definition of α among all ordinals, let T be $KP + \theta(R)^{(a)} + a < C + \neg\exists y \in\,]a,C]$ (y is an admissible ordinal) ; and choose a sentence φ such that

$$T \vdash \varphi \leftrightarrow \neg |F_\varphi(C)| < f(C)$$

where the symbol a replaces α in the $\underset{\sim}{\Sigma}$ definition of f. Arguing as in the proof of Th. I.6, one shows that F_φ is a total p.r. functor such that $f(\gamma) \leq F_{\varphi}(\gamma)$ for all $\gamma \in [\alpha,\alpha^+[$

<u>(2) → (1)</u> is obvious.

<u>(1) → (o)</u> - Assume $PAD(\alpha) = \alpha^+$; then $\hat{\alpha} = \alpha^+$ is clear, so there remains to see that α is β-definable. Let

$\alpha_1 = \min \{\gamma :$ there is a <u>total</u> p.r. functor F such that $F(\gamma) \geq \alpha \}$

By the choice of α_1, there is a total p.r. functor F_1 such that $F_1(\alpha_1) \geq \alpha$; and α_1 is <u>closed under total p.r. functors</u> (for otherwise, having $F(\gamma) > \alpha_1$ for some $\gamma < \alpha_1$ and some total p.r. functor F, we would have $F_1 \circ F(\gamma) \geq \alpha$, (contradicting the choice of α_1) since the composite $F_1 \circ F$ of two total p.r. functors is a total p.r. functor.

<u>Claim</u> - <u>There is a partial functor</u> G <u>such that</u> $G'' \alpha_1 \subset \alpha_1$ <u>and</u> $G(\alpha_1)$ <u>is not defined</u> (in other words, $\bar{G}(\alpha_1)$ is not a well ordering).

<u>Proof</u> - Otherwise, $G(\alpha_1) \in On$, for every partial p.r. G such that $G'' \alpha_1 \subset \alpha_1$.

And

$$\begin{aligned} \alpha^+ &> \sup \{G(\alpha_1) : G \text{ partial p.r. functor such that } G'' \alpha_1 \subset \alpha_1\} \\ &\geq \sup \{F(\alpha_1) : F \text{ total p.r. functor}\} \\ &\geq \sup \{F \circ F_1(\alpha_1) : F \text{ total p.r. functor}\} \\ &\geq \sup \{F(\alpha) : F \text{ total p.r. functor}\} = PAD(\alpha) \end{aligned}$$

and this contradicts $\alpha^+ = PAD(\alpha)$.

(The > relation holds because the condition $G''\alpha_1 \subset \alpha_1$ on G is $\underset{\sim}{\Delta}_o$ inside L_{α^+} ; the other relations are easily checked). The functor G of the claim allows to prove that α_1 is β-definable, by the same argument as in the proof of Th. III.8. And by the choice of α_1, $PAD(\alpha_1) = PAD(\alpha)$, so $PAD(\alpha_1) = \alpha_1^+ = \hat{\alpha}_1$, which by IV.1 implies that α is Σ definable from α_1 inside $L_{\alpha_1^+} = L_{\alpha^+}$. Then the β-definability of α_1 easily implies that of α.

<u>(o) → (o)'</u> is obvious, and <u>(o)' → (1)'</u> follows from Th. IV.2.

<u>(1)'→ (o)</u> - Clearly (1)' implies $\hat{\alpha} = \alpha^+$.

<u>Claim 1</u> - If $\hat{\alpha} = \alpha^+$, then $\gamma^+ = \alpha^+$ implies $\hat{\gamma} = \gamma^+ = \alpha^+$, for every ordinal γ.

<u>Proof</u> - If $\gamma^+ = \alpha^+ = \hat{\alpha}$, then there exists a pair (α_1, α_2) such that $\alpha_1 \leq \alpha_2 < \alpha^+$ and every $x < \gamma$ is Σ_1 definable from α_1 inside L_{α_2} ;

for we can take $\alpha_1 = \alpha$, and obtain α_2 by an argument used in the proof of III.4.
The smallest such pair (α_1, α_2) is Σ definable from γ inside L_{α^+} ; and by definition of this pair (α_1, α_2), every ordinal $x < \gamma$ is Σ definable from α_1 inside L_{α_2}, hence inside L_{α^+}. Thus every $x < \gamma$ is Σ definable from γ inside $L_{\gamma^+} = L_{\alpha^+}$, and by IV.1 $\hat{\gamma} = \gamma^+ = \alpha^+$.

Now let α_1 be the smallest ordinal such that $\alpha_1^+ = \alpha^+$; if there is no β^-- definable ordinal between α_1 and α, then by III.7 it means that for every p.r. functor G, G defined below α_1 implies G also defined at α. Moreover, G defined below $\alpha_1 \Leftrightarrow G''\alpha_1 \subset \alpha_1$, since α_1 is admissible or limit of admissible ordinals; so PAD'(α) = sup $\{G(\alpha)$:G. p.r. functor such that $G(\alpha)$ is defined$\}$ = sup $\{G(\alpha)$: G p.r. functor such that $G''\alpha_1 \subset \alpha_1\} < \alpha^+$, contradicting our assumption (1)'. So there <u>is</u> a β^- - definable ordinal between α_1 and α, and we let α_0 be the smallest one. This α_0 is closed under partial p.r. functors, hence is β-definable by Girard's result III.8. For otherwise, there is $\gamma < \alpha_0$ and a p.r. functor G such that $\alpha_0 < G(\gamma) \in 0_n$; and it is easy to see that the smallest such γ is β^- -definable and contradicts the choice of α_0.

We have thus shown that α_o is β-definable ; moreover, $\alpha_o{}^+ = \alpha^+$ implies that α is Σ-definable from α_o inside $L_{\alpha_o^+}$ (by Claim 1 applied with $\gamma = \alpha_o$). Using the β-definition of α_o, and the Σ definition of α from α_o, one easily shows that α itself is β -definable. This shows (o), and ends the proof of Th. IV.4.

IV.5 <u>Remark</u> - In Theorem IV.4, none of the two parts of condition (0) or (0') can be omitted : we have seen in III.13 that $\hat{s}_o = s_o{}^+$, although s_o is not β^--definable. And we next define an ordinal λ which is β-definable but satisfies $\hat{\lambda} < \lambda^+$: let λ be the first ordinal such that λ is not countable in L_{λ^+}. Then $\alpha < \lambda$ implies $\alpha^+ < \lambda$; for otherwise, $\alpha < \lambda < \alpha^+ \leqslant \lambda^+$ and α countable in L_{α^+} would imply λ countable inside L_{α^+} hence inside L_{λ^+}, by use of III.3. Then λ is the only ordinal such that $L_{\lambda^+} \models \varphi(\lambda)$, where $\varphi(v)$ is the Π formula

"$v \in On \wedge v$ is not countable $\wedge \forall u < v$ ($u^+ < v$ and u is countable in L_{u^+} ". Hence λ is β-definable, by III.5.

Moreover $\hat{\gamma} < \lambda^+$ is true for every $\gamma < \lambda^+$, in particular $\gamma = \lambda$. Otherwise, by III.4 every ordinal $< \lambda^+$ is Σ definable from γ inside L_{λ^+}, in particular

$$L_{\lambda^+} \models \forall\alpha < \lambda \ [\alpha \text{ is } \Sigma_1 \text{ definable from } \gamma\ ].$$

This statement is Σ by III.1.c, so by Σ-reflection, there is $\delta < \lambda^+$ such that $L_\delta \models \forall\alpha < \lambda$ (α is Σ_1 definable from γ). Then in L_δ, there is a Σ injection $f : \lambda \to L_\omega$; namely the application $f : \alpha \mapsto$ the smallest Σ_1 definition of α from γ. Since $\delta < \lambda^+$, $f \in L_{\lambda^+}$, so λ would be countable in L_{λ^+} contradicting the definition of λ .

We shall now systematically study the possible relations between α^+, PAD(α) and PAD'(α) ; Th. IV.4 settled the case $\alpha^+ =$ PAD(α) = PAD'(α). The next result yields cases where $\alpha^+ >$ PAD(α) = PAD'(α).

IV.6 <u>Theorem</u> - <u>If</u> α <u>is</u> β^*-<u>definable</u>, <u>then</u> PAD(α) = PAD'(α).

<u>Proof</u> - Assume that α is β^*-definable : there is a recursive theory T_0 such that α is the last ordinal for which T_0 has an α-model. We can assume that the symbol $\in$ is replaced in T_0 by another symbol ε. Given a p.r. functor F such that $|F(\alpha)| \in$ On, we let T be the theory KP + T_0, and ψ a sentence of $\mathcal{L}$ such that $T \vdash \psi \leftrightarrow \neg\, |F_\psi(C)| < |F(C)|$.

<u>Claim</u> - $T \vdash_\gamma \psi$ <u>for all</u> γ.

<u>Proof of claim</u> - Let M be a γ-model of T ; this implies $\gamma \leqslant \alpha$ since T contains T_0 and $T_0 \vdash_{>\alpha} \bot$. Hence $|F(\gamma)| \in$ On . So if $M \models \neg\psi$, , hence $M \models |F_\psi(\gamma)| < |F(\gamma)|$, it would imply that $|F_\psi(\gamma)| \in$ On, hence $T \vdash_{\leqslant\gamma} \psi$, in contradiction to $M \models T + \neg\psi$ Thus $M \models \psi$, and the claim is proved.

It implies that F_ψ is a total p.r. functor : ON $\to$ ON. Moreover assume $\gamma \leqslant \alpha$; since we can expand V to a γ-model of T, it implies $V \models \psi$, hence $V \models \neg\, |F_\psi(\gamma)| < |F(\gamma)|$; since $F(\gamma)$ and $F_\psi(\gamma)$ are here both well ordered, we conclude $|F(\gamma)| \leqslant |F_\psi(\gamma)|$.

As an example, if s_o is the first non β-definable ordinal, then since s_o is β^*-definable (R.III.11) and $PAD(s_o) < s_o^+$, we conclude $PAD(s_o) = PAD'(s_o) < s_o^+$. And if α is any "Π^1_1-singleton ordinal", from IV.6 and III.12 we conclude that $PAD(\alpha) = PAD'(\alpha)$.

Until now, we have no example of a β^--definable ordinal α which is not β-definable, and no example of α such that $PAD(\alpha) \neq PAD'(\alpha)$. This is provided by

IV.7 Theorem - Let λ be the first ordinal which is not countable inside L_{λ^+}, and let μ be the first ordinal $> \lambda$ which is not Σ definable from λ inside L_{λ^+} ($\mu < \lambda^+$ by R. IV.5). Then μ is β^--definable but not β^*-definable. And $PAD(\lambda) = PAD(\mu) < PAD'(\mu) < \mu^+$; so there exists a partial p.r. functor $G : ON \to ON$, such that $G(\mu)$ is defined and $G(\mu) > F(\mu)$, for every total p.r. functor F.

Proof - The theorem follows from the two claims :

Claim 1 - $\mu^+ > \hat{\mu} > \hat{\lambda} = PAD(\lambda) = PAD(\mu)$.

Claim 2 - μ is β^--definable.

Indeed, Cl. 2 and IV.3-a imply that $PAD'(\mu) = \hat{\mu}$, so by Cl. 1

$PAD(\lambda) = PAD(\mu) < PAD'(\mu) < \mu^+$. And if μ was β-definable, Cor. IV.3 would imply $PAD(\mu) = PAD'(\mu)$, which is false.

Proof of Claim 1 - If $PAD(\mu) = \mu^+$ was true, it would imply $\hat{\mu} = \mu^+$, which is false by R.IV.5. Let μ_o be the first ordinal which is not Σ definable from λ inside L_{λ^+} ; $\mu_o < \lambda$ by R.IV.5, and it is easy to see that $\mu = \lambda + \mu_o$, hence $\mu < \lambda + \lambda$. It is clear that $\lambda + \lambda < PAD(\lambda)$(*), so $\lambda < \mu < PAD(\lambda)$. The equality of Cl. 1 then follows by the

Remark - Whenever $\alpha < \gamma < PAD(\alpha)$, then $PAD(\alpha) = PAD(\gamma)$. (Indeed $\alpha \leqslant \gamma \to PAD(\alpha) \leqslant PAD(\gamma)$, since functors are increasing. And let F

(*) We can note that $\lambda + \lambda < \hat{\lambda}$ and apply Th. IV.3 ; but this is exageratly indirect, for it is an elementary fact that the application : $x \mapsto x + x$ from On to On has a canonical extension to a p.r. functor : $ON \to ON$.

be a total p.r. functor such that $\gamma < F(\alpha)$; for every total p.r. functor G, $G \circ F$ is a total p.r. functor, hence $G(\gamma) \leqslant G \circ F(\alpha) < PAD(\alpha)$, from which follows $PAD(\gamma) \leqslant PAD(\alpha)$).

Proof of Claim 2 - Consider the theory

$$T = KP + \varphi(a) + (a < C \wedge C \text{ is not } \Sigma \text{ definable from } a),$$

where $\varphi(v)$ is the Π formula considered in R.IV.5. This R.IV.5 together with the definition of μ, shows that L_{λ^+} becomes a model of T, when we interpret a by λ and C by μ. So T has a μ-model. And if M is any γ-model of T,

$$M \models a < C, \text{ so } a^{(M)} < \gamma, \text{ hence } a^{(M)} \text{ is an ordinal. Then } a^{(M)} = \lambda,$$

because T includes the β-definition of λ given in R.IV.5. Then $\lambda < \gamma$ implies that L_{λ^+} is a transitive part of M, by the truncation lemma ; so if γ was Σ_1 definable from λ inside L_{λ^+}, $\gamma = C^{(M)}$ would be Σ_1 definable from $a^{(M)} = \lambda$ inside M, in contradiction with $M \models T$. Thus $\gamma > \lambda$ and γ is not Σ_1 definable from λ inside L_{λ^+}, hence $\gamma \geqslant \mu$ by definition of μ.

We have shown that T has a μ-model and that $\gamma \geqslant \mu$ whenever T has a γ-model ; so T is a β^--definition of μ.

Remark - One would think that adding to T the sentence

$$\forall \alpha < C \quad [\alpha \text{ is } \Sigma_1 \text{ definable from a inside } L_{a^+}]$$

would force $a^{(M)} = \lambda$, $\gamma = \mu$ for any γ-model M of T ; thus μ would be β-definable. But we proved this to be false ... . What happens is like in R.III.9 : although $a^{(M)} = \lambda$, $a^{+(M)}$ might be non standard ; then although μ is not Σ_1 definable from λ in the true L_{λ^+}, it might become such in the non standard $L_{a^+}{}^{(M)}$, thus allowing T to have γ-models for $\gamma > \mu$.

The next result summarizes all possible relations between $PAD(\alpha)$, $PAD'(\alpha)$ and α^+ :

IV.8 Theorem - a - For every ordinal α, $PAD(\alpha) \leqslant PAD'(\alpha) \leqslant \alpha^+$.

- b - If $\alpha \geqslant \sigma_0$ (= first stable ordinal), $PAD(\alpha) = PAD'(\alpha) < \alpha^+$.

- c - Each one of the following three conditions on α is realized by a set of α's which is cofinal in σ_0 :

$PAD(\alpha) = PAD'(\alpha) = \alpha^+$
$PAD(\alpha) = PAD'(\alpha) < \alpha^+$
$PAD(\alpha) < PAD'(\alpha) < \alpha^+$

- d - These are the only possible cases, as

$PAD(\alpha) < PAD'(\alpha) = \alpha^+$ is never realized.

Proof - (a) follows from Th. I.5

(b) Since L_{σ_o} is a Σ submodel of L, it easily follows from I.5 that if F is a p.r. functor : ON → OL and $F(\gamma)$ is not well ordered for some $\gamma \in On$, then this happens already for some $\gamma < \sigma_o$. Hence if F is a partial p.r. functor, then F is total iff $F''(\sigma_o) \subset \sigma_o$. This implies (for $\alpha \geq \sigma_o$) :

$PAD(\alpha) = PAD'(\alpha) = \sup\{F(\alpha)$; F p.r. functor s.t. $F''\sigma_o \subset \sigma_o\} < \alpha^+$ (The "$< \alpha^+$" holds because the condition $F''\sigma_o \subset \sigma_o$ is Δ_o in L_{α^+}).

(c_1) - We have seen in R.III.6d that the set of α's for which condition (o) of Th. I.8 or IV.4 holds, is cofinal in σ_o ; by I.8 or IV.4, this set is also the set of α's such that $PAD(\alpha) = PAD'(\alpha) = \alpha^+$.

(c_2) - Because L_{σ_o} is a Σ-submodel of L and $PAD(\alpha) = PAD'(\alpha) < \alpha^+$ is true for any $\alpha > \sigma_o$, one can easily show that the same holds for a cofinal class of α's in L_{σ_o}.

(c_2) - Fix $\alpha_o < \sigma_o$; R.III.6 shows that by increasing perhaps α_o, we can assume it to be β-definable. Let $T_o(C)$ be a β-definition of α_o, and T be

KP + $T_o(a_o)$ + [a_o is countable and a is the first uncountable ordinal] + [a < C and C is not Σ_1 definable from a].

Adapting the proof of IV.7, one sees that T is a β^--definition of some ordinal $\alpha > \alpha_o$ which satisfies $PAD(\alpha) < PAD'(\alpha) < \alpha^+$.

(d) - follows from Th. IV.4, (1) ↔ (1)', so the proof is completed.

§ V - EXTENSIONS OF GIRARD'S BOUNDEDNESS THEOREM TO NON p.r. FUNCTORS

Here we consider a generalized recursive function f, and try to bound it by a functor F ; but instead of requiring that F is p.r., we set weaker requirements. The methods of Girard as well as those from § II easily yield the results exposed here.

We let δ denote an ordinal which is admissible or limit of admissible ordinals, and such that

$L_\delta \models$ every set is countable.

V.1 Theorem (Girard) - Let T be a theory whose conjunction is in $\mathcal{L}_{\omega_1\omega} \cap L_\delta$; for each sentence $\psi \in \mathcal{L}_{\omega_1\omega} \cap L_\delta$, there exists a functor F_ψ : ON $\rightarrow$ OL such that

- i - $F_\psi \upharpoonright (\mathrm{ON} \upharpoonright \omega)$ is δ-finite, and the application $\psi \mapsto F_\psi \upharpoonright (\mathrm{ON} \upharpoonright \omega)$ is δ-recursive

- ii- F_ψ commutes to $\underrightarrow{\lim}$ - hence F_ψ is a δ-finite functor

- III- for all $\gamma \in$ On $T \vdash_{\leqslant\gamma} \psi$ iff $|F_\psi(\gamma)| \notin$ On.

Proof - very similar to the proof of Th. II.1, using the extension to $\mathcal{L}_{\omega_1\omega}$ of the Henkin lemma.

V.2 Theorem - For every ordinal $\alpha \geqslant \delta$ the following are equivalent :

(o) there is a δ-finite sentence $\theta(R) \in \mathcal{L}_{\omega_1\omega}$ such that α is the only ordinal satisfying $\exists R\, \theta(R)$; and every ordinal $< \alpha$ is Σ-definable from α and from parameters $< \delta$, inside L_{α^+}.

(1) $\alpha^+ = \sup\{F(\alpha)$: F total δ-finite functor $\}$.

(2) for every α^+-recursive function f, there is a total δ-finite functor F such that $f(\gamma) < F(\gamma)$ for all $\gamma \in [\alpha, \alpha^+[$.

Proof - Similar to the proof of Th. I.8, but using V.1 instead of II.1. Note that Th. I.8 is the particular case $\delta = \omega$ of this result.

V.3 Corollary - The set of ordinals δ such that for every δ^+-recursive function there is a total δ-finite functor F such that $f(\gamma) \leqslant F(\gamma)$ for all $\gamma \in [\delta, \delta^+[$, is cofinal in $\aleph_1^{(L)}$.

Proof - For every ordinal $\alpha_o < \aleph_1^{(L)}$, let $\delta(\alpha_o)$ be the smallest ordinal $\delta \geq \alpha_o$ which satisfies our assumptions on δ. It is clear that condition (o) of Th. V.2 holds when $\alpha = \delta = \delta(\alpha_o)$. Hence condition (2) also holds, thus proving the corollary.

V.4 Theorem - Let α be an ordinal which is countable inside L_{α^+}. Then for every α^+- recursive function f, there is a total α^+-finite functor F such that $f(\gamma) < F(\gamma)$, for all $\gamma \in [\alpha,\alpha^+[$.

Proof - One applies Th. V.1 in the case $\delta = \alpha^+$, in a way similar to the use of Th. II.1 in order to prove I.6.

REFERENCES

[A - R] P. Aczel, W. Richter, Inductive definitions and reflecting properties of admissible ordinals, in Generalized Recursion theory, Fenstad, Hinman, editors, N.H. 1974.

[B] J. Barwise, Admissible sets and structures, Springer, 1974.

[G 1] J.Y. Girard, Π^1_2 -Logic Part 1, to appear in Ann. of Math. Log.

[G 2] J.Y. Girard, Proof Theoretic investigations of iterated inductive definitions, part I, to appear in the Specker volume in "L'Enseignement Mathématique".

[G 3] J.Y. Girard, Π^1_2 -Logic Part 2, to appear

[G 4] J.Y. Girard, Cours de Théorie de la Démonstration, 1979 - 1980, Université Paris VII.

[M] M. Masseron, Majoration des fonctions ω_1^{CK} -recursives par des échelles - Thèse de 3e cyle , Université Paris-Nord, 1980.

[R 1] J.P. Ressayre, Logique tous azimuts, Cours de 3e cycle, Université Paris VII, 1982.

[R 2] J.P. Ressayre, Models with compactness properties with respect to an admissible language, Ann. of Math. Log., 1977, p. 51.

[VdW] This volume

PROCEEDINGS OF THE HERBRAND SYMPOSIUM
LOGIC COLLOQUIUM '81, J. Stern (editor)

A SUPERSTABLE THEORY WITH THE DIMENSIONAL ORDER PROPERTY HAS MANY MODELS

Jürgen SAFFE
Universität Hannover

Abstract

In this paper we prove in a very direct fashion by interpretating graph theory in a superstable theory with the dimensional order property that such a theory has 2^λ non-isomorphic models of cardinality λ for every $\lambda \geqq |L| + \omega_1$.

0. INTRODUCTION AND NOTATION

All the notations in this paper follow those in [Sf 2] or [Sf 3] which are the same as [Sh 1] and [Sh 2] except the following changes : a type will be usually complete ; we omit the distinction between t and t_* ; two types $p_i \in S^{I_i}(\underline{A})$ are called weakly orthogonal iff for all realizations $\underline{B}_i$ of p_i $t(\underline{B}_1, A \cup \underline{B}_2)$ does not fork over $\underline{A}$ (if one of them is stationary this coincides with Shelah's definition) ; for the family $\underline{F}$ where $\underline{F} = \underline{F}^x_\lambda$ we omit the $\underline{F}$ and write (x,λ) for typographical reasons. We assume familiarity with the definitions and theorems of stability theory as can be found in [LP], [Po], [Sf 1], [Sf 2], [Sf 3], [Sh 1], [Sh 2] at least including the content of chapter V of [Sh 1].

In this paper we give a proof of the following theorem which is due to Shelah :

0.1. Theorem

Let T be superstable. If T has the dimensional order property then for all $\lambda \geqq |L| + \omega_1$ we have $I(\lambda,T) = 2^\lambda$.
But contrary to the proof in [Sh 2] we avoid the use of chapter VIII of [Sh 1] by interpretating graph theory in a direct way.

1. THE DEFINITIONS AND THE BASIC LEMMAS

Throughout the paper we assume the theory T bo te superstable.

1.1. Definition

T has the dimensional order property (dop in short) iff there are (a,ω)-saturated models $\underline{M}_0$, $\underline{M}_1$, $\underline{M}_2$ and $\underline{N}$ and a type $p \in S^1(\underline{N})$ such that $\underline{M}_0 < \underline{M}_1$, $\underline{M}_0 < \underline{M}_2$, $t(\underline{M}_1, \underline{M}_2)$ does not fork over $\underline{M}_0$, $\underline{N}$ is (a,ω)-prime over $\underline{M}_1 \cup \underline{M}_2$ and p is orthogonal to $\underline{M}_1$ and $\underline{M}_2$.

1.2. Remarks and example

This definition is clearly equivalent to that of Shelah in [Sh 2]. So if T does not have the dop then T satisfies some reasonable structure theorem. The typical example of such a theory is the following. Let L contain three predicates E,V,P and T has the following axioms : Any model of T is the union of the two disjoint sets E (which abbreviates "edge") and V ("vertex"), P has three arguments and we have $\models P(e, v_0, v_1)$ iff the edge e joins the verteces v_0 and v_1, between any two verteces there are infinitely many edges. The theory is clearly ω-categorical but for any higher cardinality it allows interpretating graph theory by a cardinality quantifier and so there are as many models as graphs for any cardinality $\geq \omega_1$. The proof of the theorem will show that in some sense it is the only example.

1.3. The basic lemma

Suppose T has the dop. Then there are (a,ω)-satured models $\underline{M}_0$, $\underline{M}_1$, $\underline{M}_2$ and $\underline{N}$ and $p \in S^1(\underline{N})$ according to the definition such that further the following hold : $w(\underline{M}_1, \underline{M}_0) = 1 = w(\underline{M}_2, \underline{M}_0)$ and p is regular.

Proof : Choose $\underline{M}_0$, $\underline{M}_1$, $\underline{M}_2$, $\underline{N}$ and $p \in S^1(\underline{N})$ according to the definition of dop such that $w(\underline{M}_i, \underline{M}_0)$ is minimal under all those examples. There is a regular $q \in S^1(\underline{N})$ which is not (even not weakly) orthogonal to p. By regularity q is orthogonal to $\underline{M}_1$ and $\underline{M}_2$. So wlog we can assume p to be regular.

Claim 1 : $w(\underline{M}_i, \underline{M}_0)$ is finite for $i = 1,2$. We show for $i = 2$. Choose $\bar{b} \in N$ such that p does not fork over $\underline{M}_1 \cup \underline{M}_2 \cup \bar{b}$ and $p_1 = p|\underline{M}_1 \cup \underline{M}_2 \cup \bar{b}$ is stationary. Then p_1 is orthogonal to $\underline{M}_1$ and $\underline{M}_2$. Let c realize p. Now choose $\bar{a}_2 \in M_2$ such that $t(c\bar{b}, \underline{M}_1 \cup \underline{M}_2)$ does not fork over $\underline{M}_1 \cup \bar{a}_2$, $t(c, \underline{M}_1 \cup \bar{a}_2 \cup \bar{b})$ is stationary and $t(\bar{b}, \underline{M}_1 \cup \bar{a}_2)$ is (a,ω)-isolated. As $t(\underline{M}_1, \underline{M}_2)$ does not fork over $\underline{M}_0$ we have $t(\underline{M}_1 \cup c\bar{b}, \underline{M}_2)$ does not fork over $\underline{M}_0 \cup \bar{a}_2$. Hence $t(c, \underline{M}_1 \cup \bar{a}_2 \cup \bar{b})$ is orthogonal to $\underline{M}_1$ and to $\underline{M}_0 \cup \bar{a}_2$ (see [SF 2] or [Sf 3] for a proof). Now let $\underline{M}'_2$ be (a,ω)-prime over $\underline{M}_0 \cup \bar{a}_2$. Clearly $w(\underline{M}'_2, \underline{M}_0) = w(\bar{a}_2, \underline{M}_0)$ is finite and $t(\underline{M}'_2, \underline{M}_1)$ does not fork over $\underline{M}_0$. Let $\underline{N}'$ be (a,ω)-prime over $\underline{M}_1 \cup \underline{M}'_2$ where wlog $\bar{b} \in N'$ because $t(\bar{b}, \underline{M}_1 \cup \bar{a}_2)$ is (a,ω)-isolated. The non-forking extension of $t(c, \underline{M}_1 \cup \bar{a}_2 \cup \bar{b})$ to $\underline{N}'$ is orthogonal to $\underline{M}_1$ and $\underline{M}'_2$. Hence by minimality $w(\underline{M}_2, \underline{M}_0)$ is finite.

Claim 2 : $w(\underline{M}_i, \underline{M}_0) = 1$ for $i = 1,2$.

Let $n_i = w(\underline{M}_i, \underline{M}_0)$ and suppose by way of contradiction $n_1 > 1$.

Let $(c^i_j)_{j<n_i}$ exemplify the definition of weight for $i = 1,2$. Now let $\underline{M}'_0$ be (a,ω)-prime over $\underline{M}_0 \cup c^1_0$ and $\underline{M}'_2$ (a,ω)-prime over $\underline{M}'_0 \cup \underline{M}_2$. Clearly $t(\underline{M}'_2, \underline{M}'_0 \cup \underline{M}_1)$ does not fork over $\underline{M}'_0$ and $t(\underline{M}_2, \underline{M}_0 \cup \underline{M}'_0)$ does not fork over $\underline{M}_0$. Wlog we can assume that the following hold : $\underline{M}_0 < \underline{M}'_0 < \underline{M}_1 < \underline{N}$, $\underline{M}_0 < \underline{M}_2 < \underline{M}'_2 < \underline{N}$ and $\underline{M}'_0 < \underline{M}'_2$. Now p is orthogonal to $\underline{M}_1$ and $\underline{M}_2$. By the minimality of the weights and $w(\underline{M}_1, \underline{M}'_0) < w(\underline{M}_1, \underline{M}_0)$ p is not orthogonal to some regular $q \in S^1(\underline{M}'_2)$. Then q is orthogonal to $\underline{M}_1$ and $\underline{M}_2$, hence to $\underline{M}_1$ and $\underline{M}'_0$. But then $\underline{M}_0$, $\underline{M}'_0$, $\underline{M}_2$, $\underline{M}'_2$ and q contradict the minimality of the weights. So the weight is one.

2. PROOF OF THE THEOREM

Now we turn to the proof of the theorem. Unfortunately my first version of the proof conained a gap, but this gap was filled by L. Harrington telling me a proof of S. Shelah.

So from now on we assume that the theory T has the dop. First we shall introduce some notational conventions. Let RV be an abbreviation for the rang $R(p,L,\infty)$ of [Sh 1]. Now let $\underline{M}_0$, $\underline{M}_1$, $\underline{M}_2$, $\underline{N}$ and $p \in S^1(\underline{N})$ be as in the conclusion of lemma 1.3. Choose $a^1 \in M_1 \setminus M_0$ and $a^2 \in M_2 \setminus M_0$ such that for $i = 1,2$ $\underline{M}_i$ is (a,ω)-prime over $\underline{M}_0 \cup a^i$, $t(a^i, \underline{M}_0)$ is regular and even more $t(a^i,\underline{M}_0)$ is orthogonal to all types q satisfying $RV(q) < RV(a^i,\underline{M}_0)$. Wlog we can assume in this situation that $t(a^1, \underline{M}_0) = t(a^2, \underline{M}_0)$ or these two types are orthogonal. Now the proof splits up into cases.

Case I : Using these notations among the examples for the dop there is one $t(a^i,\underline{M}_0)$ such that there exist realizations b^1, b^2 and b^{12} of $t(a^i, \underline{M}_0)$ (fixed i) satisfying : b^1, b^2 and b^{12} are pairwise independent over $\underline{M}_0$ but $t(b^{12}, \underline{M}_0 \cup b^1 \cup b^2)$ forks over $\underline{M}_0$.

In this case we can find $\bar{a}_0 \in M_2$, $\bar{b}$ and $q \in S^1(\bar{a}_0 b^{12} \bar{b})$ such that $t(b^1, \underline{M}_0)$ does not fork over $\bar{a}_0$, $t(b^1,\bar{a}_0)$ is stationary, b^1, b^2 and b^{12} are pairwise independent over $\bar{a}_0$ but $t(b^{12}, \bar{a}_0 b^1 b^2)$ forks over $\bar{a}_0$, $t(\bar{b}, \bar{a}_0\, b^{12})$ is almost orthogonal to $\bar{a}_0$ and q is regular, stationary and orthogonal to $\bar{a}_0$. For proving the theorem we can clearly assume $\bar{a}_0 = \emptyset$. If we write b^i for some index i this will implicitly mean $t(b^i,\emptyset)=t(b^1,\emptyset)$(here we already use $\bar{a}_0 = \emptyset$) (the same for double indices b^i_j). The writing of $\bar{b}^i$ will include $t(b^i, \bar{b}^i, \emptyset) = t(b^{12}\, \bar{b}, \emptyset)$. In this case q^i denotes the corresponding type to q over $b^i\, \bar{b}^i$. If we write b^{ij} in connection with some b^i and b^j this will include $t(b^{ij} b^i b^j, \emptyset) = t(b^{12} b^1 b^2, \emptyset)$.

Now let $\lambda \geqq |L| + \omega_1$ and G be a bipartite graph having λ verteces. Wlog we can assume that the verteces of G are $\{(i,j) | i = 1 \text{ or } 2,\ 0 < j < \lambda\}$ where each edge of G joins one vertex $(1,i)$ to $(2,j)$ for some i and j. For each such G we shall construct a model $\underline{M}^G$ such that $\underline{M}^G$ has cardinality λ and if $\underline{M}^G \cong \underline{M}^{G'}$ then $G \cong G'$. So let $\{b^i_j | i = 1 \text{ or } 2,\ j < \lambda\}$ be an independent set and choose suitable $b^{ij}_{\alpha\beta}$ for all b^i_α and b^j_β (corresponding indices). For all index combinations x choose a suitable $\bar{b}^x$ and let $\underline{C}^x$ be an independent set of realizations of q^x of cardinality λ. Now let $\underline{M}^G$ be (a,ω) - constructible of cardinality λ over $\{b^i_j | i = 1 \text{ or } 2,\ j < \lambda\} \cup \{b^{ij}_{\alpha\beta} | i = 1 \text{ or } 2,\ j = 1 \text{ or } 2,$ $\alpha < \lambda,\ \beta < \lambda\} \cup \{\underline{C}^i_j | i = 1 \text{ or } 2,\ j < \lambda\} \cup \{\underline{C}^{12}_{\alpha\beta} |$there is an edge in G from $(1,\alpha)$ to $(2,\beta)\} \cup \{\underline{C}^{11}_{0i} | 0 < i < \lambda\} \cup \{\underline{C}^{22}_{0i} | 0 < i < \lambda\}$. Now it suffices to prove the following claim : If $F : \underline{M}^G \to \underline{M}^{G'}$ is an isomorphism satisfying $F(b^1_0) = b^1_0$ and $F(b^2_0) = b^2_0$ then $G \cong G'$. So suppose that F is such an isomorphism. Now look at some b^i_j for some $i = 1$ or 2 and $0 < j < \lambda$. Then there exists a j' such that $t(F(b^i_j), b^i_{j'})$ (same i!) forks over $\emptyset$, because otherwise there is no possible image of $\underline{C}^i_j \cup \underline{C}^{ii}_{0j}$. So put in this case $f((i,j)) = (i,j')$. As F is an isomorphism clearly f is a bijection from the verteces of G onto the verteces of G'. If there is an edge from $(1,i)$ to $(2,j)$ in G then by construction $\dim(q^{12}_{ij}, \underline{M}^G) = \lambda$. Now $F(q^{12}_{ij})$ is orthogonal to all q^x where $x \neq (1,2,f(1,i),f(2,j))$. Hence there must be an edge in G' joining $f(1,i)$ to $f(2,j)$. By the same argument for F^{-1} and f^{-1} we get that f is an isomorphism from G onto G' as we have claimed before. So the proof of this case is finished. So in the following let us assume.

Case II : not case I.

First we shall introduce some more notations for this case. By a double use of the proof of claim 1 in lemme 1.3 we can find $\bar{a}^0 \in M_0$, $\bar{a}^1 \in M_1 \setminus M_0$, $\bar{a}^2 \in M_2 \setminus M_0$ and $\bar{b} \in N \setminus M_1 \cup M_2$ such that the following conditions are satisfied (all for $i = 1,2$) : $t(a^i, \underline{M}_0)$ does not fork over $\bar{a}^0$, $t(a^i, \bar{a}^0)$ is stationary (and regular) $a^i \in \bar{a}^i$, $t(\bar{a}^i, \bar{a}^0 \cup a^i)$ is almost orthogonal to $\bar{a}^0$, if $t(a^1, \bar{a}^0) = t(a^2, \bar{a}^0)$ then $t(\bar{a}^1, \bar{a}^0) = t(\bar{a}^2, \bar{a}^0)$, $t(\bar{b}, \bar{a}^0\, \bar{a}^1\, \bar{a}^2)$ is almost orthogonal to $\bar{a}^0\, \bar{a}^i$, p does not fork over $\bar{a}^0\, \bar{a}^1\, \bar{a}^2\, \bar{b}$ and $q = p | \bar{a}^0\, \bar{a}^1\, \bar{a}^2\, \bar{b}$ is stationary and orthogonal to $\bar{a}^0\, \bar{a}^i$. Wlog we can further assume that $RV(a^1, \bar{a}^0) \geqq RV(a^2, \bar{a}^0)$ and as in case I it is enough to prove the theorem for $\bar{a}^0 = \emptyset$. If we write a^i_j for some index j then this will implicitly mean $t(a^i_j, \emptyset) = t(a^i, \emptyset)$ and the writing of $\bar{b}^{ij}$ or $t(\bar{b}^{ij}, \bar{a}^1_i \bar{a}^2_j)$ (corresponding indices) will include $t(\bar{b}^{ij} \bar{a}^1_i \bar{a}^2_j, \emptyset) =$ $t(\bar{b}\, \bar{a}^1\, \bar{a}^2, \emptyset)$. In that case q^{ij} denotes the corresponding type to q over $\bar{b}^{ij} \bar{a}^1_i\, \bar{a}^2_j$. As we are not in case I we have the following property : if $t(a^i_*, (a^i_j)_{j<\lambda})$ forks over $\emptyset$ then for some $j < \lambda$ $t(a^i_*, a^i_j)$ forks over $\emptyset$.

Now let $\lambda \geqq |L| + \omega_1$ and G be a graph having λ verteces $\{i | i < \lambda\}$. As in

case I for any such G we shall construct a model $\underline{M}^G$ of cardinality λ. For notational simplicity let $U = \lambda \times \lambda \times \omega$. Now let $(\bar{a}^1_i)_{i<\lambda}$ and $(\bar{a}^2_j)_{j\in U}$ be independent sets of realizations of $t(\bar{a}^1, \emptyset)$ and $t(\bar{a}^2, \emptyset)$ and choose suitable $\bar{b}^{ij}$. Let $\underline{C}^{ij}$ be an independent set of realizations of q^{ij} of cardinality λ. Then let $\underline{M}^G$ be (a,ω) - constructible of cardinality λ over $\{\bar{a}^i_j | i, j\} \cup \{\bar{b}^{ij} | i, j\} \cup \{\underline{C}^{\alpha j} |$ there exists a vertex β, and edge from α to β in G and an $n < \omega$ such that $j = (\alpha,\beta,n)$ or $j = (\beta,\alpha,n)\}$. By construction $\underline{M}^G$ codes G in the following way : there is an edge from α to β iff there are infinitely many different j such that $\dim(q^{\alpha j}, \underline{M}^G) = \lambda = \dim(q^{\beta j}, \underline{M}^G)$. Now we need the following fact : if $t(a^1_*, a^1)$ does not fork over $\emptyset$ then for all a^2 and a^2_* such that $t(a^1, a^2, a^2_*)$ does not fork over $\emptyset$ and $t(a^1_*, a^2\ a^2_*)$ does not fork over $\emptyset$ (forking is only possible for $t(a^1, \emptyset) = t(a^2, \emptyset)$ because of the conditions for $t(a^i, \emptyset)$ and $RV(a^1, \emptyset) \geqq RV(a^2, \emptyset)$) we have q^{12} is orthogonal to q^{12}_{**}. The proof of this fact follows immediately from the observation that by the assumptions $t(a^1\ a^1_*, a^2\ a^2_*)$ does not fork over $\emptyset$ (remember $t(a^1, \emptyset)$ is orthogonal to all types of lower rang).
Now define for any a^1 in M^G $D(a^1, i, \underline{M}^G)$ to be $\sup\{\dim(q(a^1, a^2), \underline{M}^G) | t(a^2, a^2_i)$ forks over $\emptyset$, where $q(a^1, a^2)$ denotes a type corresponding to q$\}$. Clearly by the construction of the models $D(a^1_i, j, \underline{M}^G) = \dim(q^{ij}, \underline{M}^G)$. Suppose for a certain i $t(a^1, a^1_i)$ forks over $\emptyset$ (in M^G there are no other a^1's) then we claim that there are at most finitely many j such that $D(a^1, j, \underline{M}^G) = \lambda > D(a^1_i, j, \underline{M}^G)$. Suppose by contradiction that there are infinitely many such j. Wlog we can assume that $\dim(q(a^1, a^2_j), \underline{M}^G) = \lambda$. Choose for every such j an j' such that $q(a^1, a^2_j)$ is not orthogonal to $q(a^1_i, a^2_{j'})$ (this j' exists by the fact above). Now $(a^2_j a^2_{j'})_{j<\omega}$ is an independent set over $\emptyset$. Hence for some j $t(a^2_j a^2_{j'}, a^1_i a^1)$ does not fork over $\emptyset$. But then clearly $q(a^1_i, a^2_{j'})$ is orthogonal to $q(a^1, a^2_j)$ contradicting the assumption. Now suppose finally that F is an isomorphism from $\underline{M}^G$ to some $\underline{M}^{G'}$. We then have to show that G is isomorphic to G'. Define a map from the verteces of G to the verteces of G' as follows : for $i < \lambda$ find $j < \lambda$ such that $t(F(a^1_i), a^1_j)$ forks over $\emptyset$ and let $f(i) = j$. Now suppose that there is an edge in $\underline{M}^G$ from i to j. Then there are infinitely many different α such that $\dim(q^{i\alpha}, \underline{M}^G) = \lambda = \dim(q^{j\alpha}, \underline{M}^G)$. Hence there are infinitely many α such that $D(F(a^1_i), \alpha, \underline{M}^{G'}) = D(F(a^2_j, \alpha, \underline{M}^{G'}) = \lambda$. But this is only possible if there are infinitely many α such that $D(a^1_{f(i)}, \alpha, \underline{M}^{G'}) = \lambda = D(a^1_{f(j)}, \alpha, \underline{M}^{G'})$. So there is an edge from f(i) to f(j) in G'. Looking at F^{-1} and f^{-1} we get that f is an isomorphism from G onto G' what we wanted to prove. So the proof of the theorem is finished.

REFERENCES :

[LP] D. Lascar/B. Poizat ; An introduction to Forking ; Journal of Symbolic Logic 44 (1979), p. 330-350.

[Po] B. Poizat ; Déviation des types ; Dissertation Paris 1977.

[Sf1] J. Saffe ; Stabile Theorien ; Diplomarbeit Hannover 1980.

[Sf2] J. Saffe ; Einige Ergebnisse über die Anzahl abzählbarer Modell superstabiler Theorien ; Dissertation Hannover 1981.

[Sf3] J. Saffe ; An introduction to Regular Types ; to appear.

[Sh1] S. Shelah ; Classification Theory and the Number of Non-Isomorphic Models ; North-Holland 1978.

[Sh2] S. Shelah ; The Spectrum Problem I : κ_ε-saturated Models, the Main Gap ; Preprint Jerusalem 1980.

PROCEEDINGS OF THE HERBRAND SYMPOSIUM
LOGIC COLLOQUIUM '81, J. Stern (editor)

NUMBER THEORY AND THE BACHMANN/HOWARD ORDINAL

Ulf R. Schmerl
Mathematisches Institut der Universität München

Since the work of Gentzen and Hilbert/Bernays in the 1930's ε_0 is considered to be the characteristic ordinal of classical number theory: It is the least upper bound on the derivability of transfinite induction in number theory and the consistency of this theory can be proved by transfinite induction up to ε_0 included. This situation can also be described as follows: Let $I(\alpha)$ be the formal system which is obtained from Peano arithmetic PA when the schema of complete induction is replaced by the transfinite induction rule

$$(I_\alpha) \qquad \frac{A(0,\bar{a}) \quad A(\beta,\bar{a}) \to A(\beta+1,\bar{a}) \quad A(\lambda[f(\lambda,\bar{a})],\bar{a}) \to A(\lambda,\bar{a})}{A(\kappa,\bar{a})},$$

where A is an arithmetical formula with a (possibly empty) list $\bar{a}$ of parameters, β, λ are codes of ordinals, $(\lambda[n])_{n<\omega}$ is the fundamental sequence of a limit ordinal λ, κ is an ordinal $<\alpha$, and f is a p.r. function. Then PA and $I(\varepsilon_0)$ are connected as follows:
(i) Each Π_1^0-formula provable in PA is also provable in $I(\varepsilon_0)$ and the induction formulas in that proof can be restricted to be quantifier-free.
(ii) Each formula provable in $I(\varepsilon_0)$ is also provable in PA.
(iii) The consistency of PA is provable in $I(\varepsilon_0+1)$.
A proof of these statements is sketched below.

Gentzen's technique of establishing the consistency of a formal system by transfinite induction up to a certain ordinal and of showing transfinite induction to be derivable within the system for each smaller ordinal has since that time been successfully applied to many other theories, especially to various subsystems of analysis. The characteristic ordinal found in this way is often interpreted as a measure of the proof-theoretical strength of the considered system. However, the significance of this measure is not quite clear. It is, for example, a well-known fact that the ordinal of a system can depend on whether or not free variables are allowed in the axioms of that system (cf. Kreisel's [7]). In what follows we shall show that an ordinal analysis of number theory will yield another ordinal than ε_0 - namely $\phi\varepsilon_{\Omega+1}0$, the Bachmann/Howard ordinal - if instead of ordinary transfinite induction the following pointwise induction principle is used (i.e. the measure of proof-theoretical strength depends on the scale which is used):

$$(\mathrm{PI}_\alpha) \qquad \frac{A(0,\bar{a}) \quad A(\beta,\bar{a})\rightarrow A(\beta+1,\bar{a}) \quad A(\lambda[h(\bar{a})],\bar{a})\rightarrow A(\lambda,\bar{a})}{A(\kappa,\bar{a})}$$

where A, $\bar{a}$, β, λ, h satisfy the same conditions as in rule (I_α). Let $PI(\alpha)$ be PA with rule (PI_α) instead of complete induction. The study of this rule in connection with number theory has been suggested by Girard and Kreisel, the rule itself is due to Girard. He conjectured that (i), (ii), and (iii) also hold for pointwise induction if ε_0 is replaced by the somewhat greater Bachmann/Howard ordinal $\phi\varepsilon_{\Omega+1}0$.

The following proof of Girard's conjecture is essentially based on the fact that the induction principles (I) and (PI) have recursion counterparts in the so-called fast- and slow-growing hierarchies of number theoretic functions. These are defined as follows:

- the fast-growing Grzegorczyk/Wainer hierarchy:

$$F_0(n) = 2^n$$

$$F_{\alpha+1}(n) = F_\alpha^n(n)$$

$$F_\lambda(n) = F_{\lambda[n]}(n) \quad \text{if } \lambda \text{ is a limit ordinal}$$

- the slow-growing or pointwise hierarchy:

$$G_0(n) = 0$$

$$G_{\alpha+1}(n) = G_\alpha(n) + 1$$

$$G_\lambda(n) = G_{\lambda[n]}(n) \quad \text{if } \lambda \text{ is a limit ordinal.}$$

A central idea in our proof is to transform inductions of the form (I) or (PI) into finite inductions whose length can be bounded by terms $F_\alpha(t)$ or $G_\alpha(t)$ resp.. But before going into details we first prove the partial characterization of PA by $I(\varepsilon_0)$.

Suppose given a coding in the natural numbers of the ordinals in Bachmann's notational system which we denote by ON (cf. Bachmann [2]). ON contains 0 and Ω, the least uncountable ordinal and the countable ordinals in ON are constructed with the functions +, $\cdot$, and ϕ such that each $\alpha \in ON$, $0<\alpha<\Omega$, can uniquely be written in the form $\phi_{\alpha_1}(\beta_1)+\ldots+\phi_{\alpha_n}(\beta_n)$ with α_i, $\beta_i \in ON$, $\beta_i<\Omega$, $\phi_\alpha(\beta_1)\geq\ldots\geq\phi_\alpha(\beta_n)$, $\phi_{\alpha_i}(\beta_i)\neq\beta_i$ and $\phi_{\alpha_i}(\beta_i)\neq\alpha_i$ if $\beta_i=0$, where the indices α_i are of the form 0 or $\Omega^{\alpha_{i1}}\cdot\beta_{i1}+\ldots+\Omega^{\alpha_{ik}}\cdot\beta_{ik}$ with α_{ij}, $\beta_{ij}\in ON$, $0<\beta_{ij}<\Omega$, $\alpha_{i1}>\ldots\alpha_{ik}$. The countable ordinals of ON form an initial segment which is denoted by $\phi_{\varepsilon_{\Omega+1}}0$, the Bachmann/Howard ordinal. We also denote by +, $\cdot$, ϕ, and [ ] number theoretic functions representing the corresponding ordinal operations on the codes. Let < be an arithmetical Σ_1^0-predicate for the <-relation between ordinals. In the rules (I) and (PI) side formulas are allowed. The language of PA, $I(\varepsilon_0)$, and $PI(\phi_{\varepsilon_{\Omega+1}}0)$ is extended by function symbols F_α and G_β, $\alpha<\varepsilon_0$, $\beta<\phi_{\varepsilon_{\Omega+1}}0$. The defining equations for these functions are added as new axioms. Quantification over F- or G-terms is not allowed in PA and $I(\varepsilon_0)$, but is allowed in $PI(\phi_{\varepsilon_{\Omega+1}}0)$ over G-terms; let PA' be PA when quantification over G-terms is admitted.

1. Lemma: Let $\forall\bar{x}A(\bar{x})$ be a Π_1^0-formula provable in PA. Then there exists an ordinal $\alpha<\varepsilon_0$, a p.r. function $f(\beta,\bar{a})$ and a quantifier-free formula $B(\beta,\bar{a})$ such that $B(\alpha,\bar{a})\to A(\bar{a})$ and $B(0,\bar{a})$, $B(\beta,\bar{a})\to B(\beta+1,\bar{a})$, $B(\lambda[f(\lambda,\bar{a})],\bar{a})\to B(\lambda,\bar{a})$ are provable in PRA (primitive recursive arithmetic).

Proof: Embed a PA-derivation of $A(\bar{a})$ in an infinite derivation with ω-rule and eliminate the cuts. Then there exists an ordinal $\alpha<\varepsilon_0$ such that the height of the cut-free derivation is α. Let $B(\beta,\bar{a})$ be a reflection formula expressing the truth of the endformula of subderivations of height β for $\beta\leq\alpha$; $B(\beta,\bar{a})$ can be chosen to be quantifier-free. Then $B(\alpha,\bar{a})\to A(\bar{a})$ is provable in PRA; since $B(\beta,\bar{a})$ trivially is progressive in β, the other formulas can also be proved in PRA for some p.r. function f.

2. Proposition (PA and $I(\varepsilon_0)$):

(i) Each Π_1^0-formula provable in PA is also provable in $I(\varepsilon_0)$; the induction formulas in that proof can be chosen to be quantifier-free.
(ii) Each formula provable in $I(\varepsilon_0)$ is also provable in PA.
(iii) The consistency of PA is provable in $I(\varepsilon_0+1)$.

Proof: (i) follows immediately from lemma 1, since PRA is a subsystem of $I(\varepsilon_0)$. (ii) is proved by transfinite induction up to ordinals $\alpha<\varepsilon_0$ in PA. (iii): The reflection formula $B(\beta,\bar{a})$ of lemma 1 can be derived for $\beta=\varepsilon_0$ in $I(\varepsilon_0+1)$; it yields the consistency of PA.

In order to transform transfinite inductions of the form (I) or (PI) into finite inductions we consider the set of predecessors of a given ordinal under a list $\bar{a}$ of parameters and a function $f(\alpha,\bar{a})$ or $h(\bar{a})$: The predecessor of a successor ordinal $\alpha+1$ is α and the predecessor of a limit ordinal λ is $\lambda[f(\lambda,\bar{a})]$ or $\lambda[h(\bar{a})]$ resp.. This gives rise to the following

<u>3. Definition:</u> (i) The n-th predecessor of α under $\bar{a}$ and $f(\beta,\bar{a})$:

$$\hat{f}(\alpha,\bar{a},0) = \alpha$$

$$\hat{f}(\alpha,\bar{a},n+1) = \begin{cases} \hat{f}(\alpha,\bar{a},n)\dot{-}1 & \text{if } \hat{f}(\alpha,\bar{a},n) \text{ is not a limit} \\ \hat{f}(\alpha,\bar{a},n)[f(\hat{f}(\alpha,\bar{a},n),\bar{a})] & \text{otherwise.} \end{cases}$$

(ii) The n-th predecessor of α under $\bar{a}$ and $h(\bar{a})$:

$$\hat{h}(\alpha,\bar{a},0) = \alpha$$

$$\hat{h}(\alpha,\bar{a},n+1) = \begin{cases} \hat{h}(\alpha,\bar{a},n)\dot{-}1 & \text{if } \hat{h}(\alpha,\bar{a},n) \text{ is not a limit} \\ \hat{h}(\alpha,a,n)[h(a)] & \text{otherwise.} \end{cases}$$

$\hat{f}$ and $\hat{h}$ are p.r. if f and h are.

<u>4. Lemma:</u> (i) For each Π_1^0-formula $\forall\bar{x}A(\bar{x})$ the following holds: If $PA \vdash \forall\bar{x}A(\bar{x})$, then there exists $\alpha<\varepsilon_0$ and a p.r. function f such that $PI(\phi_{\varepsilon_{\Omega+1}}0) \vdash \hat{f}(\alpha,\bar{a},b)=0\to A(\bar{a})$.

(ii) For each formula B the following holds: If $PI(\phi_{\varepsilon_{\Omega+1}}0) \vdash B$, then there exist $\alpha_1,\ldots,\alpha_n<\phi_{\varepsilon_{\Omega+1}}0$ and p.r. functions $h_1,\ldots,h_n$ such that $PA' \vdash \forall\bar{x}_1\exists y_1[\hat{h}_1(\alpha_1,\bar{x}_1,y_1)=0]\wedge..\wedge\forall\bar{x}_n\exists y_n[\hat{h}_n(\alpha_n,\bar{x}_n,y_n)=0] \to B$.

<u>Proof:</u> (i) By lemma 1, if $PA \vdash \forall\bar{x}A(\bar{x})$, then there exist $\alpha<\varepsilon_0$, f p.r. and a quantifier-free formula $B(\beta,\bar{a})$ such that $B(\alpha,\bar{a})\to A(\bar{a})$ and $B(0,\bar{a})\wedge B(\beta,\bar{a})\to B(\beta+1,\bar{a})\wedge B(\lambda[f(\lambda,\bar{a})],\bar{a})\to B(\lambda,\bar{a})$ are provable in PRA. Hence we have

$$PRA \vdash \hat{f}(\alpha,\bar{a},b)=0\to B(\hat{f}(\alpha,\bar{a},b\dot{-}0),\bar{a})\wedge$$
$$\forall x[B(\hat{f}(\alpha,\bar{a},b\dot{-}x),\bar{a}) \to B(\hat{f}(\alpha,\bar{a},b\dot{-}x+1),\bar{a})] ;$$

by formal induction in PRA we obtain

$$PRA \vdash \hat{f}(\alpha,\bar{a},b)=0 \to \forall xB(\hat{f}(\alpha,\bar{a},b\dot{-}x),\bar{a}) \quad \text{and hence}$$

$PRA \vdash \hat{f}(\alpha,\bar{a},b)=0 \to A(\bar{a})$.

(ii) This is proved by induction on the height of the derivation of B in $PI(\phi\varepsilon_{\Omega+1}0)$. The essential step is an occurrence of an (I)-rule:

$$\frac{\Gamma\vdash F(0,\bar{a}),\Delta \quad \Gamma,F(\beta,\bar{a})\vdash F(\beta+1,\bar{a}),\Delta \quad \Gamma,F(\lambda[h(\bar{a})],\bar{a})\vdash F(\lambda,\bar{a}),\Delta}{\Gamma\vdash F(\kappa,\bar{a}),\Delta}$$

where $\kappa<\phi\varepsilon_{\Omega+1}0$. By induction hypothesis, there exist $\alpha_1,\ldots,\alpha_n < \phi\varepsilon_{\Omega+1}0$ and p.r. functions $h_1,\ldots,h_n$ such that

$$(\forall\bar{x}_i\exists y_i[\hat{h}_i(\alpha_i,\bar{x}_i,y_i)=0])_{1\le i\le n},\Gamma \vdash_{PA'} F(0,\bar{a}),\Delta$$

$$\text{"}\quad ,\Gamma \vdash_{PA'} F(\beta,\bar{a})\to F(\beta+1,\bar{a}),\Delta$$

$$\text{"}\quad ,\Gamma \vdash_{PA'} F(\lambda[h(\bar{a})],\bar{a})\to F(\lambda,\bar{a}),\Delta$$

and in the same way as in (i) we obtain

$$(\forall\bar{x}_i\exists y_i[\hat{h}_i(\alpha_i,\bar{x}_i,y_i)=0])_{1\le i\le n},\hat{h}(\kappa,\bar{a},b)=0,\Gamma \vdash_{PA'} F(\kappa,\bar{a}),\Delta.$$

In a next step we give estimations on parameters b with $\hat{f}(\alpha,\bar{a},b)=0$ or $\hat{h}(\alpha,\bar{a},b)=0$ by terms $F_\alpha(t)$ or $G_\alpha(t)$. For that purpose we have to prove some properties of the functions F_α and G_α in $PI(\phi\varepsilon_{\Omega+1}0)$ and in PA.

<u>5. Lemma:</u> For each ordinal $\alpha<\varepsilon_0$, $a\neq 0 \to F_\alpha(a)>a$ is provable in $PI(\phi\varepsilon_{\Omega+1}0)$.
<u>Proof:</u> By pointwise induction on α: $F_0(a)>a$, $F_{\alpha+1}(a)=F^a_\alpha(a) \overset{(\alpha)}{\geq} F_\alpha(a) \overset{(\alpha)}{>} a$, $F_\lambda(a)=F_{\lambda[a]}(a) \overset{(\lambda[a])}{>} a$.

<u>6. Lemma:</u> For each limit ordinal $\alpha<\varepsilon_0$, $a>1 \to F_{\alpha[k]+1}(a)\le F_{\alpha[k+1]}(a)$ is provable in $PI(\phi\varepsilon_{\Omega+1}0)$.
<u>Proof:</u> For ordinals $\alpha_i<\varepsilon_0$ let $(\alpha_0):=\alpha_0$ and $(\alpha_0,\ldots,\alpha_n,\alpha_{n+1}):=(\alpha_0,\ldots,\alpha_n+\omega^{\alpha_{n+1}})$. Then by Cantor's normal form, for each ordinal α, $0<\alpha<\varepsilon_0$, there exist uniquely determined n and $\alpha_0,\ldots,\alpha_n$ such that $\alpha=(\alpha_0,\ldots,\alpha_n,0)$. By an external induction on n and pointwise induction in $PI(\phi\varepsilon_{\Omega+1}0)$ we prove:

$$\alpha_n \# \beta = \alpha_n+\beta \wedge \beta>0 \wedge a>1 \to F_{(\alpha_0,\ldots,\alpha_n+\beta)}(a)\geq F_{(\alpha_0,\ldots,\alpha_n)+1}(a).$$

Let n=0. $\beta=1$ is trivial; $\beta\to\beta+1$: $F_{\alpha_0+\beta+1}(a)=F^a_{\alpha+\beta}(a) \overset{L.5}{\geq} F_{\alpha_0+\beta}(a) \geq F_{\alpha_0+1}(a)$; $\lambda[a]\to\lambda$: $F_{\alpha_0+\lambda}(a)=F_{\alpha_0+\lambda[a]}(a) \geq F_{\alpha_0+1}(a)$.

Now suppose that the assertion holds for all ordinals $\alpha_0,\ldots,\alpha_n,\beta<\varepsilon_0$.

Then for $(\alpha_0,\dots,\alpha_n,\alpha_{n+1})$ we have: $\beta=1$: $F_{(\alpha_0,\dots,\alpha_n,\alpha_{n+1}+1)}(a)=$

$F_{(\alpha_0,\dots,\alpha_n+\omega^{\alpha_{n+1}}+\omega^{\alpha_{n+1}}.(a-1))}(a) \geq F_{(\alpha_0,\dots,\alpha_n,\alpha_{n+1})+1}(a)$;

$\beta\to\beta+1$: analogously; $\lambda[a]\to\lambda$: trivial, since $\lambda[a]>0$.

Now let $\alpha<\varepsilon_0$ be a limit ordinal. Then there exist $\alpha_0,\dots,\alpha_{n+1}$, such that $\alpha=(\alpha_0,\dots,\alpha_{n+1},0)$ and $\alpha[k]=(\alpha_0,\dots,\alpha_n+\omega^{\alpha_{n+1}}\cdot k)$; hence

$$F_{\alpha[k]+1}(a) \leq F_{(\alpha_0,\dots,\alpha_n+\omega^{\alpha_{n+1}}.k+\omega^{\alpha_{n+1}})}(a)=F_{\alpha[k+1]}(a).$$

<u>7. Lemma:</u> For each ordinal $\alpha<\varepsilon_0$, $a>1 \to F_\alpha(a) < F_\alpha(a+1)$ is provable in $PI(\phi_{\varepsilon_{\Omega+1}}0)$.

<u>Proof:</u> By pointwise induction on α. $\alpha=0$: clear; $\alpha\to\alpha+1$: $F_{\alpha+1}(a)=$

$F_\alpha^a(a) \overset{(\alpha)}{<} F_\alpha^a(a+1) \overset{L.5}{<} F_\alpha^{a+1}(a+1)=F_{\alpha+1}(a+1)$; $\lambda[a]\to\lambda$: $F_\lambda(a)=F_{\lambda[a]}(a)<$

$F_{\lambda[a]}(a+1) \overset{L.6}{\leq} F_{\lambda[a+1]}(a+1)=F_\lambda(a+1)$.

The monotony properties of the F_α's stated in the last three lemmata have already been established by Schwichtenberg in [9].

<u>8. Proposition:</u> For each p.r. function $f(\alpha,\bar{a})$ there exists a p.r. function t and a natural number k, such that

$b>F_\alpha(t(\max(\alpha,\bar{a},k))) \to \hat{f}(\alpha,\bar{a},b)=0$ is provable in $PI(\phi_{\varepsilon_{\Omega+1}}0)$.

<u>Proof:</u> Let $|a_1,\dots,a_n|$ stand for $\max(a_1,\dots,a_n)$. For a given p.r. function $f(\alpha,\bar{a})$ let t be a monotonous p.r. function such that $1,\ \alpha[f(\alpha,\bar{a})],f(\alpha,\bar{a}),\bar{a},b < t(|\alpha,\bar{a},b|)$. For this t, take a natural number $k\geq3$ such that $t(|\alpha,\bar{a},k|)\leq F_k(|\alpha,\bar{a},k|)$. By induction on pairs (b,c) of natural numbers we prove

$$b\geq F_{\hat{f}(\alpha,\bar{a},c)}(t|\hat{f}(\alpha,\bar{a},c),\bar{a},k|) \to \hat{f}(\hat{f}(\alpha,\bar{a},c),\bar{a},b)=0.$$

(0,c): if b=0, then the assertion is trivial.

(b,c+1)→(b+1,c): by distinction of cases whether $\hat{f}(\alpha,\bar{a},c)$ is 0, a successor or a limit ordinal.

- $\hat{f}(\alpha,\bar{a},c)=0$: then $\hat{f}(0,\bar{a},b)=0$ holds trivially
- $\hat{f}(\alpha,\bar{a},c)=\hat{f}(\alpha,\bar{a},c+1)+1$: if $b+1\geq F_{\hat{f}(\alpha,\bar{a},c)}(t|\hat{f}\alpha\bar{a}c,\bar{a},k|)=$

$F_{\hat{f}(\alpha,\bar{a},c+1)+1}(..) \overset{L.5}{>} F_{\hat{f}(\alpha,\bar{a},c+1)}(..)$, then

$b\geq F_{\hat{f}(\alpha,\bar{a},c+1)}(t|\hat{f}(\alpha,\bar{a},c+1)+1,\bar{a},k|) \overset{L.7}{\geq} F_{\hat{f}(\alpha,\bar{a},c+1)}(t|\hat{f}(\alpha,\bar{a},c+1),\bar{a},k|)$

by (b,c+1) follows: $\hat{f}(\hat{f}(\alpha,\bar{a},c+1),\bar{a},b)=0$ and hence

$\hat{f}(\hat{f}(\alpha,\bar{a},c+1)+1,\bar{a},b+1)=0$.

- $\mathfrak{f}(\alpha,\bar{a},c+1)=\mathfrak{f}(\alpha,\bar{a},c)[f(\mathfrak{f}(\alpha,\bar{a},c),\bar{a})]$: if $b+1\geq F_{\mathfrak{f}\alpha\bar{a}c}(t|\mathfrak{f}\alpha\bar{a}c,\bar{a},k|)=$

$F_{\mathfrak{f}\alpha\bar{a}c[t|\mathfrak{f}\alpha\bar{a}c,\bar{a},k|]}(t|\mathfrak{f}\alpha\bar{a}c,\bar{a},k|) \overset{L.6}{\geq} F_{\mathfrak{f}\alpha\bar{a}c[|f(\mathfrak{f}\alpha\bar{a}c,\bar{a}),\bar{a},k|]+1}(..)>$

$F^2_{\mathfrak{f}\alpha\bar{a}c\,..}(t|..|)$, then

$b\geq F_{\mathfrak{f}(\alpha,\bar{a},c+1)}(F_k(|\mathfrak{f}\alpha\bar{a}c[f(\mathfrak{f}\alpha\bar{a}c,\bar{a})],\bar{a},k|)) \overset{L.7}{\geq}$

$F_{\mathfrak{f}(\alpha,\bar{a},c+1)}(t|\mathfrak{f}(\alpha,\bar{a},c+1),\bar{a},k|)$ and by $(b,c+1)$ follows

$\mathfrak{f}(\mathfrak{f}(\alpha,\bar{a},c+1),\bar{a},b)=0$, hence $\mathfrak{f}(\mathfrak{f}(\alpha,\bar{a},c),\bar{a},b+1)=0$.

<u>9. Lemma:</u> Let $\alpha \# \beta=\alpha+\beta$. Then the following equations are provable in $PI(\phi\varepsilon_{\Omega+1}0)$:

(i) $G_{\alpha+\beta}(a) = G_\alpha(a) + G_\beta(a)$

(ii) $G_{\omega^{\alpha+\beta}}(a) = G_{\omega^\alpha}(a)\cdot G_{\omega^\beta}(a)$, $G_{\omega_2(\alpha+\beta)}(a) = G_{\omega_2(\alpha)}(a)^{G_{\omega^\beta}(a)}$

$G_{\omega_{k+3}(\alpha+\beta)}(a) = (G_{\omega_{k+3}(\alpha)}(a),(..,(G_{\omega_2(\alpha)}(a),G_{\omega^\beta}(a)-1)..-1))$

where (c,d) is used to denote c^d

(iii) $G_{[\omega]_k}(a) = [a]_k$ for $a\neq o$, where $[c]_0:=1$, $[c]_{n+1}:=c^{[c]_n}$.

<u>Proof:</u> (i) and (ii) are proved by pointwise induction on β and complete induction on k, (iii) is a special case of (ii).

The following theorem gives a comparison between the functions of the slow- and the fast-growing hierarchies. A first result of this kind showing that F_{ε_0} and $G_{\phi\varepsilon_{\Omega+1}0}$ are of comparable growth was first proved by Girard in [5], other proofs can be found in [1], [3], [4], [6], and [10]. In [8] we gave a proof which can be formalized in $PI(\phi\varepsilon_{\Omega+1}0)$ and also in PA. For that purpose we extend the definition of the functions G_α to the uncountable ordinals in ON: If $\alpha=\Omega^{\alpha_1}\cdot\beta_1+...+\Omega^{\alpha_k}\cdot\beta_k$, then we inductively define

$$G_\alpha(a) := \omega^{G_{\alpha_1}(a)}\cdot G_{\beta_1}(a)+...+\omega^{G_{\alpha_k}(a)}\cdot G_{\beta_k}(a).$$

<u>10. Theorem:</u> Let $\alpha\neq 0$ and $a>1$. Then

$$G_{\phi_\alpha\beta}(a+3)> \begin{cases} F^{a\cdot G_\mu(a+3)}_{G_\alpha(a+3)+i-1}(F^{k+1}_{G_\alpha(a+3)+i}(a)) \text{ if } \beta=\phi^k_\alpha 0+\mu,\ \mu<\phi^{k+1}0 \\ F^{a\cdot G_\mu(a+3)}_{G_\alpha(a+3)+i-1}(G_{\phi_\gamma\delta}(a+3)) \text{ if } \beta=\phi_\gamma\delta+\mu, 0<\mu<\phi_\gamma\delta, \gamma>\alpha \\ F^{a\cdot G_\mu(a+3)}_{G_\alpha(a+3)+i-1}(F^{k+1}_{G_\alpha(a+3)+i}(G_{\phi_\gamma\delta}(a+3))) \text{ if } \beta=\phi^k_\alpha(\phi_\gamma\delta\cdot 2)+\mu, \\ \qquad \mu<\phi^{k+1}(\phi_\gamma\delta\cdot 2), \gamma>\alpha \end{cases}$$

is provable in $PI(\phi_{\varepsilon_{\Omega+1}}0)$, where $i=0$ for $\alpha<\Omega$ and $i=1$ otherwise.
Proof: This is proved by pointwise induction on $\phi_\alpha\beta$ using lemma 9. Due to many distinction of cases, the proof is rather lengthy but nevertheless uncomplicated; so we omit it.

11. Corollary: For each ordinal $\alpha<\varepsilon_0$, $\alpha\neq 0$, $G_{\phi_{\hat{\alpha}}0}(a+3) > F_{\alpha+i}(a)$ for $a>1$ is provable in $PI(\phi_{\varepsilon_{\Omega+1}}0)$. Here $\hat{\alpha}$ is the ordinal which is obtained from α if in its Cantor normal form ω is replaced by Ω, $i=0$ for $\alpha<\omega$ and $i=1$ otherwise.
Proof: This follows immediately from the preceding theorem.

Theorem 10 and corollary 11 also hold for "<" if on the left side $a+3$ is replaced by a and if on the right side a is replaced by $a+2$, but we don't need this direction here nor the fact that the theorem is also provable in PA. Combining the information obtained in the preceding, we can now prove the points (i) and (iii) of Girard's conjecture.

12. Theorem: Each Π_1^0-formula provable in PA is also provable in $PI(\phi_{\varepsilon_{\Omega+1}}0)$ and the consistency of PA is provable in $PI(\phi_{\varepsilon_{\Omega+1}}0+1)$. The induction formulas in these proofs can be restricted to be quantifier-free.
Proof: Let $\forall\bar{x}A(\bar{x})$ be a Π_1^0-formula which is provable in PA. By lemma 4 there exists an ordinal $\alpha<\varepsilon_0$ and a p.r. function f such that $\forall\bar{x}\exists y[\hat{f}(\alpha,\bar{x},y)=0] \rightarrow \forall\bar{x}A(\bar{x})$ is provable in PRA and hence in $PI(\phi_{\varepsilon_{\Omega+1}}0)$. By proposition 8 and theorem 10, $\hat{f}(\alpha,\bar{a},G_{\phi_\alpha 0}(t|\alpha\bar{a}k|+3))=0$ and hence $\forall\bar{x}\exists y[\hat{f}(\alpha,\bar{x},y)=0]$ is provable in $PI(\phi_{\varepsilon_{\Omega+1}}0)$. For the provability of Con(PA) the same argument as in proposition 2 applies.

In order to obtain point (ii) of Girard's conjecture, again we have to prove some properties of the functions F_α and G_α, now in PA.

13. Lemma: The statements of the lemmata 5,6, and 7 are provable in PA.
Proof: Use transfinite induction on ordinals $\alpha<\varepsilon_0$.

Since in PA transfinite induction is only available for ordinals $<\varepsilon_0$, it cannot be used to prove properties of the functions G_α for $\alpha\in ON$. For this reason we introduce reductions $\alpha\{a,n\}$ of ordinals α which are closely connected to the G_α's and which can be treated in PA.

<u>14. Definition:</u> (i) The reduction of an ordinal $\alpha \in ON$ under a natural number n:

- one-step-reduction: $0\{n\}:=0$, $(\alpha+1)\{n\}:=\alpha\{n\}+1$, $\lambda\{n\}:=\lambda[n]$
- iterated reduction: $\alpha\{n,0\}:=\alpha$, $\alpha\{n,k+1\}:=(\alpha\{n,k\})\{n\}$
- complete reduction: We write $\alpha\{n,p\}=\text{const.}$ if $\alpha\{n,p\}=\alpha\{n,p+1\}$ and $\alpha{\downarrow}n$ for $\exists x[\alpha\{n,x\}=\text{const.}]$.

(ii) The reduction of uncountable ordinals in ON:

$$(\alpha_0+\Omega^{\alpha_1}\cdot\beta)\{n\} \ : \ \begin{cases} \alpha_0+\Omega^{\alpha_1}\cdot\beta\{n\} & \text{if } \beta\neq\beta\{n\} \\ \alpha_0+\Omega^{\alpha_1\{n\}}\cdot\beta & \text{if } \beta=\beta\{n\}, \alpha_1\neq\alpha_1\{n\} \\ \alpha_0\{n\}+\Omega^{\alpha_1}\cdot\beta & \text{if } \beta=\beta\{n\}, \alpha_1=\alpha_1\{n\} \ . \end{cases}$$

<u>15. Lemma:</u> The following is provable in PA:

(i) If $\alpha \# \beta=\alpha+\beta$, then $\alpha{\downarrow}a \wedge \beta{\downarrow}a \rightarrow (\alpha+\beta){\downarrow}a$

(ii) $\alpha{\downarrow}a \rightarrow \forall k[(\alpha\cdot k){\downarrow}a]$ for additive principle numbers α

(iii) $\phi_0(\alpha){\downarrow}a \rightarrow \phi_0(\alpha+1){\downarrow}a$

(iv) $\alpha{\downarrow}a \rightarrow \phi_0(\alpha){\downarrow}a$.

<u>Proof:</u> (i) $\forall\alpha,\beta[\alpha{\downarrow}a \wedge \beta\{a,p\}=\text{const.}\rightarrow(\alpha+\beta){\downarrow}a]$ is proved by induction on p. (ii) follows immediately from (i), (iii) follows from (ii). (iv): $\alpha\{a,p\}=\text{const.}\rightarrow\omega^{\alpha}{\downarrow}a$ is proved by induction on p with application of (iii).

<u>16. Proposition:</u> For each ordinal $\alpha\in ON$,

$$PA \vdash \forall x\exists y[\alpha\{x,y\}= \text{const.}].$$

<u>Proof:</u> By induction on the length of the ordinal term α. Let $A(\alpha,a)$ be the formula $\alpha{\downarrow}a\rightarrow\forall\beta{\downarrow}a[\phi_\alpha\beta{\downarrow}a]$ and let $A_0(\alpha,a):=A(\alpha,a)$ and $A_{n+1}(\alpha,a):= \alpha{\downarrow}a\rightarrow \ \beta{\downarrow}a[A_n(\beta,a)\rightarrow A_n(\beta+\Omega^{\alpha},a)]$. By external induction on n we show that for all n

(i) $A_n(0,a)$

(ii) $A_n(\alpha,a)\rightarrow A_n(\alpha+1,a)$

(iii) $\alpha\in Lim_\omega\rightarrow[A_n(\alpha[a],a)\rightarrow A_n(\alpha,a)]$

(iv) $\alpha\in Lim_\Omega\rightarrow[\forall\xi{\downarrow}a<\Omega A_n(\alpha[\xi],a)\rightarrow A_n(\alpha,a)]$

are provable in PA.

n=0. (i): see lemma 18; (ii): from $A(\alpha,a)$ we obtain $\alpha{\downarrow}a\rightarrow\forall\beta{\downarrow}a\forall k\phi_\alpha^k(\beta){\downarrow}a$ and hence $\phi_{\alpha+1}0{\downarrow}a$, $\phi_{\alpha+1}\beta{\downarrow}a\rightarrow\phi_{\alpha+1}(\beta+1){\downarrow}a$. By formal induction on p we prove $\alpha{\downarrow}a\wedge\beta\{a,p\}=\text{const.}\rightarrow\phi_{\alpha+1}(\beta){\downarrow}a$ and hence $A(\alpha+1,a)$. (iii) is analogous to (ii). (iv): Let α Lim_Ω, then from $\forall\xi<\Omega A(\alpha[\xi],0)$ and $\phi_\alpha 0[k]{\downarrow}a$ or $\phi_\alpha(\beta+1)[k]{\downarrow}a$ it follows that $\phi_\alpha 0[k+1]{\downarrow}a$ or $\phi_\alpha\beta{\downarrow}a\rightarrow\phi_\alpha(\beta+1)[k+1]{\downarrow}a$ resp.; by induction on p with

$\beta\{a,p\}$=const. we obtain $A(\alpha,a)$.
$n\to n+1$. $(i)_{n+1}$ follows immediately from $(ii)_n$; $(ii)_{n+1}$: from $A_{n+1}(\alpha,a)$ we first obtain by induction on p with $\xi\{a,p\}$=const.:
$\alpha{\downarrow}a\to\forall\beta{\downarrow}a[A_n(\beta,a)\to\forall\xi{\downarrow}a<\Omega A_n(\beta+\Omega^\alpha\cdot\xi,a)]$; it follows:
$A_{n+1}(\alpha,a)\to[(\alpha{\downarrow}a\to\forall\xi{\downarrow}a<\Omega\forall\beta{\downarrow}a[A_n(\beta,a)\to A_n(\beta+\Omega^\alpha\cdot\xi,a)])$

$$\overset{(iv)_n}{\to}(\alpha{\downarrow}a\to[A_n(\beta,a)\to A_n(\beta+\Omega^{\alpha+1},a)])\leftrightarrow A_{n+1}(\alpha+1,a).$$

$(iii)_{n+1}$: $A_{n+1}(\alpha[a],a)\leftrightarrow[\alpha[a]{\downarrow}a\to\forall\beta{\downarrow}a[A_n(\beta,a)\to A_n(\beta+\Omega^{\alpha[a]},a)]]$

$$\overset{(iii)_n}{\to}(\alpha{\downarrow}a\to\forall\beta{\downarrow}a[A_n(\beta,a)\to A_n(\beta+\Omega^{\alpha},a)])\leftrightarrow A_{n+1}(\alpha,a).$$

$(iv)_{n+1}$: $\forall\xi{\downarrow}a<\Omega A_{n+1}(\alpha[\xi],a)\to\forall\xi{\downarrow}a<\Omega[\alpha[\xi]{\downarrow}a\to\forall\beta{\downarrow}a[A_n(\beta,a)\to$

$A_n(\beta+\Omega^{\alpha[\xi]},a)]]\overset{(iv)_n}{\to}(\alpha{\downarrow}a\to\forall\beta{\downarrow}a[A_n(\beta,a)\to A_n(\beta+\Omega^{\alpha},a)])$.

From $(i)_n,\dots,(iv)_n$ and

$$(v)_n\qquad A_{n+1}(\alpha,a)\to(\alpha{\downarrow}a\to\forall\beta{\downarrow}a[A_n(\beta,a)\to\forall\xi{\downarrow}a<\Omega A_n(\beta+\Omega^\alpha\cdot\xi,a)]),$$

which is easily proved, the assertion follows by induction on the length of the term α.

<u>17. Proposition:</u> $PA\vdash\forall\alpha<\Omega[\alpha\{a,b\}=\text{const.}\leftrightarrow\alpha\{a,b\}=G_\alpha(a)]$.
<u>Proof:</u> "$\leftarrow$" is trivial, since $G_\alpha(a)$ is finite and hence irreducible.
"$\to$": by induction on b.

<u>18. Theorem:</u> For all countable ordinals $\alpha\in ON$,

$$PA\vdash\forall x\exists y[y=G_\alpha(x)].$$

<u>Proof:</u> This is an immediate consequence of proposition 17.

<u>19. Lemma:</u> $PA\vdash\alpha>0\to G_\alpha(a+1)>0$.
<u>Proof:</u> $\alpha>0\to\alpha\{a+1,b\}>0$ is proved by induction on b.

<u>20. Definition:</u> The extension $\alpha\sqsubset\beta$ of an ordinal α:
(i) $\alpha\sqsubset\alpha+\delta$ if $\delta\neq 0$ and $\alpha\#\delta=\alpha+\delta$
(ii) $\alpha\sqsubset\phi_\gamma(\alpha)$ if $\phi_\gamma(\alpha)\neq\alpha$
(iii) $\alpha\sqsubset\beta\to\gamma+\alpha\sqsubset\gamma+\beta$ if $\gamma+\alpha=\gamma\#\alpha$ and $\gamma+\beta=\gamma\#\beta$
(iv) $\alpha\sqsubset\beta\to\phi_\gamma(\alpha)\sqsubset\phi_\gamma(\beta)$ if $\phi_\gamma(\alpha)\neq\alpha$ and $\phi_\gamma(\beta)\neq\beta$
(v) $\alpha\sqsubset\beta\to\phi_\alpha(\gamma)\sqsubset\phi_\beta(\gamma)$ if $\gamma=0$ or $\gamma=\gamma_1+1$ with $\gamma_1=\phi_\alpha(\gamma_1)=\phi_\beta(\gamma_1)$
(vi) $\alpha\sqsubset\beta\to\Omega^\alpha\sqsubset\Omega^\beta$, $\Omega^\gamma\cdot\alpha\sqsubset\Omega^\gamma\cdot\beta$ where $\alpha,\beta<\Omega$.

<u>21. Lemma:</u> The following is provable in PA:
(i) For each limit ordinal α of type ω, $\alpha[k]\sqsubset\alpha[k+1]$ and for each limit ordinal α of type Ω, $\beta\sqsubset\gamma\to\alpha[\beta]\sqsubset\alpha[\gamma]$ holds.

(ii) $\alpha \subset \beta \wedge a>1 \rightarrow G_\alpha(a)<G_\beta(a)$

(iii) For each limit ordinal $\alpha<\Omega$, $G_{\alpha[k]}(a)<G_{\alpha[k+1]}(a)$.

<u>Proof:</u> (i): By induction on the length of α. (ii): By induction on the inductive construction of the relation $\subset$. (iii) is an immediate consequence of (i) and (ii).

<u>22. Lemma:</u> For all ordinals $\alpha\in ON$, $\alpha<\Omega$ and p.r. functions $h(\bar{a})$,

$$PA \vdash b \geq G_\alpha(h(\bar{a})+1) \rightarrow \hat{h}(\alpha,\bar{a},b)=0.$$

<u>Proof:</u> By induction on pairs (b,c) of natural numbers we prove $b \geq G_{\hat{h}(\alpha,\bar{a},c)}(h(\bar{a})+1) \rightarrow \hat{h}(\hat{h}(\alpha,\bar{a},c),\bar{a},b)=0$ for all b and c.

(0,c): If b=0, then $\alpha=0$ and hence $\hat{h}(0,\bar{a},0)=0$.

$(b,c+1)\rightarrow(b+1,c)$: By distinction of cases whether $\hat{h}(\alpha,\bar{a},c)$ is 0, successor or a limit ordinal:

- $\hat{h}(\alpha,\bar{a},c)=0$; then $\hat{h}(\hat{h}(\alpha,\bar{a},c),\bar{a},b)=0$.
- $\hat{h}(\alpha,\bar{a},c)=\hat{h}(\alpha,\bar{a},c+1)+1$; if $b+1 \geq G_{\hat{h}(\alpha,\bar{a},c)}(h\bar{a}+1)=G_{\hat{h}(\alpha,\bar{a},c+1)}(h\bar{a}+1)+1$, then $b \geq G_{\hat{h}(\alpha,\bar{a},c+1)}(h\bar{a}+1) \overset{(b,c+1)}{\Rightarrow} \hat{h}(\hat{h}(\alpha,\bar{a},c+1),\bar{a},b)=0$ and hence $\hat{h}(\hat{h}(\alpha,\bar{a},c),\bar{a},b+1)=0$.
- $\hat{h}(\alpha,\bar{a},c+1)=\hat{h}(\alpha,\bar{a},c)[h(\bar{a})]$; if $b+1 \geq G_{\hat{h}\alpha\bar{a}c}(h\bar{a}+1)=G_{\hat{h}\alpha\bar{a}c[h\bar{a}+1]}(h\bar{a}+1)$ $\overset{L.21}{\geq} G_{\hat{h}\alpha\bar{a}c[h\bar{a}]}(h\bar{a}+1)=G_{\hat{h}(\alpha,\bar{a},c+1)}(h\bar{a}+1)$, then $b \geq G_{\hat{h}(\alpha,\bar{a},c+1)}(h\bar{a}+1)$ $\overset{(b,c+1)}{\Rightarrow} \hat{h}(\hat{h}(\alpha,\bar{a},c+1),\bar{a},b)=0$ and hence $\hat{h}(\hat{h}(\alpha,\bar{a},c),\bar{a},b+1)=0$.

We can now complete the proof of Girard's conjecture.

<u>23. Theorem:</u> Each formula B in the language of PA which is provable in $PI(\phi_{\varepsilon_{\Omega+1}}0)$ is also provable in PA.

<u>Proof:</u> If $PI(\phi_{\varepsilon_{\Omega+1}}0) \vdash B$ then by lemma 4 there exist ordinals $\alpha_1,\ldots,\alpha_n<\phi_{\varepsilon_{\Omega+1}}0$ and p.r. functions $h_1,\ldots,h_n$ such that

$PA' \vdash \forall x_1 \exists y_1[\hat{h}_1(\alpha_1,x_1,y_1)=0] \wedge \ldots \wedge \forall x_n \exists y_n[\hat{h}_n(\alpha_n,x_n,y_n)=0] \rightarrow B.$

By lemma 21 it follows that

$PA' \vdash \forall x \exists y[y=G_{\alpha_1}(x)] \wedge \ldots \wedge \forall x \exists y[y=G_{\alpha_n}(x)] \rightarrow B$

and hence by theorem 18, $PA \vdash B$.

There remains the question why the ordinal analysis of number theory by pointwise induction exactly yields the Bachmann/Howard ordinal. Obviously, pointwise induction as well as pointwise recursion in the slow-growing hierarchy treat limit ordinals as successor ordinals. Using the reduction $\alpha\{n,p\}$ of ordinals, this can also be carried out in number theory, with exception of ordinals like $\phi_{\hat{\alpha}}0$ whose fundamental sequences are built by strong diagonalizations. These form within $\phi_{\varepsilon_{\Omega+1}}0$ a skeleton with the structure of

ε_0 (it is for this reason that the proof of proposition 16, for example, is of great formal similarity with the usual proof of transfinite induction up to ordinals $\alpha<\varepsilon_0$ in number theory). Hence the characterization of number theory by pointwise induction preserves quite well the structure of ε_0, $\phi_{\varepsilon_{\Omega+1}}0$ appears so to speak as a "super-ε_0". What this and other examples show is that it is not the assignment of an ordinal as a set-theoretical object which gives insight into a formal theory, but the relation between the proof-theoretical means of a theory and the internal structure of a certain ordinal.

References

[1] P.Aczel, Another elementary treatment of Girard's result connecting the slow and the fast growing hierarchies of number theoretic functions, manuscript, 1980

[2] H.Bachmann, Die Normalfunktionen und das Problem der ausgezeichneten Folgen von Ordinalzahlen, Vierteljahresschrift der Naturforschenden Gesellschaft in Zürich, XCV (1950), 5-37

[3] W.Buchholz, Three contributions to the conference on "Recent Advances in Proof Theory", manuscript, 1980

[4] E.A.Cichon and S.S.Wainer, The slow-growing and the Grzegorczyk hierarchies, manuscript, 1980

[5] J.-Y.Girard, Π^1_2-Logic, to appear in Ann.Math.Logic

[6] H.R.Jervell, Homogeneous Trees, lectures held at Munic, summer-term 1979

[7] G.Kreisel, Wie die Beweistheorie zu ihren Ordinalzahlen kam, Jahresb.Dtsch.Math.Ver. 78 (1976) 177-223

[8] U.R.Schmerl, Über die schwach und die stark wachsende Hierarchie, to appear in Sitzungsber.Bay.Akad.Wiss.

[9] H.Schwichtenberg, Eine Klassifikation der ε_0-rekursiven Funktionen, Zeitschr.f.math.Logik u.Grundl.d.Math., 17 (1971) 61-74

[10] H.Schwichtenberg, Homogene Bäume und subrekursive Hierarchien, lecture held at Oberwolfach, April 1980

PROCEEDINGS OF THE HERBRAND SYMPOSIUM
LOGIC COLLOQUIUM '81, J. Stern (editor)

RELATIVE RECURSIVE ENUMERABILITY

Robert I. Soare
University of Chicago

Michael Stob
Calvin College

We show the following:

Theorem: Let C be a recursively enumerable (r.e.), nonrecursive set. Then there is a set A such that A is r.e. in C, C is recursive in A, but A does not have r.e. degree.

This theorem was first suggested to us by a conjecture of Cooper. He conjectured [1] that every Turing degree $\mathbf{a} \geq \mathbf{0}'$ which is r.e. in $\mathbf{0}'$ is r.e. in every high degree $\mathbf{d} \leq \mathbf{0}'$. (A degree $\mathbf{d}$ is <u>high</u> if $\mathbf{d}' = \mathbf{0}''$.) Our theorem, since it relativizes to any degree $\mathbf{d}$, negatively answers this conjecture. (In fact, it shows that no high degree $\mathbf{d} < \mathbf{0}'$ can have the property that any degree $\mathbf{a} \geq \mathbf{0}'$ r.e. in $\mathbf{0}'$ is r.e. in $\mathbf{d}$. The existence of some high r.e. degree $\mathbf{d}$ which fails to have this property also follows by a theorem of Jockusch and Shore [2].) Our interest in this theorem was reawakened when Jockusch and Shore showed that an index for A (as a set r.e. in C) cannot be found effectively from that of C. This led us to discover the present proof which exhibits this nonuniformity by giving indices for two sets A and B, one of which has the desired properties.

In §1 we describe a single requirement for proving the theorem and give a construction which meets this requirement. In §2 we outline a method for meeting all the requirements, assuming that C is low. In §3, we describe the method for meeting the requirements without this assumption. In §4, we give further refinements of the theorem and also discuss the relationship of this work to that of Jockusch and Shore [2]. Our notation is standard, a reference is Soare [6]. In addition, we will often identify a set A with its characteristic function and use A[x] to represent this function restricted to arguments $\leq$ x.

1. ONE REQUIREMENT.

Let C be a fixed nonrecursive r.e. set. We will construct a set A so that $A \oplus C$ has the desired properties, namely, $A \oplus C$ is r.e. in C and $A \oplus C$ does not have r.e. degree. (Obviously, C is recursive in $A \oplus C$.) We consider requirements of the following form

$$R: \qquad \Phi(W) \neq A \quad \text{or} \quad \Psi(A \oplus C) \neq W$$

where Φ, Ψ are functionals and W is an r.e. set. It is clear that if we meet the requirements R as Φ and Ψ range over all functionals and W over all r.e. sets, then $A \oplus C$ will have the desired degree provided A is r.e. in C.

Before giving a construction that meets a single (somewhat modified) requirement, we describe the basic strategy for attempting to meet the requirement R.

First we need some notation. If $\Phi_s(W_s;x)$ $(\Psi_s(A_s \oplus C_s;y))$ is defined, let $\varphi_s(x)$ $(\psi_s(y))$ be the greatest element used in the computation. (Here W_s, A_s, C_s, Φ_s, and Ψ_s are the approximations to W, A, C, Φ, and Ψ available at the end of stage s of the construction. Thus, for instance if $W_s[\varphi_s(x)] = W[\varphi_s(x)]$ then $\Phi_s(W_s;x) = \Phi(W;x)$.

To attack R we choose a "witness" x_0. We keep x_0 out of A until a stage s at which

$$(1.1) \qquad \Phi_s(W_s;x_0) = 0 \quad \text{and} \quad (\forall y \leq \varphi_s(x_0))[\Psi_s(A_s \oplus C_s;y) = W_s(y)]$$

If such a stage s does not exist, then requirement R is satisfied since either $\Phi(W;x_0) \neq A(x_o) = 0$ or $\Psi(A \oplus C;y) \neq W(y)$ for some $y \leq \varphi(x_0) = \lim_s \varphi_s(x_0)$. Suppose then that a stage s satsifying (1.1) does exist. Then at stage s+1 we enumerate x_0 in A (and restrain A through $\psi_s(\varphi_s(x_0))$). Thus we now have that $\Phi_s(W_s,x_0) \neq A_{s+1}(x_0)$. Obviously now requirement R is satisfied unless there is a stage $t > s$ so that some number $y \leq \varphi_s(x_0)$ is enumerated in W at stage t. Suppose then that such a y and such a stage t exist. Then by (1.1)

$$(1.2) \qquad W_t(y) \neq W_s(y) = \Psi_s(A_s \oplus C_s;y)$$

Now suppose that $C_s[\psi_s(\varphi_s(x_0))] = C_t[\psi_s(\varphi_s(x_0))]$; that is, suppose that C has not changed on the amount of C used in the computation $\Psi_s(A_s \oplus C_s;y)$. Then if we can remove x_0 from A at stage t we would have $A_t[\psi_s(\varphi_s(x_0))] = A_s[\psi_s(\varphi_s(x_0))]$ so that $\Psi_s(A_s \oplus C_s;y) = \Psi_t(A_t \oplus C_t;y)$ and again we have that R is satisfied since $W_t(y) \neq \Psi_t(A_t \oplus C_t;y)$ by (1.2). If we restrain A on the initial segment up to $\psi_s(\varphi_s(x_0))$ then R is satisfied forever unless $C[\psi_s(\varphi_s(x_0))]$ changes at some later stage. If C changes however we restart the procedure; using the same x_0 we search for a stage s satisfying (1.1). Two things can happen. Either we eventually satisfy R with x_0 in one of the above ways or there are infinitely many stages s and stages u > s so that $C_{u+1}[\psi_s(\varphi_s(x_0))] \neq C_u[\psi_s(\varphi_s(x_0))]$. This implies either that $\Phi(W;x_0)$ is not defined or $\Psi(A \oplus C;y)$ is not defined for some y $\leq \lim_s \varphi_s(x_0)$. In the former case, only finitely much action (and restraint on A) is required by this strategy; in the latter case infinitely much action is required (x_0 enters and leaves A infinitely often) but the restraint on A has lim inf 0.

The above construction did not guarantee that A is r.e. in C. In general, if $x_0 \notin A_s$ we may always enumerate x_0 in A at stage s. But the above construction required removing x_0 from A as well. To insure that A is r.e. in C, C must be able to tell if x_0 will ever be removed from A. The usual method of guaranteeing this is as follows. When the witness x_0 is enumerated in A at stage s we appoint a certain marker $\gamma_s(x_0)$. Then we only allow x_0 to be removed from A at a later stage t+1 if $C_{t+1}[\gamma_s(x_0)] \neq C_t[\gamma_s(x_0)]$. This insures that A is r.e. in C. However, this conflicts with our original strategy in the following ways. If stage t exists so that we now want to remove x_0 from A, we now need C to change below $\gamma_s(x_0)$. But we already noticed that if C changes below $\psi_s(\varphi_s(x_0))$, our attack is ruined and we must restart the attack. Thus, at the very least we should require $\gamma_s(x_0) > \psi_s(\varphi_s(x_0))$. Further, after stage t, we need C to change on the interval $I_0 = (\psi_s(\varphi_s(x_0)),\gamma_s(x_0)]$ so that we may remove x_0 from A and complete the attack. Now C need never change on the interval I_0. In this case x_0 must remain in A and so does not satisfy R. The solution to this problem is to use further witnesses $x_1,x_2,\ldots$ so that the corresponding intervals $I_0,I_1,I_2,\ldots$

overlap and so together, cover ω (except for finitely many integers preceding the first marker $\psi_s(\varphi_s(x_0))$). Then we argue that if C is nonrecursive, C must change on one of the intervals after it is formed. The witness x_i corresponding to the least of these intervals is the witness used to verify requirement R. Now the major difficulty arises in constructing the intervals $I_0, I_1, \ldots$ to overlap but yet so that the actions taken for the witnesses $x_0, x_1, \ldots$ do not conflict. Consider the two witnesses x_0 and x_1. Then

$$I_0 = (\psi_{s_0}(\varphi_{s_0}(x_0)), \gamma_{s_0}(x_0)] \text{ and } I_1 = (\psi_{s_1}(\varphi_{s_1}(x_1)), \gamma_{s_1}(x_1)]$$

where $s_0 < s_1$ are the stages at which we verified (1.1) for x_0 and x_1 respectively. The overlap requirement is that $\psi_{s_1}(\varphi_{s_1}(x_1)) \leq \gamma_{s_0}(x_0)$. We can only arrange this if we wait until stage s_1 to enumerate x_0 in A and hence to define $\gamma_{s_0}(x_0)$ else we would not know how large to make $\gamma_{s_0}(x_0)$. Waiting is of course acceptable since (1.1) must happen cofinitely often if R is a true requirement. There is an additional crucial difficulty here however. To verify that $\psi_{s_1}(\varphi_{s_1}(x_1)) \leq \gamma_{s_0}(x_0)$ we must have certain computations from $(A_{s_1} \oplus C_{s_1})[\psi_{s_1}(\varphi_{s_1}(x_1))]$ existing. The argument that R is satisfied for x_1 depends on these computations being preserved - that no element enters A below $\psi_{s_1}(\varphi_{s_1}(x_1))$ except possibly x_1. But undoubtedly $x_0 < \psi_{s_1}(\varphi_{s_1}(x_1))$ so that at stage s_1 we ruin our attack on x_1 by enumerating x_0 in A. This destroys our interval I_1 because the computation on x_1 used the most current information on x_0, namely that x_0 was not yet in A.

The solution to this conflict between x_0 and x_1 is to construct two sets A and B instead of just one so that the computations used in forming I_0 and I_1 do not conflict. This is done by substituting for R the requirement

R': $\quad \Phi(W) \neq A \;$ or $\; \Psi(A \oplus C) \neq W \;$ or $\; \Theta(V) \neq B \;$ or $\; \Xi(B \oplus C) \neq V$

where Φ, Ψ, Θ, and Ξ are reductions and W and V are r.e. sets. This device is the same as that used by Lachlan in his non-diamond theorem [3]. It is easy to see that if each requirement R' is met as $\Phi, \Psi, \Theta, \Xi, W, V$ range over all possible such sextuples, then one of $A \oplus C$ or $B \oplus C$ satisfies the theorem. This device overcomes the final obstruction as follows. We use $x_0, x_2, \ldots$ as witnesses which want to enter A to meet $\Phi(W) \neq A$ or $\Psi(A \oplus C) \neq W$ and $x_1, x_3, x_5, \ldots$ enter B to meet $\Theta(V) \neq B$ or $\Xi(B \oplus C) \neq V$. Then I_0 and I_1 may overlap since x_1 does not need computations using the fact that $x_0 \notin A_s$. Of course, I_0 and I_2 need not overlap so x_0 and x_2 do not interfere with each other.

To summarize, a requirement R can be met as follows. First, we could fail in our attempt to form some interval. This corresponds to the non-convergence of one of the functionals involved. This outcome results in finitely much activity taking place for requirement R and only finite restraint imposed on A and B. Second, some disagreement in computations resulting from a witness x_n may be preserved forever. This may happen because W or V fails to change after x_n is successfully removed from A or B without being later destroyed by a C change. This again is a finitary outcome. Finally, some witness x_n may have its attacks destroyed by changes in C infinitely often. Then x_n witnesses that R is satisfied by the divergence of a certain computation. Requirement R may receive attention infinitely often but the restraint imposed has finite lim inf.

We now give the complete construction.

Lemma 1.1 (One requirement). Let requirement R' be as above. Then there are sets A,B which are r.e. in C and which satisfy R'.

Proof. The construction is in stages. We will have movable markers $x_0, x_1, x_2, \ldots$ representing candidates for the witnesses mentioned above and we will use x_i^s for the position of marker x_i at the end of stage s if it is defined.

We construct A,B by enumerating integers in and out of these sets. If n is enumerated in or out of A (B) at stage s+1 then $n = x_{2i}^s$ ($n = x_{2i+1}^s$) for some i. A_s (B_s) denotes the (finite) set of integers in A (B) at the end of stage s and $A = \{n : (\exists s)(\forall t \geq s)\ [n \in A_t]\}$ $(B = \{n : (\exists s)(\forall t \geq s)[n \in B_t]\})$.

If x_i^s is defined, we may also have defined a certain auxiliary function $v(x_i,s)$. $v(x_i,s+1) = v(x_i,s)$ unless $v(x_i,s+1)$ is undefined. Similarly, $x_i^{s+1} = x_i^s$ unless x_i^{s+1} is undefined. To cancel a marker at stage s+1 means to make x_i^{s+1} and $v(x_i,s+1)$ undefined and remove x_i^s from A (B) if i is even (odd).

For the appropriate approximations to the funtionals Φ,Ψ,Θ, and Ξ at stage s we let the corresponding use functions be denoted by φ,ψ,θ, and ξ. Thus $\varphi(y,s)$ is the bound on the information used from W_s in computing $\Phi_s(W_s;y)$. We assume that $y \leq \varphi(y,s)$ if $\Phi_s(W_s;y)$ is defined and if $\Phi_s(W_s;y+1)$ is also defined, that $\varphi(y,s) \leq \varphi(y+1,s)$. Of course, similar assumptions are made for ψ, θ, and ξ. Define the length of agreement function, $\ell(s)$ by

$$\begin{aligned}\ell(s) = (\mu y)(\forall x < t)\ [&\Phi_s(W_s;x) = A_s(x)\\ &\Theta_s(V_s;x) = B_s(x)\ ,\\ &(\forall w < \varphi(y,s))[\Psi_s(A_s \oplus C_s;w) = W_s(w)],\ \text{and}\\ &(\forall w < \theta(y,s)[\Xi_s(B_s \oplus C_s;w) = V_s(w)]]\end{aligned}$$

If $y \leq \ell(s)$, let $u(y,s) = \max\{\psi(\varphi(y,s),s),\xi(\theta(y,s),s)\}$ so that $u(y,s)$ bounds all the information needed from A,B,W, and V to verify $\ell(s) \geq y$.

<u>Construction</u>

<u>stage 0</u>: Let $x_0^0 = 0$. Let $v(x_0,0)$ be undefined.

<u>stage s+1</u>: Let i be the least integer, if any, such that $(\exists y \leq v(x_i,s))[y \in C_{s+1} - C_s]$. If i exists, adopt case 1 below. Otherwise, adopt case 2.

<u>Case 1</u>. <u>Cancel</u> all markers x_j, $j > i$. If $i > 0$ <u>remove</u> x_{i-1}^s from A (B) if i-1 is even (odd) and $x_{i-1}^s \in A_s$ $(\in B_s)$. <u>Let</u> $v(x_i,s+1)$ be undefined.

<u>Case 2</u>: Let i be the greatest integer such that x_i^s is defined. If $\ell(s) < x_i^s$, go to stage s+2. Otherwise,

(1) enumerate x_{i-1}^s in A (B) if i is even (odd);

(2) let $v(x_i,s+1) = u(x_i^s,s)$; and

(3) let x_{i+1}^s be a number greater than any number used so far in the construction as a marker position or auxiliary function value.

The following fact is easy to see from the construction.

Lemma 1.2. For every s there is an i such that

i) $(\forall j < i)$, x_j^s and $v(x_j,s)$ are defined,

ii) x_i^s is defined but $v(x_i,s)$ is not defined,

iii) $(\forall j > i)$, x_j^s and $v(x_j,s)$ are not defined,

iv) $x_0^s,\dots,x_{i-2}^s$ are in A_s or B_s and x_{i-1}^s, x_i^s are in neither A_s nor B_s,

v) $x_0^s < x_1^s < \dots < x_i^s$; $v(x_0,s) < v(x_1,s) < \dots < v(x_{i-1},s)$.

Lemma 1.3. The sets A and B are r.e. in C.

Proof. We claim that A satisfies the following Σ_1^C definition

$$A = \{z : (\exists i)(\exists s)[z = x_{i-1}^s \in A_s \text{ and } C[v(x_i,s)] = C_s[v(x_i,s)]]\}.$$

For if z is enumerated in A at some stage s+1, $z = x_{i-1}^{s+1}$ for some odd i and $v(x_i,s+1)$ is defined at stage s+1 by an application of case 2 to i at stage s+1. Now z is removed from A at the first later stage $t+1 > s+1$ for which $C_{t+1}[v(x_i,s)] \neq C_t[v(x_i,s)]$ for then case 1 applies to some marker x_j, $j \leq i$. (Thus $v(x_i,s)$ plays the role of the enumeration marker $\gamma_s(x_i^s)$ mentioned in the sketch above.) If no such t exists however, then $x = x_{i-1}^s = \lim_s x_{i-1}^s$, $\lim_s v(x_i,s) = v(x_i,s)$ and z remains in A and satisfies the Σ_1^C definition above.

Lemma 1.4. There is an integer i such that

a) $(\forall j \leq i)$ [x_j^s is defined for confinitely many s], (and so $\lim_s x_j^s$ exists, and

b) $(\forall j > i)$ [x_j^s is undefined for infinitely many s].

Proof. Since $x_0^s = 0$ for every s, we may suppose for a contradiction that for every i, x_i^s is defined for almost every s. We will show that this assumption implies that C is recursive, contrary to the hypothesis of the theorem.

The assumption implies that for every i there is a stage s_i such that for all $s \geq s_i$

(1.2) $$x_i^{s_i} = x_i^s ,$$

(1.3) $$v(x_i, s_i) = v(x_i, s),$$

(1.4) $$C_s[v(x_i, s)] = C[v(x_i, s)] .$$

We show how to recursively compute stages t_i so that t_i has properties (1.2)-(1.4). By (1.3) and (1.4) and the fact that $v(x_i,s) \geq x_i$ for every i, C is then recursive.

Let $t_0 = s_0$. Supposing that we have computed t_i, we show how to compute t_{i+1}. Let us suppose for convenience that i is even. Let u be the least stage such that $(\forall s)[u \leq s \leq t_i \Longrightarrow v(x_i,s) = v(x_i,t_i)]$. Then case 2 applied to i at stage u and (1.2)-(1.4) are true for u in place of s_i. Let t_{i+1} be the least stage such that $u \leq t_{i+1}$ and

(1.5) $$W_u[\varphi(x_i^u, u)] \neq W_{t_{i+1}}[\varphi(x_i^u, u)].$$

We first argue that there is such a stage t_{i+1} and then that t_{i+1} satisfies (1.2)-(1.4) in place of s_{i+1}. If there is no such stage t_{i+1}, we have

$$0 = A_u(x_i^u) = \Phi(W_u, x_i^u) = \Phi(W, x_i^u).$$

But, $x_i^u \in A$ since $x_i^u \in A_s$ at any stage s such that $v(x_{i+1},s)$ is defined and there are cofinitely many such s. Thus for cofinitely many s, $\ell(s) < x_i^u$, contradicting the hypothesis that every marker x_j is eventually used, for if x_j^s is ever defined, $\ell(s) > x_j^s$.

Finally we show that t_{i+1} has the desired properties. First, (1.2) follows easily since $x_{i+1}^s = x_{i+1}^u$ for all $s \geq u$. To show (1.3) and (1.4) it suffices to prove

(1.6) $$(\forall s)[t_{i+1} \leq s \Longrightarrow v(x_{i+1},s) \text{ is defined}]$$

for (1.4) can only fail for some $s \geq t_{i+1}$ if $v(x_{i+1},s)$ is made undefined under case 1 at stage s. Let $s \geq t_{i+1}$ be a counterexample to (1.6); then $v(x_{i+1},s)$ is undefined. Then $A_s = A_u$. Also, since $s \geq u$, $C_s[v(x_i,u)] = C_u[v(x_i),u)] = C[v(x_i,u)]$ so that

(1.7) $$(\forall y \leq \varphi(x_i,u))[\Psi_u(A_u + C_u;y) = \Psi_s(A_s \oplus C_s;y)].$$

since $v(x_i,u) \geq \psi(\varphi(x_i,u),u)$. But then $\ell(s) < x_i^s$ since $(\exists y \leq \varphi(x_i,u))[\Psi_u(A_u \oplus C_u;y) \neq W_s(y)]$. Thus, $v(x_{i+1},s+1)$ is also undefined. Thus we have shown that if there is any counterexamnple s' to (1.6), $v(x_{i-1},s)$ is undefined for all $s \geq s'$. (Intuitively, $v(x_{i+1},s)$ becomes undefined only when $C[v(x_{i+1},s)]$ changes at which point x_i is removed from A and x_i from then on witnesses the disagreement of W and $\Psi(A \oplus C)$.) This final contradiction establishes (1.6) and hence the lemma. □

Lemma 1.5. $\Phi(W) \neq A$ or $\Psi(A \oplus C) \neq W$ or $\Theta(V) \neq B$ or $\Xi(B \oplus C) \neq V$.

Proof. Let i be as in Lemma 1.4. Fix a stage s_0 so that for every $s \geq s_0$ and for every $j \leq i$, x_j^s is defined and $x_j^s = x_j^{s_0}$ and if $j < i, v(x_j,s) = v(x_j,s_0)$. Suppose that the lemma is false so that $\Phi(W) = A$, $\Psi(A \oplus C) = W$, $\Theta(V) = B$, and $\Xi(B \oplus C) = V$. Let $s_1 \geq s_0$ be a stage so that $\ell(s) \geq x_i^{s_1}$ $(= x_i^{s_0})$ and $C_{s_1}[u(x_i^{s_1},s_1)] = C[u(x_i^{s_1},s_1)]$ and $x_i^{s_1}$ is defined but $x_{i+1}^{s_1}$ is not. There must be infinitely many such stages $s \geq s_0$ because the computations involved settle down and for such a stage s_1, $A_{s_1} = A$ and $B_{s_1} = B$. Then at stage s_1+1, $v(x_i,s_1+1) = u(x_i,s_1)$ and $x_{i+1}^{s_1+1}$ is defined and case 1 never applies to any marker x_j, $j \leq i$ at any stage $s > s_1$. Thus x_{i+1}^s is defined for all $s > s_1$, contradicting the crucial property of i. This contradiction yields the lemma. □

2. IF $C' \equiv_T \phi'$.

In this section we describe a construction based on that of §1 which proves the theorem, provided that $C' \equiv \phi'$. Of course, if $C' >_T \phi'$, then the theorem is trivial since C' is itself r.e. in C but not of r.e. degree. The assumption that

C is low will allow us to modify the construction of §1 so that the outcome is finitary in nature. This will allow us to prove the theorem using a finite injury priority argument. The technique of replacing an infinite injury priority argument by a finite injury argument using lowness is due to Robinson [4].

To prove the theorem, it suffices to construct sets A and B which are r.e. in C and which satisfy for every $i \in \omega$ the requirement

$$R_i : \quad \Phi_i(W_i) \neq A \text{ or } \Psi_i(A \oplus C) \neq W_i \text{ or } \Theta_i(V_i) \neq B \text{ or } \Xi_i(B \oplus C) \neq V_i.$$

where $\langle \Phi_i, \Psi_i, \Theta_i, \Xi_i, W_i, V_i \rangle_{i\in\omega}$ is an effective list of all sextuples of the appropriate type.

Suppose then that we apply the construction of §1 to R_0. Then Lemmas 1.4 and 1.5 together show that R_0 is satisfied. Now to prove Lemma 1.4 (and in particular (1.7)) we used the fact that if s and u are stages of the construction for which $v(x_j,s) = v(x_j,u)$ and $v(x_{j+1},s)$ and $v(x_{j+1},u)$ are undefined, then $A_s[v(x_j,s)] = A_u[v(x_j,s)]$. (This holds if j is even; substitute B if j is odd.) Thus if $v(x_j,s)$ is defined, it imposes a restraint of the same length on both A and B. If this restraint is respected by our action for other requirements, then Lemma 1.4 can be proved for R_0 with the same proof as that of §1. (We will defer discussing the proof of Lemma 1.5.) Thus our strategy for meeting R_1 should only appoint markers larger than any value $v(x_j,s)$ which is defined.

Lemma 1.4 for R_0 then guarantees that there are integers i_0 and s_0 so that $x^s_{i_0} = x^{s_0}_{i_0}$ for all $s \geq s_0$ and $x^s_{i_0+1}$ is undefined for infinitely many s. Thus if $s \geq s_0$ and $j < i_0$, $v(x_j,s) = v(x_j,s_0)$, and by b) there are infinitely many stages s such that $v(x_{i_0},s)$ is undefined. Thus, if $r(0,s)$ is the restraint imposed by R_0 at stage s, $\liminf_s r(0,s) = v(x_{i_0-1},s)$. Thus, the restraint has finite lim inf as in other infinite injury arguments. Now it could be the case that $x^s_{i_0+1}$ is undefined for almost every s; in that case, $\lim_s r(0,s)$ exists and, in fact, only finitely much action is taken for R_0 over the whole construction. On the other hand, it may be the case that $x^s_{i_0+1}$ (and hence $v(x_{i_0},s)$) is defined for infinitely many s so that $v(x_{i_0},s)$ and $r(0,s)$ are unbounded in s. Of course this latter case happens only because

$$(2.1) \qquad (\overset{\infty}{\exists} s)[v(x_{i_0},s) \text{ is defined and } C_{s+1}[v(x_{i_0},s)] \neq C_s[v(x_{i_0},s)]].$$

It is (2.1) that we prevent using the lowness of C.

The crucial tool in the Robinson technique is the following lemma. (See [5, Theorem 2.7] for a proof.)

Lemma 2.1. If C is an r.e. set such that $C' \equiv_T \phi'$ and $\{D_n\}_{n\varepsilon\omega}$ is the canonical indexing of finite sets, then there is a recursive function f such that for all j

$$(2.2) \qquad W_j \cap \{n : D_n \subseteq \overline{C}\} = W_{f(j)} \cap \{n : D_n \subseteq \overline{C}\} \text{ and}$$

$$(2.3) \qquad W_j \cap \{n : D_n \subseteq \overline{C}\} = \emptyset \Longrightarrow W_{f(j)} \text{ is finite.}$$

The lemma is used to modify our basic construction in the following way. In case 2, when we are attempting to define $v(x_i,s)$, (say to make $v(x_i,s) = y$), we use the function f supplied by the lemma to "guess" whether $C_s[y] = C[y]$. We do this by enumerating n into a fixed r.e. set W_e where $D_n = \overline{C}^s[y]$; e here depends on x_i. We can assume we know the index e by the recursion theorem. We then enumerate simultaneously $W_{f(e)}$ and C until either n appears in $W_{f(e)}$ or some number $z \leq y$ appears in C. Such must happen by Lemma 2.1. In the former case, we say that $v(x_i,s)$ is "certified" and proceed with case 2 exactly as before. However, if it is C that changes, we do not define $v(x_i,s)$ but instead proceed directly to the next stage. Now if $C_s[y] = C[y]$, then $v(x_j,s)$ is "certified" by (2.2). Thus enough certification is provided so that Lemma 1.5 can still be proved. On the other hand, there can be no i_0 such that (2.1) happens. For each time $v(x_{i_0},s)$ is defined, a new element is enumerated in $W_{f(e)}$. If (2.1) happens, $W_{f(e)}$ is infinite so by (2.3), $W_e \cap \{n : D_n \subseteq \overline{C}\} \neq \emptyset$. But this means that there is a stage s' such that $v(x_{i_0},s')$ is certified and $C_{s'}[v(x_{i_0},s')] = C[v(x_{i_0},s')]$. But then $v(x_{i_0},s) = v(x_{i_0},s')$ and (2.1) fails.

Thus, using this modified construction we have that R_0 is satisfied and only finitely many of the values of the markers and v function are ever defined. Now to meet a second requirement R_1, we use a second set of markers, say $y_0, y_1, \ldots$ and insure that y_0^s is greater than any value of x_i^s or $v(x_i,s)$ for any i. Thus the markers y_i automatically respect the restraint imposed by R_0. Otherwise, the strategy for the y_i's is the same as that for the x_i's except that whenever further work is done for R_0, y_0^s is moved to a large number and the construction for R_1 is restarted. This will happen only finitely often. This construction for R_1 does not interfere with that for R_0. Lemma 1.4 for R_0 can be proved using the fact that R_1 respects $r(0,s)$. Lemma 1.5 for R_0 is proved by arguing as in §1 that eventually long lengths of agreement would be preserved so that x_i^s is almost always defined for every i (for a contradiction of Lemma 1.4). This argument depends on the fact that for every u there are infinitely many s such that $A_s[u] = A[u]$. This is true in this case since any integer z enters and leaves A only finitely often.

Lemmas 1.4 and 1.5 are proved for R_1 exactly as for R_0 since after R_0 stops acting, R_1 behaves exactly as the requirement R of §1. It is easy to see how to extend the above construction to the infinitely many requirements R_i; it is now a relatively standard finite injury argument.

The sets A and B constructed in this way have the property that any integer x is enumerated into A (or B) only finitely many times. With a slight modification of the construction, we can insure that an integer x is enumerated into each of A and B at most once; thus A and B are each a difference of r.e. sets.

3. REMOVING THE ASSUMPTION OF LOWNESS.

We next show how to meet the requirements $\{R_i\}_{i\varepsilon\omega}$ of §2 without the assumption that C is low. The reason we wish to prove the theorem directly without this assumption on C is to indicate the uniformity that can be obtained as we explain further in §4. Also this was our original proof of the theorem. We first consider the strategy for meeting two requirements R_0 and R_1.

Recall then that the result of the activity of R_0 was to place a restraint on A and B, $r(0,s)$, at every stage s so that $\liminf_s r(0,s) < \infty$. Also Lemma 1.4 guaranteed the existence of i_0 and s_0 so that

$$(\forall s \geq s_0)\ [x_{i_0}^{s_0} = x_{i_0}^{s}] ,$$

and

$$x_{i_0+1}^{s} \text{ and } v(x_{i_0},s) \text{ are undefined for infinitely many } s.$$

The difficulty which was circumvented by the assumption of lowness in §2 was (2.1), namely the possibility that $v(x_{i_0},s)$ is defined infinitely often (and so that $\limsup_s r(0,s) = \infty$). We now describe a strategy for meeting R_1 which takes into account this possibility.

To meet R_1 we will have infinitely many strategies σ_i, $i \in \omega$, corresponding to guesses about the outcome of R_0. Strategy σ_i is associated with the guess that $i_0 = i$, i.e.,

(3.1) x_i^s is defined for cofinitely many s,

(3.2) x_{i+1}^s and $v(x_i,s)$ are undefined for infinitely many s,

(3.3) $\liminf_s r(0,s) = \lim_s v(x_{i-1},s)$.

Associated with each strategy σ_i, $i \in \omega$, is a set of markers $x_{\langle i,j\rangle}$, $j \in \omega$. The markers $x_{\langle i,j\rangle}$, $j \in \omega$, will be used to show that R_1 is satisfied if σ_i is the strategy associated with the correct guess at the outcome of R_0. Also the strategies σ_i, $i \in \omega$, together impose a restraint $r(1,s)$ so that if $R(1,s) = \max\{r(0,s),r(1,s)\}$, then $\liminf_s R(1,s) < \infty$.

Strategy σ_i acts on the markers $x_{\langle i,j\rangle}$, $j \in \omega$, as in the basic strategy of §1 with the following modifications. First, strategy σ_i only acts at stages such that $v(x_i,s)$ is undefined. (If σ_i's guess is correct, there are infinitely many such stages by (3.2).) Second, σ_i only defines $x_{\langle i,j\rangle}^s$ so that $x_{\langle i,j\rangle}^s > v(x_{i-1},s)$; this is necessary by (3.3). Thus, if $v(x_{i-1},s)$ becomes undefined, we cancel all markers $x_{\langle k,j\rangle}^s$, $k \geq i$. (If σ_i's guess is correct then $x_{\langle i,0\rangle}^s$ is redefined by this requirement only finitely often by (3.1).) These two restrictions on the activity of σ_i guarantee that σ_i respects r(0,s) at any stage such that σ_i acts, because $x^s_{\langle i,j\rangle} > r(0,s) = v(x_{i-1},s)$ for all j such that $x^s_{\langle i,j\rangle}$ is defined. Third, strategy σ_i places a modification on the activity of R_0. If $v(x_{i-1},s)$ is to be defined at stage s we require that $v(x_{\langle i,j\rangle},s) < v(x_i,s)$ for any j such that $x^s_{\langle i,j\rangle}$ is defined. This condition on the definition of $v(x_i,s)$ means that the auxiliary functions for R_0 and R_1 bear the following relationships (if all those mentioned are defined)

$$(3.4) \qquad v(x_{i-1},s) < x^s_i < x^s_{\langle i,0\rangle} < v(x_{\langle i,0\rangle},s)$$
$$< x^s_{\langle i,1\rangle} < v(x_{\langle i,1\rangle},s) < x^s_{\langle i,2\rangle} < v(x_i,s).$$

The reason for insuring (3.4) is as follows. Suppose that s+1 is a stage such that

$$(3.5) \qquad C_{s+1}[v(x_{\langle i,j\rangle},s)] \neq C_s[v(x_{\langle i,j\rangle},s)].$$

Then our basic strategy, case 1, would have us remove $x^s_{\langle i,j-1\rangle}$ from A or B and cause $v(x_{\langle i,j\rangle},s+1)$ to be undefined. But we are requiring that we can only play strategy σ_i at stages s+1 so that $v(x_i,s+1)$ is undefined (so as to respect the restraint of R_0). Now (3.4) and (3.5) together guarantee that s+1 is such a stage since $C_{s+1}[v(x_i,s)] \neq C_s[v(x_i,s)]$.

Suppose now that σ_i is the strategy with the correct guess about R_0; there is such a strategy since Lemma 1.4 still holds for R_0 because every σ_i respects r(0,s). We show that R_1 is satisfied by the action of σ_i and that $\liminf_s R(1,s) < \infty$. We then also argue that R_0 is still satisfied; namely that we can still prove Lemma 1.5 for R_0. To show that Lemma 1.4 and 1.5 still hold for the requirement R_1 and the markers $x_{\langle i,j\rangle}$, $j \in \omega$, let s_0 be a stage so that $x^s_i = x^{s_0}_i$ for all $s \geq s_0$ (so that also $v(x_{i-1},s) = v(x_{i-1},s_0)$ for all $s \geq s_0$). Notice that after stage s_0, strategies σ_k, $k < i$, are never again used so that any action (including restraint) taken by these strategies has settled down by stage s_0. Also, after s_0, $x^s_{\langle i,0\rangle} = x^{s_0}_{\langle i,0\rangle}$ since we only redefine $x^s_{\langle i,0\rangle}$ if $v(x_{i-1},s)$ becomes undefined. Let $S = \{s \geq s_0 : v(x_i,s) \text{ is undefined}\}$. By (3.2), S is infinite. Now for every $s \in S$ we have

$$(3.6)\qquad (\forall k \leq i)\,[x_k^s \in A_s \iff x_k^s \in A] \text{ and } (\forall k > i)\,[x_k^s \notin A \text{ and } x_k^s \notin A_s]$$

$$(3.7)\qquad (\forall k > i)\,(\forall j)\,[x^s_{\langle k,j\rangle} \text{ is undefined}].$$

Thus for every $s \in S$, $A_s = A$ and $B_s = B$ except for the positions of markers $x^s_{\langle i,j\rangle}$, $j \in \omega$. Thus, Lemmas 1.4 and 1.5 can be proved for R_1 and the markers $x_{\langle i,j\rangle}$, just as in §1 except that the argument takes place on the set of stages S rather than all of ω. (Of course restraint imposed by σ_i at some stage $s \in S$ remains in force at stages in $\omega - S$ until cancelled or until σ_i's guess is seen to be incorrect.) Further, for any $s \in S$, $r(0,s) = v(x_{i-1},s) = \lim_s v(x_{i-1},s)$ so that

$$\liminf_s R(1,s) = \max\{\lim_s v(x_{i-1},s),\ \liminf_{s\in S} r(1,s)\} < \infty.$$

(This description of two requirements is, so far, relatively standard. It is very similar to the minimal pair argument of Lachlan (see Soare [7] for an exposition). Basically, R_1 guesses at possible outcomes for R_0 and plays strategies for these guesses only at stages when the guesses look correct. The argument that the strategy for the correct guess succeeds takes place on the restricted "universe" of stages where that guess looks correct.)

What remains to be shown is that our strategy for meeting R_1 described above does not interfere with Lemma 1.5 for R_0. (The restraint $r(0,s)$ was imposed only so that we could prove Lemma 1.4 for R_0.) While we originally thought this point quite worrisome, the argument turns on an important standard device of the infinite injury priority method, the true stages of the enumeration of C. Let $\{C_s\}_{s\in\omega}$ be our fixed recursive enumeration of C and let c_s denote the least element enumerated in C at stage s. Then s is a true stage for (this enumeration of) C if $C_s[c_s] = C[c_s]$. The crucial observation is the following. Suppose that t is a true stage for C and let m be any marker for any requirement. If $v(m,t)$ is defined (and has not just been defined at t) then

$$(3.8)\qquad (\forall s \geq t)\,[v(m,s) = v(m,t)].$$

In fact, if we modify the construction slightly so that $v(m,s) < c_s$ if $v(m,s)$ is defined by case 2 at stage s, then (3.8) holds for all markers m such that $v(m,t)$ is defined. We assume from now on that this modification has been made. Thus, if $v(m,s)$ is undefined i.o., then $v(m,s)$ must be undefined at every true stage for C. It follows from (3.8) and the construction that for any true stage t for C and any marker m

$$(3.9) \qquad m^t \in A_t \Longrightarrow m^t \in A \text{ and } m^t \in B_t \Longrightarrow m^t \in B.$$

Now to prove Lemma 1.5 for R_0, it is enough to show that if $\Phi_0(W_0) = A$, $\Psi_0(A \oplus C) = W$, $\Theta_0(V_0) = B$, and $\Xi_0(B \oplus C) = V$ and x_i is the marker of Lemma 1.4, then there is a stage s such that $\ell(s) > x_i^s$ and such that $v(x_i,s)$ can be defined to preserve this fact forever. (This yields a contradiction of Lemma 1.4.) There are two supposed obstacles to arguing that there is such a stage s. First, R_1 is enumerating elements in and out of A and B so it could be the case that $\ell(s) < x_i^s$ for all s because R_0 is looking at computations using information about A_s which is not true about A. Second, even if R_0 sees $\ell(s) > x_i^s$, R_0 may define $v(x_i,s)$ so large that C later changes below $v(x_i,s)$ causing $v(x_i,s)$ to again become undefined. (Recall that $v(x_i,s)$ is chosen large enough so that $v(x_i,s) > x^s_{\langle i,j\rangle}$ as we stated in (2.4).)

Let s_0 be a stage such that $x_i^{s_0}$ is the final position of marker x_i and let u be the "true" use in establishing a length of agreement $\ell(s)$ larger than $x_i^{s_0}$. Note that if $x \in A$ (B) then $x \in A_s$ (B_s) for almost every s. Now let $s_1 \geq s_0$ be a true stage for C such that

$$(3.10) \qquad (\forall x \leq u)\ [x \in A\ (B,C) \Longrightarrow x \in A_{s_1}(B_{s_1}),C_{s_1})].$$

Then $A_{s_1}[u] = A[u]$, $(B_{s_1}[u] = B[u])$ by (3.9). Thus, for all sufficiently large true stages s_1, $\ell(s_1) > x_i^{s_1}$ via the <u>correct</u> computation. There is thus no obstacle towards defining $v(x_i,s_1+1)$. Furthermore, $v(x_i,s_1+1)$ is not defined to be too large since all values $v(x_{\langle i,j\rangle},s_1)$ which are defined satisfy $v(x_{\langle i,j\rangle},s_1) \leq c_{s_1}$ and, since s_1 is true, $C[c_{s_1}] = C_{s_1}[c_{s_1}]$. Thus if $v(x_i,s_1+1) = \max\{u;\ v(x_{\langle i,j\rangle},s),\ j \in \omega\}$, then $C_{s_1}[v(x_i,s_1+1)] = C[v(x_i,s_1+1)]$

and so $v(x_1,s_1+1)$ remains defined forever, giving the same contradiction to Lemma 1.4 as before.

Rather than further describe how the requirements fit together, we now give the construction. We first need some notation. Requirement R_e has markers x_α for each $\alpha \in \omega^{e+1}$. Intuitively, x_α contains guesses at requirements $R_0,\dots,R_{e-1}$. For instance, x_α guesses that the marker $x_{\langle\alpha(0)\rangle}$ is the first marker such that $v(x_{\langle\alpha(0)\rangle},s)$ is undefined infinitely often for R_0 and x_α guesses that $x_{\langle\alpha(0),\alpha(1)\rangle}$ is the least marker for R_1 such that $v(x_{\langle\alpha(0),\alpha(1)\rangle},s)$ is undefined infinitely often among those markers for R_1 which guess $\alpha(0)$ for R_0. For strings α,β, $\alpha \subseteq \beta$ means α is an initial segment of β and $\alpha < \beta$ means $(\exists e)\,[(\forall i < e)\,[\alpha(i) = \beta(i)]$ and $[\alpha(e) < \beta(e)]]$ We fix a recursive enumeration $\{C_s\}_{s\in\omega}$ of C such that $C_s - C_{s-1} \neq \emptyset$ and let $c_s = (\mu y)[y \in C_s - C_{s-1}]$. If $\beta \in \omega^{e+1}$, β^- is the string such that $(\forall i < e)\,[\beta(i) = \beta^-(i)]$ and $\beta^-(e) = \beta(e) \dot{-} 1$. Again, we will have an auxiliary function $v(m,s)$, and we will use m^s to denote the position of marker m at the end of stage s. To cancel marker m at $s+1$ means to remove m^s from A or B and to cause $v(m,s+1)$ and m^{s+1} to be undefined. Let $\ell(e,s)$ denote the length of agreement function at stage s for the functionals of R_e, which is defined just as $\ell(s)$ was in §1. Similarly, let $u(e,x,s)$ denote the bound on the amount of A,B,C,W_e,V_e used to establish $\ell(e,s) \geq x$.

<u>Construction</u>

<u>Stage 0.</u> Let $x^0_{\langle 0\rangle} = 0$.

<u>Stage s+1.</u> If

$$(\exists\alpha)\,[c_{s+1} \leq v(x_\alpha,s)] \tag{3.11}$$

adopt case 1 below. Otherwise adopt case 2.

<u>Case 1.</u> Find α so that (3.11) holds and $v(x_\alpha,s)$ is least over all such α. Cancel x_β, all $\beta > \alpha$, let $v(x_\beta,s+1)$ be undefined, all $\beta \subseteq \alpha$, remove $x^{s+1}_\beta, x^{s+1}_{\beta^-}$ from A (or B), all $\beta \subseteq \alpha$.

Case 2. In order of increasing $e \leq s$ we define α_e and decide whether R_e requires attention. If R_e requires attention we perform the indicated action and pass to the next stage, else we pass to the next $e \leq s$. Let $\alpha_{-1} = \langle\ \rangle$.

a) If $x^s_{\alpha_{e-1} \frown \langle 0\rangle}$ is undefined, R_e requires attention and the required action is: define $x^{s+1}_{\alpha_{e-1} \frown \langle 0\rangle}$ to be some number larger than any previously used (as a marker position or value of an auxiliary function).

b) Let j be the greatest integer such that $x^s_{\alpha_{e-1} \frown \langle j\rangle}$ is defined. Let $\alpha_e = \alpha_{e-1} \frown \langle j\rangle$. If $c_s, c_{s+1}, \ell(e,s) \geq u(e, x^s_{\alpha_e}, s)$ then R_e requires attention. Then let

$$v(x_{\alpha_e}, s+1) = \max\{u(e, x^s_e, s)\} \cup \{v(\beta, s) : \beta \supseteq \alpha_e\}.$$

(It will be evident from the construction that for all β, if $v(\beta,s)$ is defined then $v(\alpha_e, s+1) \geq v(\beta,s)$.) Enumerate $x^s_{\alpha_{e-1} \frown \langle j-1\rangle}$ in A (B) if $j-1$ is even (odd). Let $x^{s+1}_{\alpha_{e-1} \frown \langle j+1\rangle}$ be a number larger than any previously used.

The proofs to the next two lemmas are straightforward as described in our sketch and we omit them.

Lemma 3.1. Suppose x^s_β is defined.

(a) Suppose $\alpha \supseteq \beta$. Then x^s_α is defined. Also, if $v(x_\alpha, s)$ and $v(x_\beta, s)$ are defined then $v(x_\alpha, s) \geq v(x_\beta, s)$.

(b) Suppose that $\alpha < \beta$. Then x^s_α is defined, $v(x_\alpha, s)$ is defined, and if $v(x_\beta, s)$ is defined then $v(x_\alpha, s) \leq v(x_\beta, s)$.

(c) Suppose that $v(x_\beta, s)$ is defined and $\beta \in \omega^{e+1}$. Then $x^s_{\beta^+}$ is defined, where β^+ is such that $(\forall i < e)\,[\beta^+(i) = \beta(i)]$ and $\beta^+(e) = \beta(e) + 1$.

Lemma 3.2. Suppose that t is a true stage for C and m is any marker. Then

a) if $v(m,t)$ or $v(m,t+1)$ (m^t or m^{t+1}) is defined then $v(m,s)$ (m^s) is defined for all $s \geq t+1$;

b) if $m^t \in A_t$ (B_t) then $m^t \in A$ (B).

Lemma 3.3. There is a sequence $\beta \in \omega^\omega$ such that if $\beta_e = \beta \upharpoonright e+1$ then

a) $x^s_{\beta_e}$ is defined for confinitely many s,

b) If $\beta_e < \alpha$ then x^s_α is undefined for infinitely many s.

Proof. We define β by defining β_e by induction on e. For notation, let $\beta_{-1} = \langle\ \rangle$. Fix $e \geq -1$ and suppose that β_e has been defined; we define $\beta_{e+1} \supseteq \beta_e$. We show that there is a j such that

(3.12) $$x^s_{\beta_e \frown \langle j\rangle} \text{ is defined for confinitely many s,}$$

(3.13) $$x^s_{\beta_e \frown \langle j+1\rangle} \text{ is undefined i.o.}$$

Then it is evident that $\beta_{e+1} = \beta_e \frown \langle j\rangle$ has properties (a) and (b). Fix s_0 so that $x^s_{\beta_e}$ is defined for all $s \geq s_0$.

To see that j exists, we first show that $x^s_{\beta_e \frown \langle 0\rangle}$ is defined cofinitely often. Now if $x^{s_1}_{\beta_e \frown \langle 0\rangle}$ is defined for any $s_1 \geq s_0$, $x^s_{\beta_e \frown \langle 0\rangle}$ is defined for all $s \geq s_1$ since $x^{s+1}_{\beta_e \frown \langle 0\rangle}$ can only become undefined at stage s+1 if case 1 applies to some $\alpha < \beta_e \frown \langle 0\rangle$ (and hence some $\alpha < \beta_e$) at stage s+1. Thus, it suffices to show that $x^{s+1}_{\beta_e \frown \langle 0\rangle}$ becomes defined at some stage s+1 (by case 2a of the construction). Let $t \geq s_0$ be any true stage. Then

(3.14) $$(\forall i \leq e)\ [x^t_{\beta_i} \text{ is defined and } x^t_{\beta_i^+} \text{ is not}]$$

by Lemma 3.2. Furthermore, case 1 cannot happen at stage t+1 since for any α, if $v(x_\alpha,t)$ is defined, $v(x_\alpha,t) < c_t < c_{t+1}$, the last inequality because t is a true stage for C. We claim that at stage t+1 if $x^t_{\beta_e \frown \langle 0\rangle}$ is undefined, case 2a applies, requirement R_{e+1} receives attention, and $x^{t+1}_{\beta_e \frown \langle 0\rangle}$ is defined. To see this, notice that at stage t+1 no requirement R_i, $i \leq e$ can receive attention in case 2 for otherwise $x^{t+1}_{\beta_i^+}$ and $v(x_{\beta_i},t+1)$ are defined and $v(x_{\beta_i},t+1) < c_t$. Thus, we would have that $x^s_{\beta_i^t}$ would be defined for all $s \geq t+1$ by Lemma 3.1(c) and the fact that t is a true stage. This would be a contradiction of (b) for some $i \leq e$. Thus R_{e+1} receives attention by case 2a and hence $x^{t+1}_{\beta_e \frown \langle 0\rangle}$ is defined.

Now to prove the existence of an integer j with the properties (3.12) and (3.13), it suffices to derive a contradiction from the hypothesis that for all i

(3.15) $\qquad x^s_{\beta_e \frown \langle i\rangle}$ is defined for cofinitely many s.

The proof of (3.15) is just a careful repetition of the proof of Lemma 1.4. Namely, we show how to find recursively stages t_i such that for all $s \geq t_i$,

(3.16) $\qquad x^s_{\beta_e \frown \langle i\rangle}$ is defined,

(3.17) $\qquad v(x_{\beta_e \frown \langle i\rangle}, s)$ is defined,

(3.18) $\qquad C_s[v(x_{\beta_e \frown \langle i\rangle}, t_i)] = C_{t_i}[v(x_{\beta_e \frown \langle i\rangle}, t_i)]$.

This of course yields the contradiction that C is recursive.

Let t_i be the least stage satisfying (3.16)-(3.18) for i. We show how to find t_{i+1}. We will suppose i is even. Let t_{i+1} be the least stage such that $t_{i+1} \geq t_i$ and

(3.19) $\qquad W_{e+1,t_{i+1}}[\varphi_{e+1}(x^{t_i}_{\beta_e \frown \langle i\rangle}, t_i)] \neq W_{e+1,t_i}[\varphi_{e+1}(x^{t_i}_{\beta_e \frown \langle i\rangle}, t_i)]$

It is easy to argue as in Lemma 1.4 that there is such a stage; otherwise

$$1 = A(x^{t_i}_{\beta_e \frown \langle i\rangle}) \neq A_{t_i}(x^{t_i}_{\beta_e \frown \langle i\rangle}) = \Phi_{e+1}(W_{e+1,t_i}; x^{t_i}_{\beta_e \frown \langle i\rangle}) = \Phi_{e+i}(W_{e+1}; x^{t_i}_{\beta_e \frown \langle i\rangle})$$

so that at cofinitely many stages s, $\ell(e+1,s) < x^{t_i}_{\beta_e \frown \langle i\rangle}$. This contradicts the hypothesis that every $x_{\beta_e \frown \langle j\rangle}$ is eventually used.

Finally, we show that t_{i+1} has the desired properties. (3.16) follows from (3.18) for t_i. Let $v+1 > t_{i+1}$ be the least counterexample to (3.18) for t_{i+1} (and hence to (3.17)). We now claim that there is never a stage $s+1 > v+1$ such that we define $v(x_{\beta_e \frown \langle i+1\rangle}, s+1)$ at stage s+1 (necessarily by case 2b). This of course would serve to contradict (3.15). For at such a stage s+1 we must have that

(3.20) $\qquad$ for all $k \leq e$, $x^{s+1}_{\beta^+_k}$ is undefined,

(3.21) $\qquad \ell(e+1,s) \geq x_{\beta_e \frown \langle i+1\rangle}$

We contradict (3.21) by showing that $\ell(e+1,s) < x_{\beta_e \frown \langle i\rangle}$. The crucial observation here is that

(3.22) $\qquad A_{t_i}[v(x_{\beta_e \frown \langle i\rangle}, t_i)] = A_s[v(x_{\beta_e \frown \langle i\rangle}, t_i)]$

(and similarly for B). We will verify (3.22). Assuming (3.22) we have the desired contradiction to (3.21) for on the one hand we have

$$(3.23) \qquad (\forall y \leq \varphi_{e+1}(x_{\beta_e \frown \langle i\rangle}, t_i))[\Psi_{e+1,t_i}(A_{t_i} \oplus C_{t_i};y) = \Psi_{e+1,s}(A_s \oplus C_s;y)]$$

(by (3.22) and (3.18) for t_i); on the other hand, we have

$$(3.24) \qquad (\exists y \leq \varphi_{e+1}(x_{\beta_e \frown \langle i\rangle}, t_i))[W_{e+1,s}(y) \neq \Psi_{e+1,s}(A_s \oplus C_s;y)]$$

so that $\ell(e+1,s) < x_{\beta_e \frown \langle i\rangle}$. Finally, we must show that (3.22) holds. The only possible counterexamples to (3.22) are stages w, $t_i \leq w \leq s$, at which a marker position x_α^w enters or leaves A.

However, it cannot be the case that $\alpha < \beta_e$, else marker $x_{\beta_e \frown \langle i\rangle}$ is cancelled by such activity, contradicting the choice of t_i. If $\beta_e < \alpha$ however, $x_\alpha^w > v(x_{\beta_e \frown \langle i\rangle}, t_i)$. If $\beta_e \frown \langle i\rangle \subset \alpha$, then x_α^w can only enter or leave A_w if $v(x_{\beta_e \frown \langle i\rangle}, w)$ is undefined. Finally, if $\alpha \subset \beta_e \frown \langle i\rangle$ then $\alpha = \beta_k$ for some $k \leq e$ and (3.22) holds for these α by (3.20). □

<u>Lemma 3.4.</u> For every e, requirement R_e is satisfied.

<u>Proof.</u> Fix e. Let β_e be as in Lemma 3.3. Fix a stage s_0 so that for every $s \geq s_0$, $x_{\beta_e}^s$ is defined. Suppose the lemma is false; and

$$(3.24) \qquad \Phi_e(W_e) = A,\ \Psi_e(A \oplus C) = W_e,\ \Theta_e(V_e) = B, \text{ and } \Xi_e(B \oplus C) = V_e$$

Let x_{β_e} denote $\lim_s x_{\beta_e}^s$. Let u be the "true" use involved in establishing that the length of agreement in (3.24) exceeds x_{β_e}. Let t be a true stage such that

$$(3.25) \qquad u \leq c_t$$

$$(3.26) \qquad (\forall y \leq u)\ [y \in A \Longrightarrow y \in A_t \text{ and } y \in B \Longrightarrow y \in B_t].$$

Since t is true,

$$(3.27) \qquad (\forall y)\ [y \in A_t \Longrightarrow y \in A \text{ and } y \in B_t \Longrightarrow y \in B].$$

Certainly, for any true stage $t' \geq t$, (3.25)-(3.27) are also true so there is a true stage $t \geq s_0$ such that $\ell(e,t) \geq x_{\beta_e}$ since (3.25)-(3.27) imply that our approximations to A, B, and C are "true" through u at stage t. We claim that, for such a t, R_e requires attention at stage t+1, $x^{t+1}_{\beta^+_e}$ is defined, and so $x^s_{\beta^+_e}$ is defined for all $s \geq t+1$. This would be the desired contradiction. The proof of this claim is really the same as that of Lemma 3.3 where we showed that $x^s_{\beta_e \frown \langle 0 \rangle}$ will eventually be defined. First we note that case 1 never happens at stage t+1 if t is true. Second, we note that no requirement R_i, $i < e$, can receive attention in case 2 at any stage $t+1 \geq s_0$ such that t is true. (Otherwise $x^s_{\beta^+_e}$ is defined for all $s \geq t+1$.) Thus there is no obstacle to R_e receiving attention by case 2b at stage t+1. Therefore $x^{t+1}_{\beta^+_e}$ is indeed defined, $v(x_{\beta_e}, t+1)$ is defined, and, since t is true, $x^s_{\beta^+_e}$ remains defined forever. □

<u>Lemma 3.5.</u> A and B are r.e. in C.

<u>Proof.</u> A satisfies the following $\Sigma_1(C)$ definition.

$$A = \{y : (\exists \beta)(\exists s)\ [y = x^s_\beta \in A_s \text{ and } C[v(x_{\beta^+}, s)] = C_s[v(x_{\beta^+}, s)]]\},$$

for β^+ as defined in Lemma 3.1. □

4. FURTHER REMARKS

With the above construction, we have actually shown the following

<u>Theorem 4.1.</u> There are recursive functions f and g so that for every $e \in \omega$, if W_e is nonrecursive, then either $W_e \oplus W^{W_e}_{f(e)}$ or $W_e \oplus W^{W_e}_{g(e)}$ is not of r.e. degree.

As we mentioned earlier, the conclusion of Theorem 4.1 cannot be improved to assert the existence of a single recursive function f which always produces an index for the desired set. This is an observation of Jockusch and Shore which follows from the following theorem.

Theorem 4.2. (Jockusch-Shore, [2]): There is a total recursive function h such that for every $e \in \omega$,

(4.1) $W_{h(e)}$ is nonrecursive, and

(4.2) $W_e^{W_{h(e)}} \oplus W_{h(e)} \equiv_T \emptyset'$.

Corollary 4.3. There is no recursive function f such that for every $e \in \omega$

(4.3) W_e nonrecursive $\Longrightarrow W_{f(e)}^{W_e} \oplus W_e$ is not of r.e. degree.

Proof. Suppose there were a recursive function f such that (4.3) holds. By (4.1) and (4.2) we have that

(4.4) $W_{h(f(e))}$ is nonrecursive, and

(4.5) $W_{f(e)}^{W_{h(f(e))}} \oplus W_{h(f(e))} \equiv_T \emptyset'$.

By the recursion theorem, there is an $e \in \omega$ such that $W_e = W_{h(f(e))}$. For this e, W_e is nonrecursive so that (4.3) applies and

(4.6) $W_{f(e)}^{W_e} \oplus W_e = W_{f(e)}^{W_{h(f(e))}} \oplus W_{h(f(e))}$ is not of r.e. degree.

But (4.5) and (4.6) directly contradict each other. □

Jockusch and Shore have also proved a two-index version of Theorem 4.2.

Theorem 4.4 (Jockusch-Shore [2]). There is a total recursive function h such that

(4.7) $W_{h(e,i)}$ is nonrecursive,

(4.8) $W_e^{W_{h(e,i)}} \oplus W_{h(e,i)}$ has r.e. degree, and

(4.9) $W_i^{W_{h(e,i)}} \oplus W_e^{W_{h(e,i)}} \oplus W_{h(e,i)} \equiv_T \emptyset'$.

Jockusch and Shore have asked whether there is any function h satisfying (4.7),

(4.8), and

(4.10) $$W_i^{W_{h(e,i)}} \oplus W_{h(e,i)} \text{ has r.e. degree.}$$

The uniformity of Theorem 4.1 gives a partial answer by showing that such an h cannot be recursive.

Corollary 4.5. There is no recursive function h such that (4.7),(4.8), and (4.10) hold.

Proof. Let f and g be the recursive functions of Theorem 4.1 and suppose that an h satisfying (4.7), (4.8), and (4.9) exists. Then for any integer j, let $h'(j) = h(f(j),g(j))$. Then for each $j \in \omega$

(4.11) $$W_{h'(j)} \text{ is nonrecursive,}$$

(4.12) $$W_{f(j)}^{W_{h'(j)}} \oplus W_{h'(j)} \text{ has r.e. degree, and}$$

(4.13) $$W_{g(j)}^{W_{h'(j)}} \oplus W_{h'(j)} \text{ has r.e. degree .}$$

Let j be an integer such that $W_{h'(j)} = W_j$. Then W_j is nonrecursive by (4.11) so by Theorem 4.1, we have for this fixed j

(4.14) $$\text{either } W_{f(j)}^{W_{h'(j)}} \oplus W_{h'(j)} \text{ or } W_{g(j)}^{W_{h'(j)}} \oplus W_{h'(j)} \text{ does not have r.e. degree.}$$

But (4.14) contradicts either (4.12) or (4.13). □

We think it possible that the technique of this paper can be extended to answer the question of Jockusch and Shore, namely to refute the existence of any such function h.

In fact, there is a further uniformity in the Jockusch-Shore proof which can be used to show that there is no effective procedure for meeting a single requirement resulting from (4.3), i.e., the requirement R of section 1. This is the observation which led us to the two-set version or the theorem. (Soare had originally announced [7] the theorem of this paper, but his intended strategy to meet the requirement R of §1 used only a single set A and overlooked the difficulty mentioned in §1 that the x_1 computation uses the fact that x_0 is not in A. After seeing the Jockusch-Shore Theorem 4.2, Stob found the error and showed that the strategy would work if two sets A and B are constructed simultaneously in place of a single set A. The method for putting the requirements together to obtain Theorem 4.1 is the same as in Soare's original proof as sketched in §3. Next Soare and Stob noticed that if C is low, the method for putting the requirements together becomes a very easy finite injury argument as explained in §2.)

Jockusch and Shore actually showed that Theorem 4.2 can be made uniform in the following way.

Theorem 4.3 (Jockusch-Shore). Given e we can effectively find i, Φ, Ψ, and W so that W_i is nonrecursive, $\Phi(W) = W_e^{W_i}$ and $\Psi(W_i \oplus W_e^{W_i}) = W$.

Therefore if we could effectively meet requirements of the form

$$R: \quad \Phi(W) \neq A \quad \text{or} \quad \Psi(A \oplus W_i) \neq W$$

we could do the following. Given e we find the i, Φ, Ψ, and W of Theorem 4.3 and perform the supposed construction for the requirement R to get a set $W_{f(e)}^{W_i}$ meeting R (as A). Let e be such that $W_{f(e)}^X = W_e^X$ for all X. Then our construction produces a set $W_{f(e)}^{W_i}$ so that

$$\Phi(W) \neq W_e^{W_i} \quad \text{or} \quad \Psi(W_i \oplus W_e^{W_i}) \neq W$$

contradicting the theorem.

As a further refinement To theorem 4.1 we are able to show that $A \oplus C$ and $B \oplus C$ can always be made to have incomparable Turing degrees. This is interesting since Jockusch and Shore have shown that one cannot always insure that $A \oplus C$ and $B \oplus C$ are of comparable Turing degrees. (This follows from Theorem 4.4 above.)

One open question suggested by the above work is the following. Let $R(\underset{\sim}{a}) = \{\underset{\sim}{b} : \underset{\sim}{b} \geq \underset{\sim}{a}$ and $\underset{\sim}{b}$ is r.e. in $\underset{\sim}{a}\}$. Is $R(\underset{\sim}{a})$ order isomorphic to $R(\underset{\sim}{0})$ for every (any) r.e. degree $\underset{\sim}{a}$? This question for degrees $\underset{\sim}{a}$ not necessarily r.e. has recently been answered negatively by Shore.

REFERENCES

[1] S. B. Cooper, Sets recursively enumerable in high degrees, Notices Amer. Math. Soc. 19 (1972) A-20.

[2] C. G. Jockusch, Jr. and R. A. Shore, Pseudo jump operators I: The R.E. case, to appear

[3] A. H. Lachlan, Lower bounds for pairs of recursively enumerable degrees, Proc. London Math. Soc., 16 (1966) 537-569.

[4] R. W. Robinson, Interpolation and embedding in the recursively enumerable degrees, Ann. of Math (2) 93 (1971) 285-314.

[5] R. I. Soare, Computational complexity, speedable and levelable sets, J. Symbolic Logic, 42 (1977) 545-563.

[6] R. I. Soare, Recursively enumerable sets and degrees Bulletin Amer. Math. Soc 84 (1978) 1149-1181.

[7] R. I. Soare, Relative enumerability, Notices Amer. Math. Soc., 26 (1979) A-15)

[8] R. I. Soare, Fundamental methods for constructing recursively enumerble degrees, in Recursion Theory: its Generalizations and Applications, Proceedings of the Logic Colloquium 79, Leeds, August 1979, Ed. F.R. Drake and S.S. Wainer, London Mathematics Society, Lecture Notes 45, Cambridge University Press, 1980.

PROCEEDINGS OF THE HERBRAND SYMPOSIUM
LOGIC COLLOQUIUM '81, J. Stern (editor)

RECURSIVE DILATORS AND GENERALIZED RECURSIONS

Jacques Van de Wiele

Abstract : We establish links between Girard's notion of recursive dilator and generalized recursions like Normann's E-recursion and Hinman's (∞, 0)-recursion as well as with the concept of function uniformly Σ-definable over all admissible sets.

SECTION 1 : RECURSIVE DILATORS

We introduce briefly the notion of (recursive) dilator. For more details see [GI4].

<u>Definition 1.1.</u> - On is the class of all ordinals. An ordinal is always identified with the set of its predecessors.

<u>Definition 1.2.</u> - The category ON :

- objects : ordinals
- morphisms : strictly increasing mappings.

The set of morphisms from x to y is designed by $I(x,y)$.

<u>Definition 1.3.</u> - The category $ON < \omega$ is the full subcategory of ON with the finite ordinals as objects.

<u>Fundamental fact 1.4.</u> - In ON any ordinal is the direct limit of a system of integers.

<u>Definition 1.5.</u> - A dilator is a functor from ON to ON which preserves direct limits and pullbacks.

<u>Definition 1.6.</u> -

i) A dilator D is weakly finite iff $D(n)$ is finite for any integer n, i.e. iff D maps $ON < \omega$ into itself.

ii) If $f \in I(n,x)$ where n is an integer and x an ordinal, we encode f by the ordinal $\ulcorner f \urcorner = \langle f(0), f(1), \ldots, f(n-1), x \rangle$. (Here, $\langle\,\rangle$ is an ordinal coding of sequences of ordinals).

iii) A dilator D is recursive if it is weakly finite and if there exists some recursive function φ from ω to ω such that for all integers m and n and all $f \in I(m,n)$:

$$\ulcorner D(f) \urcorner = \varphi(\ulcorner f \urcorner).$$

Remark 1.7. - It results from 1.4. that a dilator is completely determined by its restriction to $ON < \omega$. In particular, a recursive dilator is completely determined by the associated function φ.

SECTION 2 : DILATORS AND E-RECURSION

Theorem 2.1. - If D is a recursive dilator then the function $x \mapsto D(x)$ from On to On is E-recursive.
(for E-recursion, the set recursion introduced by Normann, see [NOR]).

Proof .- We proceed by a succession of definitions and lemmas.

Definition 2.2. - ρ is the function from sets to ordinals which associates to a wellfounded relation $\prec$ its rank ; see [BAR] p. 161, with the small change that we let $\rho(\prec)$ undefined if $\prec$ is not a wellfounded relation.

Lemma 2.3. - ρ is E-(partial)-recursive.

Definition 2.4. - $<_x$

$$\mathrm{Dom}(<_x) = \{(p,f) \ / \ \exists n \in \omega \quad f \in I(n,x) \text{ and } p < D(n)\}$$

$$(p,f) <_x (p',f') \text{ iff } D(f)(p) < D(f')(p').$$

Lemma 2.5. -

$$D(x) = \rho(<_x)$$

Proof .- This results from the fact that D preserves direct limits, condition which can be written : for any ordinal x, any $y < D(x)$, there exists $n \in \omega$, $z < D(n)$ and $f \in I(n,x)$ such that $y = D(f)(z)$.

Definitions 2.6. -

i) Given $f \in I(n,x)$ and $f' \in I(n,x')$ we define $g = f \vee f'$ belonging to $I(q,x)$ by $rg(f \vee f') = rg(f) \cup rg(f')$.

ii) Let $h \in I(n,q)$ and $h' \in I(n',q)$ be such that we have the following commutative diagram :

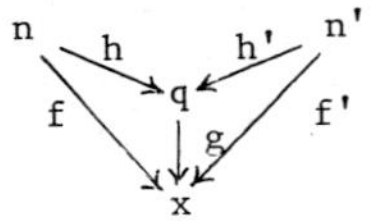

<u>Lemma 2.7.</u> - $(p,f) <_x (p',f')$ iff $D(h)(p) < D(h')(p')$.

<u>Proof</u> .- $(p,f) <_x (p',f')$ iff $D(f)(p) < D(f')(p')$
iff $D(g)(D(h)(p)) < D(g)(D(h')(p'))$
iff $D(h)(p) < D(h')(p')$.

<u>Lemma 2.8.</u> - There exists some Prim function G from ordinals to sets such that for any infinite ordinal x, $G(x) = <_x$.
(The Prim_0 and Prim functions are defined in [J-K]).

<u>Proof</u> .- Remark that $\ulcorner fvf' \urcorner$, $\ulcorner h \urcorner$ and $\ulcorner h' \urcorner$ can be obtained from $\ulcorner f \urcorner$ and $\ulcorner f' \urcorner$ by Prim functions and we have :

$$(p,f) <_x (p',f') \text{ iff } D(h)(p) < D(h')(p')$$
$$\text{iff } (\varphi(\ulcorner h \urcorner))_p < (\varphi(\ulcorner h' \urcorner))_{p'}.$$

<u>End of the proof</u> (of theorem 2.1.) .- we can define $D(x)$ by cases :
if x is infinite, $D(x) = \rho(G(x))$
if x is finite, $D(x)$ is easily obtained from φ.
The result follows from 2.5. and 2.8. and from the fact that Prim functions are E-recursive.

<u>Remarks 2.9.</u> -

i) We have not used the condition "D preserves pullbacks" in this proof.

ii) Let us mention that for a recursive dilator D, the function from On to On $x \mapsto D(x)$ is $(\infty,0)$-recursive, because a function from On to On which is E-recursive is $(\infty,0)$-recursive (Shelton's result : see [SHE]).

SECTION 3 : INDUCTIVE MODELS

<u>Preliminary 3.1.</u> - Let $\mathcal{L}$ be a language of first order with one type of object. Let $\phi(X,x)$ be a formula with a variable of unary predicate X which appears only positively. If $\mathcal{M}$ is a structure for $\mathcal{L}$, we can define the monotone operator

$\overline{\phi}_{\mathcal{M}}$ from $P(|\mathcal{M}|)$ to $P(|\mathcal{M}|)$ by :

$$\overline{\phi}_{\mathcal{M}}(A) = \{u \in |\mathcal{M}| \ / \mathcal{M} \models \phi(A,u)\}.$$

Let $D\phi_{\mathcal{M}}$ be the smallest fixed point of $\overline{\phi}_{\mathcal{M}}$. We can define $D\phi_{\mathcal{M}}$ more constructively as following :

$$I\phi^{0}_{\mathcal{M}} = \emptyset$$

$$I\phi^{a+1}_{\mathcal{M}} = I\phi^{a}_{\mathcal{M}} \cup \overline{\phi}_{\mathcal{M}}(I\phi^{a}_{\mathcal{M}})$$

$$I\phi^{a}_{\mathcal{M}} = \bigcup_{b<a} I\phi^{b}_{\mathcal{M}} \quad \text{for limit ordinal } a.$$

There exists an ordinal γ such that $I\phi^{\gamma+1}_{\mathcal{M}} = I\phi^{\gamma}_{\mathcal{M}}$ and then $D\phi_{\mathcal{M}} = I\phi^{\gamma}_{\mathcal{M}}$. The least such γ is called the closure ordinal.

Definition 3.2. -

i) $\mathcal{L}(D\phi)$ is obtained from $\mathcal{L}$ by adding a unary predicate symbol $D\phi$.

ii) $\mathcal{L}'$ is obtained from $\mathcal{L}(D\phi)$ by adding unary predicate symbols $I\phi^{a}$ for all ordinals a.

iii) If $\mathcal{M}$ is a model of a theory T in the language $\mathcal{L}$, then $(\mathcal{M}, D\phi_{\mathcal{M}}, \ldots, I\phi^{a}_{\mathcal{M}}, \ldots)$ is called an inductive model (of T relative to ϕ).

Definition 3.3. - Let A be a formula of $\mathcal{L}(D\phi)$, Δ a dilator and a an ordinal. Write $A = \mathcal{A}(D\phi^{-}, D\phi^{+})$ by separating positive and negative occurrences of $D\phi$ in A. Then define $?A_{\Delta,a}$ to be the formula of $\mathcal{L}'$: $\mathcal{A}(I\phi^{a}, I\phi^{\Delta(a)})$.

Example 3.4. - Assume that A is of the form : $\forall x \in D\phi \ \exists y \in D\phi \ A_{o}(x,y)$ where A_{o} doesn't contain $D\phi$. Then,

$$?A_{\Delta,a} = \forall x \in I\phi^{a} \ \exists y \in I\phi^{\Delta(a)} \ A_{o}(x,y).$$

Theorem 3.4. (J.Y. Girard) .- We have the following equivalence :

i) A is true in all inductive models

ii) there exists some recursive dilator Δ of the form $Id + \underline{1} + \Delta'$ such that $?A_{\Delta,a}$ is true in all inductive models for all ordinals a.

Proof .- ii) $\Rightarrow$ i). Let $(\mathcal{M},\dots)$ be an inductive model and γ the closure ordinal. The trivial condition on Δ implies $\Delta(\gamma) > \gamma$ and thus $?A_{\Delta,\gamma}$ is merely A.

i) $\Rightarrow$ ii). We give only an idea of the proof ; for complete proofs see [GI 1], [G-M] or [VDW]. We use an inductive logic which characterizes the truth in all inductive models. A proof in inductive logic depends on a dilator Δ and is conceived of as a family $(P_a)_{a\in On}$ of proofs. Locally in a, we have the following rules :

$$\frac{\Lambda,\ t\in I\phi^{a} \vdash \Pi}{\Lambda,\ t\in D\phi \ \vdash \Pi} \qquad \frac{\Lambda \vdash t\in I\phi^{\Delta(a)},\ \Pi}{\Lambda \vdash t\in D\phi,\ \Pi}$$

The cut rule can be read : $I\phi^{\Delta(a)} \subseteq D\phi \subseteq I\phi^{a}$ and we have in fact equalities because $\Delta(a) > a$. Thus the cut rule says that $D\phi$ is the smallest fixed point. Inductive logic admits a cut elimination theorem which permits for normalized proofs to reinterpret negative occurrences of $D\phi$ by $I\phi^{a}$ and positive ones by $I\phi^{\Delta(a)}$. We point out that the normalized proof of a Δ-proof is a Δ''-proof with tremendous increase of Δ : $\Delta'' >> \Delta$.

SECTION 4 : THE PRINCIPAL THEOREM

Definition 4.1. - The rank function rk from sets to ordinals is defined by :
$rk(x) = \text{Sup}\,\{rk(z)+1 \ / \ z\in x\}$.
Thus $rk(x) = $ least α $x\in V_{\alpha+1}$.

Theorem 4.2. - Let F be a function from sets to sets, unif-Σ over all admissible sets. Then there exists some recursive dilator Δ such that for any set x and any ordinal α, if $rk(x) < \alpha$ then $rk(F(x)) < \Delta(\alpha)$.

Proof .- Let $\mathcal{L}$ be the language of set theory with symbol ε for membership, let T be the Kripke-Platek theory KP and let $\phi(X,x)$ be the formula $\forall y(y \varepsilon x \longrightarrow X(y))$.
If $\mathcal{M} = (M,E)$ is a model of KP then $D\phi_{\mathcal{M}} = Wf(\mathcal{M})$, the well founded part of M for E. By Ville's lemma (truncation lemma in [BAR]), $(Wf(\mathcal{M}),\ E\restriction Wf(\mathcal{M}))$ is a model of KP.
F corresponds to a Σ_1-formula A which can be written $\exists z\ A_o(x,y,z)$ where A_o is a Δ_o-formula. By hypothesis, $M \models \forall x\exists! y\, A(x,y)$ for any admissible set M.
Let B be the formula : $\forall x\in D\phi \quad \exists y\in D\phi \quad \exists z\in D\phi\ A_o(x,y,z)$. B is true in all inductive models because these are isomorph to the transitive models of KP. Thus we may apply theorem 3.4. with $?B_{\Delta,\alpha}$ equal to :

$$\forall x\in I\phi^{\alpha}\ \exists y\in I\phi^{\Delta(\alpha)}\ \exists z\in I\phi^{\Delta(\alpha)}\ A_o(x,y,z).$$

This provides the result because if M is an admissible set then $I\phi^{\alpha}_{\mathcal{M}} = \{x\in M \ / \ rk(x)<\alpha\}$.

Theorem 4.3. - Let G be a function from On to On, uniformly Σ over all admissible sets. Then there exists some recursive dilator Δ such that for any ordinal α, $G(\alpha) \leqslant \Delta(\alpha)$.

Proof .- easy from 4.2.

SECTION 5 : CONSEQUENCES IN GENERALIZED RECURSION THEORY

Definition 5.1. - We consider the following hierarchy of constructible sets relative to a set x :

$$L_0(x) = \{x\} \cup TC(x)$$

$$L_{\alpha+1}(x) = Def(L_\alpha(x))$$

$$L_\lambda(x) = \cup_{\alpha<\lambda} L_\alpha(x) \quad \text{for limit ordinal } \lambda.$$

(Note the correction to be made in [VDW] p. 58).

Prop - def. 5.2. - If x is a set, the least admissible set containing x as an element is of the form $L_\gamma(x)$ with $\gamma \in On$. Let $\gamma = x^+$.

Lemma 5.3. - Let F be unif - Σ over all admissible sets. Then there exists some Prim predicate T and some Prim function U such that :
$F(x) = U(x^+\text{- least } y\ T(x,y))$.

Proof .- This is an adaptation of theorem 4.4. of [J-K]. See [VDW].

Theorem 5.4. - If F is a function unif - Σ over all admissible sets then F is E-recursive.

Proof .- Apply theorem 4.2. to the function $x \mapsto x^+$- least y $T(x,y)$. We obtain $F(x) = U(\Delta(rk(x)+1)\text{- least } y\ T(x,y))$ and the result follows from theorem 2.1., from the fact that rk is a Prim function and the fact that E-recursion is closed for bounded search.

Theorem 5.5. - If G is a function from On to On which is unif-Σ over all admissible sets then G is $(\infty,0)$-recursive.

Proof .- easy from 5.4. or from 4.3.

SECTION 6 : ORDINAL RECURSIONS

There are two ordinal recursions which generalize ordinary recursion on the integers ; the first one is $(\infty,0)$-recursion and the second one, which corresponds to functions unif-Σ over all admissible sets, is what I called the +-recursion in [VDW].

We can introduce +-recursion in the following way : define $\Omega_{\infty+}$ as $\Omega_{\infty\lambda}$ (see [HIN] p. 376), but replace inductive clause (4), λ-least, by inductive clause (4'), $\vec{\mu}^+$-least, viz. :

(4') if $\nu < \vec{\mu}^+$, $(b,\nu,\vec{\mu},0) \in \Omega_{\infty+}$ and $\forall\pi<\nu \quad \exists\xi>0 \quad (b,\pi,\vec{\mu},\xi) \in \Omega_{\infty+}$

then $(\langle 4,k,b\rangle,\vec{\mu},\nu) \in \Omega_{\infty+}$.

Here $\vec{\mu}^+$ stands for the least recursively regular ordinal greater than the μ_i for all $i < k$.

By denoting $\{a\}_+ (\vec{\mu}) \simeq \nu$ for all $(a,\vec{\mu},\nu) \in \Omega_{\infty+}$, we have thus :
$\{\langle 4,k,b\rangle\}_+ (\vec{\mu}) \simeq \vec{\mu}^+\text{-least } \nu \ (\{b\}_+ (\nu,\vec{\mu}) \simeq 0)$.

+-recursion is a good recursion ; in particular we have a normal form theorem.

By comparing these two recursions, we have the following strange phenomenon : +-recursion is richer than $(\infty,0)$-recursion for partial functions but they have the same (total) functions, as expressed by the following theorem which sharpens theorem 4.3. and theorem 5.5. :

Theorem 6.1. - Let F be a (total) function from On to On. Then we have the equivalences :

i) F is +-recursive (i.e. unif-Σ over all admissible sets)

ii) F is Prim_0 in a recursive dilator

iii) F is $(\infty,0)$-recursive.

Similar remarks can be made for E-recursion and a set +-recursion which corresponds to functions from sets to sets uniformly Σ over all admissible sets.

SECTION 7 : OTHER PROOFS AND FURTHER QUESTIONS

We have obtained here theorems 4.3., 5.4. and 5.5. as consequences of theorem 4.2. We have not given the most direct proofs because theorem 4.2. is certainly proof - theoretically stronger than the others. We are lead to the question : in which minimal formal systems can we prove each of these results ?

There is a Ressayre-Harrington-Simpson proof of theorem 4.2. which is based on the β-completude theorem and a gödelian self-reference argument. See [RES]. This proof does not use any cut elimination procedure and consequently does not use the functor Λ. It is shorter than the one given here, but less constructive. These two proofs of theorem 4.2. can be carried out in Π^1_1-CA or in Jäger's formalized system KPl (see [JAG]).

Masseron's direct methods in [MAS] suggest that theorem 4.3. is provable in KP. And I think that one can derive theorem 4.2. from theorem 4.3. by a relatively direct proof, probably by using Λ.

Finally different people, T. Slaman, W. Mitchell, E. Griffor, on being informed of our proof of theorem 5.4., have given an alternative proof, which is purely recursion theoretic.

REFERENCES :

[BAR] J. Barwise, Admissible sets and structures. Persp. in Math. Log. (75).

[CEN] D. Cenzer, Ordinal recursion and inductive definitions. Generalized Recursion Theory. N.H. (74).

[GI0] J.Y. Girard, Three-valued logic and cut elimination. Dissertationes Mathematicae CXXXVI.

[GI1] J.Y. Girard, A survey of Π^1_2-logic. (Hannover 79).

[GI2] J.Y. Girard, Cours de théorie de la démonstration. Chap. 6 (79-80).

[GI3] J.Y. Girard, Proof-theoretic investigations of ID. Part 1 (to appear).

[GI4] J.Y. Girard, Π^1_2-logic. Part 1, (to appear in Ann. of Math. Log.).

[G-M] J.Y. Girard and M. Masseron, Proof-theoretic investigations of ID. Part 2. (Submitted to the JSL).

[HIN] P.G. Hinman, Recursion-theoretic hierarchies. Persp. in Math. Log. (78).

[JAG] G. Jäger, Iterating admissibility in proof theory. (this volume).

[J-K] R.B. Jensen and C. Karp, Primitive recursive set functions. Proc. of Symp. in Pure Math. Vol. XIII (71).

[MAS] M. Masseron, Majoration des fonctions ω_1^{ck}-recursives par des ω-échelles prim.-réc. Thèse de 3ème cycle.

[NOR] D. Normann, Set recursion. Generalized recursion theory II (78).

[RES] J.P. Ressayre, Bounding generalized recursive functions of ordinals by prim. rec. functors (this volume).

[SHE] R. Shelton, Set recursion versus ordinal recursion. Abstracts of papers (Amer. Math. Soc.) Vol.1. No.6.

[VAU] J. Vauzeille, Interpolation et complétude en β-logique. Thèse de 3ème cycle (79).

[VDW] J. Van de Wiele, Dilateurs récursifs et récursivités généralisées. Thèse de 3ème cycle. Paris VII (81).

PROCEEDINGS OF THE HERBRAND SYMPOSIUM
LOGIC COLLOQUIUM '81, J. Stern (editor)

FUNCTORS AND ORDINAL NOTATIONS
III - DILATORS AND GARDENS

by Jacqueline Vauzeilles

INTRODUCTION

In unpublished manuscripts of 76-77 (also, Oxford conference, 1976) Jean-Yves GIRARD introduced the concept of a garden ; the terrible complexity of anything connected with gardens made it necessary to look for a simpler wiewpoint. Then, in the next years he introduced and developped the concept of a dilator. Yet, in some cases gardens may be great interest : a garden is shortly, the functorial version of a Bachmann collection in which the fundamental sequences are replaced by "fundamental flowers". In this work we prove that there is an isomorphism between the categories of gardens and dilators*. This isomorphism enables us to investigate the relation of the functor Λ introduced by Jean-Yves Girard in Π^1_2-logic with the Bachmann hierarchy, using the result of the work done in collaboration with Jean-Yves Girard on the functorial construction of the Bachmann hierarchy.

In the first part we recall definitions and some properties of gardens and dilators.

In the second part we define a functor DEC (decomposition) from the category of dilators to the category of gardens : Jean-Yves Girard has defined the separation of variables for a dilator of kind Ω in Π^1_2- logic ; if we iterate this process for each predecessor of a dilator F of kind Ω (and if F sends Ω in Ω) and if at each step we fix the value $x = \Omega$ (x being the first variable of the bilator SEP(F) (x,y)) then we obtain a family of flowers G_α indexed by the ordinals $\alpha \leqslant F(\Omega)$ of cofinality Ω. The values $G_\alpha(y)$ for $y < \Omega$ form a fundamental sequence for α, and this family of flowers forms a garden that we call DEC(F).

Conversely, in the third section we associate to each garden J a dilator SYN(J) (synthesis of J) using the inverse functor of SEP, i.e. UN (unification of variables). SYN(J) sends Ω in Ω.

Then, in part IV, we prove that the functors DEC and SYN are inverse to each other.

The functor $\mathbb{\Lambda}$ sends the category of dilators to the category of regular bilators ; then, using the above isomorphism we deduce a functor $\mathbb{\Lambda}'$ from the category of gardens to the category of regular bilators. In the section V, we show that, if we consider only regular gardens J_y, then $(\mathbb{\Lambda}'J_y)(\omega,.) = \varphi_y^J$, φ_y^J being the functor defined in (B) II. Hence the relation between Λ and the Bachmann hierarchy **.

Here we have choosen of relativizing anything to Ω but, what rôle plays Ω? :

i) Ω plays the rôle of ON : it is more simple to deal with Ω, but we can easily imagine that we replace everywhere Ω by ON ; we would have a concept of garden of type ON, with fundamental flowers for well-orders which are of cofinality ON (i.e. which are not sets) and so on ...*** Then we obtain an isomorphism between the category of dilators (instead of the category DIL Ω of dilators which send Ω in Ω) and the category of gardens of type ON (instead of type Ω).

ii) We can replace Ω (or ON as in i)) by ω or any regular cardinal : if we define the notion of garden of type α, for a regular cardinal α (i.e. we have for each ordinal of cofinality α, a flower from α to ON, etc...) then there is an isomorphism between the category of these gardens and the category DIL α of the functors F from ON $< \alpha$ to ON $< \alpha$ commuting to direct limits and to pull-backs and such that $F(\alpha)$ is well-ordered (i.e. if $\alpha \neq \omega$, F is a dilator which sends α in α). Particulary, if we have two regular cardinals α and α' with $\alpha \neq \omega$ and $\alpha < \alpha'$ we can "extend" each garden of type α in a garden of type α' using the isomorphisms : if we denote by GAR_α (resp. by $GAR_{\alpha'}$) the category of gardens of type α (resp. α') and DEC_α, SYN_α, etc... the functors decomposition and synthesis, $i_{\alpha\alpha'}$ the canonical injection from DIL α to DIL α', we have the following diagram :

$$GAR_\alpha \xrightarrow{SYN_\alpha} \text{DIL } \alpha \xrightarrow{i_{\alpha\alpha'}} \text{DIL } \alpha \xrightarrow{SYN_{\alpha'}} GAR_{\alpha'}$$

This extension could be described directly, but we will not do it. For $\alpha = \omega$, remark that DIL ω is not included in DIL α' and then the extension of a garden of type ω does not exist always (more exactly, this extension is not ever well-founded) because a functor F from ω to ω commuting to direct limits and to pull-backs and such that $F(\omega)$ is well-ordered cannot be ever extended in a dilator.
Remark that every weakly finite dilator can be representated by a garden of type ω.

iii) We can ask if it exists the concept of a garden of type α, for any ordinal α : it is certainly possible to make sense to this concept ; we can define this concept such that we have an isomorphism between a suitable category of dilators (functors from ON $\leqslant \alpha$ to ON) and the category of such gardens (at least in the case where $\alpha = \omega^{1+\beta}$) ; this is not very difficult, but we will not do it in this paper ; the difficulty is that, here, we cannot distinguish the ordinals by their cofinality but we can see easily the conditions that we must impose to obtain an isomorphism.

* An other manner for a functorial extension of a Bachmann collection is the concept of rungs and ladders (see the appendix of Π^1_2-logic) ; there is not a simple relation between the concept of garden and the one of ladder.

** The equality $(\mathbf{\Lambda}' J_y)(\omega,.) = \varphi^J_y$ for regular gardens show that we could define φ^J_y for all gardens (even non regulars).

*** If F is a such flower, for each ordinal α, $F(\alpha)$ is generally an ordinal class : we can have $F(\alpha) \geqslant ON$.

Many thanks to Jean-Yves GIRARD, who propounded this question, for his help and constant encouragements during this work.

I - The category GAR and DIL Ω

I - 1 - The category GAR

I - 1 - 1 DEFINITION .- An <u>Ω-flower</u> F is a functor from $ON < \Omega$ to ON such that :

- F commutes to direct limits and to pull-backs.
- F enjoys FL) : for all ordinals $x,y < \Omega$ $F(E_{xy}) = E_{F(x)F(y)}$.

I - 1 - 2 REMARK .- This definition shows that an Ω-flower is a flower in the sense of (G).2.4. which sends $ON < \Omega$ to ON. A <u>regular Ω-flower</u> is a particular case of Ω-flower : see (B) - 1 - 1 - 1.

I - 1 - 3 DEFINITION .- A <u>garden</u> $J_{<y}$ of height $(<y)$ and of type Ω (with $y < \Omega_2$) consists of :

i) for each ordinal $x < y$, of cofinality Ω (we denote $x \in L_\Omega$), we have a non-constant Ω-flower F_x such that :

- $x = \sup_{\alpha < \Omega} F_x(\alpha)$
- if $f \in I(\alpha,\beta)$ and $\alpha,\beta < \Omega$:

(a) if $z \in L_\Omega$ and $z \leqslant F_x(\alpha)$, for each $u < \Omega$,

$$F_x(f)(F_z(u)) \leqslant F_{\widehat{F_x(f)(z)}}(u)$$

(see the definition of ^ in (G).1.2.3.)

(b) if $g \in I(u,v)$, $u,v < \Omega$ and $Z \in L_\Omega$ and $z \leqslant F_x(\alpha)$, we have the following commutative diagram :

$$\begin{array}{ccc} F_z(u) & \xrightarrow{F_z(g)} & F_z(v) \\ \downarrow w^u & & \downarrow w^v \\ F_{\widehat{F_x(f)(z)}}(u) & \xrightarrow{F_{\widehat{F_x(f)(z)}}(g)} & F_{\widehat{F_x(f)(z)}}(v) \end{array}$$

if we set, (with $a = u$ or v) $w^a \in I(F_z(u), F_{\widehat{F_x(f)(z)}}(u))$ and, for $t < F_z(a)$, $w^a(t) = F_x(f)(t)$.

(c) if $u < F_x(\alpha)$, $\widehat{F_x(f)}(u) \leqslant^J_o F_x(f)(u)$ (we define as in (B)I-3-1 the order relations $\leqslant^J_o$ and $<^J_o$).

(d) $$\widehat{F_x(f)}(F_x(\alpha)) \leqslant^J_o F_x(\beta).$$

ii) for each ordinal x of cofinality ω (we denote $x \in L_\omega$) the set $\{t \; ; \; t <^J_o x\}$ is cofinal in x.

I - 1 - 4 DEFINITION .- A garden J_y of height y and of type Ω, is a garden $J_{<y+1}$ of height $(<y+1)$ and of type Ω. We denote by J_0 and J_1 the unique gardens of height 0 and 1.

I - 1 - 5 REMARK .- A <u>regular garden</u> (see the definition in (B) - I - 1 - 6) is a particular case of garden : in a regular garden all Ω-flowers are regular, in i) (a) we have the equality, in i) (c) and in i) (d) we have the order $\leqslant^J_*$; by remark (B)- I - 5 - 4 the properties ii) and iii) of regular gardens can be remplaced by property ii) of gardens.

I - 1 - 6 DEFINITION .-

- Let $J_{<u}$ and $J'_{<t}$ two gardens and $f \in I(x,y)$ with $x \leqslant u$, $y \leqslant t$; we say that f is a <u>compatible</u> function from $J_{<u}$ to $J'_{<t}$ iff :

i) $\forall z \leqslant x$, $z < u$, $z \in L_\Omega$, $\hat{f}(F_z(\alpha)) \leqslant F'_{\hat{f}(z)}(\alpha)$ for each $\alpha < \Omega$; and if we define $f^\alpha \in I(F_z(\alpha), F'_{\hat{f}(z)}(\alpha))$, by, for each $u < F_z(\alpha)$, $f^\alpha(u) = f(u)$, the following diagram is commutative :

$$\begin{array}{ccc} F_z(\alpha) & \xrightarrow{\quad F_z(g) \quad} & F_x(\beta) \\ \Big\downarrow f^{\alpha} & & \Big\downarrow f^{\beta} \\ F'_{\hat{f}(z)}(\alpha) & \xrightarrow{\quad F'_{\hat{f}(z)}(g) \quad} & F'_{\hat{f}(z)}(\beta) \end{array} \qquad \text{for } g \in I(\alpha,\beta)$$

ii) for each $z < x$, $\hat{f}(z) \leqslant_o^{J'} f(z)$ and $\hat{f}(x) \leqslant_o^{J'} y$.

- A compatible function from J_u to J'_t is a compatible function from $J_{<u+1}$ to $J'_{<t+1}$; if f is a compatible function from J_u to J_t, where J_u is the restriction of J_t, we say that f is compatible with J_t.

I - 1 - 7 REMARK .- In (B) - I - 3 - 3 we have defined the notion of a compatible function from a regular garden to a regular garden ; remark that in (B) - I - 3 - 3 i) we have a condition different from the one of I - 1 - 6 ; many properties proved for regular gardens and compatible functions with regular gardens still hold for gardens ; it would be fastidious to rewrite those properties but we will use them when it is necessary, referring to (B).

I - 1 - 8 LEMMA .- Let $f \in I(x,y)$ a compatible function from J_u to J'_t and let $z \in L_\Omega$ and $z \leqslant x$; then $F'_{\hat{f}(z)}(\alpha) \leqslant f(F_z(\alpha))$, for each $\alpha < \Omega$.

<u>Proof</u> .- By definition $(f^\alpha)_{\alpha<\Omega} = f$ is a natural transformation from F_z to $F'_{\hat{f}(z)}$:

by (G).2.4.10.iv) $F_z(\alpha) = (z_o ; 0, \ldots, p-2, x_{p-1})_{F_z}$ with $x_{p-1} = \sup(\alpha, p-1)$;

by (G).2.3.15. if $\alpha < \Omega$, $f^\alpha(z_o ; 0, \ldots, p-2, x_{p-1})_{F_z} = (f^p(z_o) ; 0, \ldots, p-2, x_{p-1})_{F'_{\hat{f}(z)}}$

then using (G).2.4.10.iv) we thave the result.

I - 1 - 9 DEFINITION .-

i) Let $x < \Omega$, and let $(J^z)_{z<x}$ be a family of gardens ; let, for each $z<x$, h_z the height of J^z; then it is possible to define a new garden $J = \sum_{z<x} J^z$ of height $h = \sum_{z<x} h_z$ as follows : let $y \in L_\Omega$ an ordinal such that $y \leqslant h$, then there exists $u<x$ such that $\sum_{z<u} h_z \leqslant y < \sum_{z<u+1} h_z$ if $y = \sum_{z<u} h_z$ then, since $y \in L_\Omega$, u is a successor : $u = v+1$ and we set $F_y^J = \sum_{z<v} h_z + F_v^{J^v}$; if $y > \sum_{z<u} h_z$, then $y = \sum_{z<u} h_z + v$ with $v \in L_\Omega$, and we set : $F_y^J = \sum_{z<u} h_z + F_v^{J^u}$.

ii) Let $t < \Omega$, let $(G^z)_{z<t}$ be another family of gardens of height h'_z.

and let (f^z) (with $f^z \in I(h_z, h'_{g(z)})$) be a family of compatible functions from J^z to $G^{g(z)}$, where g is a function in $I(x,t)$; then one defines a compatible function f from $\sum_{z<x} J^z$ to $\sum_{z<t} G^z$ by :

if $u<x$, $v<h_u$, $f(\sum_{z<u} h_z + v) = \sum_{z<g(u)} h'_z + f^u(v)$.

I - 1 - 10 LEMMA .- I - 1 - 9 i) and ii) are correct definitions.

Proof .- It is a trivial verification.

I - 1 - 11 REMARK .- If $x = 2$, then we obtain the definition of the sum of two (regular) gardens that we had set in (B) -I - 4 - 1.

I - 1 - 12 DEFINITION .-

i) Let x an ordinal, and $(J^z)_{z<x}$ a family of gardens such that : J^z is of height h_z and for $u \leqslant z < x$, $h_u \leqslant h_z$ and J^u is the restriction of J^z ; then it is possible to define a new garden $J_{<y} = \bigcup_{z<x} J^z$ of height $(<y)$ with $y = \sup_{z<x} h_z$ as follows : for each $t \in L_\Omega$, and $t<y$, we set $F^J_t = F^{J^u}_t$ for each u such that $t \leqslant h_u$.

ii) Let t be another ordinal and $(G^z)_{z<t}$ be another family of gardens such that : G^z is of height h'_z and for $u \leqslant z < t$, $h'_u \leqslant h'_z$ and G^u is the restriction of G^z ; let (f_z) be a family of compatible functions from J^z to $G^{g(z)}$, with $g \in I(x,t)$ and such that $f_z \in I(h_z, h'_{g(z)})$ and for each $u \leqslant z < x$, f_u is the restriction of f_z ; then one defines a compatible function $f = \bigcup_{z<g} f_z$ from $\bigcup_{z<x} J^z$ to $\bigcup_{z<t} G^z$ by : if $y' = \sup_{z<t} h'_z$, $f \in I(y,y')$ and for each $u<y$, $f(u) = f_z(u)$ if $h_z > u$.

I - 1 - 13 LEMMA .- I - 1 - 12 i) and ii) are correct definitions.

Proof .- It is a trivial verification.

I - 1 - 14 DEFINITION .- A garden J_y is <u>perfect</u> if and only if $J_y = J_1$ or $y \in L_\Omega$ and $F^J_y(0) = 0$.

I - 1 - 15 LEMMA .- A garden J_y is perfect iff it is $\neq J_0$, and if given any decomposition $J_y = J' + J''$, then $J' = J_0$ or $J'' = J_0$.

Proof .-

i) Let J_y a perfect garden ; if $y \neq 1$, then $y \in L_\Omega$ and $F_y(0) = 0$; let $J_y = J'_u + J''_v$ then $y = u+v$ and $u \overset{J}{\leqslant}_o y$ using the definition of the sum of two gardens ; hence if $y \neq 1$, either $u = y$ and $J''_v = J_0$, either, since $u \leqslant F_y(0)$, $J'_u = J_0$.

ii) Conversely, let J_y be a garden such that, for any decomposition $J_y = J' + J''$, then $J' = J_0$ or $J'' = J_0$; if $y = 1$, then $J_y = J_1$ is perfect ; we cannot have $y \in L_\omega$, because, since $\{z ; z \overset{J}{<}_o y\}$ is cofinal in y if $y \in L_\omega$ (condition ii) of a garden), we can choose $z > 0$ and $z \overset{J}{<}_o y$ and by (B) - 1 - 4 - 2 $J_y = J_z + J'_v$ (with $y = z+v$): hence a contradiction with the hypothesis since $z \neq 0$ and $v \neq 0$; if $y \in L_\Omega$, hence, since by (B)- 1 - 3 - 2 $F_y(0) \overset{J}{<}_o y$, $J_y = J_{F_y(0)} + J'_v$ (with $y = F_y(0) + v$) : hence, since $v > 0$, $F_y(0) = 0$ using hypothesis, and J_y is perfect.

I - 1 - 16 DEFINITION .- The following data define a category GAR :
objects : gardens of type Ω

morphisms from J_x to J'_y : the set $C(J_x, J'_y)$ of compatible functions $f \in I(x,y)$ from J_x to J'_y.

I - 1 - 17 LEMMA .- Let F be a flower, $f \in I(u,t)$, $x \leqslant u$ and x limit ; then $\widehat{F(f)}(F(x)) = F(\hat{f}(x))$.

Proof .- Using (G).2.4.10 we verify easily that for each $y < x$, $F(f)(F(y)) \geqslant F(f(y))$; $F(\hat{f}(x)) = \sup_{y<x} F(f(y)) \leqslant \sup_{y<x} F(f)(F(y)) = \widehat{F(f)}(F(x))$;
by (G).2.4.10.v), if $F(\omega) = (b;0,\ldots,p-2,\omega)$, for each $y \geqslant p-1$, $F(y) = (b;0,\ldots,p-2,y)$ and if $y < x$ $F(f)(F(y)) = (b;f(0),\ldots,f(p-2),f(y)) \leqslant (b;0,\ldots,p-2,\hat{f}(x)) = F(\hat{f}(x))$; hence the result.

I - 1 - 18 LEMMA .- Let T a natural transformation from the flower F to the flower G, and let y,x two ordinals such that $y \leqslant x$, then $\widehat{T(x)}(F(y)) \leqslant G(y)$.

Proof .- It is an immediate consequence of remark (G).2.4.10 and of proposition (G).2.3.15.

I - 2 - The category DIL Ω

I - 2 - 1 DEFINITION .- An Ω-dilator D is a dilator such that, for each $x < \Omega$, $D(x) < \Omega$.

I - 2 - 2 DEFINITION .- The following data define a category DIL Ω :

objects : Ω-dilators

morphisms from D to G : the set $I\Omega(D,G)$ of natural transformations from D to G.

I - 2 - 3 LEMMA .-

i) Let D be an Ω-dilator and let $D = \sum_{i<x} D_i$ be the decomposition of D in sum of perfect dilators ; then $x < \Omega$, and if $i < x$, D_i is an Ω-dilator.

ii) Let D an Ω-dilator of kind Ω ; then, for each $y < \Omega$, $SEP(D)^y$ is an Ω-dilator.

Proof .-

i) Since for each $i < x$, D_i is perfect and then $\neq \underline{0}$, we have $D_i(\omega) \geq 1$; then, since $D(\omega) < \Omega$, we conclude that $x < \Omega$; clearly D_i is an Ω-dilator.

ii) By definition of SEP, if $D = D' + D''$ with D'' perfect, $SEP(D)^y = D' + SEP(D'')^y$; then, since by i) D' is an Ω-dilator, it is sufficient to show the result when D is a perfect Ω-dilator $\neq 1$; in this case, $SEP(D)^y(x) \leq D(y+x)$ for all ordinals x,y : see the definition of SEP in (G) section 3.3 ; then, since D is an Ω-dilator, we have the result.

I - 2 - 4 DEFINITION .- An Ω-bilator B is a bilator such that, for all $x,y < \Omega$, $B(x,y) < \Omega$. We define the subcategory BIL Ω of BIL : the objects of BIL Ω are Ω-bilators.

I - 2 - 5 LEMMA .-

i) if D is an Ω-dilator of kind Ω, SEP(D) is an Ω-bilator.

ii) if B is an Ω-bilator, UN(B) is an Ω-dilator.

Proof .-

i) is an immediate consequence of I - 2 - 3 ii) ;

ii) by remark (G).3.4.15. $UN(B)(x) \leq B(x,x)$, for each ordinal x : hence the result.

I - 2 - 6 THEOREM .- Let P be a property which is defined on Ω-dilators ; we assume that the following statements are true : (all dilators considered are

Ω-dilators)

i) $P(\underline{O})$

ii) $P(D) \longrightarrow P(D+\underline{1})$

iii) if x is limit, $x < \Omega$, and for each $y < x$, $P(\sum_{i<y} D_i)$, then $P(\sum_{i<x} D_i)$ (we suppose that, for each $i < x$, $D_i \neq \underline{0}$)

iv) if D is of kind Ω, if for each $y < \Omega$, $P(SEP(D)^y)$

then $P(D)$ is true for all Ω-dilators D.

Proof .- We have the result by a trivial modification of the proof of theorem (G).3.5.1. : in this proof remplace a by Ω.

I - 2 - 7 REMARK .- The hypothesis i) - iv) indicate clearly which Ω-dilators must be considered as the predecessors of D : G is a predecessor of D (in DIL Ω) iff there exists a finite sequence $G_o = G,\ldots,G_n = D$, such that, for each $i < n$, G_i is an Ω-dilator and $G_{i+1} = G_i + G_i'$ for some Ω-dilator G_i', or $G_i = SEP(G_{i+1})^y$ for some $y < \Omega$. We say also that G is an Ω-predecessor of D.

I - 3 - Some properties of dilators

I - 3 - 1 DEFINITION .-

i) Let t be an ordinal, and let $(D_z)_{z<t}$ a family of dilators such that for all z,z' with $z \leqslant z' < t$ there exists a dilator G with $D_{z'} = D_z + G$; then it is possible to define a new dilator $\bigcup_{z<t} D_z = D$ by : $D(x) = \sup_{z<t} D_z(x)$ and if $f \in I(x,y)$ and $u < D_z(x)$, $D(f)(u) = D_z(f)(u)$.

ii) Let v be another ordinal and let $(H_z)_{z<v}$ be another family of dilators defined as above ; let (T_z) be a family of natural transformations from D_z to $H_{f(z)}$ where f is a function in $I(t,v)$ and such that, if $z \leqslant z' < t$, then there exists a natural transformation U with $T_{z'} = T_z + U$; then one defines a natural transformation $T = \bigcup_{z<f} T_z$ by : for each ordinal x, $T(x) = \bigcup_{z<t} T_z(x)$: see (G) 1 - 3 - 8 i).

I - 3 - 2 LEMMA .-

i) I - 3 - 1 i) and ii) are correct definitions

ii) If t is of cofinality ω, and if, $(D_z)_{z<t}$ is a family of Ω-dilators such that, as in definition I - 3 - 1, for $z \leqslant z'$, $D_{z'}$

extends D_z, then $\bigcup_{z<t} D_z$ is an Ω-dilator.

Proof .-

i) The verification is trivial.

ii) Since t is of cofinality ω, and since for each ordinal $x<\Omega$ and for each $z<t$, $D_z(x)<\Omega$, then $D(x) = \sup_{z<t} D_z(x) < \Omega$.

I - 3 - 3 REMARK .-

i) Using example (G).4.1.5. (iii) we have clearly, with notations and hypothesis of I - 3 - 1 : $(D, E_{D_z D}) = \lim_{z<t} (D_z, E_{D_z D_{z'}})$

ii) Using example (G).4.1.5. (iv) we have clearly, with notations and hypothesis of I - 3 - 1 ii) : $T = \bigcup_{z<f} T_z = \varinjlim T_z$.

I - 3 - 4 LEMMA .- If D and G are perfect dilators $\neq \underline{1}$, and if $T \in I(D,G)$, then for each ordinal $x = \omega^{x'}$ with $x' \neq 0$, $\widehat{T(x)}(D(x)) = G(x)$.

Proof .- As in (G).3.3.1, we have, since G is perfect, that the values $G(f)(0)$ are cofinal in $G(x)$ when f varies through $I(x,x)$; then, since $T(x)(D(f)(0)) = G(f)(T(x)(0)) \geqslant G(f)(0)$, the values $T(x)(D(f)(0))$ are cofinal in $G(x)$ when f varies through $I(x,x)$; and $\widehat{T(x)}(D(x)) = \sup_{u<D(x)} T(x)(u) \geqslant G(x)$; then, since clearly we have $\widehat{T(x)}(D(x)) \leqslant G(x)$ we have the result.

II - Decomposition of a dilator

II - 1 - THEOREM .- There exists one and only one functor DEC from DIL Ω to GAR with the following properties :

i) $DEC(\underline{0}) = J_0$ $\quad DEC(E_{\underline{0}}) = E_0$

$DEC(\underline{1}) = J_1$ $\quad DEC(E_{\underline{1}}) = E_1$ $\quad DEC(E_{\underline{01}}) = E_{01}$

ii) For each Ω-dilator G, DEC(G) is a garden of height $G(\Omega)$, and, if $T \in I\Omega(G,G')$, $DEC(T) = T(\Omega)$

iii) If $G',G'' \in DIL\,\Omega$, $DEC(G'+G'') = DEC(G') + DEC(G'')$

iv) If for each $i<x$ (with x limit) G_i is a perfect Ω-dilator, $DEC(\sum_{i<x} G_i) = \sum_{i<x} DEC(G_i)$.

v) If G is a perfect dilator, and $G \in DIL\,\Omega$, and $G \neq \underline{1}$, $(DEC(G))_{<G(\Omega)} = \bigcup_{y<\Omega} DEC(SEP(G)^y)$

$$F_G^{DEC(G)} = SEP(G)(\Omega,.).$$

Proof .- The fact that the solution is unique is easily obtained from the proof below, because, at each step, DEC is completely defined by means of some of the formulas of the above theorem. To construct DEC, satisfying i) - v) we shall proceed as Girard in (G)5 - 4 - 3 : given an Ω-dilator H, we define a subcategory $DIL\,\Omega < H$ as follows : F is an object of $DIL\,\Omega < H$ iff $I\Omega(F,H') \neq \emptyset$, for some Ω-predecessor H' of H ; define $DIL\,\Omega \leq H = DIL\,\Omega < (H+\underline{1})$; then, there exists one and only one functor DEC from $DIL\,\Omega \leq H$ to GAR enjoying i) - v) : we use an induction on H (theorem I - 2 - 6), induction hypothesis being therefore the existence of an unique functor DEC from $DIL\,\Omega < H$ to GAR enjoying i) - v). If F is in $DIL\,\Omega \leq H$, let us denote by h(F) the smallest H', for the predecessor relation, such that $I\Omega(F,H') \neq \emptyset$, H' varying through predecessors of H and H. The notation $h(F) < h(G)$, will mean that h(F) is a predecessor of h(G) ; we shall also allow the notation $h(F) \leq h(G)$ to mean $h(F) < h(G)$ or $h(F) = h(G)$; if $T \in I\Omega(F,G)$ with $G \in DIL\,\Omega \leq H$ we set $h(T) = h(G)$.

II - 1 - 1 case $H = \underline{0}$.- If $H = \underline{0}$, we set $DEC(\underline{0}) = J_0$, $DEC(E_{\underline{0}}) = E_{\underline{0}}(\Omega) = E_0$ and i) - v) are trivially verified.

II - 1 - 2 case $H = \underline{1}$.- There are two objects ($\underline{0}$ and $\underline{1}$) and three morphisms ($E_{\underline{0}}$, $E_{\underline{1}}$, $E_{\underline{01}}$) in $DIL\,\Omega \leq \underline{1}$; we set $DEC(\underline{1}) = J_1$, $DEC(E_{\underline{1}}) = E_1$, $DEC(E_{\underline{01}}) = E_{01}$; ii), iv), v), are trivial ;

iii) if $G = G' + G''$ in $DIL\,\Omega \leq \underline{1}$ then one of G' and G'' is $\underline{0}$, and then, one of DEC(G'), DEC(G'') is J_0 ; then $DEC(G'+G'') = DEC(G') + DEC(G'')$ since J_0 is neutral for the law +. We remark that in this case as in others the fact that DEC is a functor is trivial, because, using ii), $DEC(TU) = TU(\Omega) = T(\Omega)oU(\Omega) = DEC(T)oDEC(U)$.

II - 1 - 3 LEMMA .- If i) - v) hold for $DIL\,\Omega \leq H'$ and $DIL\,\Omega \leq H''$, then they hold for $DIL\,\Omega \leq (H' + H'')$.

Proof .- Let $H = H' + H''$; we extend DEC to $DIL\,\Omega \leq H$:

- if H_1 is a predecessor of H, then, either H_1 is a predecessor of H', or $H_1 = H' + H_1''$, where H_1'' is a predecessor of H'' ; so it is possible to write in both cases $H_1 = H_1' + H_1''$ with H_1' (resp. H_1'') equal to or predecessor of H' (resp. H'') ; this is still true when $H_1 = H$.

- if F is an object of $DIL\,\Omega \leq H$, then let H_1 (equal to H or predecessor of H) be such that $T \in I\Omega(F,H_1)$ for some T ; the decomposition $H = H_1' + H_1''$ induces a decomposition $F = F' + F''$, $T = T' + T''$ with $T' \in I\Omega(F',H_1')$, $T'' \in I\Omega(F'',H_1'')$: let x,y,y',y'',f, be the respective lengths of F,H_1,H_1',H_1'',T ; we write $f = f' + f''$

with $f'\in I(x',y')$, $f''\in I(x'',y'')$ $(x=x'+x'',\ y=y'+y'')$; if $F=\sum_{i<x} F_i$ if $T=\sum_{i<f} T_i$, then let $F'=\sum_{i<x'} F_i$, $F''=\sum_{i<x''} F_{x'+i}$, $T'=\sum_{i<f'} T_i$, $T''=\sum_{i<f''} T_{x'+i}$, and so, $T'\in I\Omega(F',H_1')$, $T''\in I\Omega(F'',H_1'')$.

We define $DEC(F) = DEC(F') + DEC(F'')$: this definition is possible because F' (resp. F'') is an object of $DIL\,\Omega \leqslant H'$ (resp. $DIL\,\Omega \leqslant H''$) ; we must prove that this definition is independent from the decomposition of F in sum : let $F = G' + G''$ an another similar decomposition, then, for instance, $LH(G') < LH(F')$ so, write $F' = G' + G_1'$ (with $G_1' = \sum_{i<x'-LH(G')} F_{LH(G')+i}$) ; so $G'' = G_1' + F''$ and $DEC(F) = DEC(F') + DEC(F'') = DEC(G') + DEC(G_1') + DEC(F'') = DEC(G') + DEC(G'')$ (we have used the associativity of +, and the property iii) with $F' = G' + G_1'$ in $DIL\,\Omega \leqslant H'$, with $G'' = G_1' + F''$ in $DIL\,\Omega \leqslant H''$).

If $T\in I\Omega(F,G)$ we set $DEC(T) = T(\Omega)$; given a decomposition $G = G' + G''$ (with G' in $DIL\,\Omega \leqslant H'$, G'' in $DIL\,\Omega \leqslant H''$), then, as above we write $T = T' + T''$ with $T'\in I\Omega(F',G')$, $T''\in I\Omega(F'',G'')$; then clearly $DEC(T) = DEC(T') + DEC(T'')$ and since $DEC(T')$ (resp. $DEC(T'')$) is, by induction hypothesis, a compatible function from $DEC(F')$ (resp. $DEC(F'')$) to $DEC(G')$ (resp. $DEC(G'')$), $DEC(T)$ is a compatible function from $DEC(F)$ to $DEC(G)$.

i) and ii) are trivial ; for iii) if $G = G' + G''$, then we write $G = F' + F''$, with F' in $DIL\,\Omega \leqslant H'$, F'' in $DIL\,\Omega \leqslant H''$; we distinguish two subcases :

- if $LH(G') \geqslant LH(F')$, write $G' = F' + G_1'$, so $F'' = G_1' + G''$ and $DEC(G) = DEC(F') + DEC(F'') = DEC(F') + DEC(G_1') + DEC(G'') = DEC(G') + DEC(G'')$
- if $LH(G') \leqslant LH(F')$, write $F' = G' + G_1'$, so $G'' = G_1' + G''$ and $DEC(G) = DEC(F') + DEC(F'') = DEC(G') + DEC(G_1') + DEC(F'') = DEC(G') + DEC(G'')$

iv) if $G = \sum_{i<x} G_i$ with x limit and for each i, G_i perfect Ω-dilator, write $G = F' + F''$, with F' in $DIL\,\Omega \leqslant H'$, F'' in $DIL\,\Omega \leqslant H''$; then there exists x',x'' such that $x = x' + x''$ and $F' = \sum_{i<x'} G_i$ and $F'' = \sum_{i<x''} G_{x'+i}$; if x' or x'' is equal to 0, the result is immediate by induction hypothesis ; if $x' = 0$ and $x'' = 0$, then x'' is limit and by iv) in $DIL\,\Omega \leqslant H''$, $DEC(F'') = \sum_{i<x''} DEC(G_{x'+i})$; by iii) and iv) in $DIL\,\Omega \leqslant H'$, we see easily that $DEC(F') = \sum_{i<x'} G_i$; then using the above definition of $DEC(G)$, we have : $DEC(G) = DEC(F') + DEC(F'') = \sum_{i<x'} DEC(G_i) + \sum_{i<x''} DEC(G_{x'+i}) = \sum_{i<x} DEC(G_i)$.

v) if G is a perfect Ω-dilator $\neq \underline{1}$ and if $G = G' + G''$, with G' in $DIL\,\Omega \leqslant H'$, G'' in $DIL\,\Omega \leqslant H''$, then $G = G'$ or $G = G''$, so the property is already in the hypothesis.

II - 1 - 4 COROLLARY : case $H = H' + \underline{1}$.- It is an immediate consequence of II - 1 - 2 and II - 1 - 3.

II - 1 - 5 case where H is of kind ω .- Assume that i) - v) hold in $DIL\,\Omega < H$, and define DEC(G) and DEC(T) when $h(G) = h(T) = H$; let $G = \sum_{i<y} G_i$ the decomposition of G in a sum of perfect Ω-dilators ; then y is limit : if $T \in I\Omega(G,H)$, let $f = LH(T)$, $f \in I(y,z)$ with $z = LH(H)$; if $\hat{f}(y) \neq z$, then we can write (if $H = \sum_{i<z} H_i$) $H = H' + H''$, $T = T' + E_{\underline{OH''}}$ with $H' = \sum_{i<\hat{f}(y)} H_i$ ....
Then, since $H'' \neq \underline{0}$, H' is a predecessor of H, and $h(G) \leqslant H' < H$, contrarily to the assumptions : so, $\hat{f}(y) = z$ and y is limit ; this shows that G is of kind ω. Define $DEC(G) = \sum_{i<y} DEC(G_i)$ (see definition I - 1 - 9) and observe that the formula above is valid for an arbitrary G (if $h(G) < H$ it is an immediate consequence of iii) and iv)).
If $T \in I\Omega(F,G)$ with $h(T) = h(G) = H$, let $F = \sum_{i<x} F_i$, $T = \sum_{i<g} T_i$ the decompositions in sum of F and T (for each $i < x$, F_i is a perfect Ω-dilator) we set $DEC(T) = \sum_{i<g} DEC(T_i)$ (see definition I - 1 - 9) i) and ii) are trivially verified and the fact that DEC is a functor is an immediate consequence of ii).
iii) let $G = G' + G''$; let $G = \sum_{i<x} G_i$, the decomposition of G in sum of perfect Ω-dilators and let x',x" the respective lenghts of G' and G" ; then $G' = \sum_{i<x'} G_i$ and $G'' = \sum_{i<x''} G_{x'+i}$ and as we have shown above
$DEC(G') = \sum_{i<x'} DEC(G_i)$, $DEC(G'') = \sum_{i<x''} DEC(G_{x+i})$ $DEC(G) = \sum_{i<x} DEC(G_i)$; hence the result.
iv) is immediate ; v) also is trivial.

II - 1 - 6 case where H is of kind Ω .- If H is of kind Ω, then by hypothesis, i) - v) hold in $DIL\,\Omega < H$; it is enough to treat the case H perfect : because, if H is not perfect, then write $H = H' + H''$; then $DIL\,\Omega \leqslant H'$ and $DIL\,\Omega \leqslant H''$ are subcategories of $DIL\,\Omega < H$, and since we have proved the result when H" is perfect, i) - v) hold in $DIL\,\Omega \leqslant H'$ and $DIL\,\Omega \leqslant H''$; then i) - v) hold in $DIL\,\Omega \leqslant H$, using lemma II - 1 - 3.

Let H perfect and define DEC(G) when G is in $DIL\,\Omega \leqslant H$ and $h(G) = H$; if $h(G) = H$, then $I\Omega(G,H) \neq \emptyset$ and $LH(G) \leqslant LH(H) = 1$; since we cannot have $G = \underline{0}$ and since $G \neq \underline{1}$ because $G(0) \leqslant H(0) = 0$, we can conclude that G is perfect and of kind Ω ; let $U \in I\Omega(G,H)$, then SEP(U) is a natural transformation from SEP(G) to SEP(H) and if we denote by $SEP(U)^y$ the natural transformation from $SEP(G)^y$ to $SEP(H)^y$ defined by $SEP(U)^y = SEP(U)(.,y)$, we conclude that

$I\Omega(SEP(G)^y, SEP(H)^y) \neq \emptyset$ and then $SEP(G)^y$ is in $DIL\,\Omega < H$ for each $y < \Omega$ (see lemma I-2-3) ; then $DEC(SEP(G)^y)$ is defined for each $y < \Omega$. By corollary (G)3-6-7 $SEP(G)^{y+1} = SEP(G)^y + {}_yG$ and then $DEC(SEP(G)^{y+1}) = DEC(SEP(G)^y) + DEC({}_yG)$ by induction hypothesis iii) ; using also corollary (G)3-6-7, if y is limit, $SEP(G)^y = \sum_{y'<y} {}_{y'}G$; using the decomposition of each ${}_{y'}G$ in sum of perfect dilators and induction hypothesis iii) and iv) we prove easily that $DEC(SEP(G)^y) = \sum_{y'<y} DEC({}_{y'}G)$; then, by definition I-1-12, $\bigcup_{y<\Omega} DEC(SEP(G)^y)$ is a garden $J_{<A}$ with $A = \sup_{y<\Omega} SEP(G)^y(\Omega)$ but, $SEP(G)^y(\Omega) = G(\Omega,y)$ and since the dilator $G(\Omega,.),G(E_\Omega,.)$ is a flower, $\sup_{y<\Omega} SEP(G)^y(\Omega) = SEP(G)\ (\Omega,\Omega) = G(\Omega)$ using (G)3-4-14. We set

$F^{DEC(G)}_{G(\Omega)} = SEP(G)(\Omega,.)$ and we must prove that $J_{<A}$ with the flower $F^{DEC(G)}_{G(\Omega)}$ is a garden of height $G(\Omega)$: it is this garden that we will denote $DEC(G)$; we set $K = DEC(G)$

a) clearly $G(\Omega) = \sup_{\alpha<\Omega} SEP(G)\ (\Omega,\alpha) = \sup_{\alpha<\Omega} F^K_A(\alpha)$

b) let $f \in I(\alpha,\beta)$, $\alpha,\beta < \Omega$:

- let $z \in L_\Omega$ and $z < F^K_A(\alpha)$; we must prove that for each $u < \Omega$, $\widehat{F^K_A(f)}\ (F^K_z(u)) \leqslant F^K_{\widehat{F^K_A(f)}(z)}(u)$: $F^K_A(f) = G(E_\Omega,f)$ and $F^K_A(\alpha) = G(\Omega,\alpha)$; since $z < G(\Omega,\alpha)$ and the fact that the function $G(\Omega,.)$ is increasing, there exists $\gamma<\alpha$ such that $G(\Omega,\gamma) < z \leqslant G(\Omega,\gamma+1)$ (remark that, since $z \in L_\Omega$, we cannot have $z = G(\Omega,\gamma)$ with γ limit) ; since $SEP(G)^{\gamma+1} = SEP(G)^\gamma + {}_\gamma G$ and using definition of K, $F^K_z(z) = SEP(G)^\gamma(\Omega) + F^{DEC}_v({}_\gamma G)(u) = G(\Omega,\gamma) + F^B_v(u)$ if we set $z = G(\Omega,\gamma) + v$ and $B = DEC({}_\gamma G)$; let $h \in I(\gamma, f(\gamma))$ and for $t < \gamma$, $h(t) = f(t)$; then $\widehat{F^K_A(f)}\ (F^K_z(u)) = \widehat{G(E_\Omega,f)}(F^K_z(u)) = \widehat{G(E_\Omega,h+E_1)}(F^K_z(u))$ (it is trivial consequence of the fact that $G(\Omega,.)$, $G(E_\Omega,.)$ is a flower and that $F^K_z(u) < G(\Omega,\gamma+1)$).
By (G)3-6-6 $G(E_\Omega,h+E_1) = G(E_\Omega,h) + {}_hG(\Omega)$, we have :

$\widehat{F^K_A(f)}(F^K_z(u)) = G(\Omega,f(\gamma)) + \widehat{{}_hG(\Omega)}(F^B_v(u))$.

Furthermore, $\widehat{F^K_A(f)}(z) = \sup_{t<z} F^K_A(f)(t) = \sup_{t<v} G(E_\Omega,h+E_1)\ (G(\Omega,\gamma)+t) =$

$G(\Omega,f(\gamma)) + \widehat{{}_hG(\Omega)}(v)$ and $F^K_{\widehat{F^K_A(f)}(z)}(u) = G(\Omega,f(\gamma)) + F^C_{\widehat{{}_hG(\Omega)}(v)}(u)$ if we set

$C = DEC({}_{f(\gamma)}G)$; remark that, since ${}_hG$ is a natural transformation between ${}_\gamma G$

and ${}_{f(\gamma)}G$, by induction hypothesis ii) ${}_hG(\Omega)$ is a compatible function from B to C ; then by I - 1 - 6 i) $\widehat{{}_hG(\Omega)}\,(F^B_v(u)) \leqslant F^C_{\widehat{{}_hG(\Omega)}(v)}(u)$ and hence the result.

- if $g \in I(u,c)$ with $u, c < \Omega$ and $z \in L_\Omega$, $z \leqslant F^K_A(\alpha)$ we must prove that the following diagram is commutative :

$$\begin{array}{ccc}
F^K_z(u) & \xrightarrow{\quad F^K_z(g)\quad} & F^K_z(c) \\
\Big\downarrow w^u & & \Big\downarrow w^c \\
F^K_{\widehat{F^K_A(f)(z)}}(u) & \xrightarrow{\quad F^K_{\widehat{F^K_A(f)(z)}}(g)\quad} & F^K_{\widehat{F^K_A(f)(z)}}(c)
\end{array}$$

with w^a $(a = u$ or $c)$ defined by : $w^a \in I(F^K_z(a), F^K_{\widehat{F^K_A(f)(z)}}(a))$ and for $\theta < F^K_A(a)$,

$w^a(\theta) = F^K_A(f)(\theta)$; in the sequel we use the notations of the above proof : let $b < F^K_z(u) = G(\Omega,\gamma) + F^B_v(u)$; as in the above proof we show easily that $(F^K_{\widehat{F^K_A(f)(z)}}(g) \circ w^u)(b) = (E_{G(\Omega,f(\gamma))} + F^C_{\widehat{{}_hG(\Omega)}(v)}(g)) \circ (G(E_\Omega,h) + {}_hG(\Omega))(b)$

$= (G(E_\Omega,h) + F^C_{\widehat{{}_hG(\Omega)}(v)}(g) \circ {}_hG(\Omega))(b)$; and $(w^c \circ F^K_z(g))(b)$

$= (G(E_\Omega,h) + {}_hG(\Omega)) \circ (E_{G(\Omega,\gamma)} + F^B_v(g))(b) = (G(E_\Omega,h) + {}_hG(\Omega) \circ F^B_v(g))(b)$;

hence the result since ${}_hG(\Omega)$ is a compatible function from $B = \mathrm{DEC}\,({}_\gamma G)$ to $C = \mathrm{DEC}\,({}_{f(\gamma)}G)$.

- let $z \in L_\Omega$, and $u < F^K_A(\alpha)$; we show that, if $\widehat{F^K_A(f)}(u) < z \leqslant F^K_A(f)(u)$ then $\widehat{F^K_A(f)}(u) \leqslant F^K_z(0)$: let $\widehat{F^K_A(f)}(u) < z \leqslant F^K_A(f)(u)$; then there exists $\gamma < \alpha$ such that $G(\Omega,\gamma) < z \leqslant G(\Omega,\gamma+1)$ and if we set $z = G(\Omega,\gamma) + v$, we have $F^K_z(0) = G(\Omega,\gamma) + F^B_v(0)$ with $B = \mathrm{DEC}\,({}_\gamma G)$; since $u < G(\Omega,\alpha)$, either $u = G(\Omega,\delta)$ with δ limit, or there exists δ such that $\delta+1 \leqslant \alpha$, and $G(\Omega,\delta) < u \leqslant G(\Omega,\delta+1)$, (remark that, if $u = G(\Omega,\delta+1)$, $\delta+1 < \alpha$) ;

- if $u = G(\Omega,\delta)$ with δ limit and $\delta < \alpha$, then, by I - 1 - 17

$$\widehat{F^K_A(f)}(u) = \widehat{G(E_\Omega,f)}(u) = \sup_{\mu<\delta} G(E_\Omega,f)(G(\Omega,\mu)) = \sup_{\mu<\delta} G(\Omega,f(\mu)) = G(\Omega, \hat{f}(\delta))$$

and, since $\widehat{F^K_A(f)}(u) < z$, we have $\hat{f}(\delta) \leqslant \gamma$, and then, $\widehat{F^K_A(f)}(u) \leqslant F^K_z(0)$

- if $G(\Omega,\delta) < u \leqslant G(\Omega,\delta+1)$, $\widehat{F^K_A(f)}(u) = \widehat{G(E_\Omega,f)}(u)$, and if we set $u = G(\Omega,\delta) + w$, $h \in I(\delta, f(\delta))$ defined by $h(t) = f(t)$,

$\widehat{F_A^K(f)}(u) = G(\Omega,f(\delta)) + \widehat{{}_hG(\Omega)}(w)$; then, if $f(\delta) < \gamma$, we have $F_z(0) \geqslant G(\Omega,\gamma) \geqslant G(\Omega, f(\delta)+1) \geqslant \widehat{F_A^K(f)}(u)$; if $f(\delta) = \gamma$, then,

$\widehat{{}_hG(\Omega)}(w) < v \leqslant {}_hG(\Omega)(w)$ and by induction hypothesis ${}_hG(\Omega)$ is a compatible function from $B = DEC({}_\delta G)$ to $C = DEC({}_{f(\delta)}G)$ and then

$\widehat{{}_hG(\Omega)}(w) \leqslant F_v^B(0)$; hence the result.

- let $z \in L_\Omega$ such that $\widehat{F_A^K(f)}(F_A^K(\alpha)) < z \leqslant F_A^K(\alpha)$, we show that $\widehat{F_A^K(f)}\ (F_A^K(\alpha)) \leqslant F_z^K(0)$; there exists γ such that $G(\Omega,\gamma) < z \leqslant G(\Omega,\gamma+1)$

- if $\alpha = \delta+1$, let $h \in I(\delta,f(\delta))$ such that $h(u) = f(u)$ for each $u < \delta$; for each $u < G(\Omega,\alpha)$, $G(\Omega,f)(u) = G(\Omega,h+E_1)(u)$, then $\widehat{F_A^K(f)}(F_A^K(\alpha)) = \widehat{G(\Omega,f)}(G(\Omega,\alpha))$

$$= \sup_{u<G(\Omega,\alpha)} G(\Omega,h+E_1)(u) = G(\Omega,f(\delta)) + \widehat{{}_hG(\Omega)}\ (G(\Omega))$$

if $f(\delta)+1 \leqslant \gamma$, we have $F_z^K(0) \geqslant G(\Omega,\gamma)$; hence the result ;

if $f(\delta) = \gamma$, then if we set $z = G(\Omega,f(\delta)) + v$, then, $F_z^K(0) = G(\Omega,f(\delta)) + F_v^C(0)$ (with $C = DEC({}_{f(\delta)}G)$) ; by hypothesis, $\widehat{{}_hG(\Omega)}({}_\delta G(\Omega)) < v \leqslant {}_{f(\delta)}\ G(\Omega)$, then, since by induction hypothesis, ${}_hG(\Omega)$ is a compatible function from $B = DEC({}_\delta G)$ to $C = DEC({}_{f(\delta)}G)$, $\widehat{{}_hG(\Omega)}\ ({}_\delta G(\Omega)) \leqslant F_v^C(0)$; hense the result.

- if α is limit ; by lemma I-1-17, $\widehat{G(\Omega,f)}(G(\Omega,\alpha)) = G(\Omega,\hat{f}(\alpha))$ since $\widehat{F_A^K(f)}(F_A^K(\alpha)) = \widehat{G(\Omega,f)}(G(\Omega,\alpha)) < z$ and since $G(\Omega,\gamma) < z \leqslant G(\Omega,\gamma+1)$, we conclude that $\hat{f}(\alpha) \leqslant \gamma$, and then, since $F_z^K(0) \geqslant G(\Omega,\gamma)$, the result.

Let $T \in I\Omega(D,G)$ such that $h(T) = H = h(G)$; then either $D = \underline{0}$ and $DEC(T) = T(\Omega) = E_{0G(\Omega)}$ is a compatible function from $DEC(D) = J_0$ to $DEC(G)$, or D is perfect $\neq \underline{1}$ and we set $DEC(T) = T(\Omega)$: in this case we must prove that $T(\Omega)$ is a compatible function from $DEC(D) = K$ to $DEC(G) = L$; using theorem (G)3-4-15 and theorem (G)3-6-6, we have $T(\Omega) = T(\Omega,\Omega) = \sum_{y<\Omega} {}_yT(\Omega)$;

by induction hypothesis $DEC({}_yT) = {}_yT(\Omega)$ is a compatible function from $DEC({}_yD)$ to $DEC({}_yG)$ then by I-1-12, $T(\Omega)$ is a compatible function from $(DEC(D))_{<D(\Omega)}$ to $(DEC(G))_{<G(\Omega)}$; now we prove that $T(\Omega)$ is a compatible function from $DEC(D)$ to $DEC(G)$:

- we prove that $\widehat{T(\Omega)}(F_A^K(\alpha)) \leqslant F^L_{\widehat{T(\Omega)}(A)}(\alpha)$ with $A = D(\Omega)$:

by lemma I-3-4, $\widehat{T(\Omega)}(A) = G(\Omega)$; by definition $F_A^K(\alpha) = D(\Omega,\alpha)$, and $F^L_{G(\Omega)}(\alpha) = G(\Omega,\alpha)$; then we must prove that $\widehat{T(\Omega,\Omega)}(D(\Omega,\alpha)) \leqslant G(\Omega,\alpha)$ and this is a consequence of lemma I-1-18.

- we prove that the following diagram is commutative :

$$\begin{array}{ccc} F^K_A(\alpha) & \xrightarrow{F^K_A(g)} & F^K_A(\beta) \\ \Big\downarrow f^{\alpha} & & \Big\downarrow f^{\beta} \\ F^L_B(\alpha) & \xrightarrow{F^L_B(g)} & F^L_B(\beta) \end{array}$$

with $B = G(\Omega)$ and f^{α} (resp. f^{β}) the restriction of $T(\Omega)$ to $F^K_A(\alpha)$ (resp. to $F^K_A(\beta)$) ; then $f^{\alpha} = T(\Omega,\alpha)$, $f^{\beta} = T(\Omega,\beta)$, $F^K_A(g) = D(E_{\Omega},g)$, $F^L_B(g) = G(E_{\Omega},g)$ and the result is immediate since $T(.,.)$ is a natural transformation from $D(.,.)$ to $G(.,.)$.

- we must prove that $\widehat{T(\Omega)}(D(\Omega)) \leq^L_o G(\Omega)$, but it is immediate since $\widehat{T(\Omega)}\ (D(\Omega)) = G(\Omega)$.

Now, we verify i) – v) :

i) is trivial ; ii) we have verified ii) in the above proof ;
iii) if $G = G' + G''$ is such that $h(G) = H$, then $G' = \underline{0}$ or $G'' = \underline{0}$, since H is perfect ; we achieve the proof as in II – 1 – 2 iii).
iv) is trivial because, if G is such that $h(G) = H$, we cannot have $G = \sum_{i<x} G_i$ with x limit ; v) is exactly the above definition.

II – 2 – Consequences of theorem II – 1

II – 2 – 1 LEMMA .- If H is a perfect Ω-dilator then $DEC(H)$ is a perfect garden.

Proof .- If $H = \underline{1}$, the result is obvious ; if H is perfect $\neq \underline{1}$, then it is an easy consequence of theorem II – 1 – v) and of remark (G)3 – 4 – 12 : $DEC(H)$ is a garden of height $H(\Omega)$ and $F_{H(\Omega)}(0) = SEP(H)(\Omega,0)$ and, since H is perfect $SEP(H)(\Omega,0) = 0$.

III – Synthesis of a garden

III – 1 – THEOREM .- There exists one and only one functor from GAR to DIL Ω, with the following properties :

i) $SYN(J_0) = \underline{0}$ $\qquad$ $SYN(E_0) = E_{\underline{0}}$

$SYN(J_1) = \underline{1}$ $\qquad$ $SYN(E_1) = E_{\underline{1}}$

ii) $SYN(J_y)$ is an Ω-dilator such that $SYN(J_y)(\Omega) = y$; if $f \in C(J'_x,J_y)$ $SYN(f) \in I\Omega(SYN(J'_x),SYN(J_y))$ and $SYN(f)(\Omega) = f$

iii) $SYN(J+J') = SYN(J) + SYN(J')$; $SYN(f+f') = SYN(f) + SYN(f')$

iv) If J_y is a garden of height $y \in L_\omega$, if $f \in C(J'_x, J_y)$ and $\hat{f}(x) = y$,

$SYN(J_y) = \bigcup_{z <_0 y} SYN(J_z)$ $\quad SYN(f) = \bigcup_{z <_0 x} SYN(f^z)$

(with $f^z \in C(J'_z, J_{f(z)})$ and, for each $u < z$, $f^z(u) = f(u)$).

v) If J_y is a garden of height $y \in L_\Omega$,

- let $D_\alpha = SYN(J_{F_y(\alpha)})$ for each $\alpha < \Omega$, and D the Ω-bilator defined by :

. $D(z,\alpha) = D_\alpha(z)$

. $D(h,g) = SYN(F_y(g))(h)$, with $h \in I(z,z')$, $g \in I(\alpha,\beta)$ then :

$SYN(J_y) = UN(D)$

- let $f \in C(J'_x, J_y)$ such that $\hat{f}(x) = y$; define $f^\alpha \in C(J'_{F_x(\alpha)}, J_{F_y(\alpha)})$ by $f^\alpha(u) = f(u)$, and let T the natural transformation defined by $T(x,\alpha) = SYN(f^\alpha)(x)$, then : $SYN(f) = UN(T)$.

vi) If $f \in C(J'_x, J_y)$ is such that $u = \hat{f}(x) < y$, let $J_y = J_u + J''_a$, $f = h + E_{0a}$, then $SYN(f) = SYN(h) + E_{\underline{0}\ SYN(J''_a)}$.

<u>Proof</u> .- We use an induction on y (the height of the garden), induction hypothesis being the existence of an unique functor from $GAR < y$ (J'_x is an object of the subcategory $GAR < y$ of the category GAR iff $x < y$) to $DIL\,\Omega$. The fact that the solution is unique is trivially obtained from the proof below, because, at each step, SYN is completely defined by means of some of the formulas of the above theorem.

III - 1 - 1 case $y = 0$.- We set $SYN(J_0) = \underline{0}$, $SYN(E_0) = E_{\underline{0}}$

i) - vi) are trivially verified.

III - 1 - 2 case $y = z+1$.- We have $J_y = J_z + J_1$ and we set $SYN(J_y) = SYN(J_z) + \underline{1}$; $SYN(J_y)(\Omega) = y$ let $f \in C(J'_x, J_y)$, we distinguish two cases :

- if $\hat{f}(x) = y$, then $f = h + E_1$, $x = t+1$, $h \in C(J'_t, J_z)$ and we set $SYN(f) = SYN(h) + E_{\underline{1}}$; clearly, using induction hypothesis,

$SYN(f) \in I\Omega(SYN(J'_x), SYN(J_y))$; $SYN(f)(\Omega) = SYN(h)(\Omega) + E_1 = f$

- if $\hat{f}(x) < y$, let $u = \hat{f}(x)$, $J_y = J_u + J''_a$, $f = h + E_{0a}$, we set $SYN(f) = SYN(h) + E_{\underline{0}\ SYN(J''_a)}$; since $y = z+1$, we have $a = b+1$, $J_z = J_u + J''_b$; by induction hypothesis vi), $SYN(h+E_{0b}) = SYN(h) + E_{\underline{0}SYN(J''_b)} \in I\Omega(SYN(J'_x), SYN(J_z))$, and then, we conclude that $SYN(f) \in I\Omega(SYN(J'_x), SYN(J_y))$; $SYN(f)(\Omega) = f$. We verify easily, that SYN is a functor from the category GAR $\leqslant y$ (GAR $\leqslant y$, being by definition GAR $< y+1$) to DIL Ω, using the second equality of ii) ; i) - vi) are easily verified.

III - 1 - 3 case $y \in L_\omega$.- Let $L = \{z ; z <_0 y\}$; by induction hypothesis iii) since, if $z, z' \in L$ and $z < z'$ $J_{z'} = J_z + J''_u$ (with $z' = z+u$; see (B) - I - 4 - 2) we have $SYN(J_{z'}) = SYN(J_z) + SYN(J''_u)$; then, by I - 3 - 1 and I - 3 - 2, we can set $SYN(J_y) = \bigcup_{z <_0 y} SYN(J_z)$ and $SYN(J_y)$ is an Ω-dilator (since, by induction hypothesis, for each $z \in L$, $SYN(J_z)$ is an Ω-dilator and $SYN(J_y)(\Omega) = \sup_{z \in L} SYN(J_z)(\Omega) = \sup_{z \in L} z = y$, using induction hypothesis ii) and the fact that, by definition of a garden, L is cofinal in y.

Now, we show the first equality of iii) : let $J_y = J_x + J'_u$ and $K = \{z ; z \in L$ and $z \geqslant x\}$; for each $z \in K$, there exists $v <_0^{J'} u$, such that $z = x + v$ and $J_z = J_x + J'_v$, (see (B) I - 4 - 1 and I - 5 - 3) ; by induction hypothesis iii), for each $z \in K$, $SYN(J_z) = SYN(J_x) + SYN(J'_v)$ then, as we can see easily $SYN(J_y) = SYN(J_x) + \bigcup_{v <_0^{J'} u} SYN(J'_v)$; then using the definition of $SYN(J'_u)$,

$SYN(J_y) = SYN(J_x) + SYN(J'_u)$. Let $f \in C(J'_x, J_y)$; we distinguish two cases :

- if $\hat{f}(x) = y$; let $f^z \in C(J'_z, J_{f(z)})$ defined, for each $t < z < x$, by $f^z(t) = f(t)$ (by (B) I - 3 - 8 f^z is a compatible function from J'_z to $J_{f(z)}$) ; if $z < z'$ and $z <_0^{J'} x$, $z' <_0^{J'} x$, then, by (B) I - 4 - 3 $f^{z'} = f^z + h$ for some h, and by induction hypothesis iii), $SYN(f^{z'}) = SYN(f^z) + SYN(h)$; then, by definition I - 3 - 1, we can set $SYN(f) = \bigcup_{z <_0 x} SYN(f^z)$ and $SYN(f) \in I\Omega(SYN(J'_x), SYN(J_y))$;

$SYN(f)(\Omega) = \bigcup_{z <_0 x} SYN(f^z)(\Omega) = \bigcup_{z <_0 x} f^z = f$, using induction hypothesis ii) ; if $f = g + g'$ with $g \in C(J'_u, J_v)$, $g' \in C(J'''_a, J''_b)$ (with $J'_x = J'_u + J'''_a$, $J_y = J_v + J''_b$), for each $z <_0^{J'} x$ and $u \leqslant z$, if we set $z = u + w$, $f^z = g + g'^w$ (with $g'^w \in C(J'''_w, J''_{g'(w)})$ and for $c < w$, $g'^w(c) = g'(c)$) then $SYN(f^z) = SYN(g) + SYN(g'^w)$ (using induction hypothesis iii) and since $\hat{g}'(a) = b$, we have :

$$SYN(f) = \bigcup_{z <_0^{J'} x} SYN(f^z) = SYN(g) + \bigcup_{w <_0^{J'''} a} SYN(g'^w).$$

- if $\hat{f}(x) < y$, let $u = \hat{f}(x)$, $J_y = J_u + J''_a$, $f = h + E_{0a}$; we set $SYN(f) = SYN(h) + E_{\underline{0}\ SYN(J''_a)}$; by the above proof $SYN(J_y) = SYN(J_u) + SYN(J''_a)$, and clearly $E_{\underline{0}\ SYN(J''_a)} \in I\Omega(SYN(J_0), SYN(J''_a))$; by induction hypothesis $SYN(h) \in I\Omega(SYN(J'_x), SYN(J_u))$, then we conclude that $SYN(f) \in I\Omega(SYN(J'_x), SYN(J_y))$;

$SYN(f)(\Omega) = SYN(h)(\Omega) + E_{\underline{O}\ SYN(J''_a)}(\Omega) = h + E_{Oa} = f$, using induction hypothesis ii) ; if $f = g + g'$, we show easily that $SYN(f) = SYN(g) + SYN(g')$.
The proof that SYN is a functor from $GAR \leqslant y$ to DIL Ω, is a consequence of ii) ; we have showed ii) and iii) ; iv) and vi) are exactly the above definitions ; i) and v) are trivial.

III - 1 - 4 case $y \in L_\Omega$.- By induction hypothesis $D_\alpha = SYN(J_{F_y(\alpha)})$ is an Ω-dilator for $\alpha < \Omega$; we set, for each ordinal $z \leqslant \Omega$, $D(z,\alpha) = D_\alpha(z)$, $D(h,g) = SYN(F_y(g))(h)$ with $h \in I(z,z')$, $g \in I(\alpha,\beta)$; then $D(.,.)$ is a functor from $ON \leqslant \Omega \times ON < \Omega$ to ON, as we can verify easily ; to show that D commutes to direct limits and to pull-backs, we verify that the functors $D(z,.), D(E_z,.)$ and $D(.,\alpha)$, $D(.,E_\alpha)$ commutes to direct limits and to pull-backs (see (G)2 - 1 - 9 (ii)) ; the functor $D(.,\alpha)$, $D(.,E_\alpha)$ is the functor D_α, and by induction hypothesis, it is an Ω-dilator ; let G the functor from $ON < \Omega$ to ON defined by $G(x) = D(z,x)$, $G(f) = D(E_z,f)$, we show that G is a dilator from $ON < \Omega$ to ON :

- let $(x,f_i) = \varinjlim_I (x_i,f_{ij})$ and $u < G(x)$, we show that there exists $i \in I$, $u_i \in G(x_i)$ such that $u = G(f_i)(u_i) = SYN(F_y(f_i))(z)(u_i)$ let $v = D_x(E_{z\Omega})(u)$, $v \in D_x(\Omega) = SYN(J_{F_y(x)})(\Omega) = F_y(x)$ by induction hypothesis ii) ; then there exists $i \in I$, $v_i \in F_y(x_i)$ such that $v = F_y(f_i)(v_i)$ (because, by definition of a garden, F_y commutes to direct limits) ; but, by induction hypothesis ii) $F_y(f_i) = SYN(F_y(f_i))(\Omega)$; then $v \in rg(SYN(F_y(f_i))(\Omega) \cap rg(D_x(E_{z\Omega}))$; then there exists $u_i \in D_{x_i}(z)$ such that :
$v = (SYN(F_y(f_i))(\Omega) \,\&\, D_x(E_{z\Omega}))(u_i) = (D_x(E_{z\Omega}) \circ SYN(F_y(f_i))(z))(u_i)$ (since $SYN(F_y(f_i))$ is by induction hypothesis ii) a natural transformation between D_{x_i} and D_x and using (G)4 - 2 - 9) ; hence the result.

- we show that G commutes to pull-backs ; that is : if $f \in I(x',x)$, $g \in I(x'',x)$, then $G(f) \,\&\, G(g) = G(f \,\&\, g)$; to show this equality we prove that $A = D_x(E_{z\Omega}) \circ (G(f) \,\&\, G(g))$ is equal to $B = D_x(E_{z\Omega}) \circ G(f \,\&\, g)$

$$A = D(E_{z\Omega},E_x) \circ (D(E_z,f) \,\&\, D(E_z,g)) = (D(E_{z\Omega},E_x) \circ D(E_z,f)) \,\&\, (D(E_{z\Omega},E_x) \circ D(E_z,g))$$

$$= (D(E_{z\Omega},E_x) \,\&\, (D(E_\Omega,f)) \,\&\, (D(E_{z\Omega},E_x) \,\&\, D(E_\Omega,g)) = D(E_{z\Omega},E_x) \,\&\, D(E_\Omega,f) \,\&\, D(E_\Omega,g)$$

(we have used the distributivity at left of the composition to pull-backs, theorem (G) 4 - 2 - 9, and idempotence of pull-backs).

$B = D(E_{z\Omega},E_x) \circ D(E_z,f \& g) = D(E_{z\Omega},E_x) \& D(E_\Omega,f \& g)$ by theorem (G) 4 - 2 - 9 but, $D(E_\Omega, f \& g) = SYN(F_y(f \& g))(\Omega) = F_y(f \& g) = F_y(f) \& F_y(g) =$ $(SYN(F_y(f))(\Omega) \& SYN(F_y(g))(\Omega)) = D(E_\Omega,f) \& D(E_\Omega,g)$ (we have used induction hypothesis ii), the fact that in a garden F_y commutes to pull-backs, and induction hypothesis ii)) ; hence the result.

- for each $z < \Omega$, $\alpha < \Omega$, $D(z,\alpha) = D_\alpha(z) < \Omega$, since by induction hypothesis D_α is an Ω-dilator.

- for all $x,x' < \Omega$, we show that $G(E_{xx'}) = E_{G(x)G(x')}$:
$G(E_{xx'}) = SYN(F_y(E_{xx'}))(z) = SYN(E_{F_y(x)F_y(x')})(z)$ (because in a garden F_y is a flower) ; by induction hypothesis vi) we have, if we set $F_y(x') = F_y(x) + u$, $J_{F_y(x')} = J_{F_y(x)} + J'_u$, $SYN(E_{F_y(x)\ F_y(x')}) =$ $SYN(E_{F_y(x)}) + E_{\underline{O}}\ SYN(J'_u)$; then $G(E_{xx'}) = SYN(E_{F_y(x)})(z) + E_O\ SYN(J'_u)(z)$; but, since SYN is a functor $SYN(E_{F_y(x)})(z) = E_{SYN(J_{F_y(x)})(z)} =$ $E_{D_x(z)} = E_{G(x)}$; using induction hypothesis iii), we have,
$G(x') = SYN(J_{F_y(x')})(z) = SYN(J_{F_y(x)})(z) + SYN(J'_u)(z) = G(x) + SYN(J'_u)(z)$ and we conclude that $G(E_{xx'}) = E_{G(x)G(x')}$.

- $D(z,x)$ is not constant in x : $D(\Omega,x) = SYN(J_{F_y(x)})(\Omega) = F_y(x)$ by induction hypothesis ii) and since F_y is a non constant flower, we have the result.

Now, we have proved that $D(.,.)$ is an Ω-bilator from $ON \leqslant \Omega \times ON < \Omega$ to ON and, by theorem (G) 2 - 1 - 15 we can extend it to an Ω-bilator and we set $SYN(J_y) = UN(D)$; by I - 2 - 5 $UN(D) \in DIL\ \Omega$, and by (G) 3 - 4 - 15
$SYN(J_y)(\Omega) = UN(D)(\Omega) = D(\Omega,\Omega) = \sup_{x<\Omega} D(\Omega,x) = \sup_{x<\Omega} D_x(\Omega) = \sup_{x<\Omega} SYN(J_{F_y(x)})(\Omega)$
$= \sup_{x<\Omega} F_y(x) = y$ (we have used induction hypothesis ii) and the fact that in a garden $y = \sup_{x<\Omega} F_y(x)$).

- Now we prove the first equality of iii) : let $J_y = J_x + J'_u$ by (B) - 1 - 4 - 1 $F_y^J(t) = x + F_u^{J'}(t)$, for each $t < \Omega$, and $J_{F_y(t)} = J_x + J'_{F_u(t)}$; then if we set $D_t = SYN(J_{F_y(t)})$, $H_t = SYN(J'_{F_u(t)})$ and $G = SYN(J_x)$, by induction hypothesis iii) we have $D_t = G + H_t$; if we define as above the Ω-bilators D and H,

we have, for all ordinals $t,v,D(t,v) = G(v) + H(t,v)$ and, as we can see easily using the definition of UN, $UN(D) = G + UN(H)$, that is :
$SYN(J_y) = SYN(J_x) + SYN(J'_u)$.

DEFINITION OF SYN(f) .-

- let $f \in C(J'_x, J_y)$ such that $\hat{f}(x) = y$: define $f^\alpha \in C(J'_{F'_x(\alpha)}, J_{F_y(\alpha)})$ by $f^\alpha(u) = f(u)$, for each $u < F'_x(\alpha)$, (remark that this definition is correct because $f(F'_x(\alpha)) \geqslant F_{\widehat{f(x)}}(\alpha) = F_y(\alpha)$ by I - 1 - 8 and using (B) - 1 - 3 - 8) then $SYN(f^\alpha) \in I\Omega(H_\alpha, D_\alpha)$ if we set $H_\alpha = SYN(J'_{F'_x(\alpha)})$, $D_\alpha = SYN(J_{F_y(\alpha)})$; let, for $z < \Omega$, $T(z,\alpha) = SYN(f^\alpha)(z)$; we verify that T is a natural transformation from H to D : let $g \in I(\alpha,\beta)$, $h \in I(z,z')$, then $T(z',\beta) \circ H(h,g) =$
$SYN(f^\beta)(z') \circ SYN(F_x(g))(h) = SYN(f^\beta)(z') \circ SYN(F_x(g))(z') \circ SYN(J_{F_x(\alpha)})(h)$
$= SYN(f^\beta \circ F_x(g))(z') \circ SYN(J_{F_x(\alpha)})(h) = SYN(J_{F_y(\beta)})(h) \circ SYN(f^\beta \circ F_x(g))(z)$;
in a garden, $f^\beta \circ F_x(g) = F_y(g) \circ f^\alpha$, then $T(z',\beta) \circ H(h,g) =$
$SYN(J_{F_y(\beta)})(h) \circ SYN(F_y(g))(z) \circ SYN(f^\alpha)(z) = SYN(F_y(g))(h) \circ SYN(f^\alpha)(z) =$
$D(h,g) \circ T(z,\alpha)$. We extend T by direct limits. We set $SYN(f) = UN(T)$, then $SYN(f) \in I\Omega(SYN(J'_x), SYN(J_y))$, and $SYN(f)(\Omega) = UN(T)(\Omega) = T(\Omega,\Omega) =$
$\varinjlim T(\Omega,\alpha) = \varinjlim SYN(f^\alpha)(\Omega) = \varinjlim f^\alpha = f$: we use the fact that $(\Omega, E_{\alpha\Omega}) = \varinjlim (\alpha, E_{\alpha\beta})_{\alpha \leqslant \beta < \Omega}$ and theorem (G) 3 - 4 - 15.

If $f = g + g'$ with $g \in C(J'_u, J_v)$, $g' \in C(J'''_a, J''_b)$ (with $J'_x = J'_u + J'''_a$, $J_y = J_v + J''_b$) we set $g'^\alpha \in C(F'''_a(\alpha), F''_b(\alpha))$ and for each $u < F'''_a(\alpha)$ $g'^\alpha(u) = g'(u)$; then $f^\alpha = g + g'^\alpha$, and by induction hypothesis iii) $SYN(f^\alpha) = SYN(g) + SYN(g'^\alpha)$; if we set $U(z,\alpha) = SYN(g'^\alpha)(z)$, then clearly $T = SYN(g) + U$ and $UN(T) = SYN(g) + UN(U)$; hence $SYN(f) = SYN(g) + SYN(g')$.

- if $\hat{f}(x) < y$: the proof for this case is exactly the same as in the case where $y \in L_\omega$: see III - 1 - 3.

The fact that SYN is a functor is a trivial consequence of ii) that we have verified above. We have also verified iii) ; v) and vi) are verified by construction ; i) and iv) are trivial.

III - 2 Consequences of theorem III - 1

III - 2 - 1 LEMMA .- If J_y is a perfect garden then $SYN(J_y)$ is a perfect Ω-dilator.

Proof .- If $J_y = J_1$, the result is obvious ; if J_y is perfect $\neq J_1$ it is an easy consequence of theorem III - 1 v) and of remark (G) 3 - 4 - 12 :
$D(\omega,0) = D_0(\omega) = SYN(J_{F_y(0)})(\omega) = SYN(J_0)(\omega) = 0$; then D is a perfect Ω-bilator and then $SYN(J_y) = UN(D)$ is perfect.

III - 2 - 2 LEMMA .- If for each $i < x$ $(x < \Omega)$ J^i is a perfect garden, then :
$SYN(\sum_{i<x} J^i) = \sum_{i<x} SYN(J^i)$.

Proof .- We use an induction on x ; if $x = 1$, the result is trivial ; if $x = u+1$ the result is an immediate consequence of III - 1 iii) ; if x is limit, let, for each $i < x$, h_i the height of J^i, and set
$G^j = \sum_{i<j} J^i$, $k_j = \sum_{i<j} h_j$ (for $j < x$), $k = \sum_{i<x} h_j$; clearly, $(k_j)_{j<x}$ is the maximum sequence of k (see (B) - I - 5 - 1) then if we set $J = \sum_{i<x} J^i$, by definition of $SYN(J)$ since $k \in L_\omega$, $SYN(J) = \bigcup_{i<x} SYN(G^i)$; by induction hypothesis
$SYN(G^j) = \sum_{i<j} SYN(J^i)$ then $SYN(J) = \sum_{i<x} SYN(J^i)$.

IV - Isomorphism between GAR and DIL Ω

IV - 1 THEOREM .-

i) $DEC \circ SYN = ID_{GAR}$

ii) $SYN \circ DEC = ID_{DIL}$

Proof .-

i) We show that for each garden J_y of height y, $DEC(SYN(J_y)) = J_y$ using an induction on y :

IV - 1 - 1 case $y = 0$.- $DEC(SYN(J_0)) = DEC(\underline{0}) = J_0$

IV - 1 - 2 case $y = z + 1$.- $DEC(SYN(J_y)) = DEC(SYN(J_z + J_1) = DEC(SYN(J_z)+SYN(J_1))$
$= DEC(SYN(J_z)) + DEC(SYN(J_1)) = J_z + DEC(\underline{1}) = J_z + J_1 = J_y$: we have used theorem II - 1 i) and iii), theorem III - 1 i) and iii) and induction hypothesis.

IV - 1 - 3 case $y \in L_\omega$.- Let $J'_x = DEC(SYN(J_y))$ then by II - 2 and III - 2, $x = SYN(J_y)(\Omega) = y$; let $z <^J_o y$, then $J_y = J_z + J''_u$ with $y = z + u$; then by theorems II - 1 iii) and III - 1 iii) and using induction hypothesis :
$J'_y = DEC(SYN(J_z + J''_u)) = DEC(SYN(J_z) + SYN(J''_u)) = DEC(SYN(J_z)) + DEC(SYN(J''_u))$
$= J_z + DEC(SYN(J''_u))$; so, we have proved that $z <^{J'}_o y$ and that, for each

$z <_o^J y$, J'_z coincide with J_z ; then, as by definition of a garden $\{z ; z <_o y\}$ is cofinal in y, we conclude that $J'_y = J_y$, that is, $DEC(SYN(J_y)) = J_y$.

IV - 1 - 4 case $y \in L_\Omega$

- <u>first subcase</u> : suppose that $F_y(0) = 0$, that is J_y is perfect ; then $H = SYN(J_y)$ is a perfect Ω-dilator ; as in IV - 1 - 3 we show that the height of $DEC(SYN(J_y))$ is y ; let $J'_y = DEC(SYN(J_y))$, and $D_u = SYN(J_{F_y(u)})$, for $u < \Omega$;

as in III - 1 - 4 we define the Ω-bilator D and by definition of H, $H = UN(D)$, then by theorem (G) 3 - 4 - 11 $SEP(H) = D$ and for each $u < \Omega$, $SEP(H)^u = D_u$; then $J'_{<y} = \bigcup_{u<\Omega} DEC(SEP(H)^u) = \bigcup_{u<\Omega} DEC(D_u) = \bigcup_{u<\Omega} DEC(SYN(J_{F_y(u)}) = \bigcup_{u<\Omega} J_{F_y(u)} = J_{<y}$,

using induction hypothesis ; now, we show that $F_y^{J'} = F_y^J$: $F_y^{J'}(u) = SEP(H)(\Omega,u)$ $= D(\Omega,u) = D_u(\Omega) = SYN(J_{F_y(u)})(\Omega) = F_y^J(u)$, for each $u < \Omega$; so, we have showed

that $J'_y = J_y$.

- <u>secund subcase</u> : if J_y is not perfect, we set $J_y = J_{F_y(0)} + J'_u$ and J'_u is

perfect ; then the result using theorems II - 1 iii), III - 1 iii), induction hypothesis and the first subcase.

Now we show that, for each $f \in C(J'_x, J_y)$, $DEC(SYN(f)) = f$: it is an easy consequence of II - 1 ii) and of III - 1 ii) : $DEC(SYN(f)) = SYN(f)(\Omega) = f$.

ii) we show that for each Ω-dilator G, $SYN(DEC(G)) = G$: we proceed as in the proof of II - 1 : we define $DIL\ \Omega < H$, $DIL\ \Omega \leqslant H$, we suppose that the property is true in $DIL\ \Omega < H$, and we prove it in $DIL\ \Omega \leqslant H$.

IV - 1 - 5 case $H = \underline{0}$.- $SYN(DEC(\underline{0})) = SYN(J_0) = \underline{0}$

IV - 1 - 6 case $H = H' + \underline{1}$.- $SYN(DEC(\underline{1})) = SYN(J_1) = \underline{1}$
If $G \in DIL\ \Omega \leqslant H$ and $h(G) = H$, then $G = G' + \underline{1}$, and $SYN(DEC(G)) =$ $SYN(DEC(G') + J_1) = SYN(DEC(G')) + \underline{1} = G' + \underline{1} = G$.

IV - 1 - 7 case where H is of kind ω .- Let $G \in DIL\ \Omega \leqslant H$ such that $h(G) = H$; then, as we have shown in II - 1 - 5 $G = \sum_{i<x} G_i$ with x limit is the decomposition in sum of G ; by II - 1 iv) $SYN(DEC(G)) = SYN\left(\sum_{i<x} DEC(G_i)\right)$; by I - 2 - 3, $x < \Omega$; by II - 2 - 1, for each $i < x$, $DEC(G_i)$ is perfect ; then, we can use lemma III - 2 - 2 and $SYN(DEC(G)) = \sum_{i<x} SYN(DEC(G_i)) = \sum_{i<x} G_i = G$, using induction

hypothesis since, for each $i < x$, $G_i \in DIL\,\Omega < H$.

IV - 1 - 8 case H of kind Ω

- first subcase : let H perfect Ω-dilator ; as we have shown in II - 1 - 6, if $G \in DIL\,\Omega \leqslant H$, and if $h(G) = H$, then G is perfect $\neq \underline{1}$; by definition, $SYN(DEC(G)) = UN(D)$, D being the Ω-bilator defined, for each $u < \Omega$, and each $z \leqslant \Omega$, by $D(z,u) = D_u(z) = SYN(DEC(SEP(G)^u))(z)$; by induction hypothesis, since $SEP(G)^u \in DIL\,\Omega < H$ $SYN(DEC(SEP(G)^u)) = SEP(G)^u$; then for all $u,z < \Omega$, $D(z,u) = SEP(G)(z,u)$ then $D = SEP(G)$ and $SYN(DEC(G)) = UN(D) = UN(SEP(G)) = G$ by theorem (G) 3 - 4 - 11.

- secund subcase : let $H = H' + H''$ with H'' perfect ; then, as we have proved in II - 1 - 3, if $G \in DIL\,\Omega \leqslant H$, and if $h(G) = H$, then $G = G' + G''$ with $G' \in DIL\,\Omega \leqslant H'$, and $G'' \in DIL\,\Omega \leqslant H''$; then using induction hypothesis and the first subcase : $SYN(DEC(G)) = SYN(DEC(G') + DEC(G'')) = SYN(DEC(G')) + SYN(DEC(G'')) = G' + G'' = G$ (we have also used II - 1 - iii) and III - 1 - iii)).

Now we show that, for each $T \in I\Omega(G,H)$, $SYN(DEC(T)) = T$: let $U = SYN(DEC(T))$; then, using II - 1 - ii) and III - 1 - ii), $U(\Omega) = DEC(T) = T(\Omega)$; then $U = T$.

IV - 2 - 1 COROLLARY .- The functors DEC and SYN commute to direct limits and pull-backs.

Proof .- It is an immediate consequence of the fact that they are isomorphisms.

V - The functors Λ, Λ', Λ''. Comparaison between Λ'' and φ

V - 1 - Preliminaries

V - 1 - 1 DEFINITION .-

i) A flower F is regular iff :

$$\forall x,y,z \in ON \ \forall f \in I(x,y)(z < x \longrightarrow F(z) < F(x) \text{ and } F(f)(F(z)) = F(f(z))).$$

ii) If F,G are regular flowers, if $T \in I(F,G)$, then T is regular iff :

$$\forall z,x \in ON (z < x \longrightarrow T(x)(F(z)) = G(z)).$$

iii) FL_r is a category, defined by :

objects : regular flowers.

morphisms from F to G : the set $I^r(F,G)$ of regular morphisms from F to G.

V - 1 - 2 REMARKS .-

i) If B is a regular bilator, (see definition (G) 5 - 3 - 4), then, for each ordinal x, the functor B_x defined by : $B_x(y) = B(x,y)$, $B_x(f) = B(E_x,f)$, is a regular flower ; if T is a regular morphism between the regular bilators B and D, and if we define B_x and D_x as above, then if we set $T_x(y) = T(x,y)$, T_x is a regular morphism between B_x and D_x.

ii) If F is a regular flower, its restriction to $ON < \Omega$, is a regular Ω-flower : see (B) I - 1 - 1.

V - 1 - 3 DEFINITION .- The following date define a category GAR_r :

objects : regular gardens of type Ω : see definition in (B) I - 1 - 6.

morphisms from J_x to J'_y : the set $C_r(J_x,J'_y)$ of regular compatible functions $f \in I(x,y)$ from J_x to J'_y : see definition in (B) I - 3 - 3.

V - 1 - 4 LEMMA .-

i) Let J_x a perfect garden $\neq J_1$ of height x and set for each $y < \Omega$, $J_{F_x(y+1)} = J_{F_x(y)} + {}_yJ'$; let H a perfect Ω-dilator $\neq \underline{1}$ and set for each $y < \Omega$, $SEP(H)^{y+1} = SEP(H)^y + {}_yH$; then if $H = SYN(J_x)$ (or equivalently $J_x = DEC(H)$) for each $y < \Omega$, ${}_yH = SYN({}_yJ')$ (or equivalently ${}_yJ' = DEC({}_yH)$).

ii) Let J_x a perfect garden $\neq J_1$ and define as in i), for each $y < \Omega$, ${}_yJ'$; for each function $f \in I(y,y')$ define ${}^x_fF$ by : $F_x(f+E_1) = F_x(f) + {}^x_fF$; then ${}^x_fF$ is a compatible function from ${}_yJ'$ to ${}_{y'}J'$: and if $H = SYN(J_x)$ and if we set, as in (G) 3 - 6 - 6 $SEP(H)(.,f+E_1) = SEP(H)(.,f) + {}_fH$, then $SYN({}^x_fF) = {}_fH$.

iii) Let G_z be another perfect garden $\neq J_1$ and define as in i), for each $y < \Omega$, ${}_yG'$; for each $h \in C(G_z, J_x)$ such that $\hat{h}(z) = x$, define, for each $y < \Omega$, ${}_yh$ by : let $h^y \in C(G_{F'_z(y)}, J_{F_x(y)})$, ($F'_z$ and F_x being the flowers associated respectively to z in G_z and to x in J_x) defined for each $u < F'_z(y)$ by $h^y(u) = h(u)$, then $h^{y+1} = h^y + {}_yh$; then, ${}_yh \in C({}_yG', {}_yJ')$ and if $T = SYN(h)$ and $SEP(T)(.,y+1) = SEP(T)(.,y) + {}_yT$ then $SYN({}_yh) = {}_yT$.

Proof .-

i) First remark that, since $F_x(y) <^J_o F_x(y+1)$ (see (B) 1 - 3 - 2 : here since the gardens are not necessarily regular, we have $<^J_o$ instead of $<^J_*$) we can set : $J_{F_x(y+1)} = J_{F_x(y)} + {}_yJ'$ (see (B) I - 4 - 2) ;

if $H = SYN(J_x)$ and $D_y = SYN(J_{F_x(y)})$ for each $y < \Omega$, if D is

the Ω-bilator defined as in IV - 1 - 3 by, for each ordinal z, $D(z,y) = D_y(z)$, then as we have shown in IV - 1 - 4, for each $y < \Omega$, $SEP(H)^y = D_y$; then $D_{y+1} = SYN(J_{F_x(y+1)}) = SYN(J_{F_x(y)}) + SYN({}_yJ')$

$= SEP(H)^{y+1} = SEP(H)^y + SYN({}_yJ') = SEP(H)^y + {}_yH$ (we have used

III - 1 - 3) ; hence the result.

ii) Remark that we can set $F_x(f+E_1) = F_x(f) + {}^x_fF$ because F_x is a flower ; by (B) I - 4 - 3 ${}^x_fF$ is a compatible function from ${}_yJ'$

to ${}_{y'}J'$; $D(.,f+E_1) = SYN(F_x(f+E_1)) = SYN(F_x(f)) + SYN({}^x_fF)$

$= SEP(H)(.,f+E_1) = SEP(H)(.,f) + {}_fH = SYN(F_x(f)) + {}_fH$; hence the result.

iii) By (B) I - 4 - 3 ${}_yh$ is a compatible function from ${}_yG'$ to ${}_yJ'$; if $T = SYN(f)$, then by definition of $SYN(f)$ (see III - 1 - 4) we have $SYN(h^y) = SEP(T)(.,y)$ and then : $SYN(h^{y+1}) = SYN(h^y) + SYN({}_yh)$

$= SEP(T)(.,y+1) = SEP(T)(.,y) + {}_yT = SYN(h^y) + {}_yT$; hence the result.

V - 2 The functor Λ'

V - 2 - 1 DEFINITION .- Let Λ be the functor defined by Girard in (G) 5 - 4 - 1 :

i) For each garden J, we set $\Lambda' J = \Lambda(SYN(J))$

ii) For each $f \in C(J,J')$, we set $\Lambda' f = \Lambda(SYN(f))$

V - 2 - 2 THEOREM .- Λ' that we have defined in V - 2 - 1 is a functor from GAR to BIL_r.

Proof .- It is an immediate consequence of theorem (G) 5 - 4 - 3 and of the fact that SYN is a functor from GAR to the subcategory DIL Ω of DIL.

V - 2 - 3 THEOREM .- Λ' is the unique functor from GAR to BIL_r defined by :

i) $\Lambda' J_0 = Id$ $\qquad\qquad$ $\Lambda' E_0 = E_{Id}$

ii) $\Lambda' J_1 = Id + Id$ $\qquad\qquad$ $\Lambda' E_1 = E_{Id + Id}$

iii) if $J = J' + J''$ if $f = h + g$

$\Lambda' J = (\Lambda' J') \circ_s (\Lambda' J'')$ $\Lambda' f = (\Lambda' h) \circ_s (\Lambda' g)$

iv) if $x \in L_\omega$, if $(x_k)_{k \in K}$ is the maximum sequence for x in J_x and if $f \in C(G_z, J_x)$ with $\hat{f}(z) = x$, let $(z_l)_{l \in L}$ the maximum sequence for z in G_z and set for each $l \in L$, $f^l \in C(G_{z_l}, J_{f(z_l)})$ defined by, for $u < z_l$, $f^l(u) = f(u)$: $\Lambda' J_x = \underset{\overrightarrow{K}}{\lim}^* (\Lambda' J_{x_k}, \Lambda' E_{x_k, x_l})$

$\Lambda' f = \underset{\overrightarrow{L}}{\lim} (\Lambda' f^l)$

v) if $x \in L_\Omega$, if J_x is a perfect garden, if $f \in I(v, v')$, $g \in I(z, z')$ then write $J_{F_x(z)} = \sum_{t<z} {}_tJ'$, $F_x(g) = \sum_{t<z} {}^x_{gt}F$ with ${}^x_{gt}F \in C({}_tJ', {}_{g(t)}J')$ (with the notations of V-1-4, ${}^x_{gt}F$ is exactly ${}^x_{g'}F$, with $g' \in I(t, g(t))$ defined by $g'(u) = g(u)$ for all $u < t$) ; then :

$$(\Lambda' J_x)(v, z) = \left(\prod_{t<z} (\underline{1} + \Lambda'_t J')\right)(v, 0)$$

$$(\Lambda' J_x)(f, g) = \left(\prod_{t<g} (E_{\underline{1}} + \Lambda' \, {}^x_{gt}F)\right)(f, E_0)$$

If $y \in L_\Omega$, if G_y is perfect too, if $h \in C(G_y, J_x)$, $\hat{h}(y) = x$, then set, for each ordinal $z < \Omega$, $h^z = \sum_{t<z} {}_th$ (with notations of V-1-4 iii)) ; then,

$$(\Lambda' h)(v, z) = \left(\prod_{t<E_z} (E_{\underline{1}} + \Lambda' \, {}_th)\right)(v, 0)$$

vi) Let $E_{0a} \in C_r(J_0, J''_a)$, then :

$$(\Lambda' E_{0a})(t, y)(z) = (\Lambda' J''_a)(t, z)$$

<u>Proof</u> .- We use an induction on x, the height of the garden ; the fact that the solution is unique is an immediate consequence, because, at each step Λ' is completely defined by means of some of the formulas of the above theorem.
i) and ii) are immediate using (G) 5-4-3 (ii) and (iii) and the definition of SYN(see III-1 i)).

(a) <u>case $x = z + 1$</u> ; then $J_x = J_z + J_1$ and $f = h + E_{01}$ or $f = h + E_1$; in fact we consider the more general case where $J = J' + J''$, $f = h + g$:

$\Lambda'(J'+J'') = \Lambda(SYN(J'+J'')) = \Lambda(SYN(J') + SYN(J'')) = \Lambda(SYN(J')) \circ_s \Lambda(SYN(J''))$
$= (\Lambda'J') \circ_s (\Lambda'J'')$ using theorem III - 1 iii) and theorem (G) 5 - 4 - 3 v).

If $f = h+g$, $\Lambda'(h+g) = \Lambda(SYN(h+g)) = \Lambda(SYN(h) + SYN(g)) = \Lambda(SYN(h)) \circ_s \Lambda(SYN(g))$
$= \Lambda'h \circ_s \Lambda'g$ using theorem III - 1 iii) and theorem (G) 5 - 4 - 3 v).

If $g = E_{0a}$, we prove vi) : if $J_x = J_u + J''_a$, $(\Lambda' E_{0a})(t,y)(z) =$
$(\Lambda(SYN(E_{0a})))(t,y)(z) = (\Lambda E_{\underline{0}\,SYN(J''_a)})(t,y)(z) = (\Lambda(SYN(J''_a)))(t,z) = (\Lambda'J''_a)(t,z)$,
using theorem III - 1 vi) and theorem (G) 5 - 4 - 3 (vi).

(b) <u>case $x \in L_\omega$</u> ; let $(x_k)_{k\in K}$ the maximum sequence for x in J_x ; by III - 1 - 4 iv) and using remark I - 3 - 3 i), we have :

$SYN(J_x) = \bigcup_{k\in K} SYN(J_{x_k}) = \underset{\overrightarrow{K}}{\lim}^* (SYN(J_{x_k}), E_{SYN(J_{x_k})\, SYN(J_{x_1})})$

and $\Lambda'J_x = \Lambda(SYN(J_x)) = \underset{\overrightarrow{K}}{\lim}^* (\Lambda(SYN(J_{x_k}), \Lambda(E_{SYN(J_{x_k})\, SYN(J_{x_1})}))$ using (G) 5 - 4 - 3 vii) ; as we can see using III - 1 vi) and iii),

$E_{SYN(J_{x_k})\, SYN(J_{x_1})} = SYN(E_{x_k x_1})$ and then,

$$\Lambda'J_x = \underset{\overrightarrow{K}}{\lim}^* (\Lambda'J_{x_k}, \Lambda'E_{x_k x_1}).$$

If $f \in C(G_z, J_x)$, with hypothesis of iii), we verify easily using (G) 5 - 4 - 3 vii) that $\Lambda'f = \underset{\overrightarrow{L}}{\lim}^* (\Lambda' f^1)$.

If $f \in C(G_z, J_x)$ and if $\hat{f}(z) < x$, we prove vi) as in a).

(c) <u>case $x \in L_\Omega$</u> ; we prove iii) as in a), then it is enough to treat the case J_x perfect ; with notations of v), if we set $H = SYN(J_x)$ and using notations and results of lemma V - 1 - 4, we have :

$$(\Lambda'J_x)\ (v,z) = (\Lambda H)\ (v,z) = (\prod_{t<z} (\underline{1} + \Lambda_t H))(v,0) = (\prod_{t<z} (\underline{1} + \Lambda'_t J'))(v,0)$$

$$(\Lambda'J_x)\ (f,g) = (\Lambda H)\ (f,g) = (\prod_{t<g} (E_{\underline{1}} + \Lambda_{gt} H))(f,E_0) = (\prod_{t<g} (E_{\underline{1}} + \Lambda'^{x}_{gt} F))(f,E_0)$$

If $h \in C(G_y, J_x)$ and $\hat{h}(y) = x$, using notations and results of V - 1 - 4,

$$(\Lambda'h)\ (v,z) = (\Lambda T)\ (v,z) = (\prod_{t<E_z} (E_{\underline{1}} + \Lambda_t T))(v,0) = (\prod_{t<E_z} (E_{\underline{1}} + \Lambda'_t h))(v,0)$$

(we have used (G) 5 - 4 - 3 iv)).

We prove vi) as in a) ; the others assertions of theorem are trivial.

V - 3 The functor Λ''

V - 3 - 1 DEFINITION .- We define the following functor from GAR to FL_r

i) for each garden J, $(\Lambda''J)(z) = (\Lambda'J)(\omega,z)$, $(\Lambda''J)(f) = (\Lambda'J)(E_\omega,f)$

ii) for each $f\in C(J,J')$, $\Lambda''f = (\Lambda'f)(\omega,.)$

V - 3 - 2 THEOREM .-

i) for each regular garden J_x, $(\Lambda''J_x)\upharpoonright\Omega = \varphi_x^J$

ii) if $f\in C_r(G_y,J_x)$, $(\Lambda''f)(z) = T(f,z)$, for each $z < \Omega$, φ_x^J and $T(f)$ being respectively the regular flower and the natural transformation defined in (B) - II.

Proof .- We use an induction on x, the height of the garden ; for definition and properties of φ_x^J and $T(f)$, see (B) - II'.

i) Since for all ordinals y,z and for all $f\in I(y,y')$, $g\in I(z,z')$ we have, $(\Lambda'J_0)(y,z) = z$, $(\Lambda'J_0)(f,g) = g$, then $(\Lambda''J_0)\upharpoonright\Omega = \varphi_0^J$ and $(\Lambda''E_0)(u) = T(E_0,u)$ for each $u < \Omega$.

ii) For all ordinals y,z and for all $f\in I(y,y')$, $g\in I(z,z')$, we have : $(\Lambda'J_1)(y,z) = y+z$, $(\Lambda'J_1)(f,g) = f+g$ (see (G) 3 - 4 - 16 ii)) ; then, $(\Lambda''J_1)\upharpoonright\Omega = \varphi_1^J$ and $(\Lambda''E_1)(u) = T(E_1,u)$ for each $u < \Omega$. We verify easily that for each $u < \Omega$, $(\Lambda''E_{01})(u) = T(E_{01},u)$.

iii) $x = z+1$; we consider the more general case where $J = J'+J''$, (we suppose that by induction hypothesis the result holds for J' and J'') $(\Lambda''(J'+J''))(y) = (\Lambda'(J'+J''))(\omega,y) = (\Lambda'J' \circ_s \Lambda'J'')(\omega,y) = (\Lambda'J'(\omega,(\Lambda'J''(\omega,y)))) = (\Lambda''J')(\Lambda''J''(y)) = (\varphi^{J'} \circ \varphi^{J''})(y)$ for $y < \Omega$. using induction hypothesis ; if $f\in I(y,y')$ and $y,y' < \Omega$: $(\Lambda''(J'+J''))(f) = (\Lambda'(J'+J''))(E_\omega,f) = (\Lambda'J' \circ_s \Lambda'J'')(E_\omega,f) = (\Lambda'J'(E_\omega,(\Lambda'J''(E_\omega,f)))) = (\Lambda''J')(\Lambda''J''(f)) = (\varphi^{J'} \circ \varphi^{J''})(f)$; hence we have proved that $(\Lambda''J)\upharpoonright\Omega = \varphi^J$; let $G = G'+G''$, $f = h+g$ with $h\in C_r(G',J')$, $g\in C_r(G'',J'')$ and $u < \Omega$, then $(\Lambda''(h+g))(u)$ $= (\Lambda'(h+g))(\omega,u) = (\Lambda'h \circ_s \Lambda'g)(\omega,u) = (\Lambda'h(\omega,(\Lambda'g(\omega,u)))) =$ $(\Lambda''h(\Lambda''g(u))) = (\Lambda''h \circ \Lambda''g)(u) = (T(h)\circ T(g))(u) = T(h+g)(u)$, using induction hypothesis and theorem (B) - II' - 8.
Let $E_{0a}\in C_r(J_0,J'_a)$; then if $u < \Omega$, $(\Lambda''E_{0a})(u)(t) = (\Lambda'E_{0a})(\omega,u)(t)$ $= (\Lambda'J'_a)(\omega,t) = (\Lambda''J'_a)(t)$; then if $a \leqslant x$, using induction hypothesis we have $(\Lambda''J'_a)(t) = \varphi_a^{J'}(t) = \varphi_a^{J',u}(t) = T(E_{0a},u)(t)$ using theorem (B) - II - vi) (for the notation $\varphi_a^{J',u}$ see (B) - II - 2 - 7 i)).

iv) $x \in L_\omega$; let $(x_k)_{k \in K}$ the maximum sequence for x in J_x ; then :

$$(\Lambda''J_x)(z) = (\Lambda'J_x)(\omega,z) = (\varinjlim_K{}^* (\Lambda'J_{x_k}, \Lambda'E_{x_k x_1}))(\omega,z) =$$

$$\varinjlim_K{}^* ((\Lambda'J_{x_k})(\omega,z), (\Lambda'E_{x_k x_1})(\omega,z)) =$$

$$\varinjlim_K{}^* ((\Lambda''J_{x_k})(z), (\Lambda''E_{x_k x_1})(z)) = \varinjlim_K{}^* (\varphi^J_{x_k}(z), T(E_{x_k x_1},z)) = \varphi^J_x(z)$$

for each $z < \Omega$, using induction hypothesis.

If $f \in I(y,y')$, $(\Lambda''J_x)(f) = \varinjlim_K ((\Lambda''J_{x_k})(f)) = \varinjlim_K (\varphi^J_{x_k}(f)) = \varphi^J_x(f)$,

using induction hypothesis ; hence we have proved that : $\Lambda''J_x = \varphi^J_x$.

Let $f \in C_r(G_y,J_x)$, $\hat{f}(y) = x$, and if $(y_1)_{1 \in L}$ is the maximum sequence for y in G_y we set $f^1 \in C_r(G_{y_1},J_{f(y_1)})$ defined by,

for $u < y_1$, $f^1(u) = f(u)$; then $(\Lambda''f)(z) = (\Lambda'f)(\omega,z)$

$$= (\varinjlim_L (\Lambda'f^1))(\omega,z) = \varinjlim_L ((\Lambda'f^1)(\omega,z)) = \varinjlim_L ((\Lambda''f^1)(z))$$

$$= \varinjlim_L (T(f^1,z)) = T(f,z), \text{ for each } z < \Omega.$$

Let $f \in C_r(G_y,J_x)$, $\hat{f}(y) = u < x$, $J_x = J_u + J'_a$; then we prove as in iii) that, for each $z < \Omega$, $(\Lambda''f)(z) = T(f,z)$.

v) It is enough to treat the case J_x perfect : because if J_x is not perfect, then write $J_x = J' + J''$ with J'' perfect and as in iii), we have the result if the property holds for J' and J''. Now we suppose that J_x is perfect : since J_x is regular, then, for each $y < \Omega$, $F_x(y) <^J_* F_x(y+1)$ (see (B) I - 3 - 2) and if we set as in V - 1 - 4 $J_{F_x(y+1)} = J_{F_x(y)} + {}_yJ'$, then obviously we have the decomposition of ${}_yJ'$ in ${}_yJ' = J_1 + {}_yD$, for some garden ${}_yD$, (because $F_x(y) + 1 \leqslant^J_o F_x(y+1)$) ; if $f \in I(y,y')$, with hypothesis and notations of V - 1 - 4 ii), we have $F_x(f+E_1) = F_x(f) + {}^x_fF$ and since J_x is regular we have the following decomposition of ${}^x_fF$: ${}^x_fF = E_1 + {}^x_fF'$ for some compatible function ${}^x_fF'$ from ${}_yD$ to ${}_{y'}D$ (it is a consequence of (B) I - 4 - 3 and because $F_x(f+E_1)(F_x(y)) = F_x(y'))$; if $h \in C_r(G_z,J_x)$, with notations and hypothesis of V - 1 - 4 iii), then since G_z, J_x, h are regular, we have a decomposition of ${}_yh$ in : ${}_yh = E_1 + {}_yh'$ for some compatible function ${}_yh'$ (it is a consequence of (B) I - 4 - 3 and because $h^{y+1}(F^G_z(y)) = F^J_x(y)$, F^G_z and F^J_x, being the regular flowers associated respectively to z in G_z and to x in J_x). Then with the above notations :

$(\underline{1}+\Lambda'_y J')(\omega,u) = (\underline{1}+\Lambda'(J_1+{}_yD))(\omega,u) = 1+((\Lambda' J_1)\ o_s\ (\Lambda'_y D))(\omega,u)$

$= 1+\Lambda' J_1(\omega,(\Lambda'_y D(\omega,u))) = 1+\omega+(\Lambda'_y D)(\omega,u) = \omega+(\Lambda'_y D)(\omega,u)$

$= (\Lambda'_y J')(\omega,u) = (\Lambda''_y J')(u)$; in a similar way we show that :

$(E_{\underline{1}}+\Lambda'\ {}^x_f F)(E_\omega,g) = (\Lambda''\ {}^x_f F)(g)$ and $(E_{\underline{1}}+\Lambda'\ {}_y h)(\omega,u) = (\Lambda''_y h)(u)$;

then we verify easily by induction on y that :

$$(\prod_y(\underline{1}+\Lambda'_t J'))(\omega,u) = (\Lambda'' J_{F_x(y)})(u)$$

$$(\prod_f(E_{\underline{1}}+\Lambda'\ {}^x_{ft}F))(E_\omega,g) = (\Lambda'' F_x(f))(g)$$

$$(\prod_{E_y}(E_{\underline{1}}+\Lambda'_t h))(\omega,u) = (\Lambda'' h^y)(u)$$

and hence $(\Lambda'' J_x)(z) = (\Lambda'' J_{F_x(z)})(0) = \varphi^J_{F_x(z)}(0) = \varphi^J_x(z)$ for each $z<\Omega$, using induction hypothesis and definition of φ^J_x ; if $f\in I(z,z')$, $(\Lambda'' J_x)(f) = (\Lambda'' F_x(f))(E_0) = (\Lambda'' F_x(f))(0) = T(F_x(f),0) = \varphi^J_x(f)$ (with $z, z'<\Omega$) using induction hypothesis and definition of φ^J_x ; let G_y be another regular garden and $h\in C_r(G_y,J_x)$ with $\hat{h}(y)=x$, then $(\Lambda'' h)(z) = (\Lambda'' h^z)(0) = T(h^z,0) = T(h,z)$ for $z<\Omega$, using induction hypothesis and definition of $T(h)$; if $h\in C_r(G_y,J_x)$ and $\hat{h}(y)=u<x$, $J_x = J_u + J'_a$, then we prove as in iii) that for each $z<\Omega$, $(\Lambda'' h)(z) = T(h,z)$.

V - 3 - 3 COROLLARY .- $(\Lambda'' G_{\epsilon_{\Omega+1}})(0)$ is the usual Howard ordinal ; ($G_{\epsilon_{\Omega+1}}$ is defined in (B) - I - 2.)

Proof .- It is an immediate consequence of corollary (B) - 2 - 5 - 1 and of the above theorem.

PROCEEDINGS OF THE HERBRAND SYMPOSIUM
LOGIC COLLOQUIUM '81, J. Stern (editor)

ON THE CONSISTENCY STRENGTH OF PROJECTIVE UNIFORMIZATION

W. Hugh Woodin[1]

Department of Mathematics
California Institute of Technology
Pasadena, California 91125
U.S.A.

Let $\mathbb{R}$ denote the set of real numbers. We identify $\mathbb{R}$ with ω^ω, the collection of infinite sequences of natural numbers. We assume familiarity with the fundamentals of descriptive set theory in the context of determinacy. We restrict attention to the projective sets and will assume at most Projective Determinacy ($\underset{\sim}{PD}$), thus we work in ZFC throughout this paper. By way of a reference see [8].

Recall that a subset $A \subseteq \mathbb{R} \times \mathbb{R}$ can be uniformized if there is a partial function $f: \mathbb{R} \to \mathbb{R}$ such that for all $x \in \mathbb{R}$, $f(x) \in A_x$ if $A_x \neq \emptyset$. $A_x = \{y \mid (x,y) \in A\}$.

We are interested in two consequences of Projective Determinacy ($\underset{\sim}{PD}$) for the projective sets:

1) Every projective set has the property of Baire and is Lebesque measurable.

2) Every projective subset of the plane can be uniformized by a projective function (i.e. a function whose graph is projective).

(1) and (2) are seemingly in conflict in that (1) implies a failure of the Axiom of Choice at the projective level while (2) attempts to retain some aspect of the Axiom of Choice.

For both (1) and (2) to hold it would seem that some underlying structure is present. At the same time an important test of the naturalness of the hypothesis of $\underset{\sim}{PD}$ is if it is equivalent to its analytical consequences for the projective sets.

A natural question therefore arises. Assume (1) and (2). Can one show $\underset{\sim}{PD}$?

We show the following:

<u>Theorem 1</u>. Assume (1) and (2). Then for every real x, $x^\#$ exists i.e. $\underset{\sim}{\Pi}^1_1$-Determinacy holds.

The proof yields much more ($x^\dagger$) though it seems short of showing $\underset{\sim}{\Sigma}^1_2$-Determinacy. The proof also yields simple proofs of some of the known consequences of $\underset{\sim}{PD}$.

Throughout this paper we use the following notation. A formula φ is

---

[1]Research partially supported by NSF Grant MCS 8021468

$\Sigma^1_n(\Pi^1_n)$ if it is $\Sigma^1_n(\Pi^1_n)$ is the usual sense but <u>without</u> real parameters. φ is $\utilde{\Sigma}^1_n(\utilde{\Pi}^1_n)$ if it is $\Sigma^1_n(\Pi^1_n)$ possibly with real parameters. In special cases we will say that φ is $\Sigma^1_n(x_0)(\Pi^1_n(x_0))$ by which we will mean that φ is $\utilde{\Sigma}^1_n(\utilde{\Pi}^1_n)$ with real parameter x_0. Similarly a set of reals, A, is $\Sigma^1_n(\Pi^1_n)$ if it can be defined by a $\Sigma^1_n(\Pi^1_n)$ formula etc. Further A is $\Delta^1_n(\utilde{\Delta}^1_n\ \Delta^1_n(x_0))$ if it can simultaneously be defined by a $\Sigma^1_n(\utilde{\Sigma}^1_n\ \Sigma^1_n(x_0))$ formula and a $\Pi^1_n(\utilde{\Pi}^1_n\ \Pi^1_n(x_0))$ formula.

Finally most of our remarks concerned with forcing will be of the 'two valued' form i.e. we pretend generic objects exist. This is for conceptual simplicity (we hope) and the patient reader should have no trouble restating everything in its proper form.

§1 <u>Proof of Theorem 1</u>. We obtain sharps via failure of the covering lemma (see [2]). Suppose M is a standard model of ZFC. M is Σ^1_n-correct in V if for any Σ^1_n-formula $\varphi(x_1 \ldots x_m)$ and reals $c_1 \ldots c_m$ in M, $M \models \varphi(c_1 \ldots c_m)$ if and only if $\varphi(c_1 \ldots c_m)$ is true in V. We say that M is Σ^1_n-absolute for set generic extensions (more simply, M is Σ^1_n-absolute) if for any G_1 set-generic over M and any G_2 set-generic over $M[G_1]$, $M[G_1]$ is Σ^1_n-correct in $M[G_1,G_2]$. This we regard as an internal statement of M with truth independent of whether or not the generic extensions exist in V, of course if M is countable there is no difficulty. To say that G is set-generic over M is to say simply that G is a generic filter over M on some partial order that occurs as a set in M, as opposed to being a generic filter on a partial order that is a class over M.

<u>Lemma 1</u>: Assume M is a model of ZFC that is Σ^1_3-absolute. Then $M \models \forall x \in \mathbb{R}[x^\# \text{ exists}]$.

<u>Proof:</u> Work in M. Choose a real y such that $y^\#$ does not exist, we work toward a contradiction. $\aleph_\omega$ is a cardinal κ in L[y]. By the covering lemma κ^+ in the sense of L[y] is $\aleph_\omega^+$. We shall need some mild coding. A real x codes a countable ordinal α if in some canonical way x codes a wellorder of ω of length α. For x coding a countable ordinal we denote the ordinal coded by α_x. Similarly x codes a standard model N, if x codes a relation E on ω such that $(\omega,E) \cong (N,\in)$. Provided all is done in a reasonable fashion the predicate P(x) iff 'x codes a countable

ordinal' is Π^1_1. Similarly Q(x) iff 'x codes a standard model of $ZF^- + V = L[y]$' is $\Pi^1_1(y)$ i.e. Π^1_1 with y as a parameter.

Force with the standard Levy conditions to collapse $\aleph_\omega$ to become countable and in the generic extension M[g] choose $z \in \mathbb{R}$, a code for $\aleph^M_\omega$.

Thus in M[g], $\aleph_1$ is a successor cardinal in L[y] to $\aleph^M_\omega = \alpha_z$. But this is a $\underset{\sim}{\Pi}^1_3$ statement in the parameters z, y:

$\forall x\ \exists t$ [x codes a countable ordinal $\Rightarrow$

t codes a standard model, N, of '$ZF^- + V = L[y]$',

$\alpha_x \in M$, $\alpha_z \in M$ and $N \models \bar{\bar{\alpha}}_x \leq \bar{\bar{\alpha}}_z$]

The matrix (the part in brackets) is easily seen to be $\Sigma^1_2(z,y)$. Thus the statement is $\Pi^1_3(z,y)$.

M is Σ^1_3-absolute thus the statement must remain true in any generic extension of M[g]. This is absurd since collapsing the $\aleph_1$ of M[g] makes the statement false. This completes the proof of lemma 1. ⊣

It in fact now follows by a theorem of Martin-Solovay [6] that Σ^1_3-absoluteness is equivalent to the existence of $S^\#$ every set S.

Assuming (1) and (2) we will build for each n a standard model M of ZFC, M Σ^1_n-correct and Σ^1_n-absolute.

We shall need a few more lemmas.

It will be convenient to recall now two variants of uniformization. Suppose $A \subseteq \mathbb{R} \times \mathbb{R}$. A can be **-uniformized if there is a borel function f: $\mathbb{R} \to \mathbb{R}$ such that on a comeager set $A_x \neq \emptyset \to f(x) \in A_x$. Similarly A can be μ-uniformized if there is a borel function f: $\mathbb{R} \to \mathbb{R}$ such that for almost all x, $A_x \neq \emptyset \to f(x) \in A_x$ i.e. $\mu(\mathbb{R} \setminus \{x \mid A_x \neq \emptyset \to f(x) \in A_x\}) = 0$. μ denotes Lebesque measure. For **-uniformization we can require f to be continuous on a comeager set, for μ-uniformization we can require f to be continuous off a set of measure 0.

Lemma 2. Assume (1) and (2). Suppose x_0 is a Cohen or random real over V. Then V is Σ^1_n-correct in $V[x_0]$ for each n.

Proof: We do the Cohen case. The case for random is similar.

All we actually need for this result is **-uniformization for projective sets (μ-uniformization in the random case). **-uniformization is an easy consequence of (1) and (2).

Suppose c is a Cohen real over V. We show that V is Σ^1_3-correct in V[c]. The general case follows by induction on formula complexity.

Suppose φ is $\underset{\sim}{\Sigma}^1_3$ and $V[c] \models \varphi$. $\varphi = \exists x\ \forall y\ \psi(x,y)$ where ψ is $\underset{\sim}{\Sigma}^1_1$. Choose in V[c] a real x_0 such that $V[c] \models \forall y\ \psi(x_0,y)$. Choose in V a term τ for x_0. τ may be chosen as a borel function g such that $V[c] \models \forall y\ \psi(g(c),y)$. Assume φ is false in V. Thus $V \models \neg\varphi$. $\neg\varphi = \forall x\ \exists y\ \neg\psi(x,y)$. Therefore $V \models \forall z\ \exists y\ \neg\psi(g(z),y)$. Let $A = \{(z,y) \mid \neg\psi(g(z),y)\}$. By **-uniformization choose a borel function f such that $\{z \mid (z,f(z)) \in A\}$ is comeager. Choose a comeager borel set $B \subseteq \{z \mid (z,f(z)) \in A\}$. Hence $V \models \forall z[z \in B \rightarrow \neg\psi(g(z),f(z))]$. $\neg\psi$ is $\underset{\sim}{\Pi}^1_1$ therefore $\forall z[z \in B \rightarrow \neg\psi(g(z),f(z))]$ is $\underset{\sim}{\Pi}^1_1$ in the additional parameters, borel codes for B, f, g. Thus $V[c] \models \forall z[z \in B \rightarrow \neg\psi(g(z),f(z))]$. But $c \in B$ therefore $V[c] \models \neg\psi(g(c),f(c))$ which contradicts $V[c] \models \forall y\ \psi(g(c),y)$.

This completes the proof of lemma 2. ⊣

A projective sequence of reals is a projective relation that defined a well-ordered sequence of reals.

Lemma 3. Assume (1) and (2). Then there is no uncountable projective sequence of (distinct) reals.

This lemma is a consequence of a more general fact:

Lemma 4. Assume ZFC. Suppose S is an uncountable sequence of reals and that c is a Cohen real over V. Then in V[c] there is no real r random over L(S,c).

Proof. Let λ denote the length of S. Choose in L(S) a sequence $\langle\sigma_\alpha\rangle\ \alpha < \lambda$ of pairwise almost disjoint subsets of ω i.e. $\alpha \neq \beta \rightarrow \sigma_\alpha \cap \sigma_\beta$ is at most finite. Fix an enumeration $\langle I_k\rangle\ k < \omega$ of the open intervals in $\mathbb{R}$ with rational endpoints, such that each interval is repeated infinitely often.

Suppose c is Cohen over L(S). We work in L(S,c). Regard c as a function

$c: \omega \to \omega$. Define $f, g \in \omega^\omega$ by $f(n) = c(2n)$, $g(n) = c(2n+1)$. Thus f, g are mutually generic over L(S). Let $g_\alpha = g \restriction \sigma_\alpha$. For each $\alpha < \lambda$, $1 \leq n < \omega$, we construct an open set $O^n_\alpha \subseteq \mathbb{R}$ such that $\mu(O^n_\alpha) \leq 1/n$. Fix n, α. Define a sequence $k_1 \cdots k_m \cdots$ of integers inductively as follows:

$$k_1 = \min\{k \mid g_\alpha(k) = n \text{ and } \mu(I_{f(k)}) < 1/n\}$$

$$k_{m+1} = \min\{k > k_m \mid g_\alpha(k) = n \text{ and } \mu(I_{f(k_1)} \cup \cdots \cup I_{f(k_m)} \cup I_{f(k)}) < 1/n\}$$

Set $O^n_\alpha = \cup I_{f(k_m)}$

Claim: If $\sigma \subseteq \lambda$, $\sigma \in L(S)$ and σ is infinite then for each n, $\bigcup_{\alpha \in \sigma} O^n_\alpha = \mathbb{R}$.

Proof. Routine. The key point is that if $p \in \omega^{<\omega}$ is a Cohen condition then extending p to get p^* by adding zeros does not increase information about any of the O^n_α. Further if σ_α, σ_β are disjoint off the domain of p then for each n further commitments about O^n_α and O^n_β are completely independent of one another.

Let $C^n_\alpha = \mathbb{R} \setminus O^n_\alpha$. Thus for each α, $\mathbb{R} \setminus \bigcup_n C^n_\alpha$ has measure zero. For each n, α let τ^n_α be the canonical term in L(S) for C^n_α. By the last claim the following is true in L(S);

* Given any infinite $\sigma \subseteq \lambda$, any Cohen condition p and any $n \geq 1$,

$$p \not\Vdash \bigcap_{\alpha \in \sigma} \tau^n_\alpha \neq \emptyset.$$

Claim: * is true in V.

Proof. Consider the following tree defined in L(S):

$$T = \{\langle p, n, \alpha_1, \ldots, \alpha_m\rangle \mid \text{The } \alpha_i \text{ are distinct and } p \Vdash [-n,n] \cap \bigcap_i \tau^n_{\alpha_i} \neq \emptyset\}$$

The ordering on T is by extension. Any counterexample to this claim provides an infinite branch through T and conversely. For the converse one uses the fact that $C^n_\alpha \cap [-n,n]$ is compact. Thus T is wellfounded in L(S) and therefore by absoluteness T is wellfounded in V.

We now finish the proof of lemma 4. Suppose c is Cohen over V and in V[c] there is a real, r, random over L(S,c). For each $\alpha < \lambda$, $n > 1$, $r \in \bigcup C^n_\alpha$ since $\mathbb{R} \setminus \bigcup C^n_\alpha$ has measure zero. Choose in V a term τ for r. For each α choose p_α, n_α such that $p_\alpha \Vdash \tau \in \tau^{n_\alpha}_\alpha$. λ is uncountable so choose p, n and an infinite $\sigma \subseteq \lambda$

such that $\alpha \in G \to p = p_\alpha$, $n = n_\alpha$. Thus $p \Vdash \bigcap_{\alpha\in\sigma} \tau_\alpha^n \neq \emptyset$, a contradiction. ⊣

Using lemma 4 we can now prove lemma 3. Suppose S is an uncountable projective sequence (of distinct reals) and φ is the $\underset{\sim}{\Sigma}{}^1_m$ formula that defines S. We may assume that S has length ω_1. It follows from lemma 2 that if x is Cohen or random over V then φ defines the same sequence in V[x]. Suppose z is a real. The set of reals in L(S,z) is projective since by reflection $y \in L(S,z) \to y \in L(S \restriction \alpha, z)$ for some $\alpha < \omega_1$. From this it follows that for all z there is a real r random over L(S,z). To see how suppose not. Choose z_0 such that there is no real random over $L(S,z_0)$. This is a projective statement in z_0 and therefore by lemma 2 it must hold in V[r] for r random over V. But such an r is random over $L(S,z_0)$ a contradiction. Thus for all z there is a real random over L(S,z). Finally by lemma 2 this must hold in V[c] for c Cohen over V i.e. in V[c] there is a real random over L(S,c), a contradiction.

We now can construct M. Fix a positive integer m. We build a standard model of ZFC, N_m, that is Σ^1_{m+1}-correct, Σ^1_{m+1}-absolute.

Choose a Π^1_m set $A \subseteq \mathbb{R} \times \mathbb{R}$ that is universal for $\underset{\sim}{\Pi}{}^1_m$ subsets of $\mathbb{R}$. Let $\varphi^*(x,y)$ be a Π^1_m formula that defines A. Let f be a projective function that uniformizes A. Assume f is $\underset{\sim}{\Pi}{}^1_n$. If A has been chosen in a reasonable fashion then given any Π^1_m formula $\varphi(x\ x_1 \ldots x_k)$ f effectively induces a skolem function f_φ for φ i.e. for all reals $z_1 \ldots z_k$, $\exists x\ \varphi(x\ z_1 \ldots z_k) \to \varphi(f_\varphi(z_1 \ldots z_k)\ z_1 \ldots z_k)$.

Let $H_{\omega_1} = \{a \mid a \text{ is hereditarily countable}\}$. We define from f a partial function $F\colon H_{\omega_1} \to H_{\omega_1}$. N_m will be $L_{\omega_1}(F) = L(F) \cap H_{\omega_1}$. Choose some canonical coding of elements of H_{ω_1} by reals (for instance code $a \in H_{\omega_1}$ by coding the countable structure $(b,\in)$, b the transitive closure of a, and identifying a as a subset of b). Done properly the set of codes is Π^1_1. Given a code x we denote by a_x that element in H_{ω_1} coded by x. Suppose R(a,b) is a relation on H_{ω_1}. We say that R is projective in the codes if the induced relation on the reals $R^*(x,y) \leftrightarrow$ 'x is a code $\wedge$ y is a code $\wedge$ $R(a_x,a_y)$' is projective. Similarly a partial function $G\colon H_{\omega_1} \to H_{\omega_1}$ is projective in the codes if the induced relation on reals, G^*, is projective where $G^*(x,y) \leftrightarrow G(a_x) = a_y$.

Suppose Q is a partial order, $Q \in \in H_{\omega_1}$. Assume that forcing with Q is non-trivial for instance assume Q is atomless and separative.

Let $\tau_1 \ldots \tau_k$ be terms in V^Q for reals such that $\tau_1 \ldots \tau_k \in H_{\omega_1}$. For simplicity one may as well assume that each term τ_i is of the canonical form $\{(q,t) \mid q \in Q, t \in \omega^{<\omega}$ and $q \Vdash \tau_i$ extends $t\}$. Suppose $\varphi(x\ x_1 \cdots x_k)$ is a Π^1_m formula in $k+1$ variables. Let g be a generic filter on Q. $\tau_1 \cdots \tau_k$ define reals in V[g], $T_1(g) \cdots T_k(g)$. Suppose $V[g] \models \exists x\ \varphi(x\ \tau_1(g) \cdots \tau_k(g))$. Q is countable therefore V[g] = V[c] for some real, c, Cohen over V. Thus by lemma 2, $V[g] \models \varphi(f_\varphi(\tau_1(g) \cdots \tau_k(g))\tau_1(g) \cdots \tau_k(g))$. Let τ be the canonical term for $f_\varphi(\tau_1(g) \cdots \tau_k(g))$. Define $F_\varphi(Q\ \tau_1 \cdots \tau_k) = \tau = \{(q,s) \mid q \in Q,\ s \in \omega^{<\omega}$ and $q \Vdash f_\varphi(\tau_1(g) \cdots \tau_k(g))$ extends $s\}$. If τ does not exist we view F_φ as undefined. Thus F_φ is a partial function from $H^{k+1}_{\omega_1} \to H_{\omega_1}$. We reinterpret F_φ as a partial function from $H_{\omega_1} \to H_{\omega_1}$. Set $F = F_{\varphi^*}$ where φ^* is the Π^1_m formula that defined the universal set A that f uniformizes. We need to show that F is projective in the codes. The key point is that Q is countable and any term $\tau \in V^Q$ for a real is equivalent to $\tau^* = \{(q,t) \mid q \Vdash \tau$ extends $t\}$ a term in canonical form. The appropriate forcing analysis becomes projective in the codes since quantifiers can be restricted to range over terms in canonical form. In particular if $\psi(x_1 \ldots x_k)$ is $\underset{\sim}{\Pi}^1_n$ ($\underset{\sim}{\Sigma}^1_n$) then the relation $R(q,\tau_1 \cdots \tau_k) \leftrightarrow$ '$\tau_1 \cdots \tau_k$ are terms in canonical form and $q \Vdash \psi(\tau_1 \cdots \tau_k)$ ' is $\underset{\sim}{\Pi}^1_n$ ($\underset{\sim}{\Sigma}^1_n$) in the codes with an additional parameter, a code for Q.

Thus F is projective in the codes. In fact the corresponding relation $R(a,b) \leftrightarrow a \in F(b)$ is $\underset{\sim}{\Sigma}^1_{n+1}$ in the codes given that f is $\underset{\sim}{\Pi}^1_n$.

L(F) (= L[R]) denotes the minimal model of ZFC containing the ordinals, closed under F with $F \restriction L(F) \in L(F)$. L(F) has a definable wellordering in the parameter F. $L_{\omega_1}(F) = L(F) \cap H_{\omega_1}$ is projective in the codes further there is a wellordering of $L_{\omega_1}(F)$ that is projective in the codes.

Claim: $L_{\omega_1}(F) \models$ ZFC

Proof. Suppose not. Then the power set axiom must fail. Assume α is a countable ordinal and $P(\alpha) \cap L_{\omega_1}(F) \notin L_{\omega_1}(F)$. Therefore $P(\alpha) \cap L_{\omega_1}(F)$ has size

x_1. This yields via a code for α an uncountable projective sequence of reals. But this contradicts lemma 3.

Claim: $L_{\omega_1}(F)$ is Σ^1_{m+1}-correct.

Proof. Immediate. Suppose $\varphi(x\ x_1 \cdots x_\kappa)$ is Π^1_m, $z_1 \cdots z_\kappa$ are reals in $L_{\omega_1}(F)$ and $\exists x\ \varphi(x\ z_1 \dots z_\kappa)$ holds in V. Then $f_\varphi(z_1 \dots z_\kappa)$ is defined and an element of $L_{\omega_1}(F)$.

Claim: $L_{\omega_1}(F)$ is Σ^1_{m+1}-absolute.

Proof. Suppose $Q_1 * Q_2$ is an iteration defined in $L_{\omega_1}(F)$. Let g_1 be generic over V for Q_1, g_2 be generic over $V[g_1]$ for Q_2. By an argument identical to that for the previous claim, $L_{\omega_1}(F)[g_1]$ is Σ^1_{m+1}-correct in $V[g_1]$ and $L_{\omega_1}(F)[g_1,g_2]$ is Σ^1_{m+1}-correct in $V[g_1 g_2]$. But by lemma 2 $V[g_1]$ is Σ^1_{m+1}-correct in $V[g_1 g_2]$ hence $L_{\omega_1}(F)[g_1]$ is Σ^1_{m+1}-correct in $L_{\omega_1}(F)[g_1 g_2]$. This proves the claim.

Thus $N_m = L_{\omega_1}(F)$ is as desired. If x is a real let $N^x_m = L_{\omega_1}(F,x)$ i.e. N_m relativised to x. Theorem 1 follows immediately. For every real x, N^x_2 is Σ^1_3-absolute. Hence by lemma 1 $N^x_2 \models x^\#$ exists but N^x_2 is Σ^1_3-correct therefore $x^\#$ exists in V.

This completes the proof of Theorem 1.

The proof of Theorem 1 provides for an interesting corollary.

Suppose $a \subseteq \omega$ is infinite. Let $[a]^\omega$ denote the set of infinite increasing sequences of elements of a. Recall that $A \subseteq \omega^\omega$ is Ramsey if there exists an infinite $a \subseteq \omega$ such that $[a]^\omega \subseteq A$ or $[a]^\omega \subseteq \omega^\omega \setminus A$.

Corollary. Assume (1) and (2). Then every projective set is Ramsey and every uncountable projective set contains a perfect subset.

Proof. We prove that every projective set is Ramsey. Fix $n \geq 1$. We prove that every Σ^1_n set is Ramsey, the argument easily relativises to prove that every $\underset{\sim}{\Sigma}^1_n$ set is Ramsey. Fix a Σ^1_n formula $\psi(x)$ that defines $A_\psi \subseteq \omega^\omega$. To say that A_ψ is Ramsey is no worse than Σ^1_{n+2}. Let $m = n+1$. We will show that $N_m \models A_\psi$ is Ramsey. It will then follow by the Σ^1_{m+1}-correctness of N_m that $V \models A_\psi$ is Ramsey.

N_m is Σ^1_{m+1}-absolute. We work in N_m.

Recall Mathias forcing. Conditions are pairs (t,σ) where $t \subseteq \omega$ is finite and $\sigma \subseteq \omega$ is infinite. The ordering of conditions is given by $(t,\sigma) \leq (t',\sigma')$ if $t' \subseteq t$, $\sigma \subseteq \sigma'$ and $t \setminus t' \subseteq \sigma'$. The generic object, g, may be viewed as a subset of ω. Let f_g denote the corresponding element of $[\omega]^\omega$. Mathias forcing has two very nice features (see Mathias [7]). First if $g \subseteq \omega$ is Mathias generic over V then any infinite $g' \subseteq g$ is Mathias generic over V. Second if φ is any statement in the forcing language φ may be decided by a condition of the form $(\emptyset,\sigma)$.

Choose (in N_m) a Mathias condition $(\emptyset,\sigma)$ that decides $\psi(f_g)$. Assume with no loss in generality that $(\emptyset,\sigma) \Vdash \psi(f_g)$.

Let g be Mathias generic over N_m, $g \subseteq \sigma$. Thus for all infinite $g' \subseteq g$ $N_m[g'] \models \psi(f_{g'})$. Hence by the Σ^1_{m+1}-absoluteness of N_m, $N_m[g] \models \forall z \in [g]^\omega [\psi(z)]$. Therefore $N_m[g] \models A_\psi$ is Ramsey. This again by absoluteness implies $N_m \models A_\psi$ is Ramsey.

That every uncountable projective set contains a perfect subset follows by a similar argument.

§2 Some applications. We use the techniques of the proof of Theorem 1 to sharpen and simplify certain proofs involving Projective Determinacy. As an example we have already produced a simpler proof that $\underset{\sim}{\mathrm{PD}}$ implies every projective subset of ω^ω is Ramsey.

Our goal is to provide sharp analogs (at higher levels of the projective hierarchy) of the following theorem of Solovay. A $\underset{\sim}{\Sigma}^1_n$ sequence is a $\underset{\sim}{\Sigma}^1_n$ relation that defines a wellordered sequence of distinct reals.

Theorem (Solovay [9]) Assume every $\underset{\sim}{\Sigma}^1_2$ sequence is countable then

1) Every $\underset{\sim}{\Sigma}^1_2$ set is measurable.

2) Every $\underset{\sim}{\Sigma}^1_2$ set has the property of Baire.

3) Every $\underset{\sim}{\Sigma}^1_2$ set is Ramsey.

We outline a proof of:

Theorem 2 Assume $\underset{\sim}{\Delta}^1_{2m}$-Determinacy and that every $\underset{\sim}{\Sigma}^1_{2m+2}$ sequence is countable then

1) Every $\underset{\sim}{\Sigma}^1_{2m+2}$ set is measurable.

2) Every $\underset{\sim}{\Sigma}^1_{2m+2}$ set has the property of Baire.

3) Every $\underset{\sim}{\Sigma}^1_{2m+2}$ set is Ramsey.

For simplicity we restrict our attention to the fourth level of the projective hierarchy i.e. m = 1. The general case is identical.

We shall need to generalize the notion of Σ^1_4-absoluteness to include standard models N in which the Power Set Axiom may fail i.e. N $\models$ ZFC - Power Set. We simply use the same definition, forcing can be defined in this restricted setting.

We construct $N_3 \subseteq H_{\omega_1}$ as we did in the proof of Theorem 1 working now only with $\underset{\sim}{\Delta}^1_2$ -Determinacy. We will not be guaranteed that $N_3 \models$ ZFC however it will follow as before that $L(N_3) \cap H_{\omega_1} = N_3$. N_3 will be Σ^1_4-correct and Σ^1_4-absolute (this makes sense). Further we will show that N_3 is Σ^1_4 in the codes and that if x_0 is a real generic over N_3 then $N_3^{x_0} = N_3[x_0]$. $N_3^{x_0}$ denotes N_3 relativised to x_0. In particular $N_3[x_0]$ is Σ^1_4-correct for reals, x_0, generic over N_3. From this Theorem 2 follows immediately since if $\underset{\sim}{\Sigma}^1_4$-sequences are countable then $N_3 \models$ ZFC.

Before proceeding we attempt to clarify what is going on. $N_1^x = L_{\omega_1}[x] = L[x] \cap H_{\omega_1}$. Thus N_3^x is an analog of $L_{\omega_1}[x]$ at the Σ^1_4-level of the projective hierarchy. From $\underset{\sim}{\Delta}^1_2$-Determinacy it follows that Π^1_3 sets admit Π^1_3 scales (see [8]). Let T_3 be the tree of a Π^1_3 scale on a universal Π^1_3 set. $L[T_3]$ has been proposed by Moschovakis as an analog of L at the Σ^1_4-level. Indeed it satisfies many of the basic requirements (see [8]). $L[T_3]$ is Σ^1_4-correct as is $L[T_3,x]$ for any real x. $L[T_3]$ can be used to prove Theorem 2 though the hypothesis needed is that $\mathbb{R} \cap L[T_3,x]$ is countable for each real, x. Harrington and Kechris [3] have shown that the reals in $L[T_3,x]$ are the reals of a $\underset{\sim}{\Sigma}^1_4$-sequence (actually a $\Sigma^1_4(x)$-sequence). Thus the hypothesis of $\mathbb{R} \cap L[T_3,x]$ being countable appears to be identical to assuming that $\underset{\sim}{\Sigma}^1_4$-sequences are countable. The difficulty is that the analysis of [3] is quite complicated and requires $\underset{\sim}{\Delta}^1_4$-Determinacy.

It actually follows from the results of Harrington and Kechris that $N_3^x = L[T_3,x] \cap H_{\omega_1}$. The main point of this section is that $N_3^x(N_{2m+1}^x)$ can be defined and fully analyzed using only $\underset{\sim}{\Delta}^1_2$-Determinacy ($\underset{\sim}{\Delta}^1_{2m}$-Determinacy).

We shall need some basic consequences of $\underset{\sim}{\Delta}^1_2$-Determinacy. They all can be found in [8].

3) (Kechris) $\underset{\sim}{\Sigma}^1_3$ sets are measurable and have the property of Baire.

4) (Kechris) ** uniformization for $\underset{\sim}{\Pi}^1_2$ sets and μ-uniformization for $\underset{\sim}{\Pi}^1_2$ sets.

5) (Martin) Every uncountable $\underset{\sim}{\Sigma}^1_3$ set contains a perfect subset.

6) (Martin, Moschovakis) Every Π^1_3 set admits a Π^1_3 norm.

7) (Moschovakis) Every Π^1_3 set can be uniformized by a Π^1_3 function.

We first observe that lemma 2 holds in a restricted form.

Lemma 5. Assume $\underset{\sim}{\Delta}^1_2$-Determinacy. Suppose c is Cohen over V. Then V is Σ^1_4-correct in V[c].

Proof. This follows from (3) and (4) via an argument identical to that in the proof of lemma 2. ⊣

As an immediate corollary to lemma 5 we get:

Lemma 6. Assume $\underset{\sim}{\Delta}^1_2$-Determinacy. Suppose c is Cohen over V. Then V[c] $\models$ $\underset{\sim}{\Delta}^1_2$-Determinacy.

Proof. $\underset{\sim}{\Delta}^1_2$-Determinacy is a Π^1_4 statement. ⊣

Both lemma 5 and lemma 6 hold for random generic extensions as well.

The next lemma provides for a slight generalization of (4) above.

Lemma 7. Assume $\underset{\sim}{\Delta}^1_2$-Determinacy. Let $\varphi(x,y)$ be a $\underset{\sim}{\Pi}^1_3$ formula. Suppose that for any c Cohen over V, V[c] $\models \exists y\, \varphi(c,y)$. Then the subset of the plane defined by φ can be ** uniformized.

Proof. Choose a borel function f: $\mathbb{R} \to \mathbb{R}$ such that for any c Cohen over V, V[c] $\models \varphi(c,f(c))$ (i.e. choose a term for y). Thus by a now familiar argument $\{z \mid \varphi(z,f(z))\}$ is comeager. Hence f ** uniformizes the $\underset{\sim}{\Pi}^1_3$ subset of the plane defined by φ. ⊣

Lemma 7 provides a sufficient condition for ** uniformization of a $\underset{\sim}{\Pi}^1_3$ set in the context of $\underset{\sim}{\Delta}^1_2$-Determinacy. It can also be shown to be a necessary condition. The point is that $\underset{\sim}{\Delta}^1_2$-Determinacy cannot (modulo inconsistency) by itself prove ** uniformization for $\underset{\sim}{\Pi}^1_3$ sets. This is analogous to ZF cannot prove ** uniformization for $\underset{\sim}{\Pi}^1_1$ sets (it fails in L).

We now can build N_3 much as we did in the proof of Theorem 1. Choose a Π^1_3 set $A \subseteq \mathbb{R} \times \mathbb{R}$ that is universal for $\underset{\sim}{\Pi}^1_3$ subsets of $\mathbb{R}$. Choose A canonically in the usual fashion. Let f be a Π^1_3 function that uniformizes A. We must be more careful in our choice of f. Choose a Π^1_3 formula $\varphi(x,y)$ that in V[c] for c Cohen over V, defines a Π^1_3 function that uniformizes A. By lemma 5 and homogeneity of Cohen forcing, $\varphi(x,y)$ defines a Π^1_3 function in V that uniformizes A. Let f be such a Π^1_3 function.

Following the construction in section 1, define from f a partial function $F: H_{\omega_1} \to H_{\omega_1}$. Set $N_3 = L(F) \cap H_{\omega_1}$. Thus N_3 is Σ^1_4-correct and Σ^1_4-absolute. Define the relativised versions of N_3 by $N_3^x = L(F,x) \cap H_{\omega_1}$ for x a real.

We wish to show that N_3 is Σ^1_4 in the codes and that for reals, x_0, generic over N_3, $N_3^{x_0} = N_3[x_0]$.

It is usual to define a set of reals as thin if it does not contain a perfect subset. We need that there is a largest thin Σ^1_4 set, it is called C_4. However since we wish to work only with $\underset{\sim}{\Delta}^1_2$-Determinacy we must be more careful in defining the notion of thin. The analogous problem at the second level concerns thin Σ^1_2 sets in the context of only ZFC. ZFC proves that if $\mathbb{R} \neq \mathbb{R} \cap L$ then there is a largest thin Σ^1_2 set namely $\mathbb{R} \cap L$. However with the definition of thin given above, in L there is no largest thin Σ^1_2 set. Thus if one wishes to work uniformly one needs a slightly sharper notion than the notion of thin.

Assume $\underset{\sim}{\Delta}^1_2$-Determinacy. We say that a set of reals is Σ^1_4-rigid if it can be defined by a Σ^1_4 formula $\varphi(x)$ which defines the same set in V[c] for c Cohen over V. We define the notions of $\underset{\sim}{\Sigma}^1_4$-rigid, Π^1_3-rigid and $\underset{\sim}{\Pi}^1_3$-rigid similarly. We are taking advantage of lemma 5. Also notice that by lemma 5 $\underset{\sim}{\Pi}^1_3$-rigid sets cannot contain perfect subsets. There is no need to define $\underset{\sim}{\Sigma}^1_3$-rigid as it coincides with being countable (by (5)).

C_4 is the largest Σ^1_4-rigid set and this is provable just assuming $\underset{\sim}{\Delta}^1_4$-Determinacy. C_4 is analogous to $\mathbb{R} \cap L$ at the Σ^1_4-level.

In particular assuming $\underset{\sim}{\Delta}^1_2$-Determinacy our notions of Σ^1_4-rigid, Π^1_3-rigid coincide with the usual notion of thinness just in case $\mathbb{R} \neq C_4$.

Any Σ^1_4-rigid set is the projection of a Π^1_3-rigid set since Π^1_3 sets can be uniformized by Π^1_3 functions. To see this use lemma 5. Suppose A is Σ^1_4-rigid. Thus in V[c] A is Σ^1_4 and is therefore the projection of a Π^1_3 set $B \subseteq \mathbb{R} \times \mathbb{R}$. Choose in V[c] a Π^1_3 function, f, that uniformizes B. Cohen forcing is homogeneous thus $f \in V$ and is therefore Π^1_3-rigid.

Thus to find C_4 it suffices to find a Σ^1_4-rigid set containing all Π^1_3-rigid sets, for given such a set C, C_4 may be defined by $\{x \mid x$ is recursive in y for some $y \in C\}$. A definition for C can be motivated by the observation that if S is a Π^1_3-rigid set and $\leq$ is a 'nice' Π^1_3 norm on S (a Π^1_3 norm chosen in V[c] will suffice) then for each $x \in S$, $\{y \mid y \leq x\}$ is countable. This is because $\{y \mid y \leq x\}$ is $\Delta^1_3(x)$ and $\underset{\sim}{\Delta}^1_2$-Determinacy proves that uncountable $\underset{\sim}{\Delta}^1_3$ sets contain perfect subsets.

We give a definition of C. It is due to Moschovakis (see [8]).

Choose canonically $W_3 \subseteq \omega \times \mathbb{R}$, Π^1_3 and universal for Π^1_3 subsets of $\mathbb{R}$. Define $P_3 \subseteq \mathbb{R}$ by $x \in P_3 \leftrightarrow (x(0),x') \in W_3$ where $x' = (x(1), x(2), \ldots)$. Let $\leq$ be a (nice) Π^1_3 norm on P_3. Define C by $x \in C \leftrightarrow \exists m\, [\{y \mid m^\frown y \leq m^\frown x\}$ is countable$]$ $\leftrightarrow \exists m\, \exists z\, \forall y\, [m^\frown y \leq m^\frown x \rightarrow y$ is recursive in $z]$ where $m^\frown x = (m, x(0), x(1), \ldots)$.

Thus C is Σ^1_4-rigid and contains every Π^1_3-rigid set. C is in fact Π^1_3-rigid (see Kechris [5]). It is therefore C_3 the largest Π^1_3-rigid set.

For arbitrary reals x we denote by C^x, C^x_4 the relativised versions of C, C_4. Observe that C^x_4 is Σ^1_4-correct for each x. This follows from (7).

<u>Lemma 8</u>. Assume $\underset{\sim}{\Delta}^1_2$-Determinacy. For each real x, $\mathbb{R} \cap N^x_3 = C^x_4$.

<u>Proof</u>. We show $\mathbb{R} \cap N_3 = C_4$. The general case is similar.

Let $\varphi(x,y)$ be any reasonable Σ^1_4-formula such that $\varphi(x,y) \leftrightarrow y \in C^x_4$ (choose φ that works in V[c] for c Cohen over V).

For countable ordinals α, a generic code of α is a real, g_α, coding a generic collapse of α i.e. g_α codes the prewellordering on ω induced by a generic enumeration of α in length ω. Let Q_α be the corresponding (countable) notion of forcing. $Q_\alpha \in L$ hence $Q_\alpha \in N_3$.

Let F be the partial function $F: H_{\omega_1} \to H_{\omega_1}$ used to define N_3. F is Σ^1_4 in the codes. Let R be the relation $R(a,b) \leftrightarrow a \in F(b)$. R is Σ^1_4 in the codes. $L(F) = L(R)$. Let α be a countable ordinal and suppose g_β is a generic code of $\beta = \alpha + 1$. The forcing for $g_\beta(Q_\beta)$ is countable hence $V[g_\beta] = V[c]$ for some c Cohen over V. It follows by lemma 5 and lemma 6 that L(F) is the same computed in V or $V[g_\beta]$. Therefore since $C_4^{g_\beta}$ is Σ^1_4-correct (in $V[g_\beta]$), $L_\alpha(R)$ can be computed inside $C_4^{g_\beta}$. Fix $x_0 \in L_\alpha(R)$ x_0 a real in V. Hence $x_0 \in C_4^{g_\beta}$ in $V[g_\beta]$ for any generic code of β. Thus in V^{Q_β} $[[\varphi(g_\beta,x_0)]]^{Q_\beta} = 1$. Let, in V, A = $\{x \mid \exists\beta\ [[\varphi(g_\beta,x_0)]]^{Q_\beta} = 1\}$. By an analysis similar to that used to show that F is projective in the codes, A is Σ^1_4. We claim that A is Σ^1_4-rigid. Suppose not. Therefore for some countable ordinal β and any c Cohen over V we have that in V[c], $\{x([[\varphi|g_\beta,x)]]^{Q_\beta} = 1\} \cap V \neq \{x \mid [[\varphi(g_\beta,x)]]^{Q_\beta} = 1\} \cap V[c]$. Add g_β a generic code of β to V[c]. Therefore in $V[c,g_\beta] = V[g_\beta][c]$, $C_4^{g_\beta} \cap V[g_\beta] \neq C_4^{g_\beta} \cap V[c,g_\beta]$ which implies that $C_4^{g_\beta}$ is not $\Sigma^1_4(g_\beta)$-rigid a contradiction. Hence A is Σ^1_4-rigid and contained in C_4. But $x \in N_3 \to x \in L_\alpha(R)$ for some $\alpha \to x \in A$. Thus $\mathbb{R} \cap N_3 \subseteq A \subseteq C_4$.

We need to show that $C_4 \subseteq \mathbb{R} \cap N_3$. Suppose S is Π^1_3-rigid. It suffices to show $S \subseteq N_3$. Choose a Π^1_3 norm, $\leq$, on S in V[c]. S is Π^1_3-rigid hence $\leq$ is a Π^1_3 norm on S in V which prewellorders S in length $\leq \omega_1$. For each $x \in S$ $\{y \mid y \leq x\}$ is countable and $\Delta^1_3(x)$ hence $N_3 \cap S$ is an initial segment of S relative to $\leq$. Suppose $S \not\subseteq N_3$ and choose β, length $(\leq \restriction N_3 \cap S) < \beta \leq$ length $(\leq \restriction S)$. Add a generic code, g_β, of β to V. N_3 is the same defined in V or $V[g_\beta]$ ($=V[c]$). Further $N_3[g_\beta] = N_3^{g_\beta}$ in $V[g_\beta]$ so $N_3[g_\beta]$ is Σ^1_4-correct in $V[g_\beta]$. N_3 and $N_3[g_\beta]$ compute the same S, otherwise $N_3[g_\beta] \models$ 'S contains a perfect set'. But in this case $V[g_\beta] \models$ ' S contains a perfect set' since $N_3[g_\beta]$ is Σ^1_4-correct in $V[g_\beta]$. Thus S is not Π^1_3-rigid in $V[g_\beta]$ a contradiction. Since N_3, $N_3[g_\beta]$ compute the same S, $N_3[g_\beta] \models$ 'length $(\leq \restriction S) < \beta$' but 'length $(\leq \restriction S) < \beta$' is $\Pi^1_4(g_\beta)$ therefore $V[g_\beta] \models$ 'length $(\leq \restriction S) < \beta$' and again we have a contradiction. Thus $S \subseteq N_3$ and therefore $C_4 \subseteq N_3$. This completes the proof of lemma 8. ⊣

We get as a corollary to lemma 8:

Lemma 9. Assume $\underset{\sim}{\Delta}^1_2$— Determinacy. N_3 is Σ^1_4 in the codes.

Proof. Suppose g_α is a generic code of α. N_3 is the same computed in V or $V[g_\alpha]$ and in $V[g_\alpha], N_3[g_\alpha] = N_3^{g_\alpha}$. Further by lemma 8 $\mathbb{R} \cap N_3^{g_\alpha} = C_4^{g_\alpha}$ in $V[g_\alpha]$. Hence $a \in N_3 \leftrightarrow \exists\alpha[$a has a real code in $C_4^{g_\alpha}$ for g_α a generic code of $\alpha] \leftrightarrow$ $\exists\alpha\ \left[[[\exists z[z \in C_4^{g_\alpha} \wedge z \text{ codes } a]]]^{Q_\alpha} = 1\right]$. This yields via a standard computation a Σ^1_4 definition of N_3 in the codes. Notice that this definition does not involve f the Π^1_3 function used to define N_3. This provides an alternate view of the fact that N_3 is independent of the choice of f. ⊣

Lemma 10. Assume $\underset{\sim}{\Delta}^1_2$-Determinacy. Assume z is Cohen over N_3. Then $\mathbb{R} \cap N_3^z = \mathbb{R} \cap N_3[z]$ i.e. $\mathbb{R} \cap N_3[z] = C_4^z$.

Proof. We know from the construction of N_3 that if c is Cohen over V then $N_3^c = N_3[c]$ in V[c]. This lemma is a partial improvement of this.

For each x, C_4^x is $\Sigma^1_4(x)$-rigid. Hence it is the projection of a $\Pi^1_3(x)$-rigid subset of $\mathbb{R} \times \mathbb{R}$. Choose a Π^1_3 formula $\varphi(x,y,z)$ that uniformly in x defined such a set i.e. for each x, $A^x = \{(y,z) \mid \varphi(x,y,z)\}$ is a $\Pi^1_3(x)$-rigid set with projection C_4^x. Choose $\varphi(x,y,z)$ so that it works in V[c] for c Cohen over V. Reinterpret A^x as a subset of $\mathbb{R}$, $\varphi(x,y,z)$ as a Π^1_3 formula $\varphi(x,y)$.

Similarly choose a Π^1_3 formula $\psi(x,y,t)$ such that for each x, $\psi(x,y,t)$ defines a Π^1_3 norm, $y \leq^x t \leftrightarrow \psi(x,y,t)$, on A^x. Choose $\psi(x,y,t)$ so that it works in V[c]. Thus by previous arguments, for each $t \in A^x$, $\{y \mid y \leq^x t\}$ is countable. Notice then that for each x the induced norm $\leq^x$ on A^x has length at most ω_1.

For $x \in C_4$, A^x projects onto $C_4^x = C_4$. Consider A^0. A^0 is Π^1_3-rigid hence $A^0 \subseteq C_4$. Therefore A^0 and C_4 have the same cardinality. We may assume by taking unions that $A^0 \subseteq A^x$ for each real x.

There are two cases to consider.

Case 1. C_4 has size ω_1.

Let z_0 be Cohen over N_3. A^0 is cofinal in A^{z_0} relative to the norm $\leq^{z_0}$ since A^0 is of size ω_1 Assume $\mathbb{R} \cap N_3[z_0] \neq C_4^{z_0}$. Hence $A^{z_0} \not\subseteq N_3[z_0]$. Choose $t_0 \in A^0$

such that $\{y | y \leq^{z_0} t_0\} \nsubseteq N_3[z_0]$.

We abbreviate y is recursive in x by $y \in d(x)$ i.e. for each x let $d(x) = \{y | y$ is recursive in $x\}$.

Suppose c is Cohen over V. Choose a term τ (with boolean value one) for a real x in V[c] such that $V[c] \models \forall y\, [y \leq^c t_0 \rightarrow y \in d(x)]$. Choose τ as a borel function h. Thus (with boolean value one) $V[c] \models \forall y\, [y \leq^c t_0 \rightarrow y \in d(h(c))]$. Hence in $V, \{z | \forall y\, [y \leq^z t_0 \rightarrow y \in d(h(z))]\}$ is comeager since it is $\Pi^1_3(t_0,h)$. Therefore as a set in the codes, $\{\langle h,B\rangle | h$ is a borel function, B is a comeager borel set, $z \in B \rightarrow \forall y\, [y \leq^z t_0 \rightarrow y \in d(h(z))]\}$ is nonempty and $\Pi^1_3(t_0)$. $t_0 \in A^0 \subseteq C_4$ and C_4 is Σ^1_4-correct therefore C_4 contains such an h,B. But z_0 is Cohen over N_3 hence $z_0 \in B$ since B is a comeager borel set in N_3. Thus in V, $\forall y\, [y \leq^{z_0} t_0 \rightarrow y \in d(h(z_0))]$. $h \in N_3$ so $h(z_0) \in N_3[z_0]$ and therefore $\{y | y \leq^{z_0} t_0\} \subseteq N_3[z_0]$ a contradiction.

<u>Case 2.</u> C_4 is countable.

Let $\beta = \aleph_1^{N3}$. Add to V a generic code of β, g_β. g_β is generic over N_3 (for the collapse of $\aleph_1$) and in $N_3[g_\beta]$ there is a comeager borel set, B, of reals Cohen over N_3. Further in $N_3[g_\beta]$ there is a borel function h such that
$N_3[g_\beta] \models z \in B \rightarrow C_4^z \subseteq d(h(z))$, which may be rewritten as
$N_3[g_\beta] \models \forall z\, \forall y\, [z \in B \vee y \notin C_4^z \vee y \in d(h(z))]$. This in $N_3[g_\beta]$ is $\Pi^1_4(h,B)$. g_β is generic over V therefore $N_3^{g_\beta} = N_3[g_\beta]$ in $V[g_\beta]$ so $N_3[g_\beta]$ is Σ^1_4-correct in $V[g_\beta]$. Therefore $V[g_\beta] \models z \in B \rightarrow C_4^z \subseteq d(h(z))$. Now suppose z_0 is Cohen over N_3 in V. Then z_0 is Cohen over $N_3[g_\beta]$. Thus in $V[g_\beta]$, $z_0 \in B$ and $h(z_0) \in N_3[g_\beta, z_0]$ i.e. $C_4^{z_0} \subseteq N_3[g_\beta, z_0]$. But g_β is generic over V hence $C_4^{z_0} \subseteq N_3[z_0]$.

This finishes the proof of lemma 10.

It is clear by the proof that lemma 10 also holds for N_3^x given any x i.e. z Cohen over $N_3^x \rightarrow \mathbb{R} \cap N_3^x[z] = C_4^{z,x}$.

<u>Lemma 11.</u> Assume $\underset{\sim}{\Delta}^1_2$-Determinacy. Suppose x, x_0 are reals and that x_0 is generic over N_3^x. Then $N_3^x[x_0] = N_3^{x,x_0}$.

<u>Proof.</u> We do the case of x_0 generic over N_3. As usual the general case is

identical.

Let $F\colon H_{\omega_1} \to H_{\omega_1}$ be the partial function used to define N_3. Let R be the relation $R(a,b) \leftrightarrow a \in F(b)$. Hence $L(F) = L(R)$. We have already observed that if z_α is a code of α (a countable ordinal) then $L_\alpha(R)$ can be computed inside $C_4^{z_\alpha}$.

Let x_0 be a real, generic over N_3 for a partial order $Q \in N_3$. Let λ denote the cardinality of Q in N_3. λ is a countable ordinal β in V. Add to V a generic code, g_α, of a countable ordinal $\alpha > \beta$. $N_3[g_\alpha] = N_3^{g_\alpha}$ in $V[g_\alpha]$. Further $N_3[g_\alpha][x_0] = N_3[g_\alpha][z_0]$ for some z_0 Cohen over $N_3[g_\alpha]$. Hence by the appropriate version of lemma 10, $\mathbb{R} \cap N_3[g_\alpha][z_0] = C_4^{g_\alpha, z_0}$. Thus $L_\alpha(R,x_0) \subseteq N_3[g_\alpha][x_0]$ which may be rewritten as $L_\alpha(R,x_0) \subseteq N_3[x_0][g_\alpha]$. But g_α is generic over V hence $L_\alpha(R,x_0) \subseteq N_3[x_0]$. This must hold for every countable α hence $L_{\omega_1}(R,x_0) \subseteq N_3[x_0]$. But $N_3^{x_0} = L_{\omega_1}(R,x_0)$ therefore $N_3^{x_0} = N_3[x_0]$. ⊣

Theorem 2 (m = 1) follows immediately from lemma 9 and lemma 11. We prove a sharper version of the Ramsey case. This answers a question of Kechris [4].

Lemma 12. Assume $\underset{\sim}{\Delta}^1_{2m}$-Determinacy and that $\underset{\sim}{\Sigma}^1_{2m+2}$-sequences are countable. Then $\underset{\sim}{\Sigma}^1_{2m+2}$ sets are Ramsey. Further this is true in an effective way i.e. if A is Σ^1_{2m+2} and $[x]^\omega \subseteq A$ for some infinite $x \subseteq \omega$ then there is an infinite $y \subseteq \omega$, $[y]^\omega \subseteq A$, such that y is Δ^1_{2m+2}. Thus Δ^1_{2m+2} subsets of ω^ω have Δ^1_{2m+2} homogeneous sets.

Proof. We do the case of m = 1 and use notation as in the proof of the corollary in §1.

Suppose $A \subseteq \omega^\omega$ is $\underset{\sim}{\Sigma}^1_4$. Then A is $\Sigma^1_4(x_0)$ for some real, x_0. Let $\varphi(y)$ be a $\Sigma^1_4(x_0)$ formula that defines A. $N_3^{x_0} \models$ ZFC since $\underset{\sim}{\Sigma}^1_4$-sequences are countable. Choose (in $N_3^{x_0}$) a Mathias condition of the form $(\emptyset,\sigma)$ to decide $\varphi(f_g)$. Recall that for infinite $z \subseteq \omega$, f_z denotes the corresponding element of $[\omega]^\omega$. Assume with no loss in generality that $(\emptyset,\sigma) \Vdash \varphi(f_g)$ in $N_3^{x_0}$. Choose g Mathias generic over $N_3^{x_0}$, $g \subseteq \sigma$. Hence for all infinite $g' \subseteq g$, g' is Mathias generic over $N_3^{x_0}$ and $N_3^{x_0}[g'] \models \varphi(f_{g'})$. But by lemma 11, $N_3^{x_0}[g'] = N_3^{x_0,g'}$ hence $N_3^{x_0}[g']$ is Σ^1_4-correct, thus $\varphi(f_{g'})$ is true in V and therefore $[g]^\omega \subseteq A$.

Now assume A is Σ^1_4 and let $\varphi(y) = \exists z\ \psi(z,y)$ be a Σ^1_4 formula that defines A. Suppose $x \subseteq \omega$ is infinite and that $[x]^\omega \subseteq A$.

$N_3^x \models$ ZFC. We work in N_3^x. Since $[x]^\omega \subseteq A$ in V we have that in N_3^x, $(\emptyset,x) \Vdash \varphi(f_g)$ which is to say $(\emptyset,x) \Vdash \exists z\ \psi(z,f_g)$, ($\Vdash$ for Mathias forcing). Choose a term, τ, for z i.e. such that $(\emptyset,x) \Vdash \psi(\tau,f_g)$. By the nice properties of Mathias forcing one can find $x' \subseteq x$, a continuous function $h:[x']^\omega \to \mathbb{R}$, such that $(\emptyset,x') \Vdash \psi(h(f_g),f_g)$.

Return to V. Choose g Mathias generic over N_3^x, $g \subseteq x'$. Hence in V, $\forall s \in [g]^\omega\ \psi(h(s),s)$. Thus $\{(z,h) \mid z \subseteq \omega$ is infinite, $h: [z]^\omega \to \mathbb{R}$ is continuous and $\forall s \in [z]^\omega\ \psi(h(s),s)\}$ is nonempty. But it is Π^1_3 as a set in the appropriate codes. Hence by (7) it contains a Π^1_3-singleton (a real y such that $\{y\}$ is Π^1_3), $y = (z^*,h^*)$. Therefore $[z^*]^\omega \subseteq A$ and $z^* \subseteq \omega$ is Δ^1_4. ⊣

Finally we note that the general version (arbitrary m) of lemma 5 has as a corollary:

<u>Corollary to lemma 5.</u> Assume 'ZFC + $\underset{\sim}{\Delta}^1_{2m}$-Determinacy' is consistent. Then so is 'ZFC + $\underset{\sim}{\Delta}^1_{2m}$-Determinacy + ¬CH'. (See Becker [1] for related results).

<u>Proof</u>. (m = 1). Assume $V \models$ ZFC + $\underset{\sim}{\Delta}^1_2$-Determinacy. It follows by lemma 5 that V is Σ^1_4-correct in V[G] where V[G] is a generic extension of V obtained by adding any infinite number of Cohen reals. Hence as for lemma 6, $V[G] \models \underset{\sim}{\Delta}^1_2$-Determinacy. ⊣

This completes our analysis of N_3^x. In many ways N_3^x is the analog of $L_{\omega_1}[x] = H_{\omega_1} \cap L[x]$ at the fourth level. Certainly lemmas 8, 9, and 11 vindicate this point of view (given that C_4 is the largest Σ^1_4-rigid set). One could hope that a better analogy exists so that in addition lemma 11 holds for arbitrary x_0 (i.e. $N_3^x[x_0] = N_3^{x,x_0}$ for all x_0). This cannot happen. In fact lemma 9 and lemma 11 serve to in some sense characterize N_3^x. All in all, as has been well documented (see [8]), $\underset{\sim}{\Delta}^1_2$-Determinacy plays a role in the theory of $\underset{\sim}{\Sigma}^1_4$ sets analogous to that of ZFC in the theory of $\underset{\sim}{\Sigma}^1_2$ sets.

§3 Remarks and further problems. The main problem of course is to prove Projective Determinacy assuming certain of its nondeterministic consequences. The metatheorem should be that assuming $\underset{\sim}{PD}$ is the only way to build a structure theory for the projective sets that extends the structure theory of the $\underset{\sim}{\Sigma}^1_1, \underset{\sim}{\Pi}^1_1$ sets provided by ZFC. This kind of problem has merit independent of the consistency of determinacy. For example should (heaven forbid) measurable cardinals be inconsistent then (1) and (2) cannot both hold. This then could become a deep theorem of ZFC about the projective sets.

To prove that $\underset{\sim}{PD}$ follows from (1) and (2) may seem overly ambitious nevertheless we conjecture this is true. In any case it may be easier to augment (1) and (2) with more of the consequences of $\underset{\sim}{PD}$, for instance (6) - (7) of §2 generalized to the appropriate levels. We note that the entire analysis in §2 (all of the lemmas) needs nothing more of $\underset{\sim}{\Delta}^1_2$-Determinacy beyond (3) - (7).

REFERENCES

[1] Becker, H., Partially playful universes, Cabal Seminar (1976-1971). Vol. 689, Lecture Notes in Mathematics, Springer-Verlag (1978), 55-90.

[2] Devlin, K. J., and Jensen, R. B., Marginalia to a theorem of Silver, Logic Conference, Kiel 1974, Vol. 499, Lecture Notes in Mathematics, Springer, Berlin, 115-142.

[3] Harrington, L. A. and Kechris, A. S., On the determinacy of games on ordinals, Ann. Math. Logic, 20(1981), 109-154.

[4] Kechris, A. S., Effective Ramsey theorems in the projective hierarchy, this volume.

[5] Kechris, A. S., The theory of countable analytical sets, Trans. Amer. Math. Soc., **202** (1975), 259-297.

[6] Martin, D. A. and Solovay, R. M., A basis theorem for Σ^1_3 sets of reals, Ann. of Math. 89 (1969), 138-160.

[7] Mathias, A. R. D., Happy families, Ann. Math. Logic, 12 (1977), 59-111.

[8] Moschovakis, Y. N., Descriptive Set Theory, North Holland (1980).

[9] Solovay, R. M., On the cardinality of $\underset{\sim}{\Sigma}^1_2$ sets of reals, Foundations of Mathematics, Symposium papers commemorating the 60th birthday of Kurt Gödel, Springer-Verlag (1966), 58-73.